VACUUM AND THIN FILM TECHNOLOGY

Proceedings of an international symposium held at the University of Uppsala, Sweden, 1976

VACUUM AND THIN FILM TECHNOLOGY

Proceedings of an international symposium held at the University of Uppsala, Sweden, 31st August–3rd September 1976

edited by

J. YARWOOD

The Polytechnic of Central London, England

and

P. ANDERSSON

Uppsala Universitet Institute of Technology, Sweden

PERGAMON PRESS

OXFORD · NEW YORK · TORONTO · SYDNEY
PARIS · FRANKFURT

U.K.	Pergamon Press Ltd., Headington Hill Hall, Oxford OX3 0BW, England
U. S. A.	Pergamon Press Inc., Maxwell House, Fairview Park, Elmsford, New York 10523, U.S.A.
CANADA	Pergamon of Canada, Suite 104, 150 Consumers Road, Willowdale, Ontario M2 J1P9, Canada
AUSTRALIA	Pergamon Press (Aust.) Pty. Ltd., P.O. Box 544, Potts Point, N.S.W. 2011, Australia
FRANCE	Pergamon Press SARL, 24 rue des Ecoles, 75240 Paris, Cedex 05, France
FEDERAL REPUBLIC OF GERMANY	Pergamon Press GmbH, 6242 Kronberg-Taunus, Pferdstrasse 1, Federal Republic of Germany

First edition 1978

Library of Congress Catalog Card No. 77-81976

Published as a special issue of the journal Vacuum, *Volume 27, No. 3/4 and supplied to subscribers as part of their subscription.*

Printed in Great Britain by A. Wheaton & Co. Ltd., Exeter

ISBN 0 08 022112 2

Contents

	85	Preface
		1. Vacuum and thin films—measurements
J Yarwood	**87**	International vacuum standardization
G Kuus, A J de Ridder and A Pools	**93**	Measuring technique for partial hydrogen pressures in nitrogen–neon atmospheres
S H Norrman and T G Andersson	**97**	An evaporation cryostat for electrical and optical measurements on ultrathin metal film under ultra high vacuum conditions
H Norström, S Berg and L P Andersson	**99**	Fast volume determination using a differential capacitance manometer
P Braun, W Färber, G Betz and F P Viehböck	**103**	Effects in Auger electron spectroscopy due to the probing electrons and sputtering
R N Castellano, M R Notis and G W Simmons	**109**	Composition and stress state of thin films deposited by ion beam sputtering
C W B Martinson, P J Nordlander and S-E Karlsson	**119**	AES investigations of the interface between substrate and chromium films prepared by evaporation and ion plating
D Kropman, M Vinnal and P Putk	**125**	Influence of defects in the silicon wafer on the properties of silicon oxide films formed by reactive sputtering
J Liday, R Harman and M Veselý	**129**	Depth chemical profiles of MNOS structures measured by AES
S R Veljković and P Bibović	**141**	Growth and composition of Al clusters on NaCl
		1. Vacuum and thin films—coating processes and pumps
L N Alexandrov, F L Edelman and V V Voskoboinikov	**145**	Silicon films growth in vacuum by pyrolysis of silane
A Aronson and S Weinig	**151**	Inline production magnetron sputtering
G Sørensen and J L Whitton	**155**	Plating with ion accelerators
M Khorossany and R K Fitch	**159**	Energy distribution of the ions produced by saddle field ion sources
H Sanchez, E Posada, J Panqueva, R Paéz, I Jaramillo and G Fritsch	**163**	Deposition of thin chromium fllms on substrates heated between room temperature and 300°C in the uhv chamber of an annular cryopump
H Schwarz and U Leal de Castro	**167**	Extremely uniform thin films by vacuum deposition from a 'tube source'
J Henning and H Lotz	**171**	Turbomolecular pumps with corrosion-resistant oil for the lubrication and cooling of the rotor bearings
J Visser, B Symersky and J M Geraerts	**175**	A versatile cryopump for industrial vacuum systems
		2. Electrical properties of thin films—metal–semiconductor contacts
K Eisele and M Schulz	**181**	Metal–semiconductor junctions—related to deposition of thin films
S Berg, L P Andersson, H Norström and E Grusell	**189**	Substrate surface damages by rf-sputtering
M Hafiz, E Mgbenu, P S Tove, H Norde and S Petersson	**193**	Investigations of metal contacts to amorphous evaporated Ge films
P Glaser, Á Herman and Gy Vágó	**197**	Formation of various metal–semiconductor junctions by means of ion evaporation
H Norde and P A Tove	**201**	Behaviour of amorphous Ge contacts to monocrystalline silicon
E Mgbenu, S Petersson and P A Tove	**209**	Surface resistant measurement as an aid in controlling the fabrication of silicides

Contents

2. Electrical properties of thin films—dielectrics

D S Campbell	**213**	Deposition methods for dielectric films and their electrical properties
K Kristiansen	**227**	A statistical approach to the analysis of dielectric breakdown strength of thin insulating films
J S Sandved and K Kristiansen	**235**	Glow discharge polymerization of organic silicones (E-10)
F Morin	**241**	Pulsed photovoltaic effect in thin film MIS devices
I Lundström, M S Shivaraman and C Svensson	**245**	Hydrogen sensitive MOS structures
B Lalevic, N Annamalai, B Leung, G Taylor and W Slusark Jr	**249**	Dielectric dispersion and polarization effects in thin transition metal oxide films
Vidyardhi Nanduri	**255**	Behaviour of ceramic insulator materials during initiation of trigger discharge in TVG

2. Electrical properties of thin films—compounds

J G Anderson	**263**	Polycrystalline semiconductor thin films
R Hill and I A S Edwards	**277**	The stoichiometry of surfaces in cadmium sulphide thin films
H W Schock, G Bilger, W H Bloss, G H Hewig and F Pfisterer	**281**	Technology of large area Cu_2S–CdS solar cells
D May	**287**	Joule heating in ZnS electroluminescent films
Y Danto and J Salardenne	**293**	ac electrical properties of magnesium and aluminium fluoride thin films
E Sciri and R Chabicovsky	**299**	Conduction phenomena in Nb–Nb_2O_5Au thin film structures fabricated by various techniques
L Żdanowicz, G Pocztowski and W Żdanowicz	**305**	Size effects in thin films of cadmium arsenide

2. Electrical properties of thin films—special topics

L N Aleksandrov and R N Lovyagin	**311**	Epitaxial germanium–silicon structures obtained in high pure vacuum
A Rogalski, J Rybiński and M J Małachowski	**317**	Electrical and piezoresistive properties of $Pb_{1-x}Sn_xTe$ thin films
F d'Heurle, C Alliota, J Angilello, V Brusic, J Dempsey and D Irmischer	**321**	The deposition by evaporation of Cu–Al alloy films
T G Andersson and S H Norrman	**329**	Structural and electrical properties of discontinuous gold films on glass
A J Tosser, C R Tellier and J Launey	**335**	Thickness dependent effects of thermal ageing in thin conductive films
S Hellström and H Wesemeyer	**339**	Nonlinearity measurements of thin films
S Husa	**345**	1/f noise voltage in a thin film structure
A Masoero, P Mazzetti and A Stepanescu	**349**	Current noise in W_2O_{5-x} films due to surface adsorption of water

3. Mechanical properties of thin films

R F Bunshah	**353**	Mechanical properties of PVD films
J Gouault, M Hubin, G Richon and B Eudeline	**363**	The electromechanical behaviour of a full component (dielectric and Cu/Ni constantan alloy) for thin film strain gauge deposited upon steel-substrate
E Söllner, E Benes, A Biedermann and D Hammer	**367**	Properties of heavily preloaded quartz crystals for thin film thickness monitors
P J Nordlander, K Å B Andersson, H T G Nilsson and S B M Hagström	**373**	Influence of condensation surface temperature on macrostructure in evaporated chromium films
K Å B Andersson, S-E Karlsson and N Ohmae	**379**	Morphologies of rf sputter-deposited solid lubricants

4. Optical properties of thin films

H A Macleod	**383**	The monitoring of thin films of optical purposes
C J vd Laan and H J Frankena	**391**	Monitoring of optical thin films using a quartz crystal monitor
L E Flordal and R Kivaisi	**399**	Vacuum evaporated thin films for solar collectors
C Misiano and E Simonetti	**403**	Co-sputtered optical films
O S Heavens	**407**	Use of lateral waves in detecting surface absorption
L Zdanowicz and T Kwiecień	**409**	Some optical properties of thin Cd_3As_2 films

Preface

These proceedings contain 56 of the 80 originally invited and contributed papers presented at the Symposium on Vacuum and Thin Film Technology held in Uppsala, Sweden, 31 August–3 September 1976. The Symposium was organized by The University of Uppsala, The University of Linköping and The Swedish Vacuum Society and sponsored by IUVSTA, EPS, ITFC and related national Swedish societies.

More than 200 participants representing 25 different countries from all parts of the world took part in the symposium.

Many of the basic ideas in the field of vacuum and vacuum technology are well known but as reflected from the papers given at the Symposium there is now work going on towards the realization of these ideas in new and improved technological applications. In the papers it is also evident that the relatively new vacuum coating processes such as rf-sputtering and ion-plating will strongly influence both the production and the related properties of thin films.

The outline of the Proceedings follows closely the format used for the symposium. The Proceedings are divided into four parts:

1. Vacuum
2. Electrical properties of thin films
3. Mechanical properties of thin films
4. Optical properties of thin films.

The parts for Vacuum and Electrical properties are subdivided. Each part, as well as each subdivision, starts with an invited survey paper before the contributed papers are presented.

Our task as Editors of the Proceedings was considerably eased by the friendly co-operation of authors and referees at the Symposium. We will therefore express our deep thanks to all the authors, who were so diligent in getting their manuscripts to us on a very tight schedule, and all the referees who provided the authors and us with excellent constructive remarks.

We also want to express our sincere thanks for the financial assistance given to the Symposium and the Proceedings by The Swedish Board for Technical Development, The Swedish Government, The Ministry of Education and The Swedish National Science Research Council.

1. Vacuum and thin films measurements

1.1 International vacuum standardisation

J Yarwood, *Department of Physics, School of Engineering and Science, The Polytechnic of Central London, W1M 8JS, England*

An account of the activity of the International Standards Organization (ISO) since the early 1960s towards setting up agreed standards in vacuum science and technology. This has involved 19 countries in all, but chiefly Belgium, Britain (who run the secretariat), France, Germany, Holland and the USA. Reference is also made to work by other Standards organizations in this connection. The main topics have been terminology, subject classification, the measurement of the performance characteristics of the main types of vacuum pump, the calibration of vacuum gauges, vacuum flange dimensions, vacuum couplings, graphical symbols and leak detection. A critical review stresses some of the basic technical problems and considers present and future progress.

1. Introduction

Since the early 1960s, the International Standards Organization (ISO) have been undertaking work towards setting up agreed standards in vacuum science and technology. The ISO technical committee concerned is ISO/TC112 and the secretariat for this activity has been in Britain at the British Standards Institution (BSI) with division of specific responsibilities amongst various countries: thus, Belgium is dealing with terminology; Britain has been responsible for the measurement of the performance of vacuum pumps; France has been concerned with flanges and fittings; Germany has been involved with vacuum gauges; Holland has taken on subject classification; the USA have contributed to leak detection and aspects of gas tightness. This division of labour is not rigid: occasionally, the secretariat in Britain (after discussion with members in other countries) asks another country's standards organization to undertake a specific task. There are some 19 countries involved in all (most of them are sleeping partners) and a substantial majority of them have to give their approval to a draft ISO document before it is finally accepted and published from the central ISO office in Geneva. Arriving at International Standards is a very slow business.

During the same period, indeed, for longer, the BSI (and other national standards organizations) have also produced several vacuum standards themselves but, generally, and especially during the past decade, BSI have been concerned that their standards are within the ISO setting.

Pneurop (a co-ordinated assembly of manufacturers of compressors, vacuum pumps and pneumatic tools from 12 European countries—including Britain) have also produced standards concerned with vacuum equipment, especially pumps. Considerable use has been made by ISO of Pneurop documentation to avoid duplication of effort. Indeed, Pneurop, being more directly related to industrial concerns, has made considerably more progress on vacuum standards than ISO—which is not solely due to the advantages of commercial enterprise as against international bureaucracy, there have been other factors involved! Regrettably, the USA seem to be 'going it alone' on vacuum standards, presumably (and not surprisingly) because they have wearied of the slow progress of ISO; recently the American Vacuum Society (AVS) has brought out a booklet entitled 'Standards of the American Vacuum Society' prepared by the AVS Standards Committee under the chairmanship of Professor Hablanian; you can buy a copy for US $10. It may not be so excellent a treatise as is promised by ISO activity, but at least it has been done!

2. General vacuum standards

One of the standards dealing with general aspects produced so far is British Standard (BS2951) 'Glossary of terms used in vacuum technology, Part 1: Terms of general application' which is divided into six main sections: (1) General terms; (2) Vacuum operations; (3) Pumps and associated terms; (4) Manometers and gauges; (5) Vacuum systems and their components; (6) Leak detection. The ISO have made considerable use of this British Standard. Their activity (based in Belgium) has resulted in Draft International Standard ISO/DIS3529/1 'Vacuum technology—vocabulary—Part 1: General Terms' and, at a more preliminary stage of preparation: 'Part 2: Pumps and related terms' and 'Part 3: Manometers and vacuum gauges', and at an even more preliminary stage 'Part 4: Vacuum systems, components and accessories' and 'Part 5: Leak detection'. These ISO proposals (not yet published) are in English, French and German (the three official languages of ISO are English, French and Russian but, so far, little co-operation with the USSR has been established). The ISO terminology is more up-to-date than the British (1974–6 as against 1969) in that the names of more recent technical devices have been included. The chief problems have been, first, to classify both vacuum pumps and vacuum gauges in accordance with their fundamental principles of operation and, second, to ensure precise translations. To give one example of translation problems: 'manomètre à déflection électrostatique' is reasonably translated into 'ionisationsvakuumeter mit abgelenktem Ionenstrahl' but the proposed English term 'bent beam gauge' is unhappy, to say the least.

The most recent British work on vacuum terminology (published by BSI) is the 'Glossary of terms used in vacuum technology, Part 2: Terms of specific application'. There are three main sections: (7) Vacuum coating and allied techniques (comprising five sub-sections: 'vacuum coating processes'; 'generation and sources of coating material'; 'characteristics and types of thin films'; 'instrumentation'; and 'cleaning and removal of material from surfaces'—all together amounting to 42 entries); (8) Vacuum processing and fabrication of materials (comprising 37 entries) and (9) Vacuum instrumentation for surface studies (30 entries). It is hoped that this documentation from BSI on the terminology involved in the applications of vacuum may also act as a spur to international work in Belgium.

A 'subject classification' document, prepared in Holland, was accepted at an ISO meeting in October 1973 in Scheveningen, subject to scrutiny to ensure that the use of terms in it agrees with that in ISO vocabulary documents. The scrutiny has been done but the document has not yet been published.

After several years and following recent meetings between ISO TC112, ISO TC10 and ISO TC145, a draft International Standard (ISO/DIS3753) entitled 'Graphical symbols for vacuum technology' has been submitted but it has not yet been finally approved or published.

3. Flanges

Considerable progress has been made in arriving at standard sizes for vacuum flanges (a responsibility of AFNOR, France). There has resulted: 'Draft international standard ISO/DIS 1609.2, Vacuum technology—flange dimensions', 'International standard ISO2861/1: quick-release couplings—dimensions—Part 1: clamped type' and 'quick-release couplings—dimensions—Part 2: screwed type'.

Whilst paying tribute to the logical French in actually establishing these useful international standards it is to be noted that dimensions based on a well-worked out metric system do not cause so much controversy in committee and correspondence as do attempts to reach agreement about a laboratory-based procedure for measuring a vacuum component performance.

4. Pumps

Deliberations about the measurement of the performance characteristics of vacuum pumps has resulted in two international standards: ISO R1607 and ISO R1608, both concerned with the measurement of the volume rate of flow (pumping speed) where R1607 is about positive displacement pumps (mechanical rotary pumps) and R1608 is devoted to vapour vacuum pumps.

The test dome for the measurement of the volume rate of flow is specified to be cylindrical. For positive displacement pumps (defined as mechanical pumps which discharge gas to the atmosphere and achieve an inlet pressure below 100 Pa in one stage) to minimize fluctuations of the residual gas pressure being measured, this dome is to have a volume at least five times the volume swept by the vacuum pump during one compression cycle and where the diameter D of the test dome for pumps of various sizes (swept volumes) is as in Table 1.

Table 1. Volume of test dome for positive-displacement pump measuring system.

Swept volume (litre)	Test dome Volume (litre)	Diameter D (mm)
0 to 0.26	1.3	100
0.26 to 1.1	5.4	160
1.1 to 4.2	21	250
4.2 to 17	84	400
17 to 65	325	630
65 to 260	1300	1000

Clearly, with such test dome diameters, an adapter is needed to connect the dome to the pump inlet. This adapter is to be 0.5 D in height and be in the form of a right cone of vertex angle 45°.

For vapour vacuum pumps (R1608) it is necessary to satisfy the following main criteria in arriving at a suitable test dome.

As far as possible, the molecules of the residual gases within the dome should be of a known kind and in a state of true chaos. The former demands the admission of a measured throughput Q of a known gas (dried air, usually) at a pressure p in the dome at least 10 times the ultimate pressure achievable by the pump; the molecular chaos—no gas streaming—requires an extremely large test dome, which must be simulated to avoid excessive expense and unwieldy plant design. Figure 1 illustrates the test dome recommended by ISO: for convenience it is cylindrical and of diameter D the same as that of the inlet mouth of the pump. Indeed, ISO recommend that, wherever practicable, for all kinds of vacuum pump (other than positive displacement types) this practice is preserved regarding the test dome diameter D. To ensure that the intrinsic volume rate of flow of the pump (i.e. as when connected to an infinitely large chamber) is realized, the location of the vacuum gauge head, its tubulation dimensions and the arrangement and positioning of the gas inlet (Figure 1) are all specified. Although purists may (and do) argue that this recommended practice may not

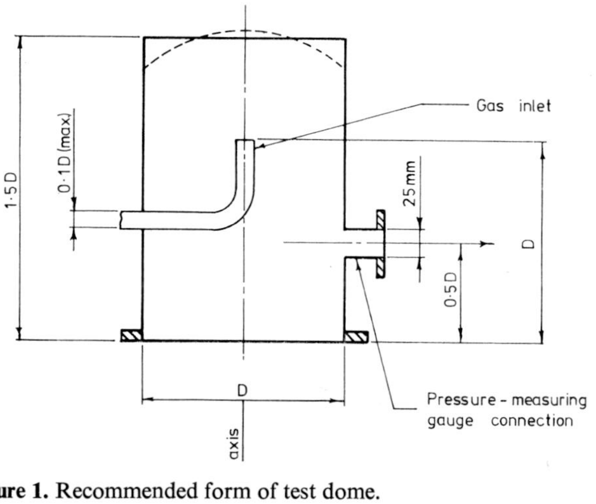

Figure 1. Recommended form of test dome.

correspond with the ideal case where the molecules entering the pump inlet have a Maxwell–Boltzmann distribution of speeds and a Knudsen cosine law distribution of directions, it is emphasized that a close approximation to the ideal is achieved and that the prime purpose of this standard is to ensure that all manufacturers and laboratories follow the same recommendations.

For practical purposes, the volume rate of flow (pumping speed) S of a given pump for a given gas (usually, dried air) is taken to be given by

$$S = Q/p \qquad (1)$$

where Q is the volumetric throughput of the gas and p is the equilibrium pressure measured by a suitable, calibrated vacuum gauge at the specified position in the test dome and under specified conditions of operation.

The problem of arriving at a standard for measuring the throughput Q of a gas at low pressures has not yet been resolved internationally but a considerable amount of work on this topic has been undertaken. So far, this has been based on work at the National Physical Laboratory (NPL), England, and has resulted in the BSI 'UK proposal for gas flow measurements using a constant pressure liquid displacement technique in the range 5×10^{-7} to 1×10^{-3} Pa m^3 s^{-1}'. It is planned to extend this range, particularly below 5×10^{-7} Pa m^3 s^{-1}, to determine the recommendations needed at very low pressures and for vacuum gauge calibration procedures. The methods of gas throughput measurement proposed are simple; the argument is about the best choice and the accuracy involved. A useful guide to throughput measurement in the range 10^{-4} to 10^{-10} Pa m^3 s^{-1} is the NPL Divisional Report MOM 14, October 1975.

A draft BSI standard giving the procedure for measuring the ultimate pressure of positive displacement pumps was prepared in 1974 but has not yet been published. Clearly, two facets are involved here: (a) the use of a gauge (such as a McLeod gauge) for the determination of the permanent gas pressure, where a liquid-nitrogen chilled cold trap is inserted between the pump inlet and the gauge; (b) the use of a gauge of which the sensitivity is independent of the nature of the gas or vapour (such as a diaphragm gauge) to measure the total pressure, with no cold trap in line. Certainly, most manufacturers of vacuum pumps prefer (a) to (b), because it puts the pump's performance in a much more favourable light and, anyway, procedure (b) demands the use of a vacuum gauge costing a thousand pounds or more. One must contend, however, that these problems do not excuse excessive tardiness in making an international standards recommendation.

Sputter-ion pumps have been the subject of much controversy from the aspect of the measurement of their performance characteristics. Organizations involved and contacted have been CERN, BSI, Pneurop and ISO. The outstanding problem here is that the performance of a sputter-ion pump is much affected by its history of usage. The 'Draft British Standard Sputter-ion Pumps—Part 1: Measurement of performance characteristics' has become a draft ISO standard DIS3556/1—still not accepted and published. This recommendation deals only with pumps for which the volume rate of flow exceeds $10\ l\ s^{-1}$ and the procedure involves two test domes (Figure 2), (a) for the measurement of ultimate pressure; (b) for the determination of the volume rate of flow against pressure. It can be argued with considerable justification that the idea of the ultimate pressure of a sputter-ion pump is meaningless: strictly scientifically speaking, this is true: but the manufacturers like to quote ultimate pressures.

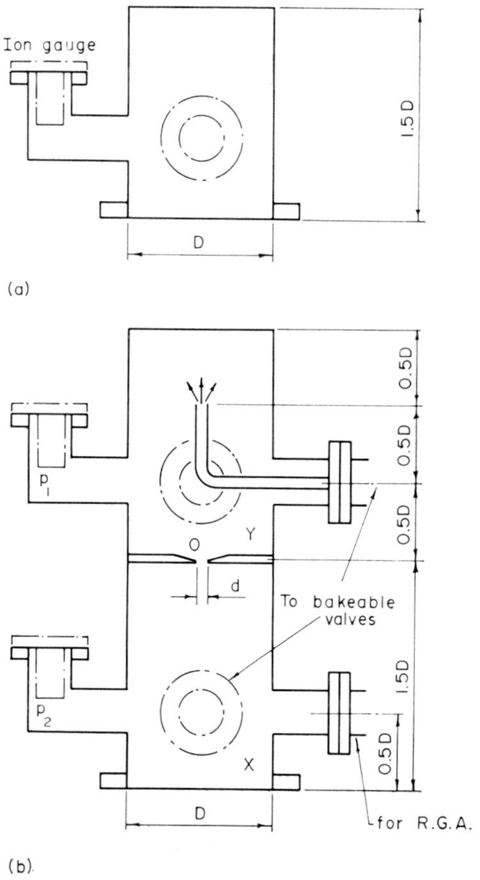

Figure 2. Test domes for measurement of performance of sputter-ion pumps: (a) for ultimate pressure measurement; (b) for the determination of volume rate of flow against pressure.

As sputter-ion pump performance is much affected by the history of the pump, the practice of a 'regenerative bake-out' is adopted. In this practice, the test dome and the sputter-ion pump are evacuated by a suitable roughing pump to below the manufacturer's recommended starting pressure and then baked for 4 h at 300°C (or the maximum temperature recommended by the manufacturer, where this is lower). The normal operating voltage is then applied to the sputter-ion pump, the roughing pump is isolated by a suitable vacuum valve and the sputter-ion pump and test dome are baked for an additional 10 h. The ionization gauge electrodes are degassed by electron bombardment, the bake-out oven is switched off and the ultimate pressure is recorded 48 h afterward.

To measure the rate of flow of gas during the determination of the characteristic of the volume rate of flow (pumping speed) against pressure, the second test dome is used (Figure 2b). The gas inlet to the upper section Y of this test dome is via a bakeable control valve to establish within Y a measured pressure p_1 whilst the pressure p_2 in the lower section X is also measured, identical ionization gauges being used.
Then,

$$Q = U(p_1 - p_2)$$

where Q is the measured throughput of dried air and U is the calculated molecularconductance of the orifice (separating sections X and Y) of diameter d.

Three main features of this standard are:

(a) To measure the ultimate pressure obtained within the test dome after regenerative bake-out.

(b) To determine the relation between the volume rate of flow for dried air and the pressure, after regenerative bake-out.

(c) To repeat (b) but after stabilization of the sputter-ion pump with dried air. This involves first the regenerative bake-out procedure and the establishment of the ultimate pressure as in (a), then a total quantity of dried air which is in excess of 4 S Pa litre is pumped where S is the nominal volume rate of flow (pumping speed) in litre per second. After pumping this quantity of dried air, the pump is allowed to reach its ultimate equilibrium pressure without bake-out, and is regarded as stabilized.

From work originally undertaken in France and originally prepared for Pneurop, there has also emanated a 'proposal for a draft International Standard for methods of measurement of the performance characteristics of turbomolecular pumps'. This draft has not yet been accepted and published as an International Standard. Two pressure ranges are selected:

(a) 1–10^{-4} Pa; (b) below 10^{-3} Pa approximately. For (a) the procedure is much the same as for the vapour pump; for (b) the test dome is like that used for sputter-ion pump performance measurement, involving a cylinder with two sections separated by an orifice of calculated molecular conductance. In the example of (b) shown in Figure 3 an addition is the right-angle tube in the lower section for *in situ* calibration of the vacuum gauges used (ionization gauges or residual gas analysers).

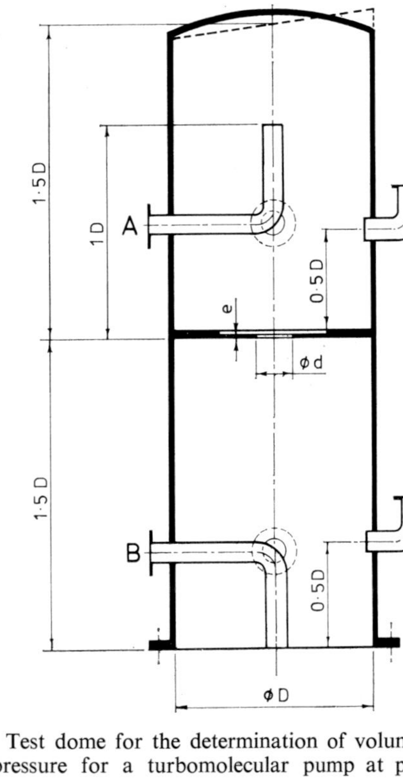

Figure 3. Test dome for the determination of volume rate of flow against pressure for a turbomolecular pump at pressures below 10^{-3} Pa.

This follows a statement from Pneurop 'one may ensure vacuum gauge agreement by fitting at B (Figure 3) a gas admission pipe directed to the pump orifice in the lower part of the dome. The adjustable valve for gas admission in this pipe line should be opened so as to obtain the desired pressure approximately. After equilibrium, gauges at points p_1 and p_2 should give the same indications, otherwise the corrections required can be deduced therefrom'. The present draft ISO recommendation is that this additional right-angle tube be optional as it increases the difficulty of construction of the test dome.

Problems of the effects of history of usage do not aggravate measurements on turbomolecular pumps as they do with sputter-ion pumps.

Two additional aspects of measurement dealt with in this draft standard are (a) the compression ratio for the turbomolecular pump, defined as 'the ratio of the backing pressure to the inlet pressure for a given gas' together with the 'compression ratio for zero gas throughput' defined as 'for a given gas under the conditions that the throughput of that gas to the pump inlet is zero and that, in the backing region, the partial pressure of that gas is at least 90% of the total pressure', and (b) 'the ultimate operational pressure' defined as 'the ultimate pressure obtained under specified conditions of operation of the pump and the test system'.

5. Gauges

Standards at a reasonably advanced state of preparation which are concerned with the calibration of vacuum gauges are:

ISO/DIS3567: a draft specification for vacuum gauges calibrated by direct comparison with a reference gauge (Figure 4). Recent discussions between the author and leaders of the NPL vacuum standards laboratory emphasize that this standard should be given priority over other methods of vacuum gauge calibration because very few laboratories in the world are interested in, or have the budget for, undertaking experimental studies on the absolute calibration of vacuum gauges. Admittedly, absolute calibration needs to be done somewhere—but the clear road to progress is surely to confine this work to national standards laboratories (such as NPL) who hardly need an ISO specification to tell them how to go about such tasks.

ISO/DIS3570/1: a draft specification—standard methods for calibration—Part 1: pressure reduction by continuous flow in the pressure range 10^{-1}–10^{-5} Pa. This is an interesting document on an absolute calibration procedure; as intimated above, it should in no way receive priority in ISO documentation schemes of the future.

In this case of pressure reduction by continuous flow, an orifice of calculated molecular conductance U separates cylindrical chambers (Figure 5) and the orifice is circular (Figure 6) with a cross-sectional area less than 0.001 of the inner surface of an inscribed sphere in the calibration chamber (the one beyond the orifice, regarded from the pumping system).

The principle is that a given gas of accurately measured throughput Q is admitted and pumped away through a gas sink of accurately known volume rate of flow S. Then,

$$p = Q/S$$

where p is the pressure of gas in the calibration chamber.

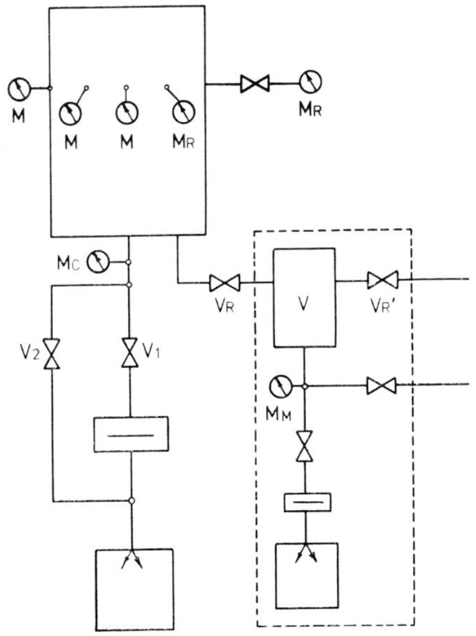

Figure 4. System for calibration of a vacuum gauge by direct comparison with a reference gauge.

M_R: reference gauge. A valve is needed if the calibration gas is condensable and if the gauge cannot be used under these conditions.
M: gauge to be calibrated.
M_C: ionization gauge, for monitoring chamber pressure.
M_M: monitor gauge for pressure in the buffer volume.
V_1: pump adjustment valve.
V_2: by-pass to be used if the calibration gas is condensable.
V_R, V_R': gas inlet adjustment valves.
V: buffer volume.

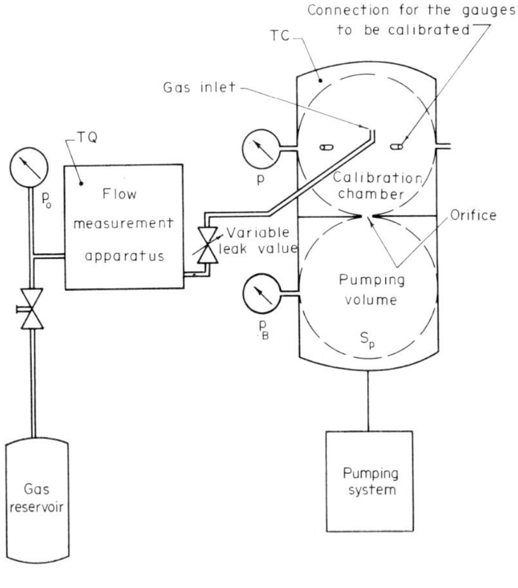

Figure 5. System for standard method of calibration of a vacuum gauge by means of pressure reduction utilizing continuous flow and an orifice of known molecular conductance.

The volume rate of flow S in the calibration chamber is determined from

$$S = U/(1 + U/S_P)$$

where S_P is the effective volume rate of flow of the pumping system.

To find S_P, either:

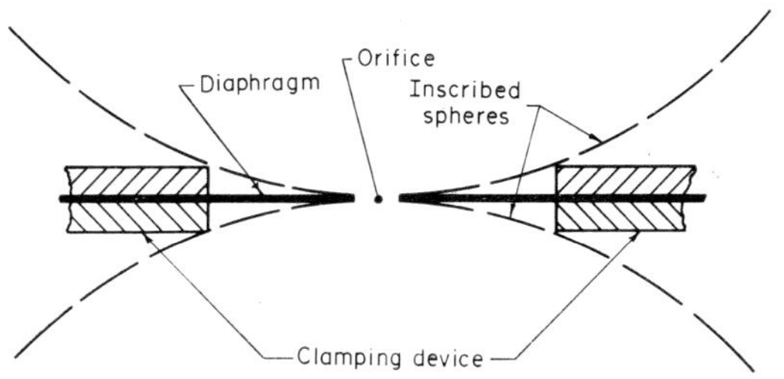

Figure 6. Circular orifice of calculated molecular conductance used in system depicted in Figure 5.

(a) the pressure increase Δp in the calibration chamber and the pressure increase Δp_B in the volume between the orifice and the pumping system are both measured when additional gas is introduced into the apparatus. Then,

$$U/S_P = \Delta p_B/\Delta p$$

(Note: if two similar linear vacuum gauges of the same sensitivity are used, they do not have to be calibrated) or

(b) the pressure p in the calibration chamber is recorded by a linear pressure gauge (not necessarily calibrated) at constant throughput Q of the given gas for a series of values of the molecular conductance U. If S_P is constant, straight line plots obtained for various values of Q should intersect at a single point on the axis of $1/U$, where

$$1/U + 1/S_P = 0$$

A draft proposal to ISO from the UK has been actively discussed for a standard based on work carried out at the NPL: 'pressure reduction by expansion into a succession of evacuated vessels in the range 10^5 Pa–10^{-6} Pa'. This excellent work aimed at absolute calibration is essential, but should it be given priority in ISO documentation and publication? I suggest not.

6. Miscellaneous

Further draft standards are:

(a) Draft international standard ISO/DIS3530, 'Vacuum technology—mass spectrometer-type leak detector calibration', and

(b) Draft British standard D62/666, 'Methods of proving the gas tightness of vacuum or pressurized plant'.

Other vacuum standards activities under discussion are 'the positioning of gauges in industrial vacuum systems', 'the measurement of the critical backing pressure of vapour pumps', 'the measurement of the water vapour tolerance of positive displacement pumps'.

7. Conclusions

Much has been done over several years by BSI, ISO and the standards organizations in France, Germany, the Netherlands and the USA in an endeavour to set up vacuum standardization. But the established internationally accepted results are too meagre. The relevant industry in most countries has lost interest because of the time and frustrations involved and current economic difficulties. There is now surely a real need to take some short cuts, to hack away at the fantastic amount of red tape and to set limited objectives of real value in this international standardization work concerned with vacuum equipment. In arriving at limited objectives, it is recommended that drastic pruning of the programme is undertaken and where survival of a standardization project is decided on a strictly utilitarian basis. Who is going to decide utility? I don't consider it good practice now to go back to all those earnest standards sub-committees; if we do, experience indicates they'll try to bring the whole lot back again! Utility here must surely be decided by the user: the industry and the government laboratories. Arrogantly assuming that such bodies need guidance on priorities, a list in order of merit, with notes, follows:

(A) ISO/DIS1609.2, concerned with 'Vacuum technology—flange dimensions' should be published and not remain much longer as a draft.

(B) Vacuum terminology is of crucial importance; the work of an international character in Belgium should be expedited, extended to include specific applications, and receive first priority as an ISO publication.

(C) A set of internationally agreed symbols for vacuum components is a considerable requirement. The committee work (which has gone on spasmodically for at least 10 years) is complete: publication is needed.

(D) Vacuum gauges: the document ISO/DIS3567 'a draft specification for vacuum gauges calibrated by direct comparison with a reference gauge' should be completed and published as soon as practicable. Consideration should be given to calibrating hot-filament ionization gauge control units (see AVS Standards Book).

(E) Vacuum pumps: ISO publications R1607 and R1608 are available; priority should be given to international agreement about gas flow measurements in the range 10^{-10}–10^{-3} Pa m^3 s^{-1}; the laboratory work at NPL forms a good basis for this.

(F) The measurement of the ultimate pressures of positive displacement pumps. This activity, which has progressed to a fairly advanced stage in committee, deserves some priority in publication internationally.

(G) Draft international standard ISO/DIS3530 'Vacuum technology—mass spectrometer-type leak detector calibration', needs to be up-dated, agreed and published.

(H) A technical report agreed internationally on 'the positioning of gauges in industrial vacuum systems'. This activity, already begun, should be completed and published. It should be concerned virtually exclusively with those vacuum systems (primarily those operating at molecular pressures) where misleading results are obtained if the vacuum gauges are positioned unwisely.

(I) Roots pumps: recommended that acceptance specification given by Pneurop be used.

(J) Consideration should be given to extending upwards the range of pressures involved in standards so far: most work has been involved with total pressures of below 1 Pa. Much industrial activity on vacuum technology involves pressures between 1 and 10^5 Pa (the atmospheric pressure, approx.) and this activity has been sadly neglected.

Other vacuum standardization work, some of which is particularly interesting to the scientist, but which should not be given high priority in international circles but is probably better completed on the national scale, includes: Specification of methods of measuring the performance of cryopumps, sputter-ion pumps and turbomolecular pumps; absolute methods of calibrating vacuum gauges; subject classification; further aspects of the measurement of the performance of positive displacement pumps such as critical backing pressure and water vapour tolerance.

1.2 Measuring technique for partial hydrogen pressures in nitrogen–neon atmospheres

G Kuus, A J de Ridder and A Pools, *N V Philips' Gloeilampenfabrieken, Lighting Division, Eindhoven, The Netherlands*

Gaseous impurities inside the discharge tube of a high pressure metal iodide lamp can have a large influence on the electrical and radiation properties of the lamp. Since the main gaseous impurity is hydrogen this gas must be eliminated from the discharge tube. One way of eliminating this gas is to use a hydrogen gettering material inside the outer bulb of the lamp since, under operational conditions hydrogen can diffuse through the quartz wall of the discharge tube into the outer bulb.

Because most types of high pressure metal iodide lamps have a mixture of a noble gas and nitrogen inside the outer bulb the gettering material must react only with the hydrogen and not with the nitrogen. For this reason the getter has to be selective. In order to investigate the selective gettering action of different chemical compounds a simple measuring technique was used to measure continuously the hydrogen partial pressure inside the nitrogen–noble gas atmosphere.

Most investigators use a mass-spectrometer or a gas chromatograph to register the change of the partial hydrogen pressure during the lifetime of the lamp. A much better technique is the use of a nickel diffusion leak combined with a manometer. A higher accuracy can be reached with this method and it is possible to add or pump away the hydrogen without changing the nitrogen or noble gas atmosphere.

The selective gettering action of a Zr–Al alloy has been examined by this measuring technique and the results of this investigation are given.

Introduction

During the last few years there has been an increasing interest in the metal halide gas discharge lamp[1] because of the fact that good colour rendition can be combined with a high efficacy.

One of the problems with this type of lamp is to keep the initial luminous efficacy and the ignition properties constant over the life time of the lamp.

It is known from other studies[2, 3] that these properties are influenced by the gas-atmosphere. Desorption of gas from the different lamp components such as the glass-wall and the electrodes will initiate chemical transport reactions. The most serious reaction is the transport of tungsten from the electrodes to the wall of the discharge tube. This transport proceeds under the influence of water or the dissociation products hydrogen and oxygen and causes a blackening of the discharge tube wall. As well as this transport reaction, the presence of this molecular gas give rise to an increase in ignition voltage of the lamp. For this reason good bakeout processes under high vacuum or a noble gas atmosphere are necessary in order to keep the lamp properties, such as the ignition voltage and luminous efficacy, constant during the life time.

Instead of this usually long degassing processes getters can be used to absorb the contaminating gases. Because the main gaseous impurity is hydrogen and the quartz wall of the discharge tube is permeable for this gas under the operating conditions of the lamp, the getter can be placed inside the outer bulb of the lamp. Since this outer bulb is filled with a mixture of neon and nitrogen the getter has to absorb selectively the hydrogen out of the nitrogen–neon atmosphere. This paper deals with such a getter as well as with a measuring technique by which it is possible to measure the change of the hydrogen pressure during the life time of the lamp.

Measuring technique

Depending on the outgassing process of the different lamp components the hydrogen pressure inside the outer bulb will rise to an equilibrium pressure between 1–10^3 Pa. From other studies[2, 3] it is known that for a pressure of hydrogen above 1 Pa there is a marked influence on the lamp quality. Thus the measuring technique has to detect hydrogen in the above-mentioned partial hydrogen pressure range in the presence of a nitrogen–neon gas mixture at a pressure of 5.10^5 Pa.

Using a mass-spectrometer or a gas chromatograph it is quite possible to measure 1 Pa of hydrogen inside this noble gas–nitrogen mixture with sufficient accuracy. However, for both techniques it is necessary to remove, from time to time, a quantity of the gas atmosphere for analysing. Especially for small lamps this can influence the lamp properties.

Of course it is possible to compensate for the gas loss but it is simpler to use a stationary system in which the nitrogen–neon mixture and the hydrogen can be separated by exploiting the difference in their diffusion properties. It is known that hydrogen and nitrogen as well as neon can be separated by means of a palladium or nickel diffusion leak.[4, 5] This separation method is used, for example, with success in order to obtain high purity hydrogen.

In 1960 Young and Whetten[6] published an article about the quantitative measurement of small amounts of hydrogen in other gases by means of a diffusion leak. They found that it was possible to determine a partial pressure of 10^{-5} Pa of hydrogen in the presence of 10^5 Pa nitrogen gas. For the present application this measuring technique is very interesting.

The experimental arrangement used is shown schematically in Figure 1. Because the degassing of the different lamp components takes place very slowly a relatively long response time

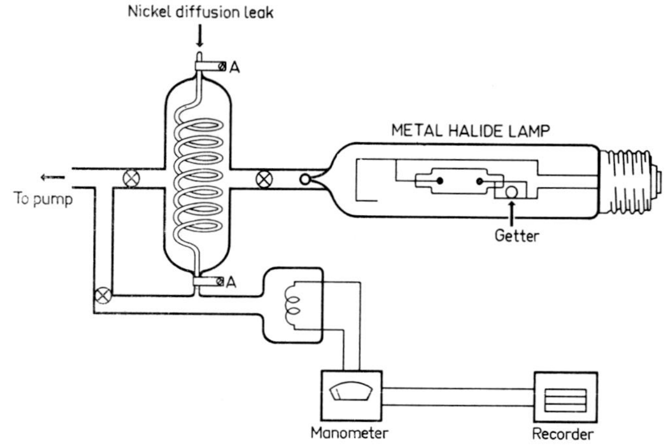

Figure 1. Technique for measuring the partial hydrogen pressure as a function of the operating time of the lamp.

for the measuring technique is acceptable. For this reason nickel can be used instead of the expensive palladium or palladium–silver alloy. The inside of the nickel tube is connected to a diffusion pump and a Pirani manometer. The nickel tube is about 100 cm in length, 0.1 mm thick and 2.0 mm in diameter. The tube is in the form of a helix to obtain a compact unit. The heating of the tube is realized by an electrical current supply which is connected to the metal–glass feedthrough, indicated in Figure 1 by A.

After reducing the nickel tube with hydrogen and degassing the glass ware at about 400°C for several hours the diffusion leak is attached to the outer envelope of the lamp and the partial hydrogen pressure can be measured as a function of the operating time of the lamp. By means of this technique different gettering materials are examined. The examination done with one special alloy will be discussed in the next section.

Selective gettering of hydrogen

Zirconium metal or alloys are used on a large scale as getter material inside vacuum systems as well as gas filled lamps. We have examined with the diffusion leak method the selective property of a zirconium aluminium alloy commercially available as St 101 getter made by the SAES firm.[7]

When this metal alloy is introduced inside the outer bulb of the lamp there is a thin oxide layer on the alloy surface. This oxide layer has to be removed in order to obtain a sufficiently high gettering rate. For vacuum tubes for example this is achieved by high frequency heating of the getter at a temperature of about 900°C in vacuum. At this temperature the oxide layer diffuses into the metal lattice. When the zirconium–aluminium alloy getter is placed in a nitrogen atmosphere this activation process is not possible because zirconium will react immediately with nitrogen to form zirconium nitride. High frequency heating of the getter before the introduction of nitrogen will give an active surface but as soon as nitrogen is introduced a protective layer of adsorbed nitrogen or even a nitride layer is formed. This nitrogen layer as well as the oxide layer forms a diffusion barrier for hydrogen. During our experiments with the zirconium–aluminium alloy it was found that this diffusion barrier can be broken by the hydrogen impurity itself. When hydrogen is brought into the nitrogen–neon atmosphere by means of the diffusion leak there is a certain hydrogen pressure

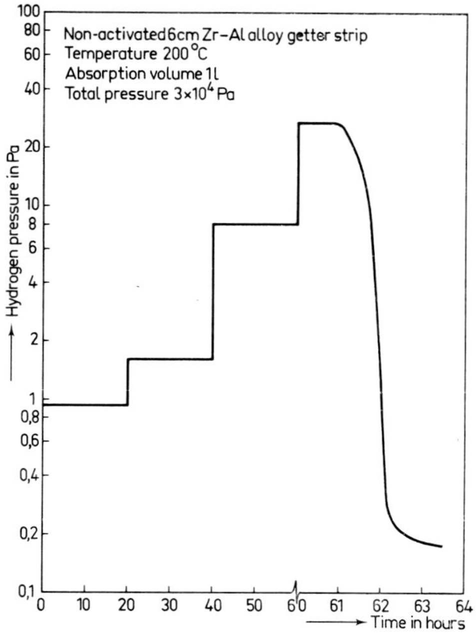

Figure 2. Absorption of hydrogen as a function of time measured in an atmosphere of nitrogen–neon.

at which the gettering action of the alloy starts and depending on the getter capacity the pressure will decrease below 1 Pa (Figure 2). So a starting pressure is necessary to obtain a good gettering action with this particular alloy. Experiments done at other temperatures have shown that this starting pressure depends on the temperature. At lower temperatures a higher starting pressure is necessary in order to activate the getter. The explanation of this effect might be a cracking of the surface layer by the hydrogen.

In 1961 Van Vucht[8] published an article about this cracking process. He found that, under certain circumstances, hydrogen can crack the oxide or nitride layer. The mechanism of this cracking can be explained by the fact that hydrogen is able to penetrate through surface imperfections and microcracks. At the interface between oxide layer and metal this hydrogen forms

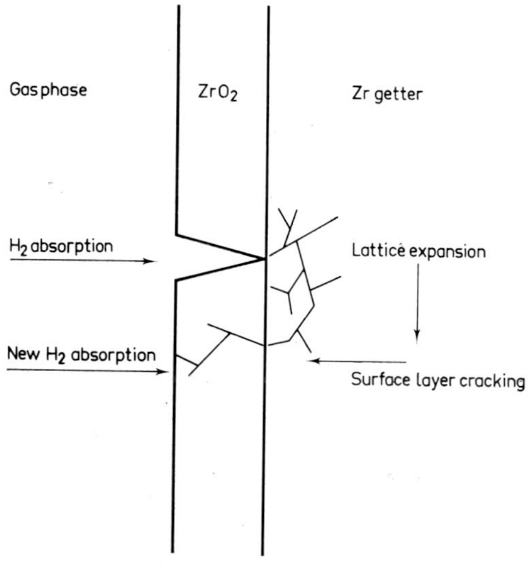

Figure 3. Getter activation by means of hydrogen.

a solid solution or even a metal hydride which gives rise to metal lattice expansion and cracking of the surface layer occurs. Through these new cracks hydrogen can diffuse more easy and crack formation goes on until the surface layer is completely broken.

It was interesting to note that during the experimental time no detectable reaction takes place with nitrogen to stop crack formation. So the reaction with hydrogen must be much faster than with nitrogen.

In Figure 3 the cracking mechanism is given schematically.

Conclusion

The experimental data presented in this paper show that the diffusion leak measuring technique can be used to measure the selective gettering action of different materials inside lamps.

From the examination done with the zirconium–aluminium alloy it may be concluded that this particular alloy has only selective gettering properties in a certain hydrogen pressure range. When the hydrogen pressure inside the lamp remains just below the starting pressure no hydrogen will be adsorbed.

References

[1] P C Drop, E Fischer, F Oostvogels and G A Wesselink, *Philips Techn Rev*, **35**, 1975, 347.
[2] W C Gungle and J F Waymouth, *J Illum Eng Soc*, **1**, 1972, 222.
[3] G Kuus, *Philips Techn Rev*, **35**, 1975, 354.
[4] O M Katz and E A Gulbransen, *Rev Sci Instrum*, **31**, 1960, 615.
[5] J R Young, *Rev Sci Instrum*, **34**, 1963, 374.
[6] J R Young and N R Whetten, *Rev Sci Instrum*, **31**, 1960, 1112.
[7] P della Porta, *J Vac Sci Technol*, **9**, 1971, 537.
[8] J H N van Vucht, *Philips Res Repts*, **16**, 1960, 263.

1.3 An evaporation cryostat for electrical and optical measurements on ultrathin metal film under ultra high vacuum conditions

S H Norrman and T G Andersson, *Department of Physics, Chalmers University of Technology, Fack, S-402 20 Göteborg, Sweden*

An evaporation cryostat for electrical and optical measurements on ultrathin metal films under ultra high vacuum conditions is discussed. The main feature of this cryostat is that the electrical and optical measurements can be carried out simultaneously, or individually in situ *or after deposition of the films in the temperature region 1–77 K. Some physical properties of ultra-thin films, which give the motivation for the construction, are also discussed.*

1. Introduction

The growth of a metal film on to a glass substrate is characterized by several steps. These are: first nucleation and initial growth, later coalescence growth and finally the formation of a metallic continuous film at large scale coalescence (LSC). Thicker films have metallic electrical and optical properties while island films have properties characterized by the small metal particles.

Experimental works on discontinuous films show that the conduction can be represented by an activation energy, which depends strongly on the film thickness.[1,2] On the other hand, theoretical calculations predict no simple dependence of the conduction on the activation energy.[3–5]

The optical transmission through a film is strongly affected by the thickness and therefore its structure. In a discontinuous film the optical properties are best characterized by an absorption peak.[6] This peak is not present in a continuous film which have bulk metal properties.[7]

Recently the growth of gold films have been investigated by measuring the resistance of the films during and after their condensation on glass[8,9] in the temperature region 77–320 K. In order to give a detailed description of the growth and structure of ultrathin metal films from resistance measurements or e.g. the optical transmission both electrical and optical properties must be accurately known.

Measurements within the temperature region 1–77 K is anticipated to give new information about: (i) the electric conduction and its activation process and (ii) the optical transmission and reflectance. Furthermore, it will be investigated: (iii) if it is possible to monitor the film growth by electrical and/or optical measurements at low temperatures and (iv) how the thickness at LSC depends on the temperature less than 77 K. These four topics give a solid motivation for construction of an evaporation cryostat.

2. Design principles

Experiments on cold samples are often restricted by condensed contamination.[10] Consequently the measurements must be performed in ultra high vacuum. One of the starting requirements was to construct the cryostat in such a way that it would fit to an already existing uhv-system. A schematic diagram of the cryostat adapted to the bell jar is shown in Figure 1.

Since the time of pumping and baking of the uhv-system together with the subsequent cooling of the cryostat is rather extensive, several substrates ought to be mounted in the system each operation cycle. It is also necessary, when studying the electrical and optical properties that the substrate is maintained at a constant temperature for a sufficiently long time.

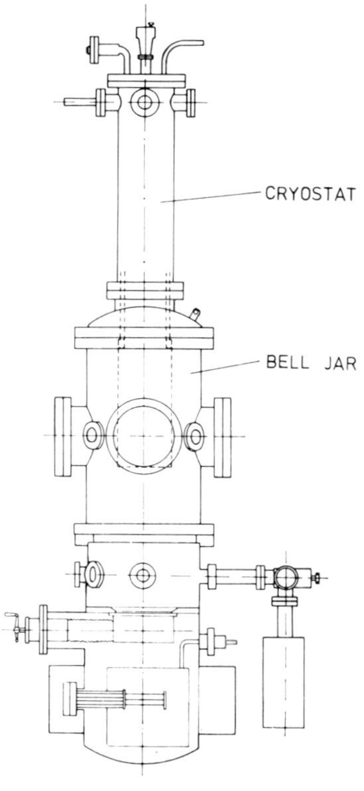

Figure 1. A schematic diagram of the evaporation system with the cryostat in position. The nitrogen radiation shield of the cryostat is indicated inside the bell jar by the dashed lines.

A cryostat has been designed and is in course of construction according to the design principles above. The following main features were set up for the cryostat in order to fulfil the desired principles:

(i) be possible to carry out electrical and optical measurements simultaneously or individually during as well as after deposition,

(ii) be able to operate down to a minimum substrate temperature of approximately 1 K. It should also be possible to vary and maintain the temperature in a controlled way from outside the uhv-chamber in the temperature region 1 –77 K with an accuracy of ± 0.5 K for at least 30 min,

(iii) work under ultra high vacuum conditions, where the pressure during deposition must not exceed $4 \cdot 10^{-9}$ torr,

(iv) be capable of handling eight substrates, whose position can be changed from outside the uhv-chamber. The rotation of the holder must not change the temperature on the substrates. Furthermore, there must be a good thermal contact between the substrates and the substrate holder.

An overall photograph of the cryostat is shown in Figure 2. The cryostat is a ^{4}He pumped all-metal cryostat of Dewar type. The construction consists of the following, partly conventional, main elements:

(i) an outer vacuum jacket,

(ii) a liquid nitrogen container of stainless steel cooling a copper radiation shield surrounding the inner parts of the cryostat,

(iii) a stainless steel liquid ^{4}He container cooling another copper heat shield,

(iv) a pumped liquid ^{4}He pot of oxygen-free high-conductivity (OFHC) copper bringing down the temperature of itself and the substrate holder,

(v) the substrate holder, which is rotatable and handles eight substrates and

(vi) an axle with accessories, which sustains the substrate holder and makes it rotatable. The substrate holder is rotated via the axle from a feedthrough on top of the cryostat.

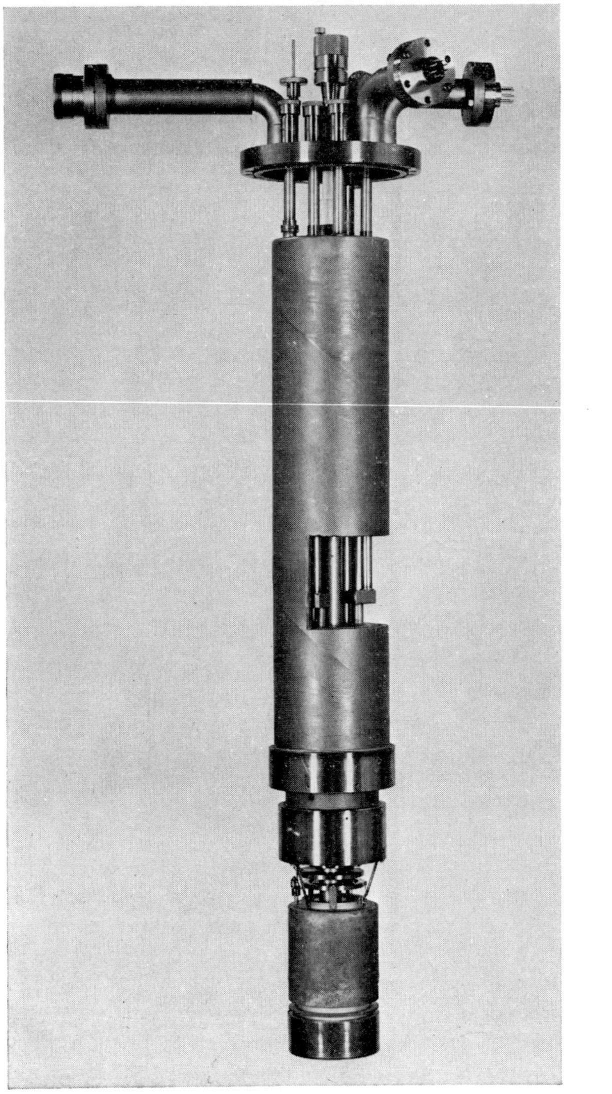

Figure 2. An overall photograph of the cryostat with the vacuum jacket and the two radiation shields removed.

3. Discussion

A cryostat has been designed and is in course of construction to match the listed features in Section 2. It is a ^{4}He pumped evaporation cryostat suitable for electrical and optical measurements on insulating and transparent substrates. It works under ultra high vacuum conditions and it is therefore completely bakeable to 300°C without having removable or cooled thermometers or other parts. The temperature of the substrates may be varied and maintained at any temperature from 1 to 77 K with an estimated accuracy better than ± 0.5 K for at least 30 min.

The substrates are in turn put into position with the help of an axle, which is rotatable half a revolution in both directions from outside the uhv-chamber, a 'stop-location' arrangement, and a rotational feedthrough.

Several advantages compared with earlier constructed cryostats are easily recognized. We consider the main new feature of this cryostat being the simultaneous (and/or individual) measurements of electrical and optical properties of ultrathin films during and/or after deposition. The technique to monitor electrical and optical properties *in situ* has usually been performed in separated cryostats. As far as it is known to the authors there has not earlier been constructed an uhv-evaporation cryostat for simultaneous *in situ* measurements of electrical and optical properties of ultrathin films.

Later a complete report will be given concerning the detailed construction of the cryostat. In the report also the operation and performance will be discussed.

References

[1] C A Neugebauer and M B Webb, *J Appl Phys*, **33**, 1962, 74.
[2] L L Kazmerski and D M Racine, *J Appl Phys*, **46**, 1975, 791 and 2658.
[3] R Kiernan and D W Stops, *Nature*, **224**, 1969, 907 and *J Vac Sic Techn*, **9**, 1971, 433.
[4] R M Hill, *Le Vide* (Suppl), **147**, 1970, 247.
[5] A Barr and R D Finney, *Thin Solid Films*, **24**, 1974, S11.
[6] V V Truong and G D Scott, *J Opt Soc Am*, **66**, 1976, 124.
[7] C G Granqvist and O Hunderi, *Phys Rev* B, (in press).
[8] T Andersson, *J Phys D: Appl Phys*, **9**, 1976, 973.
[9] T Andersson, *J Appl Phys*, **47**, 1976, 1752.
[10] R A George, I Roberts and E G Wilson, *J Phys E: Sci Instrum*, **8**, 1975, 35.

1.4 Fast volume determination using a differential capacitance manometer

H Norström, S Berg and L P Andersson, *Institute of Physics, University of Uppsala, Uppsala, Sweden*

The reliable pressure readings with a capacitance manometer are the basis for a new type of instrument designed for accurate and rapid measurements of capillary volumes. The basic measuring principle is based on expansion of air of atmospheric pressure enclosed in a known volume into a pre-evacuated cylinder so that the enclosed volume will cause a reduced pressure in the cylinder. The same is done with the air in an unknown volume into another identical pre-evacuated cylinder. A differential capacitance manometer is then used to measure the difference in pressure between the two cylinders with a high degree of accuracy. This difference in pressure is shown to be a linear function of the difference in volume between the unknown and the known volume. The accuracy and repeatability of the instrument is better than 0.5%.

1. Introduction

The need of rationalization in precision glass thermometer manufacturing was the original reason for designing this instrument.

A glass thermometer contains the bulb with the mercury and capillary tube intended for the thermal expansion of the mercury. A certain bulb filled with mercury in combination with a specific capillary tube will give the desired expansion length difference L_0 for a 100°C difference in temperature.

The inner diameters of the capillary tubes are known from the glass tube manufacturing company, but deviations of several per cent from predicted values do frequently occur.

The glass bulbs, however, are made manually by glass blowing. With this technique it is not possible to achieve better accuracy than $\pm 10\%$ in the predetermination of the manufactured bulb volume.

It is thus obvious that such a glass bulb, in combination with a standard capillary tube will result in a thermometer with a predicted expansion length not better than $\pm 10\%$. Each manufactured thermometer, therefore has to be individually calibrated. Further problems arise when scales are to be attached to the thermometer. The manufacturer has to have a set of some 20 different scales to match the different expansion lengths obtained.

If the bulb volume could be determined within 0.5% accuracy, it is possible to combine this bulb with a capillary tube that gives the desired expansion length L_0. The possibility of making such combinations eliminates the individual calibrations in the manufacturing process.

2. Description

At room temperature and atmospheric pressure the air (oxygen and nitrogen) behaves as a perfect gas obeying Boyle's law[1]

$$P_0 V_0 = P_1 V_1$$

This is the basic principle of the instrument designed.* The unknown volume (V_x) in a glass bulb contains air of atmospheric P_0. This air volume is expanded into a large pre-evacuated cylinder with a volume $V_1 (V_1 \gg V_x)$. The pressure in the cylinder will then be

$$P_1 = \frac{P_0}{V_1} \cdot V_x$$

In another branch of the vacuum system a glass bulb containing a known volume V_0 of air at atmospheric pressure P_0 is expanded into another pre-evacuated cylinder with volume V_2. The pressure in this cylinder will then be

$$P_2 = \frac{P_0}{V_2} \cdot V_0$$

Now a differential capacitance manometer[2] (MKS Baratron 170M-6 instrument supplied with a type 145AH-10 head) is connected between the two cylinders thus indicating the pressure

$$P = P_1 - P_2$$

This pressure reading P can by electronic compensation be processed into a reading

$$u_2 = k(V_x - V_0) \qquad \text{(appendix)}$$

u_2 is then a signal that tells the difference in volume between an unknown volume V_x and a well-known volume V_0. The technique of calibration of the instrument is pointed out in the appendix.

A schematic diagram of the mechanical set-up of the system is shown in Figure 1. S1–S8 are electrically regulated valves that can be either closed ('0') or open ('1'). The mechanical pump (Edwards EDM2) has a pumping speed of 2 $m^3\ h^{-1}$ and the total internal volume to be evacuated is about 0.5 l. This situation gives a total pumping time less than 1 min.

The valves are operated according to Table 1.

* Swedish pat 741209–9, US pat 618161, German pat 2542838.

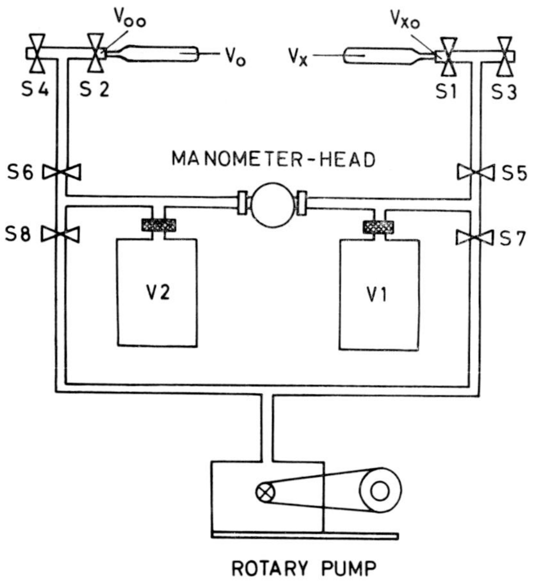

Figure 1. Schematic of the mechanical set-up of the instrument.

Table 1.

Sequence	S1	S2	S3	Valve S4	S5	S6	S7	S8
Start	1	1	1	1	0	0	1	1
Evacuate	0	0	0	0	1	1	1	1
Read	1	1	0	0	1	1	0	0

The three operating sequences of the instrument are the

'start' sequence in which the unknown glass bulb V_x is attached to the instrument. V_0 is constantly fitted to the instrument. The purpose of *S*3 and *S*4 is to maintain atmospheric pressure in V_x and V_0 in this sequence. Then comes the

'evacuate' sequence in which all the internal volume of the instrument is evacuated to a pressure better than 0.1 torr by the rotary pump. Finally comes the

'read' sequence in which the two volumes V_x and V_0 of air are expanded into the instrument and the corresponding pressure difference between the two branches is measured.

3. Experimental results

The actual volumes of the thermometer bulbs measured by this instrument have been in the range of 0.3–1.0 cm^3. To ensure the required accuracy, the instrument must be able to resolve a volume difference of less than 5 mm^3.

It should be pointed out that the actual volume of air of atmospheric pressure expanded into the instrument is $V_x + V_{x0}$, where V_{x0} is the internal volume of the valve *S*1 at atmospheric pressure during the sequence 'evacuate'. The volume V_{x0} has been measured to be about 0.3 cm^3. The valve *S*2 has a similar added volume V_{00}. In the differential pressure sensing technique however, these two volumes almost compensate each other and cause only a small residual 'error pressure'. Experiments have verified that this 'error pressure' is a design parameter depending primarily on the valves (*S*1 and *S*2) used. This 'error pressure' also turns out to be constant and can therefore easily be eliminated as shown in the appendix. Further advantage of the differential measuring system is that the instrument is insensitive to the pressure reached (about 0.1 torr) during the 'evacuation' sequence. Non-symmetry of pumping speeds in the two branches of the system can cause a pressure difference of at most 0.005 torr at the end of this sequence. This difference adds a negligible error to the value measured during the 'reading' sequence.

The instrument is designed to give a pressure difference of 1 torr between the two branches if the unknown volume deviates 0.35 cm^3 from the known standard volume $V_0 = 0.70$ cm^3. In this pressure region the overall error of the gauge head is better than 0.1%.

Figure 2 shows the correlation between measured (with this instrument) volumes V_x in fraction of V_0 and the same fraction (V_x/V_0) obtained from the manufactured complete thermometer.

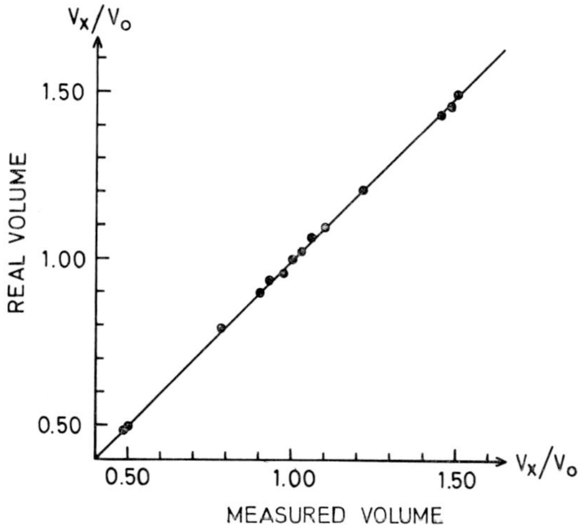

Figure 2. Normalized measured bulb volumes versus normalized real bulb volumes. The normalizing factor being the known standard volume V_0.

The correlation is very good. The deviations from perfect correlation have been found to be mainly due to errors in the supplied diameter data of the capillary tubes, which thus at this moment limit the accuracy of predetermination of thermal expansion length of the thermometer.

4. Conclusion

Volume determination using a differential capacitance manometer has proved to be very reliable. The symmetrical design of the two branches in the instrument has reduced the influence of 'extra added volumes' to the measured volume to a minimum since such volumes compensate each other. A further advantage is the independence of variations in the limiting vacuum reached by the mechanical vacuum pump. The instrument was designed to measure the volume of glass thermometer bulbs but it is expected to be used soon in a wide variety of applications.†

† The instrument is commercially available under the name 'ITM-761-A Capillary Bulb Measurement Unit' and exported by Qvintus Inc, Box 408, 126 04 Hägersten, Sweden.

The developement of this instrument was financially supported by the Swedish Board of Technical Development.

Appendix

P_0 = atmospheric pressure

V_x = unknown volume to be measured

V_{x0} = internal volume of valve S1 at atmospheric pressure during sequence 'evacuate'

V_1 = extra volume to which V_x and V_{x0} are expanded during sequence 'read'

P_1 = pressure reached in V_1 during sequence 'read'

V_0 = thermometer glass bulb with known internal volume

V_{00} = internal volume of valve S2 at atmospheric pressure during sequence 'evacuate'

V_2 = extra volume to which V_0 and V_{00} are expanded during sequence 'read'

P_2 = pressure reached in V_2 during sequence 'read'

The volumes $V_x + V_{x0}$ of atmospheric pressure P_0 are expanded into the volume V_1. Thus

$$P_0(V_x + V_{x0}) = P_1(V_x + V_{x0} + V_1) \tag{1}$$

but

$(V_x + V_{x0}) \ll V_1$ is always valid so

$$P_1 = \frac{P_0}{V_1}(V_x + V_{x0}) \tag{2}$$

The other branch behaves identically, thus

$$P_0(V_0 + V_{00}) = P_2(V_0 + V_{00} + V_2) \tag{3}$$

and

$(V_0 + V_{00}) \ll V_2$, so

$$P_2 = \frac{P_0}{V_2}(V_0 + V_{00}) \tag{4}$$

The differential capacitance manometer reads the pressure P

$$P = P_1 - P_2 \tag{5}$$

Combination of equation (2), (4) and (5) gives

$$P = \frac{P_0}{V_1} \cdot V_x + P_0\left(\frac{V_{x0}}{V_1} - \frac{V_0}{V_2} - \frac{V_{00}}{V_2}\right) \tag{6}$$

Here

$$P_0\left(\frac{V_{x0}}{V_1} - \frac{V_{00}}{V_2}\right)$$

is the earlier mentioned 'error pressure'.

Equation (6) shows that P is a linear function of V_x. The measured value P is available from the manometer unit as an analogue voltage u. Thus

$$u = k \cdot V_x + l \tag{7}$$

where k and l are instrument constants. By electronically adding a constant voltage u_1

$$u_1 = -(kV_0 + l) \tag{8}$$

to u one obtains a value

$$u_2 = u_1 + u = k(V_x - V_0) \tag{9}$$

The value of u_1 is easily found by applying a volume V_0 in the position of V_x and during the 'read' sequence adjust to zero output signal.

Now u_2 can be amplified to desired value to obtain full scale reading during sequence 'read' with a glass bulb of known volume 1.50 V_{00} in position of V_x.

References

[1] A Roth, *Vacuum Technology*, North Holland, Amsterdam, 1976.
[2] N G Utterback and T Griffith Jr, *Rev Sci Instrum*, **37**, 1966, 866.

1.5 Effects in Auger electron spectroscopy due to the probing electrons and sputtering

P Braun, W Farber, G Betz and F P Viehböck, *Institut für Allgemeine Physik, University of Technology, Vienna A-1040, Austria*

The analysing electron beam and sputtering used for cleaning and depth profiling can alter the chemical composition of composite materials in the near-surface region. Therefore, in a correct analysis several effects caused by the incident ion- or electron beam have to be considered besides the general problem of quantitative AES. No general formalism can be given as these effects depend on the type of material investigated. Examples are given which arise in analysing oxides, carbonates, silicates, homogeneous and heterogeneous alloys.

In oxides, the chemical bonding between oxygen and metal is broken up under the influence of the electron beam and oxygen is desorbed. Spectroscopying carbonates, the probing electrons evolve CO_2 which is desorbed and only the remaining oxide can be detected. For silicates, changes in the surface composition due to the electron beam as well as the ion beam were observed. Sputtering of alloys results in preferential sputtering changing the surface composition in accordance to different sputtering yields of the constituents. This is shown for one-phase as well as two-phase alloy systems. For the two-phase system Ag–Cu a further effect is observed which can be explained by pronounced surface diffusion of one constituent, namely silver. Finally for materials of a low concentration of heavy atoms in a matrix of light atoms the knock-in effect is observed.

1. Introduction

Auger transitions evolve electrons of core levels and often also valence band electrons. Therefore, the line shape as well as the electron energy is strongly determined by the chemical environment.[1, 2] If a strong bonding between the atoms in a compound exists, the electron levels may shift several electron volts. The electron levels of electronegative elements are shifted to lower binding energies, whereas the levels of electropositive elements are shifted to higher binding energies in accordance to the charge transfer. Auger electrons out of these levels show a corresponding chemical shift in their kinetic energy. Furthermore, changes in the electronic density of state in the valence band caused by the chemical environment alter the shape of Auger peaks of transitions involving valence band electrons. This was demonstrated by several authors by oxidizing clean metal surfaces.[3, 4]

1.1. Effects due to the probing electrons. In many cases interactions can occur between the solid surface and the probing electron beam. One can distinguish three effects[5]:

(a) Electron impact desorption, where the probing electron beam can cause desorption of quite strongly bound gas molecules. Desorption of weakly bound oxygen atoms has an initial probability of about 10^{-2} atoms per electron.[6] Using a typical primary beam current in the order of 10^{-5} A physisorbed oxygen atoms desorb in less than 1 s. Therefore, for beam currents of this magnitude only chemisorbed oxygen can be detected.

(b) Electron stimulated adsorption of residual gas molecules can take place. Gas molecules which would only be weakly bound to a surface dissociate under the electron beam on or close to the surface. Depending on the type of dissociated species and the target atoms it is possible that one of the components binds much more strongly to the surface than the original gas molecule. For example, oxygen arising near a previously clean silicon surface results in oxygen build-up previously inside the area irradiated by the electron beam.[7] Reduction of the residual gas pressure can reduce this electron beam effect.

(c) Electron induced dissociation or decomposition is probably most important. Many compounds are really damaged by electron irradiation and care must be taken, especially in a quantitative analysis of such materials. Examples are alkali halides which are sensitive to the probing electrons, causing preferential desorption of halogen atoms[8] and SiO_2 which shows a partial decomposition to elemental Si.[9]

1.2. Effects due to sputtering. Sputtering is essential in AES for cleaning contaminated samples, after being introduced into the uhv system, and also for composition vs depths profiling of thin films. AES is sensitive only to the topmost 2–5 atomic layers. Therefore, surface composition changes of alloys and compounds due to sputtering are of great importance, if quantitative analysis is strived for. While sputtering of pure metals is understood fairly well, sputtering of alloys and compounds is not fully understood yet. Selective or differential sputtering and related effects have to be accounted for a quantitative in depths analysis with AES and also surface diffusion probably enhanced by ion bombardment can be found.

Considering the effect of differential sputtering of alloys one has to distinguish different cases:

(a) One-phase systems consisting of continuous series of solid solutions like Ag–Au, Au–Cu, or Cu–Ni. For such systems we cannot assume *ad hoc* that the sputtering yields of the pure constituents are preserved for the alloy. There are not many experimental results available in literature. Only the system Cu–Ni was studied more extensively by different authors. Tarng and Wehner[10] found that the surface becomes enriched in Ni by ion bombardment of a 55 Cu–45 Ni alloy. For ion energies between 200–500 eV the ratio of sputtering yields for the two components is the same as for the pure elements. For lower energies, especially near the sputtering threshold, large deviations occur. Shimizu *et al.*[11] obtained similar results for a series of different Cu–Ni alloys.

(b) Multiphase systems. In this case where different phases are present in the surface it is obvious to expect a preferential sputtering effect. Especially in a two-phase system like Ag–Cu, where the solid solubility at room temperature of one constituent in the other is negligible, preferential sputtering takes place in accordance to the different sputtering yields. For instance, by sputtering an Ag–Cu eutectic and using an optical emission spectroscopy technique Anderson[12] found an enrichment in Cu in a thick surface layer. Similar conditions can be stated for MgO/Au cermets.[13] Sputtering by 500 eV Ar ions a depletion of Au is caused by the low sputtering yield of the MgO.

In all mentioned examples a preferred sputtering of one component was interpreted preliminarily by the different sputtering of the constituents. Surfaces which contain 'difficult to sputter' species tend to become covered with cones.[14] These difficult to sputter species accumulate in steps or scratches and are mixed together with sputter deposited substrate atoms and this mixture has a low sputtering yield.

A more complete picture of changes in surface composition due to sputtering involves furthermore other effects especially knocking-in of heavy atom impurities in a matrix of a light element and amorphization of the crystal structure. For example the Auger peak ratio S/Mo of a freshly cleaved single crystal MoS_2 sample is reduced by a factor of 2.5 after sputtering away a few monolayers.[22]

2. Experimental results

2.1. Carbonates. Auger analysis of carbonates reveals a very surprising effect. A large number of different carbonates were analysed, but in none of them a C-peak could be detected.

The following carbonates have been analysed: a calcite crystal ($CaCO_3$), a pearl whose composition is 92% aragonite ($CaCO_3$) and 8% organic compounds, and samples of pressed powder: $MgCO_3$, $MnCO_3$, $SrCO_3$, and $BaCO_3$. Figure 1 shows an Auger spectrum of $BaCO_3$ after sputter cleaning the sample. A rough estimate of the C:Ba Auger peak heights ratio according to the stochiometric composition $BaCO_3$ and using the Handbook of AES[15] one should expect a ratio C:Ba equal 1. The energy region of the C-peak is shown enlarged by a factor of 10 in Figure 1. Therefrom follows that the peak heights ratio C:Ba is less than 0.01 in the analysed surface layer. Also spectra taken during sputtering did not show any C on the surface. One sample, the pearl, was also analysed with ESCA, which shows carbon distinctly.

The sputter yield of C has to be larger than that of Ba by at least two orders of magnitude to explain this effect by differential sputtering of C. This, however, is most unlikely, as the sputter yields of all pure elements measured are within one order of magnitude.[14]

More likely the electron beam decomposes the carbonate and CO or CO_2 desorbes from the surface. The part of the surface exposed to the electron beam becomes a metal oxide which is in addition deficient in oxygen. This is evident from a semi-quantitative analysis of the Auger peak heights of oxygen and metal and furthermore supported by experiments with oxides, like SiO_2, Al_2O_3 which also show the electron beam induced decomposition.[9,16]

2.2. Silicates. Three different silicates: K Al Si_3O_8, Na Al Si_3O_8, and Ca $Al_2Si_2O_8$ were studied. These three materials, which form a continuous series of solid solutions with each other and have similar crystal structure and chemical composition, were Auger analysed during sputtering and after sputtering. Table 1 summarizes the results obtained. Using the corresponding chemical formula for each silicate one can define

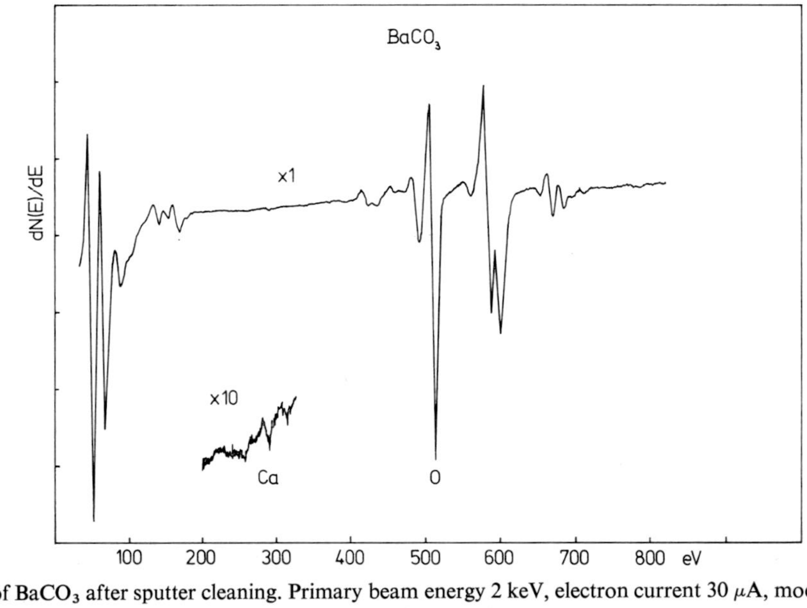

Figure 1. Augerspectrum of $BaCO_3$ after sputter cleaning. Primary beam energy 2 keV, electron current 30 μA, modulation amplitude 4 Vpp.

Table 1. Auger yields for three silicates in relation to aluminium calculated from measured Auger peak heights during and after sputtering.

	Al	Si	O	K	Na	Ca	Calc formula
K Al Si_3O_8							
during sputtering	1	0.37	2.1	3.2	—	—	$K_{0.2}$ Al $Si_{1.2}O_{0.8}$
10 min after sputtering	1	0.37	2.3	0.85	—	—	$K_{0.05}$Al $Si_{1.2}O_{0.9}$
Na Al Si_3O_8							
during sputtering	1	0.35	1.9	—	0.21	—	$Na_{0.2}$Al $Si_{1.2}O_{0.7}$
10 min after sputtering	1	0.37	2.1	—	<0.05	—	$Na_{<0.04}$Al $Si_{1.2}O_{0.8}$
Ca $Al_2Si_2O_8$							
during sputtering	1	0.43	3.4	—	—	7.5	$Ca_{0.8}Al_2Si_{0.9}O_{1.3}$
10 min after sputtering	1	0.39	3.2	—	—	9.0	$Ca_{0.9}Al_2Si_{0.85}O_{1.2}$
Data from Handbook[15]	1	0.92	21	16.8	1.2	9.4	

relative Auger yields per atom relative to aluminium calculated from the measured Auger peak heights. As can be seen the relative Auger yield for silicon is nearly the same for all silicates investigated and independent if the data were taken during or after sputtering. For oxygen the relative Auger yield is equal only for K Al Si_3O_8 and Na Al Si_3O_8 whereas for Ca $Al_2Si_2O_8$ with a slightly different chemical composition the Auger yield is somewhat higher, but again independent of sputtering. However, the relative Auger yields of potassium and sodium decrease drastically after stopping sputtering and the yield of calcium increases.

Furthermore, Table 1 contains the relative Auger yields as taken from data given in the Handbook of AES.[15] These data are derived from the pure elements aluminium, silicon, calcium and the compounds MgO, KCl, and WO_3Na_7. Deviations up to 50% can be expected[17] due to matrix effects on electron transport properties, chemical effects on peak shapes and surface topography. However, differences up to a factor 10, as for oxygen, cannot be explained in this way. Changes in the surface composition must have taken place due to differential sputtering and/or the influence of the incident probing electron beam.

Using the relative Auger yield, from the Handbook[15] one can calculate the altered surface composition which is also listed in Table 1. Under this assumption silicon, oxygen, potassium and sodium are found to be largely depleted relatively to Al at the surface. Besides of possible differential sputtering there is strong evidence that the electron beam itself changes the near surface composition.

Figure 2 shows the changes in the surface composition of a Ca $Al_2Si_2O_8$ sample under ion bombardment, without, and under resumed ion bombardment. After stopping sputtering, the surface becomes slowly enriched in aluminium and calcium and depleted in silicon. If the changes would be due to annealing after differential sputtering, only the few topmost layers should be involved. However, as can be seen in Figure 1, a layer exists with an altered surface composition of a thickness of about a few hundred Angströms. This indicates diffusion processes probably enhanced by the electron irradiation.

The corresponding experiments on K Al Si_3O_8 and Na Al Si_3O_8 show also long-time changes after sputtering under the influence of the electron beam, however of a different behaviour. For example, in the case of K Al Si_3O_8, K, Al decrease and Si increases with time, in opposition to the behaviour of Ca, Al, Si in $CaAl_2Si_2O_8$. Electron microprobe experiments with similar materials, namely binary oxide glasses of Na_2O and K_2O with SiO_2 show also changes in the surface composition with time due to the primary electron beam.[18]

2.3. Binary alloys. The binary alloy systems Ag–Au, Au–Cu and Ag–Cu have been studied in view of surface composition changes due to sputtering.[19, 20] For that purpose a series of

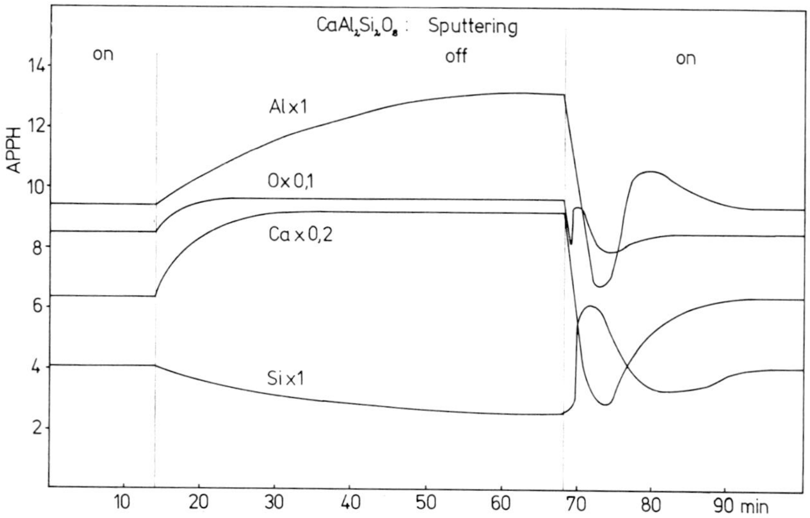

Figure 2. Changes in surface composition of Ca $Al_2Si_2O_8$ under ion bombardment, without, and after ion bombardment resumed again.

alloys with different and known bulk composition of each binary system has been sputtered with 1 keV Ar^+ ions under 20° to the sample surface and simultaneously analysed by AES. Ag–Au and Au–Cu are one-phase systems with a continuous series of solid solutions over the whole concentration range, whereas Ag–Cu is a two-phase system with an eutectic point at 39.9 at.% Cu.[21]

To observe changes in the surface composition of an alloy due to sputtering effects it is necessary to start from a clean surface with a composition equal to the bulk composition. For solid solutions as in the Ag–Au and Au–Cu systems this was achieved by scribing or cleaving the samples in the uhv-system. Unfortunately, these methods are not valid for obtaining a surface composition equal to the bulk in the two-phase Ag–Cu system.[20]

(a) One-phase systems: Ag–Au and Au–Cu.

For the one-phase systems Ag–Au and Au–Cu differential or selective sputtering was found in agreement with the sputter yield ratios of the pure elements. The sputtering yields for Ag, Cu, and Au under 1 keV Ar^+ ion bombardment are: 3.8; 3.6; and 3.6 resp.[14] So for Ag–Au alloys an enrichment of Au on the surface under ion bombardment was observed and is largest in the topmost surface layer, decreasing with depth. The changes in composition with depth were detected by comparing Auger electrons with different escape depths. The ratio of the sputtering yields for Ag and Au estimated from the observed enrichment is constant over the whole concentration range and in qualitative agreement with the data given in ref 14. For the Au–Cu system no changes in the surface composition occur due to the equal sputter coefficients of pure Au and Cu, and therefore sputtering, cleaving and scribing of the samples yield the same results.

(b) Two-phase system: Ag–Cu.

A quite different behaviour shows the two phase system Ag–Cu consisting of almost pure Ag and Cu crystallites at room temperature. The grain size determined by SEM was about 1 μm. To obtain an equilibrium surface composition under ion bombardment a removal of at least 1 μm thickness is necessary, equivalent to the size of crystallite.[13] This is contrary to the one-phase systems, where equilibrium was reached after removal of a few monolayers.

Figure 3 shows the normalized Auger peak heights $Ag_N = [Ag_m/(Ag_m + Cu_m)] \times 100$ for silver versus the bulk composition Ag_b after cleaving, scribing and sputtering. Ag_m and Cu_m are the measured Auger peak heights of the 351/356 eV Ag and the 58/60 eV Cu Auger transitions. Because of the different tensile strength of the Ag and Cu crystallites, cleaving and scribing are not valid methods to obtain reference data for the bulk composition. This can be seen from Figure 3, where cleaving and scribing result in different curves, which is not the case for one-phase systems. Due to the lack of reference data for the bulk composition differential sputtering cannot be properly shown. However, from the form of the sputtering curve in Figure 3, especially in the vicinity of the eutectic point, it is obvious that sputtering has also created a surface composition different from the bulk.

Furthermore, changes in the surface composition due to surface diffusion were observed after sputtering Ag–Cu

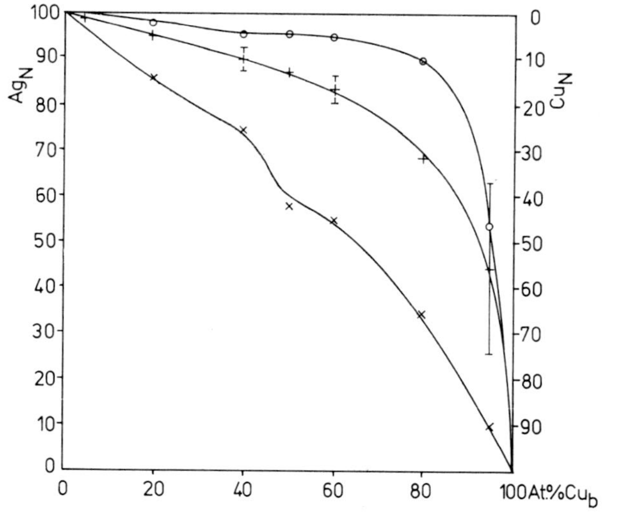

Figure 3. Normalized Auger peak heights Ag_N, Cu_N versus the bulk composition after cleaving (○), scribing (+), and sputtering (×).

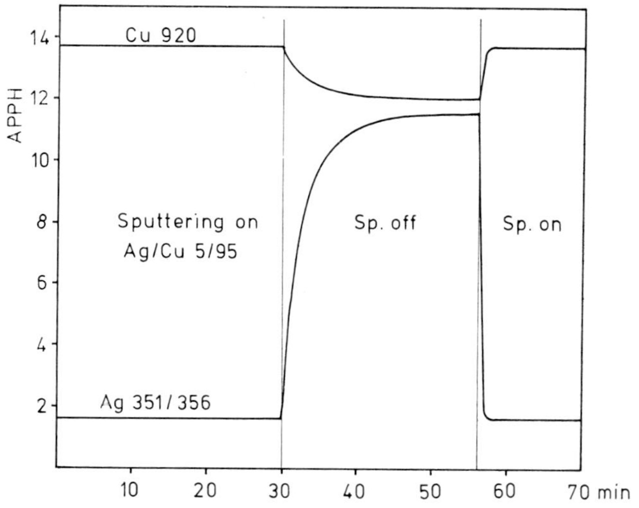

Figure 4. Changes of the Ag and Cu Auger peak heights for a 5 at.% Ag–95 at.% Cu sample vs time.

samples. Figure 4 shows for a 5 at.% Ag–95 at.% Cu sample the changes of the Ag and Cu Auger peak heights with time. Stopping sputtering after a prolonged bombardment time, the Ag signal increases rapidly by a factor of about 8 within 15 min. Starting sputtering again the changes are immediately reversed, Because the solubility of Ag in Cu and vice versa is negligible. annealing after sputtering through bulk diffusion can be excluded. Furthermore, it was established that the effect is not caused by heating of the probing electron beam. The relative increase of Ag, when sputtering is stopped, is inversely proportional to the Ag bulk concentration. Assuming surface diffusion, a simple calculation shows that the surface becomes covered by about 1 to 2 atomic layers of Ag independent of the bulk composition.

There were no changes in surface composition detected after stopping the ion bombardment for the one phase Ag–Au and Au–Cu alloys.

2.4. Knock-in effect. Besides of differential sputtering composition changes at the surface and especially at interfaces can be caused by knock-in effects. This is demonstrated by sputtering a Si/Ag sample with Ar^+ ions of different energies. The sample was made by simultaneous sputter deposition of silicon and silver on a Mo target. The estimated content of silver by using the Handbook of AES[15] was about 10 at.%. Figure 5 shows the change of the silver concentration by switching the ion energy from 2 keV to 500 eV and vice versa. The incident primary ion beam creates collision cascades, resulting in a variation of atomic composition in the near surface layer. If a few heavy atoms (Ag) are embedded in a matrix of light element (Si) the collision kinetics will result in a depletion of the heavier silver atom near the surface, their enrichment in a certain depth, depending on the primary ion energy. Under continuous sputtering with 500 eV Ar^+ ions this zone of changed composition will be established in a certain depth from the surface. Switching now the argon ion energy from 500 eV to 2 keV, initially the depletion increases resulting in a decreased silver Auger signal. However, after steady-state composition are reached the silver concentration has to become nearly the same as before preventing an ever increasing silver concentration. Small changes in the silver concentration may be explained by change in differential sputtering between 500 eV and 2 keV. Switching back to 500 eV one obtains a depths profile of the enriched silver zone produced by the 2 keV argon ions folded with the much less composition changes caused by the 500 eV ions. From the sputter time needed to reach the maximum in the silver concentration after switching back to 500 eV and using the sputter coefficient of silicon one can roughly estimate the depth of the depletion zone to be about 10–15 Å.

3. Conclusion

It was shown that sputtering and electron beam effects cannot be neglected in any attempt for quantitative AES. Furthermore, these effects are very specific for each type of material analysed and, therefore, no general method to deal with them can be given. Electron beam effects are inherent to AES, but can be reduced by lowering the beam current density to a certain degree. Sputtering is not a part of AES itself, but is essentially for cleaning the samples and composition vs depth profiling of thin films. Nevertheless, these changes can give useful information about sputtering mechanism. Considering the examples given it is not always possible to distinguish distinctly, whether ion or electron beam effects alone are responsible for the obtained concentration changes. However, even an ideally clean surface of a composed sample may exhibit a surface composition different from the bulk due to minimization of the surface free energy.

Nevertheless, the relative errors in a quantitative analysis can be reduced drastically by using standards with compositions close to the samples analysed which are treated in the experiment in the same way.

Acknowledgement

We are grateful to the Austrian 'Fonds zur Förderung der wissenschaftlichen Forschung', project No 2559, and the 'Jubiläumsfond der Gemeinde Wien' which supported this work in part financially.

References

1 K Siegbahn *et al* ESCA; *Atomic, Molecular, and Solid State Structure Studied by Means of Electron Spectroscopy*, Almqvist & Wiksells, Uppsala (1967).
2 C C Chang, *Surf Sci*, **25**, 1971, 53.
3 M A B Withaker, *J Phys C*, **5**, 1972, L 102.
4 W Färber and P Braun, *Surf Sci*, **41**, 1974, 195.
5 T E Gallon and J A D Matthew, *Rev Phys Technol*, **3**, 1972, 31.
6 T E Madey and J T Yates Jr, *J Vac Sci Technol*, **8**, 1971, 525.
7 J P Coad, H E Bishop and J C Rivière, *Surf Sci*. **21**, 1970, 253.
8 P W Palmberg and T N Rhodin, *J Phys Chem Solids*, **29**, 1968, 1917.
9 C C Chang, in: *Characterization of Solid Surfaces*, Edited by P F Kane and G B Larrabee, Plenum Press, New York (1974).
10 M L Tarng and G K Wehner, *J Appl Phys*, **42**, 1971, 2449.
11 H Shimizu, M Ono and K Nakayama, *Surf Sci*, **36**, 1973, 817.
12 G S Anderson, *J Appl Phys*, **40**, 1969, 2884.

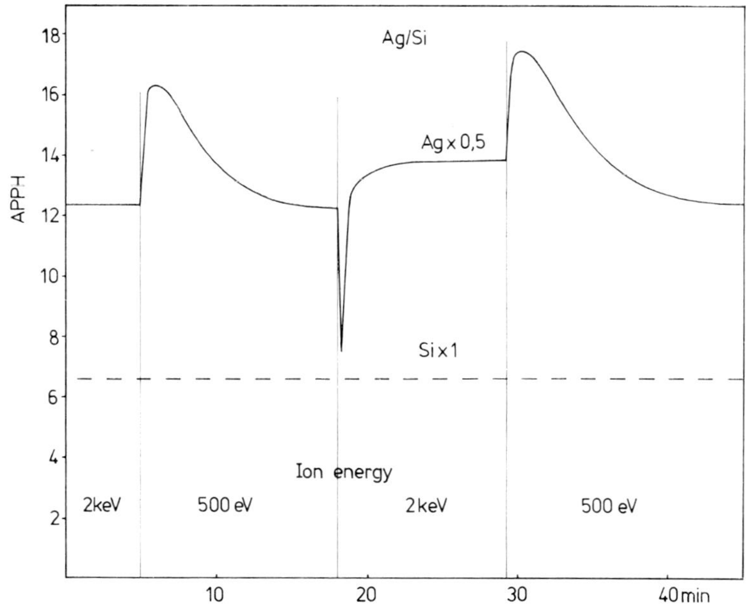

Figure 5. Knock-in of Ag in Si matrix by changing the primary Ar^+ ion energy.

[13] V E Henrich and J C C Fan, *Surf Sci*, **42**, 1974, 139.
[14] G K Wehner, in: *Methods and Phenomena 1, Method of Surface Analysis*, Edited by A W Czanderna, Elsevier, New York (1975).
[15] P W Palmberg *et al*, *Handbook of Augerelectron Spectroscopy, Physical Electronics Ind*, Minnesota (1972).
[16] G Betz, P Braun and W Färber, to be published.
[17] P W Palmberg in: *Methods and Phenomena 1, Method of Surface Analysis*, Edited by A W Czanderna, Elsevier, New York (1975).
[18] L F Vassamillet, V E Caldwell, *J Appl Phys*, **40**, 1969, 1637.
[19] W Färber, G Betz and P Braun, *Nucl Instr Meth*, **132**, 1976, 351.
[20] W Färber, Thesis, Vienna, 1975.
[21] M Hansen, *Constitution of binary alloys*, McGraw–Hill, New York (1958).
[22] A Preisinger, T Tortschanoff, F Viehböck and W Weissmann, *Jap J Appl Phys Suppl* 2, Pt 2, 1974, 791.

1.6 Composition and stress state of thin films deposited by ion beam sputtering

R N Castellano, *Bell Laboratories, Allentown, PA, U.S.A.*

M R Notis and G W Simmons, *Lehigh University, Bethlehem, PA. U.S.A.*

The ion beam sputtering technique offers several advantages over conventional sputtering systems. This technique operates at a lower pressure and substrate temperature than conventional sputtering. In addition, the angle of deposition which is easily varied with ion beam sputtering is essentially fixed in dc and rf diode sputtering. As a result of these advantages, many of the parameters which effect film stress, resistivity, and grain size can be varied independently.

Several properties of ion beam sputtered Ni, Al, Ni_3Al and Au thin films have been evaluated as a function of ion beam current density, target material, and the angle and distance of the substrate from the target.

The grain size and stress were found to vary with the angle of deposition. There is an apparent correlation between electrical resistivity and the oxygen content in the films. Both properties depend upon the grain size. The stress levels of the films are shown to be influenced by the oxygen content. These experimental results are discussed in light of models proposed to explain the origin of stress in thin films.

1. Introduction

Little research has been done which details the physical principles and technology underlying thin film preparation with a focused ion beam sputtering technique.[1–4] In this novel method, a modification of ion beam milling, material sputtered from one surface is deposited on another as a thin film. A plasma is generated in an ion gun and a collimated beam of energetic ions is accelerated at high potential toward the target. The plasma is contained in the ion gun so that the sample is not in a plasma environment. As a result, operating parameters (deposition rate, angle of deposition, temperature) that can affect thin film properties can be varied easily and independently. In addition, the substrate and target are at ground potential so that secondary electrons from the target are not accelerated to high energy, thus reducing heating of the substrate.

In contrast, normal sputtering techniques[5] suffer several limitations which can effect film properties: (1) 75–90% of the energy of the bombarding ions is dissipated in the form of heat. As a result, substrates are usually clamped to a water cooled holder. It is difficult to maintain good heat transfer between the holder and the substrate so that the substrate surface temperature can get as high as 250°C. (2) The system operates at 10^{-2} torr compared with 10^{-4} torr for ion beam sputtering. At this higher pressure, the likelihood of residual gas incorporation into the film is increased. (3) With diode sputtering, the plasma is generated between the target and substrate. In order that the plasma is sustained, the angle of deposition cannot be readily varied.

Thermal evaporation[5] offers the advantage of low pressure deposition so that films of high purity and low gas incorporation are obtained. However, films of high melting point materials and alloys with correct stoichiometry are difficult to make.

The purpose of this study is to evaluate the properties of thin films prepared by the ion sputtering technique. Thin films of Ni, Al, Ni_3Al, and Au have been prepared by this technique. Several properties of the thin films have been evaluated including: (1) intrinsic stress, resistivity, and purity as a function of target material, ion beam current density, and the angle (θ_s) and distance (d_2) of the substrate from the target, (2) alloy composition as a function of target material and current density, and (3) crystal structure and grain size as a function of target material and substrate angle. The deposition rate has been evaluated as a function of time, current density, target material, θ_s, and d_2. These results are compared with published data on films deposited by conventional deposition systems (evaporation, dc and rf sputtering).

2. Experimental

2.1. Ion deposition system. The ion beam sputtering system (Technics, Inc, Alexandria, VA) shown in Figure 1 consists of a liquid N_2 trapped, diffusion pumped, vacuum system, dc power supply, and ion gun. The vacuum chamber measures 30 cm in diameter × 45 cm high and the ion gun measures 20 cm in diameter × 20 cm long. The ion gun shown schematically in Figure 2, is a low voltage Kaufman design[6] whereby electrons from a heated filament are accelerated to the anode and ionize a noble gas such as argon while *en route*. A magnetic field inside the gun deflects the electrons in a cycloidal path for increased collision and ionization. Nearly monoenergetic argon ions are then ejected from the gun by an ion extraction–focusing system and a collimated accelerated ion beam, 5.5 cm in diameter, is produced. A heated filament at ground potential between the gun and target supplies low energy electrons to the target, space-charge neutralizing the beam, and preventing positive charge buildup on insulators. The target is contained

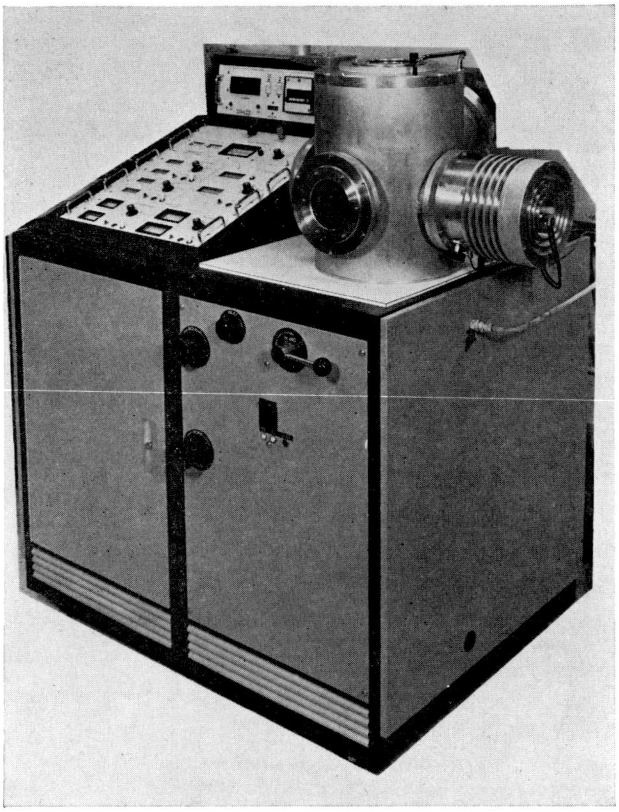

Figure 1. Photograph of ion beam milling system.

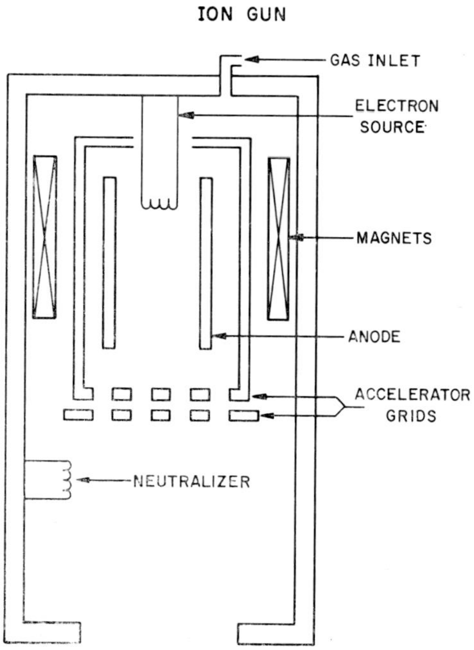

Figure 2. Schematic diagram of ion gun.

in the vacuum chamber at ground potential so that secondary electrons ejected from the target are not accelerated to high energy, thus reducing heating of the substrate. By a momentum transfer mechanism, an atom on or near the surface of the target is ejected, provided the energy transferred to the atom in the collision process exceeds the chemical binding energy of that atom and that the momentum imparted to the atom is directed away from the surface.

Operational parameters of the system are variable; the ion beam current densities, measured with a Faraday cup, range from 0.1 to 1.0 mA/cm^2 at accelerating voltages of 150–3000 V. The operational pressure of Ar inside the vacuum chamber is typically 7×10^{-5} torr after an initial pumpdown to 2×10^{-6} torr. The mean free path at this pressure is approximately 250 cm, large compared to the gun–target distance of 10 cm. Little scattering of the impinging ions occurs so that a well-collimated beam results.

2.2. Target–substrate geometry. The target–substrate geometry is shown in Figure 3. Bulk Al and Ni plates of 99.99% purity and electron beam-evaporated Au on a copper plate have been used as targets in this study. The Ni_3Al is a composite target consisting of Ni foil epoxy-bonded to an Al plate in a wedge-shaped configuration of alternating Ni–Al wedges meeting at

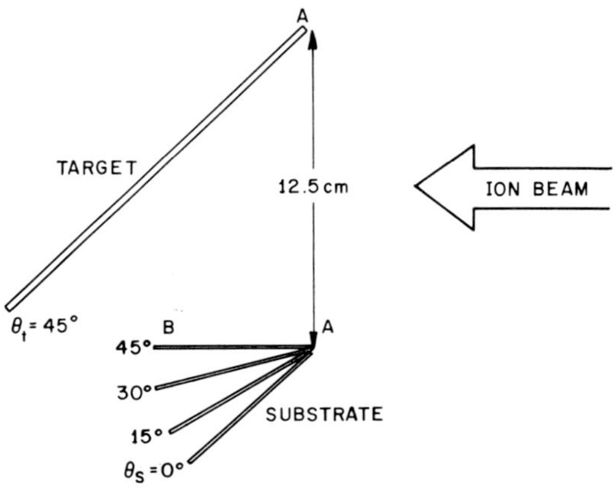

Figure 3. Diagram of target–substrate geometry.

the center of the plate to compensate for the gaussian profile of the beam with respect to ion current. The composition of the target was estimated from the ratio of the relative areas of Ni and Al on the target, the deposition rates of Ni and Al determined in this study, and the beam uniformity. For all films made in this study, the angle of incidence of the substrate with respect to the target was 0° (parallel), 15°, 30° and 45° and the target-substrate distance at point A was maintained at either 12.5 cm or 15.0 cm. As the angle of inclination varied from 0° to 45°, the target–substrate distance measured at point B decreased to a distance of 5.7 cm or 8.3 cm. The substrates were (100) oriented silicon wafers (Western Electric Co.), 3.8 cm in diameter and coated with 5000 Å of thermally grown SiO_2 to prevent any epitaxy of the deposited film. The plates were held solely by gravity since any mechanical clamp will interrupt film deposition and change the stress field, and were rotated at 150 rpm to assure thickness uniformity which in all cases was ±2%.

2.3. Film purity and composition. The purity of thin films of Ni and Ni_3Al deposited at $\theta_s = 0°$ and Al at $\theta_s = 0°$ and 45° were measured by means of electron microprobe analysis (Material Analysis Co). In addition, the composition of the Ni_3Al measured at several spots on the films was determined by this technique.

The oxygen content of the films was measured with the electron microprobe and by Auger electron spectrometry (AES) (Physical Electronics).

2.4. X-ray measurements. The crystalline structure of Al and Ni_3Al thin films was measured using X-ray diffraction analysis.[12] Utilizing a crystal monochromator with Cu Kα radiation, these thin film samples were scanned at 0.2° min^{-1} using a 3° beam slit and a 0.02° receiving slit. The grain sizes of all thin films, prepared at various substrate angles were calculated from diffraction peak broadening measurements.

2.5. Thickness-resistivity measurements. The film thickness was determined by chemically etching an unmasked section of each film and measuring the step height with a profilometer (Taylor–Hobson Talysurf). Sheet resistance measurements were made with a 4 point probe (Veeco); when multiplied by the film thickness, the resistivity in μm Ω-cm was calculated.

2.6. Stress measurements. The stress of thin films of Al, Ni and Au as a function of substrate angle and of Ni and Ni_3Al as a function of thickness has been measured for films deposited at current densities of 0.5 mA/cm² and 0.8 mA/cm² at accelerating voltages of 2.0 and 1.5 keV respectively. These stresses can be responsible for film rupture, loss of adhesion, and substrate cracking. Stresses in thin films consist of two major components; thermal stress due to the difference in coefficients of thermal expansion of the film and substrate, and intrinsic stress which is a fundamental result from the structure and growth of the thin film. Intrinsic stress can be either compressive or tensile depending on the material and conditions during growth and is normally observed by a deformation of a thin substrate. A compressive stress will bend the substrate so that the film surface is convex, and a tensile stress so that the surface is concave.

The stress has been measured with an optically-levered laser beam technique.[8] A schematic diagram is given in Figure 4. The wafer is translated across a He–Ne laser beam with the normal to the sample nearly parallel with the laser beam. The radius of curvature of the plate induced by the stress of the film is then determined. The contour of the substrate is measured prior to deposition to isolate information about the influence of the film alone on the curvature of the wafer. The film stress can then be calculated by use of the equation:[7]

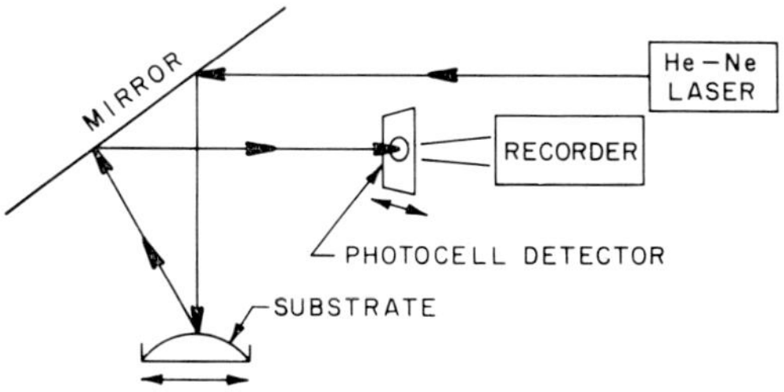

Figure 4. Schematic diagram of optically-levered laser beam technique.

$$S = \frac{ED^2}{6(1-\gamma)Rt} \tag{1}$$

where E is Young's Modulus of the substrate (1.302×10^{12} dynes cm^{-2} for (100) Si), γ is Poisson's ratio of the substrate (0.279), D is the thickness of the substrate (3×10^{-2} cm), R is the radius of curvature induced in the substrate due to the stress in the film (in cm), and t is the thickness of the film in kÅ.

The substrate is scanned in precise increments and the reflected laser beam position is detected by a photocell with a slit width of 0.05 cm that is connected to a micrometer. The radius of curvature (R) can then be calculated by

$$R = (2L)\frac{\Delta x}{\Delta d} \tag{2}$$

where $L = 2.24 \times 10^2$ cm is the distance between the sample and photocell, Δx is the increment of beam scans across the sample (0.635 cm in this study) and Δd is the change in reflection of the laser beam (measured on the micrometer) for each value of Δx. Substituting all constants into equation (1) we obtain:

$$S(\text{dynes cm}^{-2}) = \frac{9.52 \times 10^{10} \times \Delta d}{t} \tag{3}$$

3. Results and discussion

3.1. Film purity and composition. Electron microprobe analysis has been used to determine film purity. The source of the impurities Fe, Mn, Cr, and Ni totalling 1.3 at.% is attributed to the sputtering by the primary beam of the stainless steel Faraday cup that is used to monitor the current density of the ion beam. A small amount of Al (1.2 at.%) contamination also comes from a sputtering of the walls of the ion gun.

No oxygen was detected in the Al films at $\theta_s = 0°$ as indicated in Table 1. However, the oxygen content of films deposited at $\theta_s = 45°$ varies between 0.24–1.16 at.% for several Al samples. The latter films are visually dull in appearance. Hill and Hoffman[9] and Orlowski *et al.*[10] have noted that oblique incidence of deposition yields a more porous film which is highly sensitive to deposition conditions and susceptible to adsorption effects. Also, the apparent oxidation at 45° may be associated with the bombardment of the film by the primary ion beam.

Oxygen was also detected in Ni and Au films as indicated in Table 1. There is not an obvious correlation between oxygen content and grain size for the Ni and Au films. However, the two sigma limits for the oxygen content for Au are large introducing a high uncertainty in the quantitative values. For several Ni samples a direct correlation has been determined; the larger the grain size (fewer grain boundaries) the less oxygen present. It is probable therefore, that the oxygen atoms detected are concentrated in the grain boundaries.

The phase diagram[11] for the Al–Ni system is given in Figure 5. The Ni_3Al phase forms within the range of 72.5–77.0 at.% Ni although there are reported discrepancies as to the equilibria involving its formation and homogeneity.[12,13] Electron microprobe analysis gives a composition of Al ~25.5 ± 1.0 at.% and Ni ~74.5 ± 1.0 at.% measured on several samples at various spots on the film.

Table 1.

Sample	Substrate angle (degrees)	Current density (mA/cm^2)	Grain size (A)	Oxygen detected (at.%)	Resistivity ($\mu m\ \Omega$-cm)
Al	0	0.5	1125	ND*	20
	0	0.8			15
	45	0.5	450	0.24–1.16 ± 0.1	140
	45	0.8	750	2.74	35
Ni	0	0.5	200–350	1.00–2.04 ± 0.2	30
	0	0.8			20
	45	0.5	300–425	0.13–0.17 ± 0.1	25
	45	0.8			25
Au	0	0.5	450	0.21 ± 0.14	12
	45	0.5	525	0.55 ± 0.36	15
Ni_3Al	0	0.5		ND	110
	0	0.8			100

* ND = not detected.

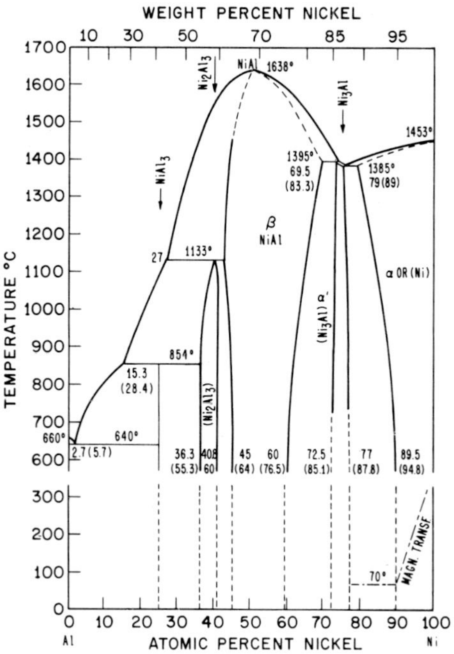

Figure 5. Phase diagram of Al–Ni system.

3.2. X-ray measurements. The grain size of the films as a function of substrate angle has been calculated by measuring the half-widths of line profiles of the X-ray diffraction peaks. The grain size is determined semi-empirically by [14]

$$D = \frac{57.3\ K\lambda}{\beta \cos \theta} \tag{4}$$

where D is the grain size, K is a constant (close to unity), which depends on the crystal plane corresponding to the peak measured, λ is the wavelength of the X-ray source, β is the half-width of the diffraction peak, and θ is the Bragg angle. All films deposited in this study are highly oriented with the (111) plane parallel to the surface of the substrate. The calculated grain size of these films is listed in Table 1.

Ni_3Al has a face-centered cubic structure of the Cu_3Au type with $a_0 = 3.561$ Å.[15] The presence of this structure was confirmed by employing X-ray diffraction techniques. No other phase or element was detected.

3.3. Temperature measurements. Target and substrate temperatures measured by an iron–constantan thermocouple at various ion beam current densities are plotted in Figure 6. For

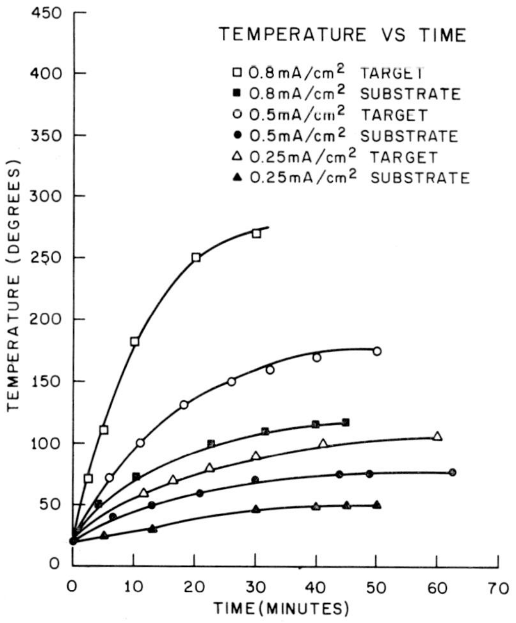

Figure 6. Substrate and target temperatures vs time at various ion beam current densities.

all values given, the thermocouple is in intimate contact wth the back of the target and substrate at point A (Figure 3). It is noted that steady-state temperatures do not exceed 125°C, even at relatively high ion beam current densities.

3.4. Deposition rate and profile. The thickness vs time for Ni films deposited at a current density of 0.8 mA/cm^2 at $\theta_s = 0°$ and for Ni_3Al films deposited at 0.5 mA/cm^2 at $\theta_s = 0°$ is linear as indicated in Figures 7 and 8.

As shown in Figure 9, deposition rate is also a linear function of current density and also found to be nearly independent of substrate angle. However, for films deposited at 0.5 mA/cm^2 at $\theta_s = 45°$, the rate decreases by a factor of 4 for Al and 1.5 for Ni and Au. A plausible explanation of the decrease is the slight sputtering of the film by the primary ion beam.

Figure 10 shows the angular variation of film thickness, indicating that the shape of the experimental curve fits nearly exactly a cosine cubed distribution. A cosine distribution is plotted for comparison. This angular dependence has nearly circular symmetry radially about the cosine lobe maximum. The maximum film thickness has been found to be normal to the center of the target (center of beam impingement). This effect is independent of target material and ion beam current density. Neither the target nor substrate were in motion during these measurements.

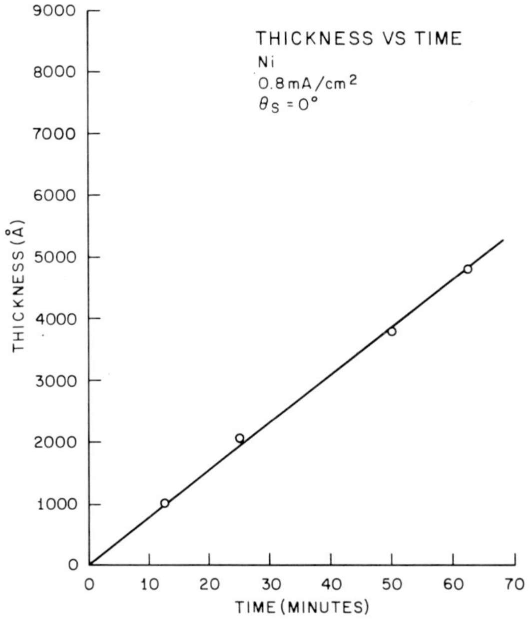

Figure 7. Thickness vs time for Ni.

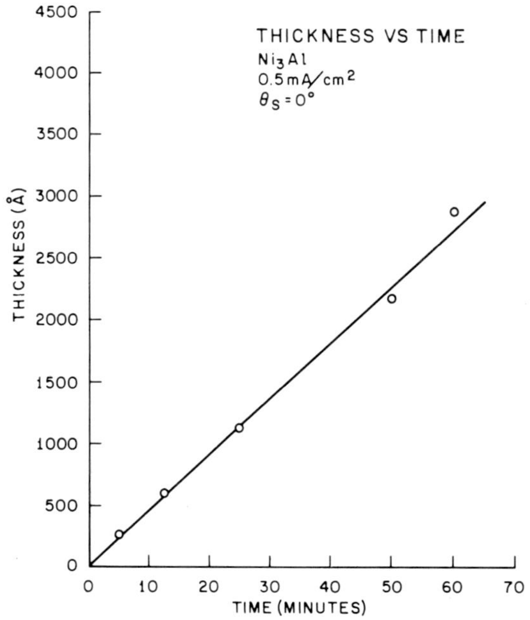

Figure 8. Thickness vs time for Ni_3Al.

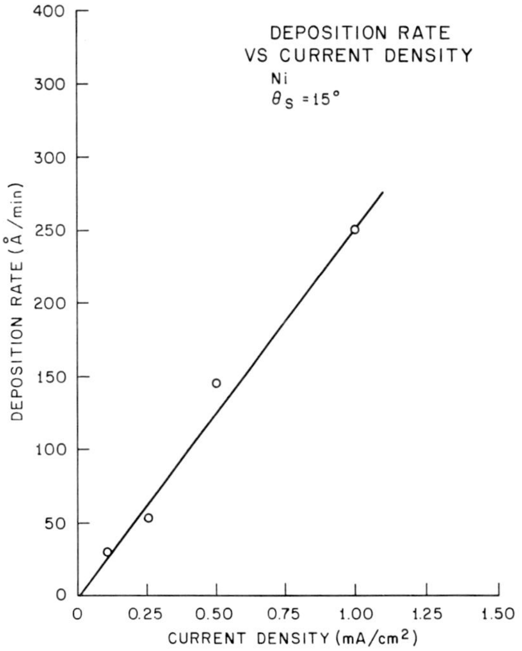

Figure 9. Deposition rate vs ion beam current density for Ni.

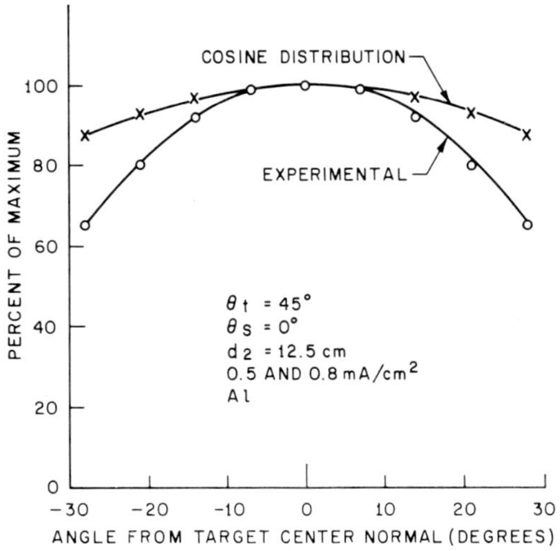

Figure 10. Angular variation of film thickness.

3.5. Resistivity. The high electrical resistivity of thin films is due to the scattering of conduction electrons by the high density of lattice imperfections (vacancies, dislocations, and grain boundaries) normally found in thin films.[16] Average resistivity values of films deposited at substrate angles of 0° and 45° at 0.5 and 0.8 mA/cm^2 current density are listed in Table 1. The higher resistivity values can be correlated to smaller grain size (grain boundary scattering) and higher levels of oxygen present.

3.6. Stress. The measured stress of Al films as a function of substrate angle for 0.5 mA/cm^2 and 0.8 mA/cm^2 is shown in Figures 11 and 12. Nominal film thickness is 4000 Å for most samples. The negative sign represents tensile stress and the positive sign represents compressive stress. In both cases, the stress reaches a maximum at $\theta_s = 15°$ and becomes increasingly less tensile at more oblique angles as shown by the solid line.

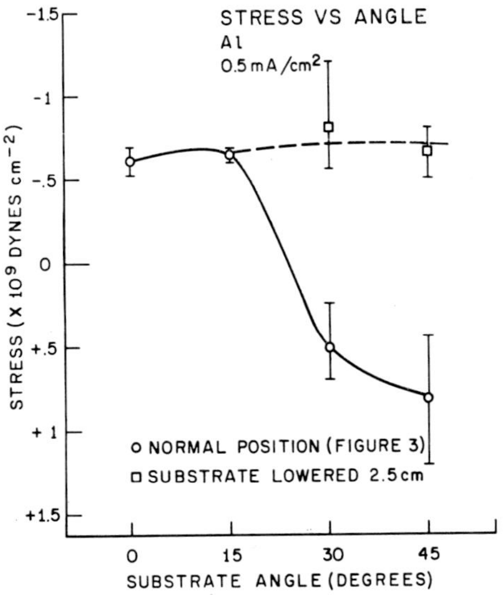

Figure 11. Stress vs substrate angle for Al (0.5 mA/cm^2).

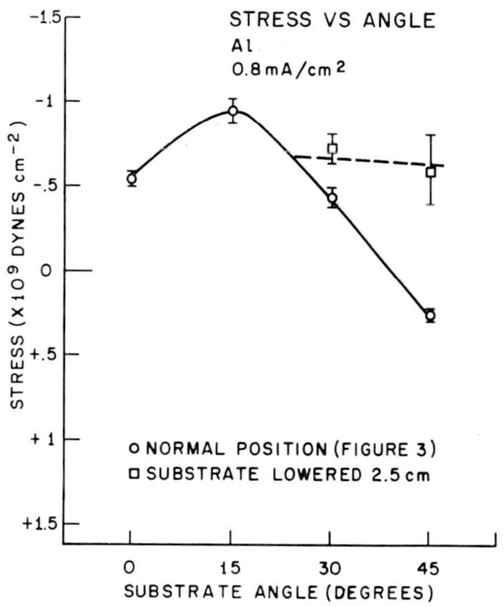

Figure 12. Stress vs substrate angle for Al (0.8 mA/cm^2).

The stress changes to compression for films deposited at 0.5 mA/cm^2 at 30° and 45° and at 45° for films deposited at 0.8 mA/cm^2. When the substrate is lowered 2.5 cm at $\theta_s = 30°$ and 45° to minimize secondary electron bombardment and eliminate direct impingement by the primary beam, the stress values are all tensile and approximately independent of substrate angle as shown by the dashed line. The target–substrate mean distance is then essentially constant over all values of θ_s. The magnitude of the stress values is approximately 2.5 times lower than that reported by Rymer[17] for vacuum evaporated films.

This compressive stress is probably due to the oxygen content measured in films, as previously noted. This is in agreement with Murback and Wilman[18] who found compressive stresses in Al films whenever the chamber pressure exceeded 10^{-4} torr during deposition. Similarly, Halliday *et al*[19] observed compressive stress in Cu films aged in air resulting in Cu_2O formation. As previously mentioned, Hill and Hoffman[9] have noted that oblique incidence of deposition yields a more porous film which is highly sensitive to deposition conditions and susceptible to adsorption effects. They also have observed an angle of incidence dependence for SiO films where the stress changes from tensile to compressive at oblique angles. The range in stress values at oblique angles (Figures 11 and 12) is attributed to the range in oxygen levels in the films.

An additional contribution to the compressive stress measured in films deposited at oblique angles of incidence may be the decrease in the grain size of Al films as the substrate angle is increased. This correlation between grain size and internal stress has been reported by Sun *et al.*[20]

Stress vs substrate angle data for 3000 Å Ni thin films deposited at 0.5 mA/cm^2 and 0.8 mA/cm^2 are plotted in Figures 13 and 14. Again a maximum stress level is observed at $\theta_s = 15°$

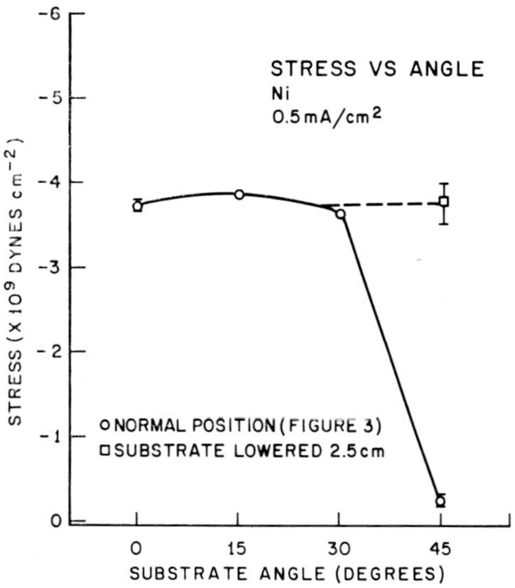

Figure 13. Stress vs substrate angle for Ni (0.5 mA/cm^2).

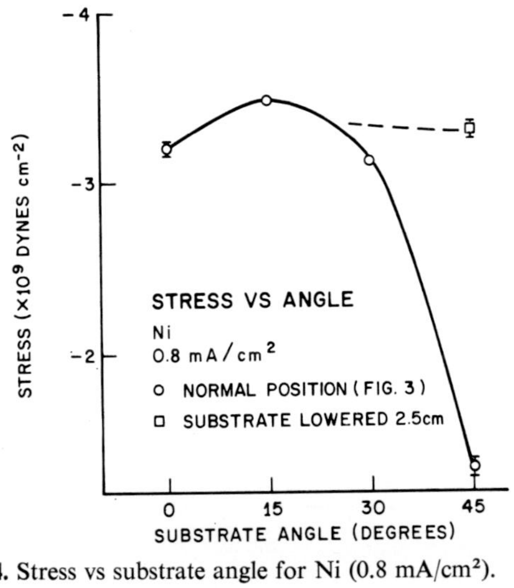

Figure 14. Stress vs substrate angle for Ni (0.8 mA/cm^2).

followed by a large decrease in stress level at $\theta_s = 45°$. The stress level in Ni films is five times greater than in Al films.

Figure 15 is a plot of stress vs thickness for Ni films deposited at a current density of 0.8 mA/cm^2 at $\theta_s = 0°$. The stress increases to a maximum value of 3.0×10^9 dynes cm^{-2} at a thickness of 5300 Å. This is approximately 2.5 times lower than data published by Klockholm and Berry for evaporated Ni films.[21]

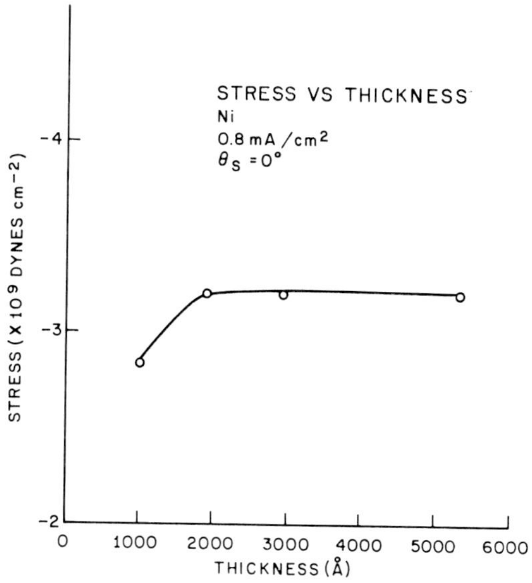

Figure 15. Stress vs thickness for Ni.

In order to understand the influence of oxygen incorporation on stress, thin films of Au (3500 Å) were deposited at different θ_s values. Gold was chosen because it is not susceptible to oxidation at these operating parameters. Since the grain sizes vs θ_s for Au films were found to be similar, the trend toward increased compressive stress (Figure 16), also observed with Ni and Al, can be correlated to higher O_2 levels and higher film porosity at oblique angles of deposition. Stress levels of these films are a factor of 10 less than reported by Klockholm and Berry.[21]

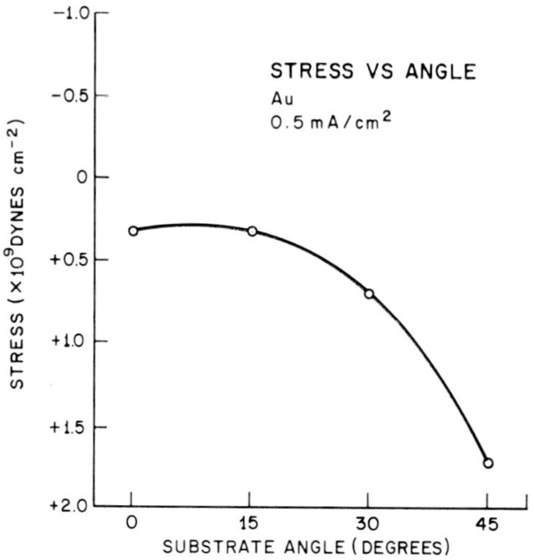

Figure 16. Stress vs substrate angle for Au.

Stress vs thickness data for the Ni_3Al thin films deposited at 0.5 mA/cm^2 and 0.8 mA/cm^2 at $\theta_s = 0°$ is given in Figures 17 and 18. A large stress is calculated which decreases with increasing film thickness. In neither case does the stress level become thickness independent within the range of values tested, in contrast to the Ni films which reached a plateau at 1800 Å.

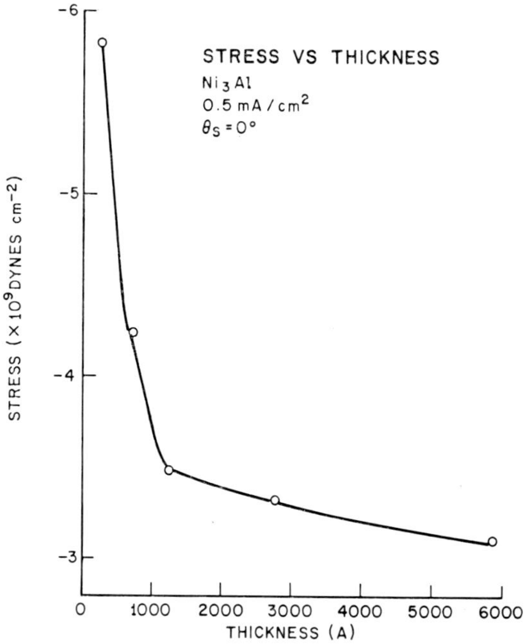

Figure 17. Stress vs thickness for Ni_3Al (0.5 mA/cm^2).

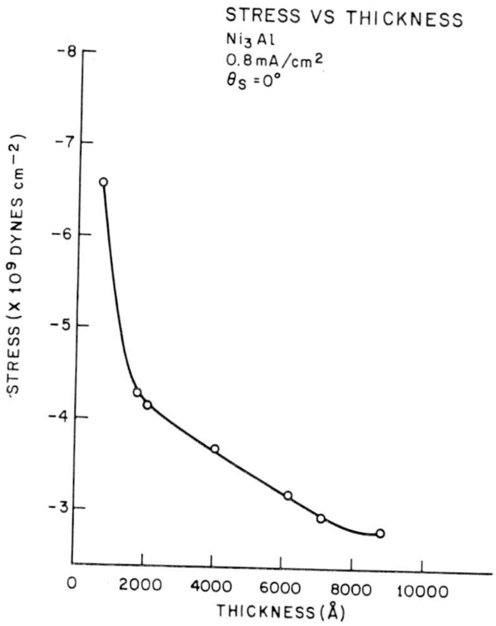

Figure 18. Stress vs thickness for Ni_3Al (0.8 mA/cm^2).

Klockholm and Berry[21] suggested that a major contribution to the intrinsic stress (referred to as growth stress) originates from the annealing and attendant shrinkage of disordered material buried behind the advancing surface of the growing film. The shrinkage of the film into a more ordered state results in an isotropic stress. This predicts that a stress × thickness vs thickness (*St* vs *t*) curve should be approximately linear in all cases where the above model is the dominant contribution to the total intrinsic stress. The annealing rate of disordered material, *r*, is assumed to follow a simple Arrhenius relationship. At low substrate temperatures such that $r < R_f$ (R_f is the deposition rate), a large amount of disorder is buried behind the growing surface so that a large intrinsic stress is observed.

St vs *t* curves for Ni_3Al deposited at ion beam current densities of 0.5 and 0.8 mA/cm² are given in Figures 19 and 20. Both curves are linear in agreement with the above model indicating that the stress is isotropic throughout the film. This is also true for Ni shown in Figure 21. Experimental data

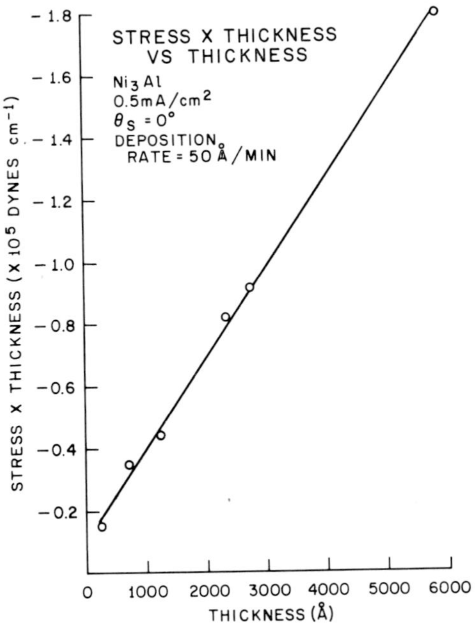

Figure 19. Stress × thickness vs thickness for Ni_3Al (0.5 mA/cm²).

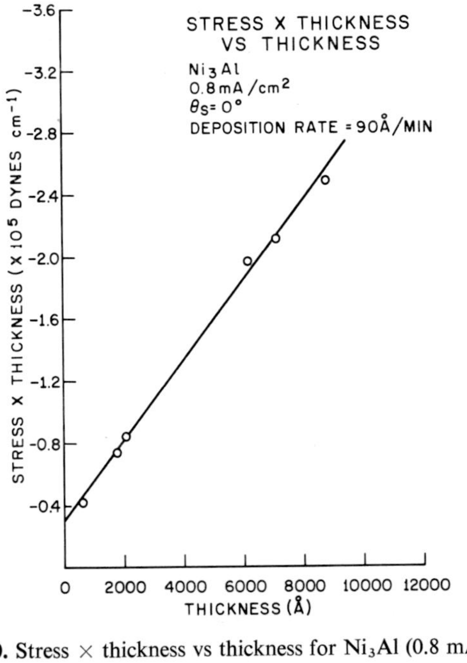

Figure 20. Stress × thickness vs thickness for Ni_3Al (0.8 mA/cm²).

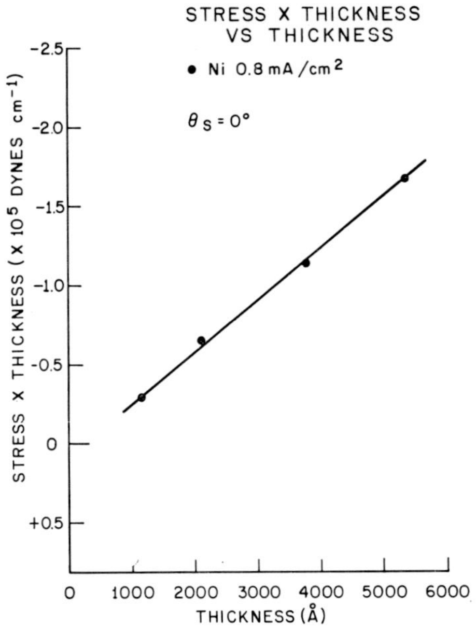

Figure 21. Stress × thickness vs thickness for Ni.

indicate that the stress is at least isotropic in the plane of the substrate since the radius of curvature of the substrate is constant regardless of the direction of the laser scan across the wafer.

In addition to the intrinsic stress contributions to the stress values described here, there is a thermal stress present induced into the film by substrate heating which can be calculated by:[15]

$$S_{th} - (\alpha_f - \alpha_s)\,\Delta T\, E_f \tag{5}$$

where α_f and α_s are the average coefficients of expansion for the film and substrate respectively, ΔT is the temperature of the substrate during film deposition minus the temperature during measurement, and *E* is Young's modulus of the film. Calculations indicate that for temperature rises shown in Figure 6, S_{th} is approximately 57% of the total Ni stress and essentially 100% of the total Al stress of films deposited at a current density of 0.8 mA/cm². S_{th} comprises 24% of the total Ni stress, and essentially 100% of the total Al and Au stress levels of films deposited at a current density of 0.5 mA/cm².

4. Conclusion

Several properties of thin films deposited by ion beam sputtering have been evaluated. Low stress levels have been obtained and determined to be a function of film material and substrate angle. By properly controlling the substrate temperature to eliminate thermal stress and by varying the substrate angle, films with very low stress levels could be deposited. Grain size could also be varied by adjustment of the angle of deposition. This could be useful in grain boundary studies where control and variability of grain size is necessary. Also, the increased porosity of films at oblique angles could be utilized in catalysis applications.

Thin films of multi-element alloys in which the elemental constituents have greatly different vapor pressures can be deposited with correct stoichiometry from a composite target of the elements. Since secondary electron heating is virtually eliminated, films can be sputtered on a variety of substrates such as plastic and paper with no need to control the substrate temperature. Thus ion beam sputtering could be considered as a viable tool for the thin film industry.

Acknowledgements

We are indebted to Dr R C Sundahl for his helpful comments and suggestions for this manuscript and to Dr R C Sun for many stimulating discussions. We thank D R Wonsidler and R J Holmes for their invaluable assistance in electron microprobe and X-ray diffraction analyses. M R Notis is appreciative to The National Science Foundation for its support under grant DMR-76-02445.

References

[1] S Aisenberg and R Chabot, *J Vac Sci Technol*, **6**, 1970, 112.
[2] K L Chopra and R M Randlett, *Rev Sci Instrum*, **38**, 1967, 1147.
[3] P H Schmidt, R N Castellano and E G Spencer, *Solid State Technol*, **15**, 1972, 27.
[4] W A Scaife, P R Hanley and K H Purser, *4th International Conference of the Nuclear Target Development Society*, Argonne, Ill. (1975).
[5] R Glang, in *Handbook of Thin Film Technology*, Edited by L I Maissel and R Glang, McGraw–Hill, New York, Chapter 1 (1970).
[6] H R Kaufman and P D Reader, *Electrostatic Propulsion*, Edited by D B Langmuir, E Stuhlinger and J M Sellen, Jr, Academic, New York, p. 3 (1961).
[7] K L Chopra, *Thin Film Phenomena*, McGraw–Hill, New York, Chapter 5, (1969).
[8] N N Axelrod, H J Levinstein and W Royer, internal memorandum, Bell Laboratories, (1972).
[9] A E Hill and G R Hoffman, *Br J Appl Phys*, **18**, 1967, 13.
[10] B A Orlowski, W E Spicer and A D Baker, *Thin Solid Films*, **34**, 1976, 31.
[11] M Hansen, *Constitution of Binary Alloys*, pp 118–21, McGraw–Hill (1958).
[12] I Iitaka, *Tetsu-to-Hagane*, **10**, 1924, 1.
[13] W O Alexander and N B Vaughan, *J Inst Metals*, **61**, 1937, 247.
[14] P Scherrer, *Gottinger Nachrichten*, **2**, 1918, 98.
[15] S Rosen and J A Goebel, *Trans AIME*, **242**, 1968, 722.
[16] C A Neugebauer, in *Handbook of Thin Film Technology*, edited by L I Maissel and R Glang, McGraw–Hill, New York, Chapter 8 (1970).
[17] T B Rymer, *Proc R Soc*, **A235**, 1956, 274.
[18] H P Murbach and H Wilman, *Proc Phys Soc*, **B66**, 1953, 905.
[19] J S Halliday, T B Rymer and K H R Wright, *Proc R Soc*, **A225**, 1954, 548.
[20] R C Sun, T. C. Tisone and P D Cruzan, *J Appl Phys*, **46**, 1975, 112.
[21] R Klokholm and B S Berry, *J. Electrochem Soc*, **115**, 1968, 823.

1.7 AES investigations of the interface between substrate and chromium films prepared by evaporation and ion plating

C W B Martinson, P J Nordlander and S-E Karlsson, *Department of Physics and Measurement Technology, Linköping University, 581 83 Linköping, Sweden*

Auger electron spectroscopy (AES) combined with ion etching technique was used to determine the composition profile of chromium films and the interface between films and nickel and aluminum substrates. The chromium films were deposited by either evaporation or ion plating and depending on used technique differences concerning both contamination in the film and in mean etching rate were obtained. In situ *cleaning of substrates before evaporation was found to reduce the amount of contamination at the interface but only to a rather limited extent.*

Introduction

In order to improve and understand adhesion and other properties of thin films and coatings on different substrates a knowledge of the composition and its variation in the film and in the interface between the film and the substrate is of great value. Auger Electron Spectroscopy (AES) in combination with sputter etching is a very favourable method to determine this composition profile. In this paper, a preliminary AES study of the composition variation in thin chromium films and in the interface between the films and nickel and aluminium substrates is presented. Films have been prepared by evaporation and by ion plating at two different bias voltages. Although care was taken in order to keep as many parameters as possible constant between the different preparations of films, the results obtained must be regarded as qualitative and no quantitative comparisons between composition profile variations due to different preparation methods can be given. Further investigations are, however, in progress.

Method

Determination of composition profiles using AES combined with sputter etching has been described in many recent review articles.[1-3] In summary the atoms and molecules on the surface are removed layer by layer by bombardment of the surface with inert gas ions. The energy of the ions are in the range of a few hundred to some thousand electron volts. The chemical analysis of the surface is simultaneously or intermittently performed by AES. Since only a few atomic layers (5–30 Å depending on electron energy) are probed by AES a detailed knowledge of the composition variation can be obtained in this way. In interpreting the obtained composition profiles one must be aware of different phenomena that if not taken into proper account may give rise to erroneous conclusions. Such complicating phenomena are:

(1) Ion beam effects, such as preferential sputtering, different erosion rates in different grains due to angular dependence of the sputtering yield, and knock-on effects that is ion implantation of primary ions and surface atoms in the lattice.
(2) Electron beam effects, such as stimulated decomposition and desorption. If simultaneous ion etching and AES analysis are performed the sputtering yield may be changed.[4]
(3) Crater and alignment effects. If the electron beam diameter is not much smaller than that of the ion beam and/or it is not centred at the bottom of the etched crater the depth resolution deteriorates and the thickness of the coating may be exaggerated.
(4) Surface roughness and local contamination may cause shadowing effects (compare Figure 2).
(5) Auger peak shape variations may be quite large for instance due to different amounts of partial oxidation (compare Figure 7). If the relative peak-to-peak heights in the differentiated Auger spectrum are taken as measures of the relative element occurrence, erroneous results may be obtained.

In general the depth resolution is limited to about 10% of the sputtered depth due to the effects mentioned above, but in favourable cases such as sputtering on amorphous films a depth resolution of 3% may be obtained.[2]

Experimental

The AES experiments were performed in a commercial scanning Auger spectrometer[5] with a base pressure of about 10^{-7} Pa (10^{-9} torr). For the energy analysis of the Auger electrons a cylindrical mirror analyser (CMA) with an energy resolution of 0.25% was used. The analyser contains an internal coaxial electron gun. The primary electron beam current used was 2 μA at 3 keV giving an electron beam diameter on the order of 10 μm or less.

The spectrometer is equipped with a carousel sample holder mounted on a precision manipulator. This makes it possible to analyse up to 15 samples consecutively under equal conditions. The incidence angle of the primary electrons is 30°. The ion gun was mounted to give an incidence angle of about 70° in order to optimize the sputtering yield.[6] The ion etching used for profiling was performed with argon gas (99.9995%) at a pressure of 7×10^{-3} Pa (5×10^{-5} torr). The ion beam, with a current density of 70 μA cm^{-2} at 3 keV, was slightly rastered to ensure uniform current density in the area examined by the analyser. By using a beam probe, based on the Faraday cup principle, the ion beam was centred around the much thinner electron beam.

During profiling the ion gun and the electron gun were operated alternatively. At each time the guns were turned on and off in the same manner and special care was taken to assure stable measuring conditions. The currents from both guns were maintained constant within 1%. A modulation with an amplitude of 2 V peak-to-peak was used to obtain the differentiated spectra. All spectra were recorded with a sweep rate of 10 eV s^{-1}.

To obtain the composition profiles the measured Auger peak-to-peak heights were approximately corrected for elemental sensitivity by using the data given by Palmberg *et al.*[7] The composition profiles were normalized by assigning a value of 100 to the signal from the substrate far from the interface.

Specimen preparation

The coating apparatus used in this study as well as the evaporation details are described elsewhere.[8] To make ion plating[9] and

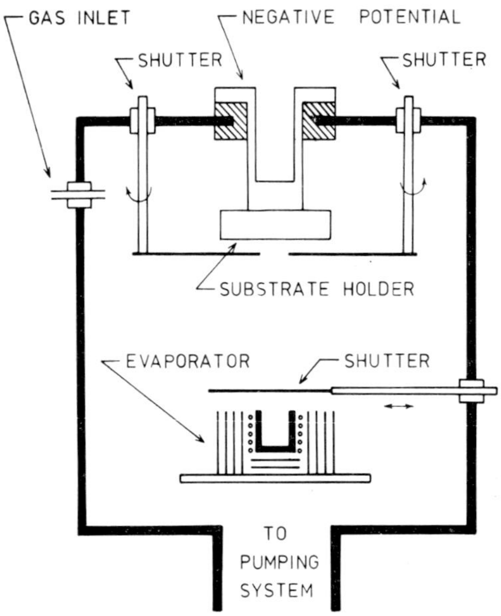

Figure 1. Schematic figure of the coating apparatus. Cr is evaporated from a ZrO crucible. The substrates, Ni and Al, are mounted in a substrate holder which during sputter cleaning and ion plating is connected to a negative potential.

Figure 2. SEM micrograph showing a sputter etched Ni substrate after that the Cr film had been removed. However, in the shadow of a bigger Cr particle a part of the original Cr film remains unaffected.

sputter cleaning of the substrates possible the substrate holder is insulated from the chamber. To protect the diffusion pump from the relatively high argon pressure used in these processes a conduction restriction is introduced between the deposition chamber and the diffusion pump. This restriction lowers the pumping speed and, therefore, the partial pressures of gases other than argon may be higher in the deposition chamber during ion plating than during evaporation.

The Ni and Al substrates had a diameter of 30 mm and a thickness of 0.2 mm and 1 mm respectively. After electrolytic polishing (perchloric acid solution) the substrates were ultrasonically cleaned in trichloroetylene, acetone and alcohol. They were placed in a special substrate holder where up to six substrates could be coated in the same deposition cycle.

After evacuating to a pressure of 10^{-4} Pa (10^{-6} torr) or lower the chamber was backfilled with 99.9997% argon gas to a pressure of about 2.6 Pa (20 mtorr). The substrates were sputter cleaned in a self sustained glow discharge, i.e. a negative potential was connected to the substrate holder. The used voltage was 1.5 kV and the standard cleaning time was 20 min.

When examining the influence of sputter cleaning time, two shutters directly below the substrate holder (Figure 1) were used to allow the glow discharge to reach the different substrates successively. In this way substrates could be sputter cleaned during different periods (or not cleaned at all) and still be simultaneously coated.

When ion plating, the shutter, just above the evaporation source, was withdrawn, directly when the sputter cleaning was ready, i.e. the pressure in the deposition chamber as well as the bias voltage on the substrate holder were kept constant during the deposition. However, in some ion plating experiments the bias potential was changed to another value when the shutter was withdrawn. This procedure ensures that no gases will readsorb and thus contaminate the substrate surface before the start of deposition. In evaporation the chamber was re-evacuated to a pressure lower than 10^{-3} Pa (10^{-5} torr) before the deposition was started.

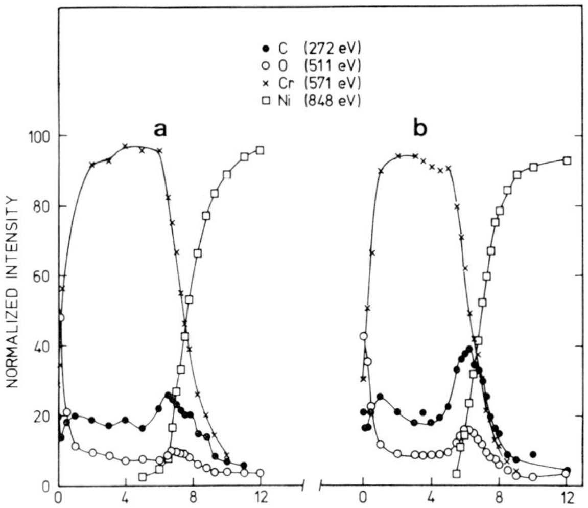

Figure 3. Composition profiles of evaporated Cr films on Ni substrates with different pretreatment. In (a) the substrate was sputter cleaned for 8 min before evaporation while in (b) no sputter cleaning was performed.

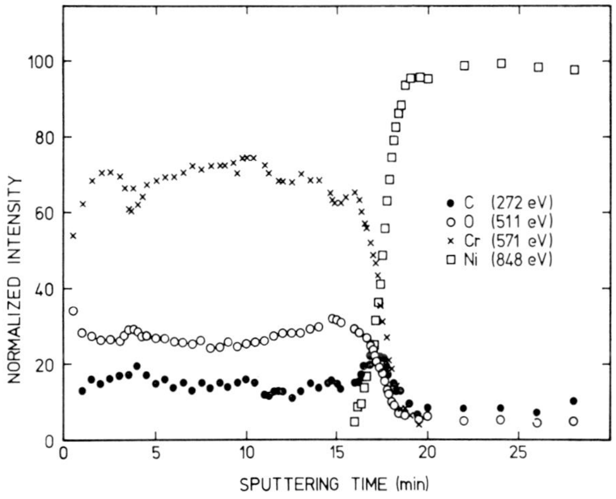

Figure 4. Composition profile of an ion plated Cr film on a sputter cleaned Ni substrate. During coating the cathode potential was −1.5 kV.

Results and discussion

In the presented composition profiles, elemental composition is shown as a function of sputter etching time. It would, of course, be more desirable to give it as a function of depth. As mentioned above several uncertainties are then introduced but an estimate of the mean etching rate can be given. A measure of this rate for some evaporated Cr films was performed. Studied samples were cooled in liquid nitrogen and then immediately bent whereby the Cr films were broken into flakes. The flake thickness was determined in a scanning electron microscope. The mean etching rate could then be calculated by combining the time to sputter through a film with the corresponding film thickness. A value of 50 Å min^{-1} was obtained. Ion plated Cr films could not be broken, presumably due to better adhesion between film and substrate as well as higher ductility of the film. However, a rough estimate could be obtained in another way. A spat Cr particle was fortunately found on a Ni sample where the Cr film had been etched away (Fig. 2). A part of the original ion plated Cr film was left

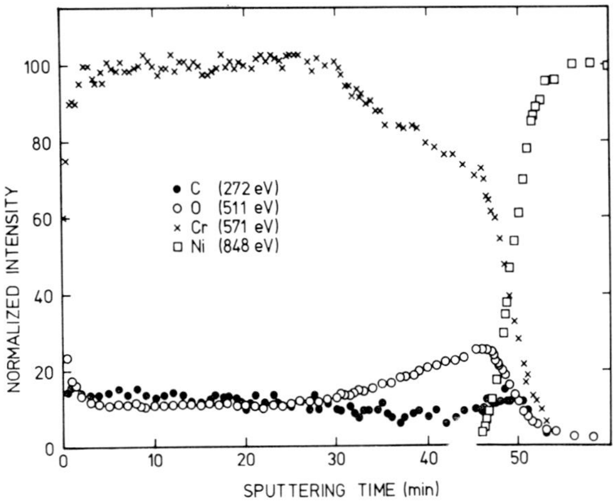

Figure 5. Composition profile of an ion plated Cr film on a sputter cleaned Ni substrate. During coating the cathode potential was −3 kV.

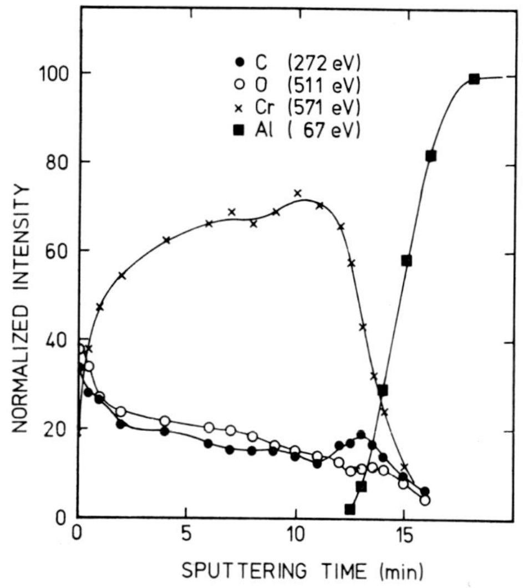

Figure 6. Composition profile of an ion plated Cr film on a sputter cleaned Al substrate. During coating the cathode potential was −1.5 kV.

unaffected under the Cr particle due to shadowing effects. Although the film had been removed after 50 min, the etching had been continued up to 70 min. Consequently, the step at the film edge consists of contributions both from the Cr film and the Ni substrate which are indistinguishable. Sputtering yields for Ni and Cr are about the same for Ar ions at 500 eV (1.45 resp. 1.18).[10] Even if the yield ratio Ni/Cr would be as high as 2.5 when using 3 keV Ar ions at least half of the step seen in Figure 2 is Cr. This gives a lower limit of the mean etching rate of 140 Å min^{-1} for the ion plated film. The difference in mean etching rate for ion plated and evaporated films may be related to differences in hardness[11] or in microstructure. *In situ* sputter cleaning of Ni substrates was found to reduce the oxygen and carbon content in the interface between evaporated films and substrates (Figure 3). The reduction was similar for all the sputter cleaning periods used, 8, 16 and 32 min.

In evaporated films the carbon and oxygen concentrations form peaks at the interface. These peaks may be caused by gases readsorbed on the substrate before evaporation and/or by gettering during beginning of the evaporation process.

A composition profile of an ion plated (−1.5 kV) Cr film on a Ni substrate is shown in Figure 4. All investigated ion plated

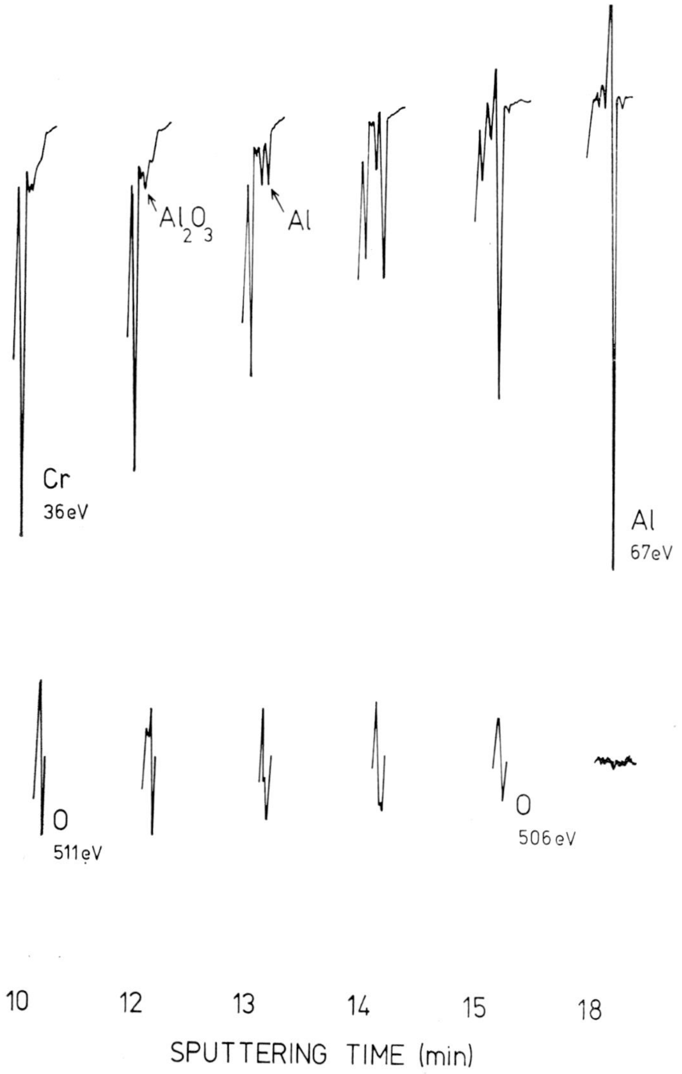

Figure 7. Cr, Al and O Auger lines at different positions at the interface between the ion plated Cr film and the Al substrate used in Figure 6.

films coated at a cathode potential of −1.5 kV have a higher oxygen content than the evaporated films. There are several possible reasons for this difference. The pressures during evaporation and ion plating differ by several orders of magnitude which increases collision cross-section and thus the reaction probability in the case of ion plating. Heating of the source will cause outgassing in the coating chamber. This may lead to a higher oxygen partial pressure during ion plating since the pumping speed is reduced there, compared to that during evaporation. Furthermore, the current of ions and electrons may enhance the reactivity. The temperature of the substrate surface may also be of importance. During sputter cleaning both dislocation density and substrate surface temperature increases, favouring diffusion processes. Due to the differences in coating processes as described above, a higher surface temperature may be maintained when ion plating than when evaporating.

Ion plated films with a composition more similar to evaporated films were obtained by changing the deposition parameters. Figure 5 shows the composition profile of a thicker Cr film on a Ni substrate ion plated at a cathode potential of −3 kV. The oxygen peak at the interface has a tail which, in contrast to evaporated films, falls off rather slowly into the film.

In this study mainly Ni substrates have been used but some depositions were also performed on Al substrates. Ion plating at a cathode potential of −1.5 kV seems to cause a larger width of the interface between film and substrate for Al than for Ni. A ratio of two between the widths can be deduced from Figures 4 and 6. This relation indicates that the penetration depth for Cr in Al is larger than in Ni. Other possible explanations are differences in sputtering yield, surface roughness and porosity of the substrates.

Information about the chemical nature of the elements in the interface can also be obtained from the Auger peaks. In Figure 7 some of the Cr, Al and O Auger lines recorded to obtain the composition profile of the Cr film on Al substrate shown in Figure 6 are presented.

In the region of the low energy Cr and Al lines a small peak around 46 eV corresponding to Cr in Cr oxide is seen before the interface (sputtering time 10 min). The strong line at 36 eV corresponding to Cr metal is also slightly depressed compared to that from pure metal. After further sputtering a line at 53 eV corresponding to Al in Al_2O_3 shows up but it is soon followed by a line at 67 eV. The latter line corresponding to Al metal grows rapidly and reaches a stable level after passing through the interface (sputtering time 18 min). At the latter time the O line has disappeared. The transition from Cr oxide to Al oxide can also be followed by studying the O line which shifts from 511 eV to 506 eV when passing through the interface.

Acknowledgement

We are grateful to Mr K Å B Andersson and Mr H T G Nilsson for assistance in parts of the experiments.

References

[1] C C Chang, in *Characterization of Solid Surfaces*, p 509, Edited by P F Kane and O Larrabee, Plenum, New York (1974).
[2] A Joshi, L E Davis and P W Palmberg, in *Methods of Surface Analysis*, p 159, Edited by A W Czanderna, Elsevier, New York (1975).
[3] N J Taylor, in *Techniques of Metals Research*, Vol VII, Part 1, p 117, Edited by R F Bunshah, Wiley, New York (1972).
[4] T Smith, *Surf Sci*, **55**, 1976, 601.
[5] Varian, CMA/LEED Console Fast Cycle System, 981–2001.
[6] H Oechsner, *Appl Phys*, **8**, 1975, 185.
[7] P W Palmberg, G E Riach, R E Weber and N C MacDonald, *Handbook of Auger Electron Spectroscopy*, Physical Electronics Industries, Edina, MN. (1972).
[8] P J Nordlander, K Å B Andersson, H T G Nilsson, S B M Hagström, presented at *Int Symp on Vacuum and Thin Film Techn*, Uppsala, Sweden, 31 August–3 September 1976.
[9] D M Mattox, *6th Int Vacuum Congress*, Kyoto, Japan, 25–29 March 1974.
[10] G K Wehner, General Mills Report No 2309, 1962.
[11] S Komiya and K Tsuruoka, *J Vac Sci Technol*, **1**, 1976, 520.

1.8 Influence of defects in the silicon wafer on the properties of silicon oxide films formed by reactive sputtering

D Kropman, M Vinnal and P Putk, *Institute of Cybernetics of the Estonian Academy of Sciences, Tallinn 200 001, USSR*

It is well known that the properties of SiO_2 films, formed by reactive sputtering in oxygen, sensitively depend upon the details of chemical pretreatment of the Si surface and on the conditions of oxide formation. The influence of bulk defects in the silicon wafer has, however, not been adequately studied. Bombardment of the silicon surface with electrons and ions during the initial stages of the oxide formation is accompanied by extensive interaction with the surface defects, leading to changes in the charge of these defects and properties of the deposited film in general. If vacancies are the dominant defects, then a positive charge is created in the oxide film. If both vacancies and stacking faults are present, then both positive and negative charge can appear in the SiO_2 film.

Introduction

Thermal oxidation leads to defect formation on silicon surfaces. These defects can be of various different types, such as stacking faults, dislocations, clusters of vacancies,[1,2] and precipitates. Interaction of these defects with charged particles during SiO_2 surface film formation through reactive sputtering in a glow discharge can have decisive influence upon the properties of the Si–SiO_2 bonding layer thus formed.[3,4] Investigation of these interactions is of practical importance for the protection of silicon surfaces in various solid-state devices and provides a new approach for the study and differentiation of the various surface defects in silicon.

Experimental

n-Type silicon with 15 $\Omega\cdot$cm specific resistance was used. The wafers were mechanically polished and chemically treated. The defects were artificially created by oxidation at 1100 to 1200°C temperature in a three-stage process in dry–wet–dry oxygen, or in wet oxygen only. Thickness of the oxide layers thus created was 0.5 μm. The oxidation procedure was repeated four times with the oxide film etched away before each treatment and before the reactive sputtering in a two-electrode apparatus with 1 to 1.7 kV cathode potential and 1 mA cm^{-2} current density at 2×10^{-2} torr oxygen pressure. The film deposition rate was 15 Å min^{-1} and final thickness 1500 Å. Silicon surface defects were brought out by selective etching with the Sirtle reagent. Effective charge in the Si–SiO_2 layer was measured from the C–V curves, using the MOS capacitance technique at 1 MHz frequency.

Results and discussion

The structural defects, formed in silicon during oxidation, are stacking faults, dislocations, clusters of vacancies, and precipitates.[1,2] At 1100–1150°C mainly stacking faults and dislocations are formed, while at 1200°C the clusters of vacancies and precipitates predominate. Repeated oxidations at 1100–1150°C temperature diminish (Figure 1), and those at 1200°C enhance the density of defects formed.

The influence of oxidation temperature on the charge in the thermal SiO_2 film depends upon the number of oxidation cycles used (Figure 2). If the silicon wafer has been treated only once or twice, then a rise in oxidation temperature leads to a reduced charge, three oxidations provide independence of temperature and charge in wafers that have been treated four times passes through a maximum. The relations between the charge in the sputtered oxide film and the temperature of oxidation before the final treatment, $Q(T)$, are plotted in Figure 3. It is readily apparent that the high-temperature pretreatment influences charges, created in the low-temperature oxide. In case of a high oxidation temperature (1200°C), and even at lower temperatures for the four-fold-treated wafers, the charge in the sputtered oxide was positive. At 1100° and 1150°C, both the magnitude and sign of the charge in SiO_2 depend upon the total number of oxidation cycles, and can be either positive or negative.

The latter complicated oxidation temperature dependence of the charge in the sputtered oxide is caused by the different nature of structural defects created on the silicon surface. At 1200°C pretreatment temperature, with clusters of vacancies as the predominant type of surface defects, the resulting positive charge in the sputtered oxide film may be caused by the creation of positively charged centres by association of the vacancies with oxygen atoms (A—centres). Of course, other surface defects can also play a role in the oxide charge formation, especially through creation of partial bonds between silicon atoms. Surface recombination processes are also enhanced in the vicinity of defects.[5] Ion bombardment in plasma during oxide deposition can lead to bond rupture at these defects with the formation of acceptor centres. Repeated oxidations lead to a reduction in the density of stacking faults, but the density of vacancies increases. The complicated dynamics of the changing densities of various kinds of surface defects can lead to sign changes for the charge in sputtered oxide films.

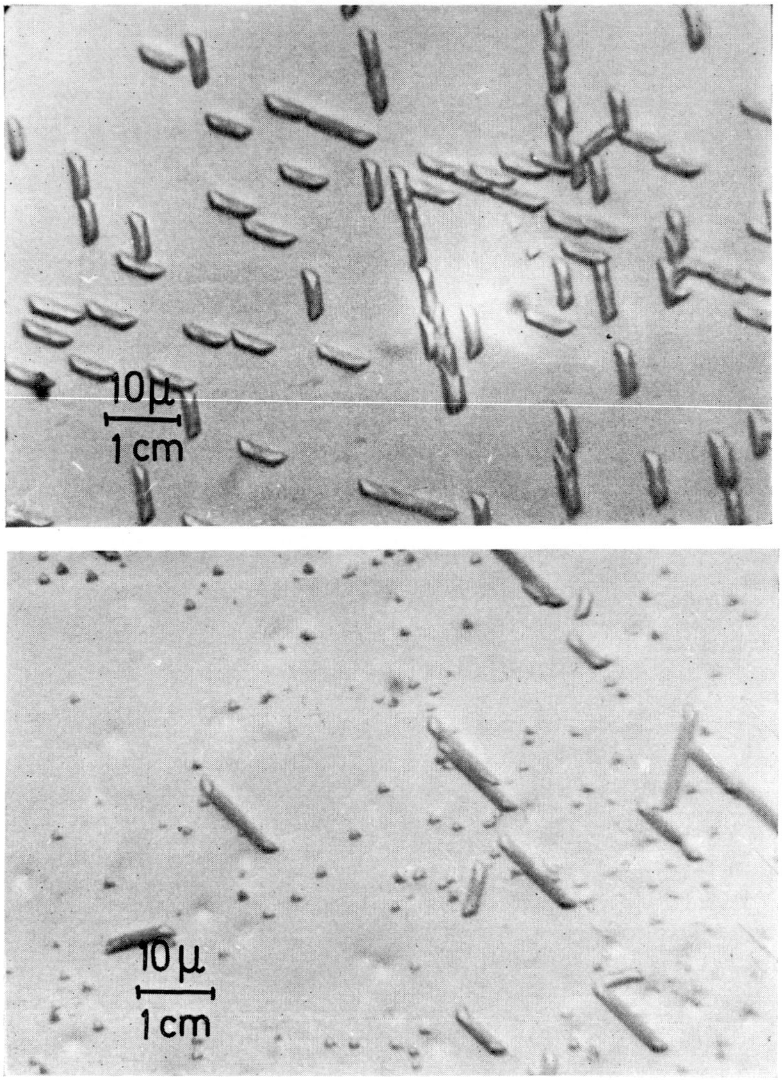

Figure 1. Silicon surface after oxidation at 1150°C; a—first oxidation, b—third oxidation.

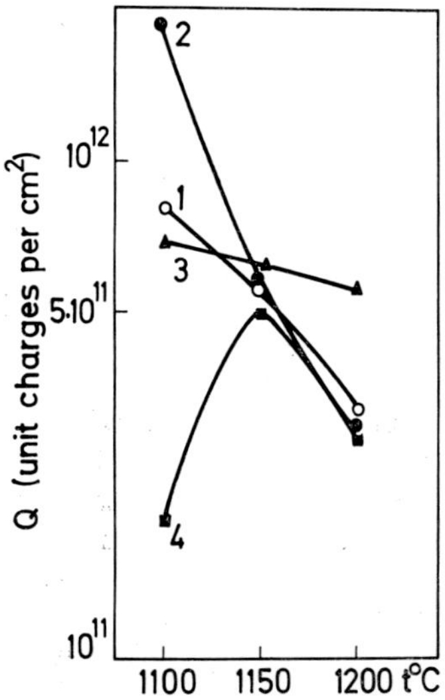

Figure 2. Relation between the charge in the thermal oxide and oxidation temperature; 1, 2, 3, 4—number of oxidations.

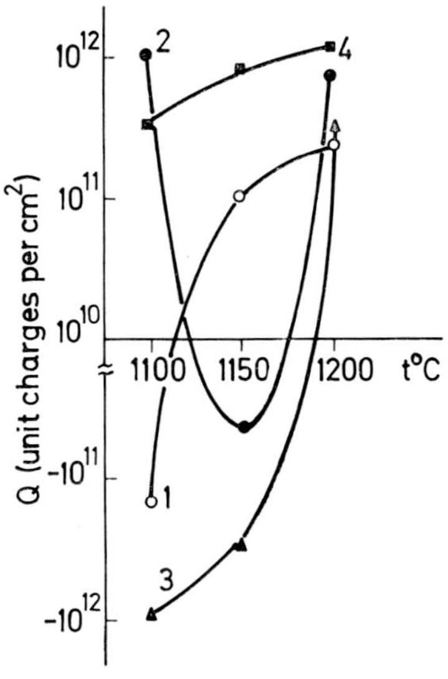

Figure 3. Relation between the charge in the oxide film, deposited by reactive sputtering in oxygen atmosphere, and the temperature during oxidative pretreatment; 1, 2, 3, 4—number of oxidations before sputtering.

It must be stressed that in case of oxidation in wet oxygen, the relation between the oxidation temperature and charge in the sputtered oxide, is changed. These changes parallel those in defect formation and must be caused by the interaction between charged particles and surface irregularities. The relation between oxidation temperature, number of treatments, and the resulting charge also depends upon the glow discharge voltage—an effect, which is best rationalized through the different nature of defects with different mean bond energies between silicon atoms.

It follows from the Figures 2 and 3 that the charges in the thermal and reactively sputtered oxides are related for the lower temperature range from 1100 to 1150°C and do not correlate for higher oxidation temperatures between 1150 and 1200°C. Thus, the defects generated or annealed in the first case must influence the charge formation in both kinds of oxides, which is not the case for the higher temperatures. If the formation of a positive charge in the latter case of high pretreatment temperature can be assigned to a high density of vacancies in the silicon surface, then the decrease of charge in the thermal oxide, prepared at this temperature, can be caused by stresses, created by the different thermal expansion coefficients of Si and SiO_2. This assumption is partially confirmed by the accompanying broadening of the SiO_2 absorption band at 1100 cm^{-1} in the infrared.

The properties of oxide films, deposited by reactive sputtering, thus provide useful information about the nature of surface defects in the silicon wafers used.

References

[1] H J Queisser and P G G Van Loon, *J Appl Phys*, **35**, 1964, 3076.
[2] K V Ravi, *J Electrochem Soc*, **121**, 1974, 1090.
[3] A G Revesz, *J Vac Sci Technol*, **6**, 1969, 596.
[4] D Kropman, P Langovits, M Vinnal and B Meyler, *Thin Solid Films*, **34**, 1976, 91.
[5] D W Small and R F Pierret, *Appl Phys Lett*, **27**, 1975, 148.

1.9 Depth chemical profiles of MNOS structures measured by AES

J Liday, R Harman and M Veselý, *Slovak Technical University, Microelectronics Department, 880 19 Bratislava, Vazovova 5, Czechoslovakia*

A review of recent work on measurements of surface and chemical depth profiles of materials of great importance in modern metal oxide and metal nitride oxide semiconductor technology.

1. Introduction

The important SiO_2/Si structure has earlier been studied using Auger Electron Spectroscopy (AES) and the results have been published in several papers. Weber[1] analysed the SiO_2 layer on Si which was covered by P or Cl before oxidation. Chang *et al*[2] measured the chemical depth profile of SiO_2 layers doped with 0.5 at.% of P.

Inoue *et al*[3] studied diffusion of P and B into the SiO_2/Si and polycrystalline $Si/SiO_2/Si$ structures, by chemical depth profiling. Chou *et al*[4] studied the escape depth of Auger electrons from SiO_2 by measuring the chemical profile of SiO_2/Si structures doped with P. A comprehensive Auger analysis of the SiO_2/Si and the MOS interface has recently been published.[5–7] The interface of Al_2O_3/Al has been studied.[8] For the chemical depth profile analysis of the MNOS structure, which is of great importance for semiconductor technology, appears the short information recently published.[9]

2. The preparation of MNOS sample

The wafer of *p*–Si (111) was used for the preparation of $Al/Si_3N_4/SiO_2/Si$ structure. The Si wafer was oxidized in dry oxygen at $T = 1100°C$. A silicon nitride was then created by the CVD method at $T = 830°C$. The Al layer was evaporated using *e*-beam heating at $p \sim 10^{-6}$ torr and at the wafer temperature $T \sim 100°C$. The thickness of the layers on the *p*-Si (111) as measured by ellipsometry were as follows: $d(Al) = 1030 \pm 5$ Å, $d(Si_3N_4) = 720 \pm 5$ Å, $d(SiO_2) = 320 \pm 5$ Å.

3. The chemical depth profile measurement by AES

For the analysis we used a Varian 120 LEED–AES system with a CMA analyser and a PHI Ion gun model 04-131 F. After introducing the sample into the analysing chamber, the system was evacuated without baking to $p \sim 10^{-9}$ torr. Then the vacuum system was degased at $T \sim 220°C$ for 20 h. The lowest total pressure was $p \sim 1 \times 10^{-10}$ torr. A mass spectrum at this pressure taken by a Finnigan Quadrupole gas analyser Spectrascan 750 is shown in Figure 1. We introduced spectral clean argon into the analysing chamber thus increasing the total pressure to $p = 5 \times 10^{-5}$ torr. A mass spectrum at this pressure is shown in Figure 2.

Every analysing cycle was composed of: sputtering of surface layer by ion gun at $p_{Ar} \sim 5 \times 10^{-5}$ torr, rotating the sample holder in front of the energy analyser and taking the Auger electron spectrum at the same pressure $p_{Ar} \sim 5 \times 10^{-5}$ torr. A focused ion beam with energy of Ar^+, $E_i \sim 950$ eV, bombarded an area with the diameter ϕ 4–5 mm on sample surface. For the AES we used a glancing incidence gun; $E_p \sim 1800$ eV, $I_p \sim 15$ μA. We measured and analysed mainly low energy Auger transitions. Because the investigated sample was exposed for a long time to the atmospheric condition an aluminum oxide Al_2O_3 was formed on the Al surface. This is detected by the Auger spectrum shown in Figure 3, taken at the beginning of the analysis.

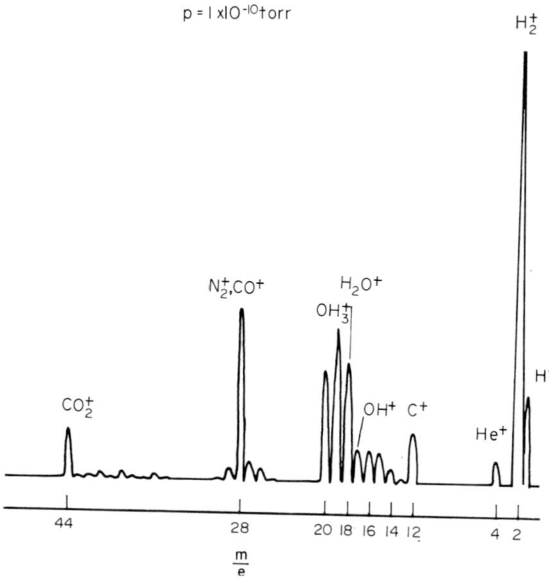

Figure 1. Mass spectrum of residual gases in the uhv analysing chamber at $p \sim 1 \times 10^{-10}$ torr.

The investigated sample thus represents a structure $Al_2O_3/Al/Si_3N_4/SiO_2/Si$.

4. Experimental results and discussion

Results and discussion are divided according to the structure interfaces as they follow from the sample surface into the bulk silicon.

4.1. AES of the Al_2O_3/Al structure. The Auger spectrum shown in Figure 3 was taken immediately after fine cleaning of the surface by an ion bombardment dose $I \cdot t \sim 30$ μA min. In

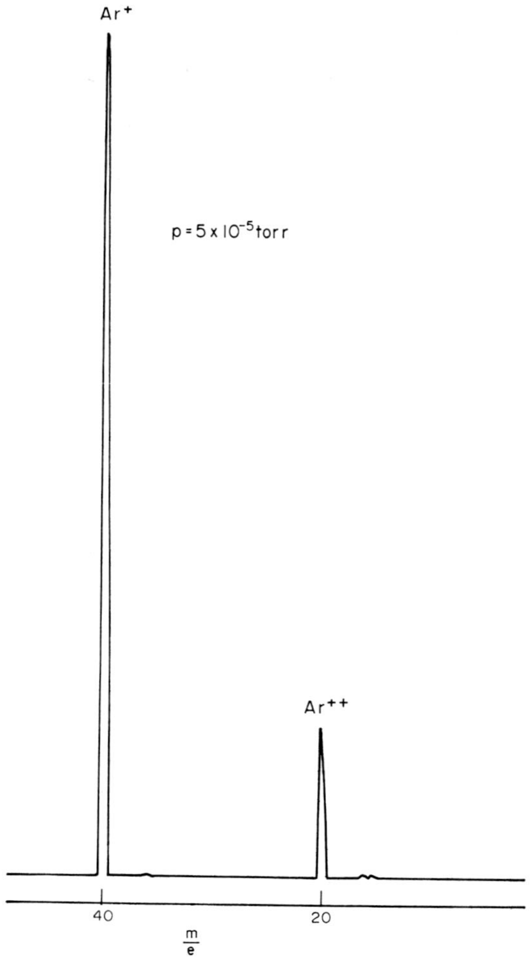

Figure 2. Mass spectrum of gases in the analysing chamber at $p \sim 5 \times 10^{-5}$ torr.

the spectrum besides C, O and Ar we identify S, F, Fe, Ni and peaks at 38 and 56 eV. Previously[8] Al peaks have been found at 67, 51, 41 eV and a small peak at 83 eV. Dufour *et al*[10] published a peak at 52 eV for Al_2O_3 and small spectrum bend at 34 eV. For elemental Al they published peaks at 65 eV (main peak), at 51 eV a bend at 37 eV. Quinto and Robertson[11] published for elemental Al peaks at 67 eV (main peak), 52, 39 and 81 eV. Palmberg *et al*[12] published for elemental Al peaks at 68 eV (main peak) and 84 eV.

When we compare our AES measurements with data from[10–12] for elemental Al, it seems to us that the Al spectra[10] are shifted and peaks are thus also shifted about 2–3 eV to the lower values. We detected peaks at 38 and 56 eV which correspond to the Auger peaks from Al in Al_2O_3. Main O peak from Al_2O_3 lies at 509 eV.

With the continuous sputtering and AES measurement of the Al_2O_3/Al structure, we see new peaks at the beginning of the interface, corresponding to elemental Al. Furthermore, we can detect peaks at 38 and 55 eV (Al in Al_2O_3) and 507 (O in Al_2O_3), as shown in Figure 4. The peaks of elemental Al at 56 and 38 eV are shifted to the values 52 and 41 eV respectively, going through the interface. The Auger spectrum of a clean Al surface is shown in Figure 5. This demonstrates that the peaks of elemental Al lie at 67 eV (main peak), 41, 52 and 83 eV. The intensity of the C peak was in Al_2O_3 decreased, but it is in fact unchanged in the Al layer. A small signal of Fe was detected extending from the surface to the interface. The signal again appeared in Al at the previous value. The small N and Na peaks disappeared after the sputtering dose $I \cdot t \sim 28$ μA min. A depth concentration profile of Al_2O_3/Al is shown in Figure 6.

The great changes of the peak-to-peak values at 56, 67, 83 eV and (507–509) eV lie between 45 and 130 μA min, representing the distance of about 60 Å. The shift of 56 eV peak to the 52 eV value, is observed here. This layer (about 60 Å thick)

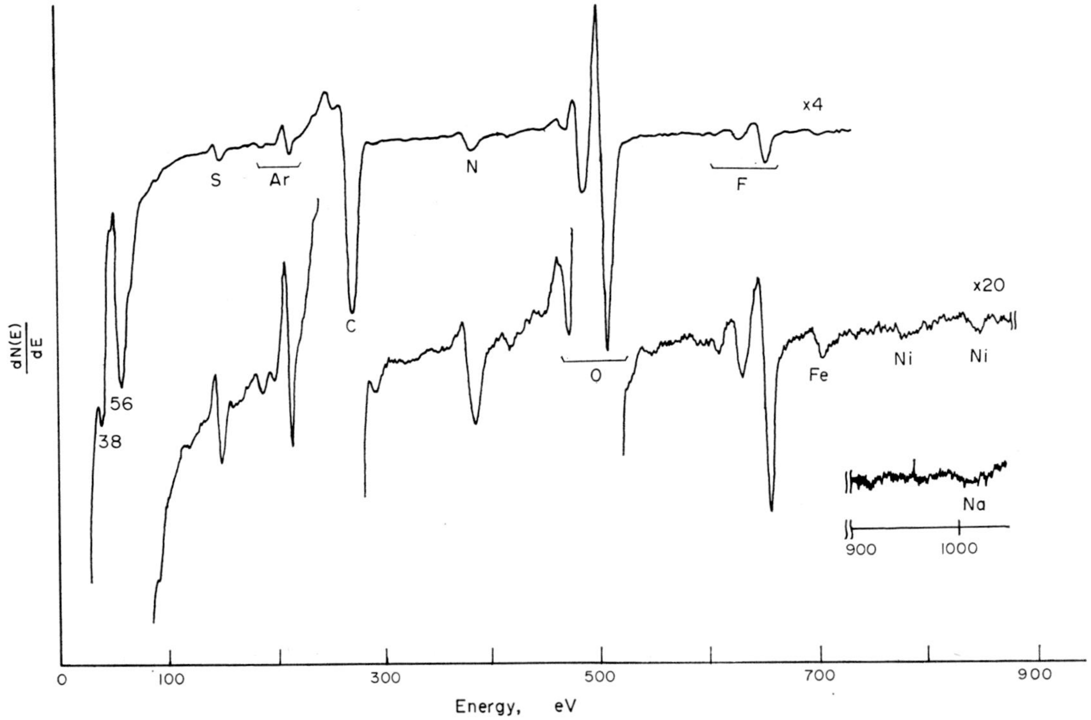

Figure 3. An Auger electron spectrum of an oxidized aluminum surface.

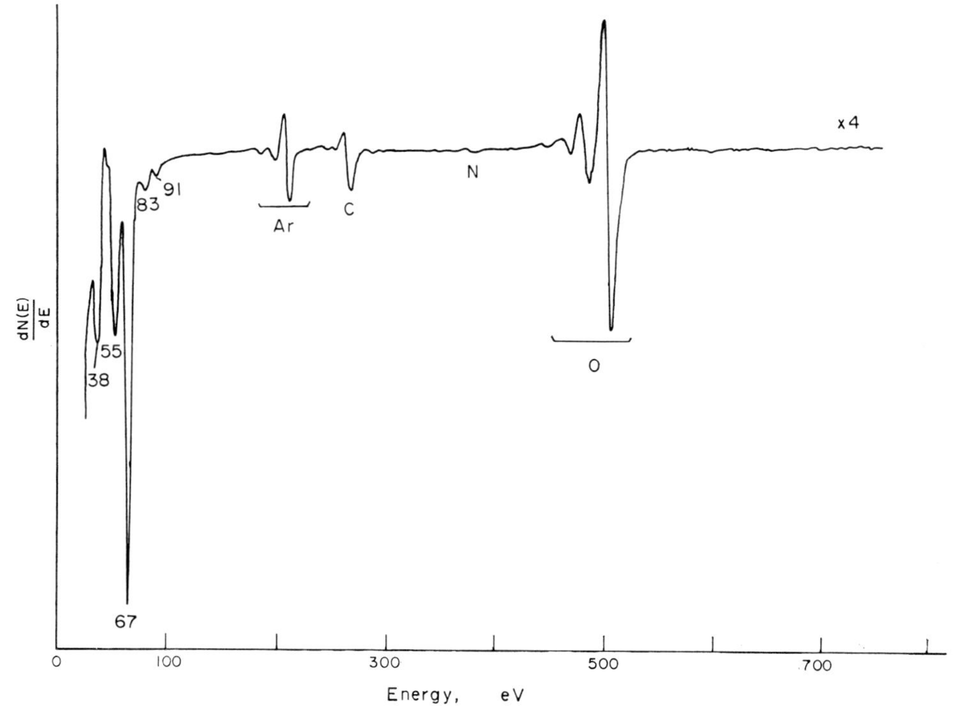

Figure 4. An Auger electron spectrum of an analyzed sample in depth of about 50 Å from the surface.

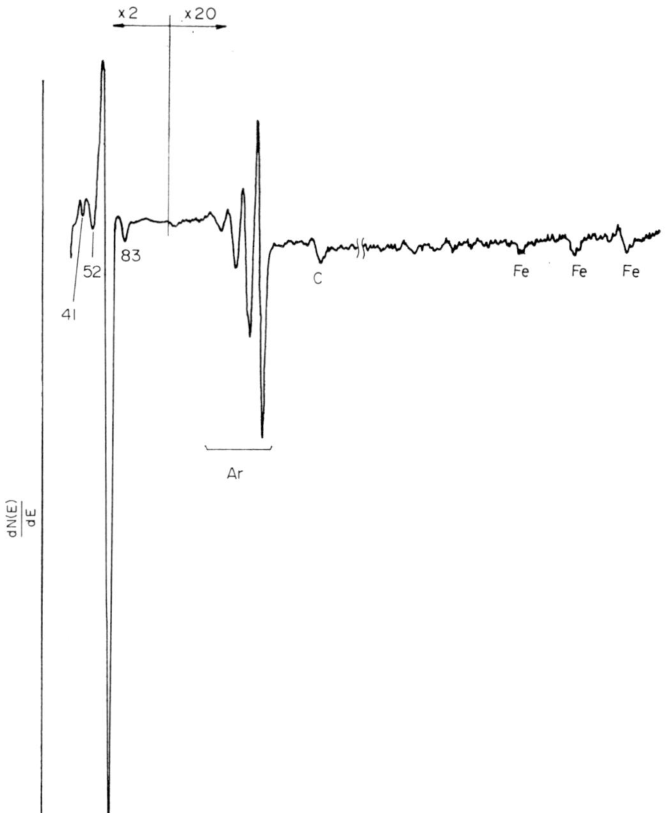

Figure 5. An Auger electron spectrum of clean aluminum layer in MNOS structure, depth ~ 500 Å.

represents an 'interface' containing a measurable amount of Al, Al_2O_3 and eventually the other Al and O compounds. The centre of this interface lies about 60 Å from the sample surface. It is necessary to take into account that the surface after sputtering is not absolutely flat and smooth, which is caused mainly by an uneven initial surface, by non-homogeneous ion density and by the presence of non-monoenergetic sputtering ions. Moreover, the uneven crater supports the affects of the different sputtering yields of Al_2O_3, Al, Si_3N_4, SiO_2, Si, if they are creating the macroscopic uneven areas in crater during the sputtering. The dissociation probability of Al_2O_3 and the motion of oxygen atoms (ions) into the bulk[13] as a result of ion interaction with surface in this case is very small. According to the above discussion, we conclude that the mean thickness of Al_2O_3 on Al is about 60 Å and that it was created by the diffusion of oxygen into Al and subsequent oxidation over a long period of time.

4.2. AES of the Al/Si_3N_4 structure. After the next sputtering of Al there appeared at $I \cdot t$ 1400 μA min in Auger spectrum a new peak at 92 eV, which belongs to elemental Si and N, O peaks, as shown in Figure 7. Later we can see the shift of 92 eV peak into 91 eV position. At $I \cdot t \sim 1490$ μA min a peak at 83 eV disappeared and a strong signal at 90 eV was created, as shown in Figure 8.

Further, the position of 90 eV peak was shifted to the value 87 eV and his amplitude was increased, as seen in Figure 10. The peak at 87 eV represents the Auger electrons emitted from Si in Si_3N_4. Heckingbottom and Wood[14] published for the main peak from Si in Si_3N_4 the value 86 eV. According to Holloway[15] the main peak of Si in Si_3N_4 lies at 87 eV. The concentration depth profile of an interface Al/Si_3N_4 at $I \cdot t \sim 1400$–1670 μA min is shown in Figure 9. The greatest changes of peak-to-peak values at 67 eV (Al), 87 eV (Si in Si_3N_4) and 376 eV (N in Si_3N_4) lie in the range 1470–1630 μA min, which is equivalent to a thickness of about 120 Å. In this depth of crater the effect of various inhomogenities on the interface profile identification is greater than in the previous interface. The measured values indicate that this interface is wider. In this case there exists a real possibility for the movement of Al into Si_3N_4 layer, as a result of sputtering ion–surface interaction, that is shown in Figure 10, taken at the distance 1420 Å ($I \cdot t \sim 2100$ μA min) from the surface, it means almost in the centre of Si_3N_4 layer. The presence of oxygen signal at 504 eV in the Si_3N_4 layer shows the probability that silicon oxides are also present.

4.3. AES of Si_3N_4/SiO_2 structure. The depth concentration profile of Si_3N_4/SiO_2 interface and near SiO_2 range ($I \cdot t \sim 2500$ —3180 μA min) is shown in Figure 11 where the changes of peak-to-peak heights at 87 (Si in Si_3N_4) -92 eV, 77, 62–63 eV (Si in SiO_2), 377–380 eV (N in Si_3N_4) and 504–506 eV (O in SiO_2) are drawn. The characterization of interface range is now more difficult. This interface, according to the great changes of O peak and the creating of 62–65 eV peak, lies between $I \cdot t \sim 2650$ and 2870 μA min, which represents the width of about 150 Å. These concentration profile dependencies characterize a possibly uneven surface of crater in the depth 1750 Å and possibly more complicated diffusion and chemical processes at the Si_3N_4/SiO_2 interface which have occurred during the sample preparation. An Auger spectrum from the interface ($I \cdot t \sim 2660$ μA min) is shown in Figure 12. An Auger spectrum taken at $I \cdot t \sim 2940$ μA min, where the peak at 77 eV (Si in SiO_2) is already visible, is shown in Figure 13. An Auger spectrum from the SiO_2 layer, at $I \cdot t \sim 3000$ μA min, where are well identified the peaks at 62–63 eV, 77 eV/(Si in SiO_2), also 92 eV, 108 eV (el Si), is shown in Figure 14. This spectrum shows that the SiO_2 is partially dissociated as a result of ion bombardment during sputtering and electron bombardment during the analysis.

4.4. AES of SiO_2/Si structure. A depth concentration profile of SiO_2/Si interface ($I \cdot t \sim 3180$–3780 μA min) is shown in Figure 15. The interface lies more precisely between the values $I \cdot t \sim 3280$ and 3600 μA min. At $I \cdot t \sim 3700$ μA min a bend occurs at 55 eV characteristic for clean Si. To this interface range corresponds the width of about 220 Å. The Auger spectra taken at $I \cdot t \sim 3280$, 3415 and 3700 μA min are shown in Figures 16–18, respectively. The spectrum in Figure 18 contains characteristic peaks for elemental Si LVV transitions (35–36 eV, 45–46 eV, 55, 74, 92, 108 and 138–139 eV). Moreover, we detected N, O and Fe. N peaks vanished after a dose of about 5500 μA min, O peaks vanished after a dose of about 6900 μA min. The spectrum of Si peaks is unchanged in these depths.

To check the influence of sputtering during the depth profile analysis, we made this analysis of the SiO_2 structure only, prepared in identical conditions to a structure MNOS, but the sample production cycle was over after Si oxidation in dry oxygen, at $T \sim 1100$°C, $d_{SiO_2} \sim 320$ Å. The analysis was performed like in the MNOS case, only for sputtering we used ion beam energy 450 eV for Ar^+. An Auger spectrum of the SiO_2 surface before sputtering is shown in Figure 19. The peaks at 62 and 77 eV correspond to Si in SiO_2. A depth concentration profile from the surface and in the interface SiO_2/Si is shown in Figure 20. When we compare a similar interface in MNOS structure we see now the interface width is narrower, $I \cdot t \sim 540$–600 μA min represents 100 Å. The interface is more clearly defined by the sharp changes of peak-to-peak values.

5. Conclusions

From the analysis of Al_2O_3/Al/Si_3N_4/SiO_2/Si and SiO_2/Si we come to these conclusions:

(a) Pure Al exposed to the atmospheric conditions was oxidized to a depth of about 30–90 Å.

(b) The characteristic Auger peaks corresponding to the Auger electrons emitted from Al atoms bonded in Al_2O_3 lie at 38 and 56 eV.

(c) The Auger peaks from the pure Al not bonded in compounds lie at 41, 52, 67 eV (main peak) and 83 eV.

(d) Main characteristic Auger peak of Si bonded in Si_3N_4 lies at 87 eV.

(e) At the Al/Si_3N_4 interface may arise chemical reactions from thermal processes and argon ion bombardment, under the above discussed conditions, in which were formed elemental (not bonded) Si peaks at 90, 92 eV.

(f) Under the foregoing conditions the probability of the dissociation of Si_3N_4 does not exist as the result of argon ion and electron bombardment.

(g) The Al atoms or ions probably move into the Si_3N_4 layer as a result of argon ion bombardment during sputtering,

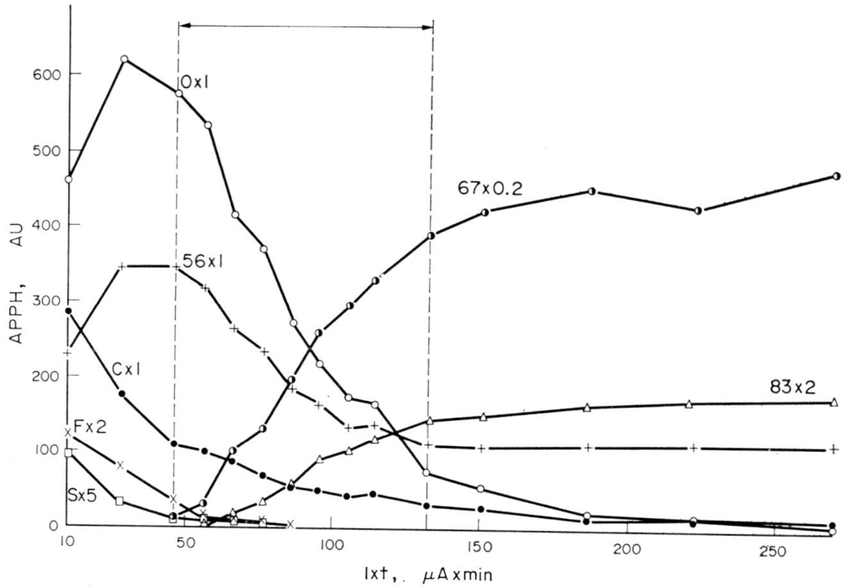

Figure 6. An Auger electron peaks depth profile of Al_2O_3/Al structure.

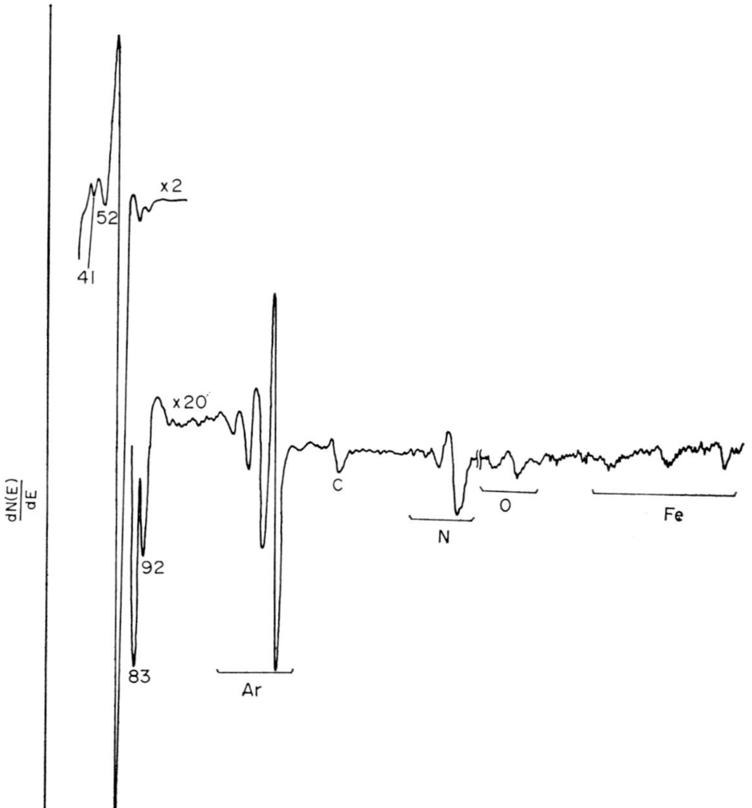

Figure 7. An Auger electron spectrum from the beginning of Al/Si_3N_4 interface; depth ~950 Å.

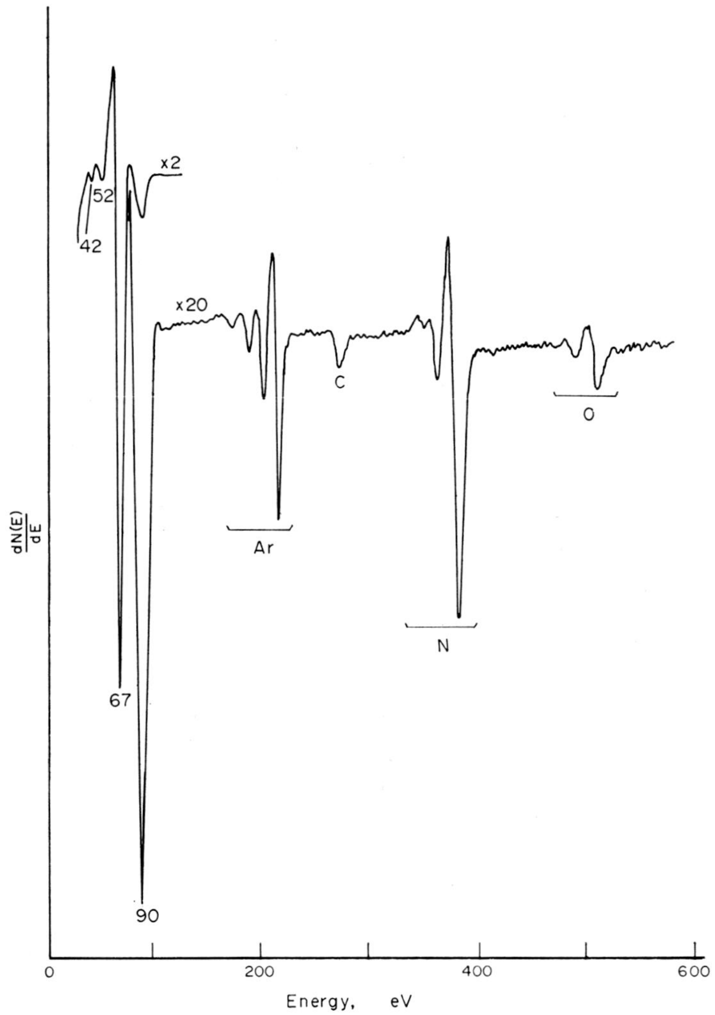

Figure 8. An Auger electron spectrum from the Al/Si_3N_4 interface; depth ~ 1000 Å.

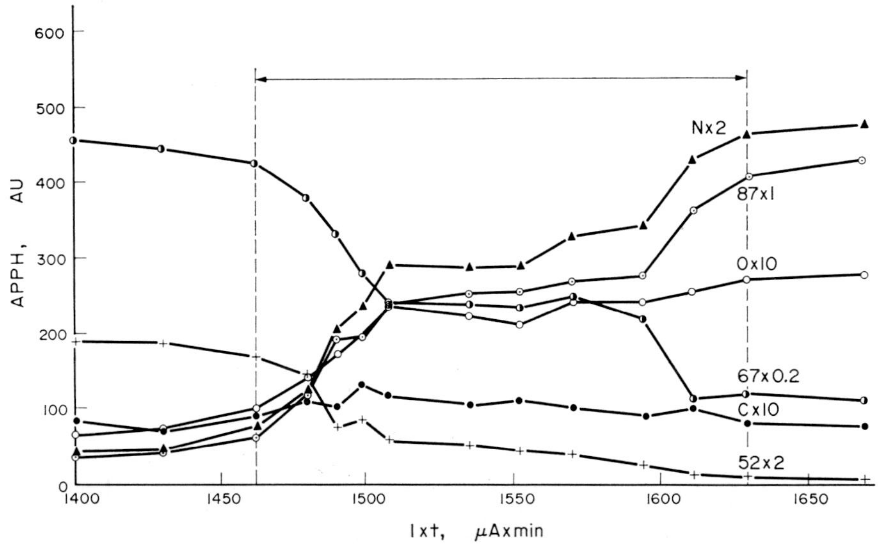

Figure 9. An Auger electron peaks depth profile of Al/Si_3N_4 structure.

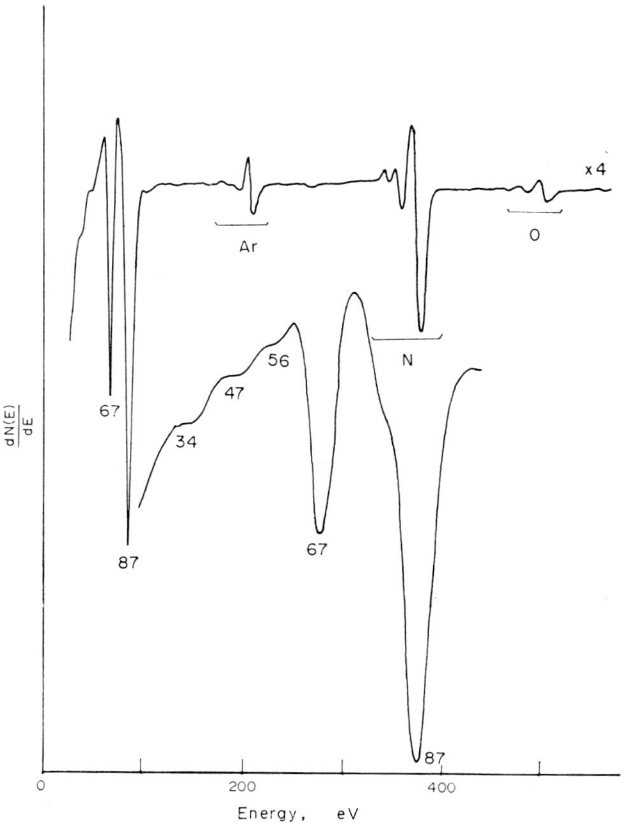

Figure 10. An Auger electron spectrum from the Si_3N_4 layer; depth ~ 1420 Å.

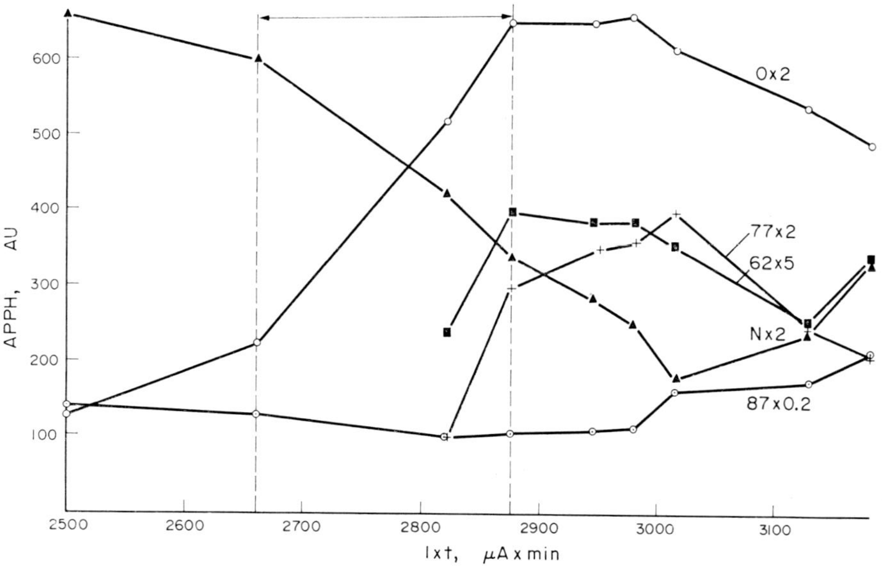

Figure 11. An Auger electron peaks depth profile of Si_3N_4/SiO_2.

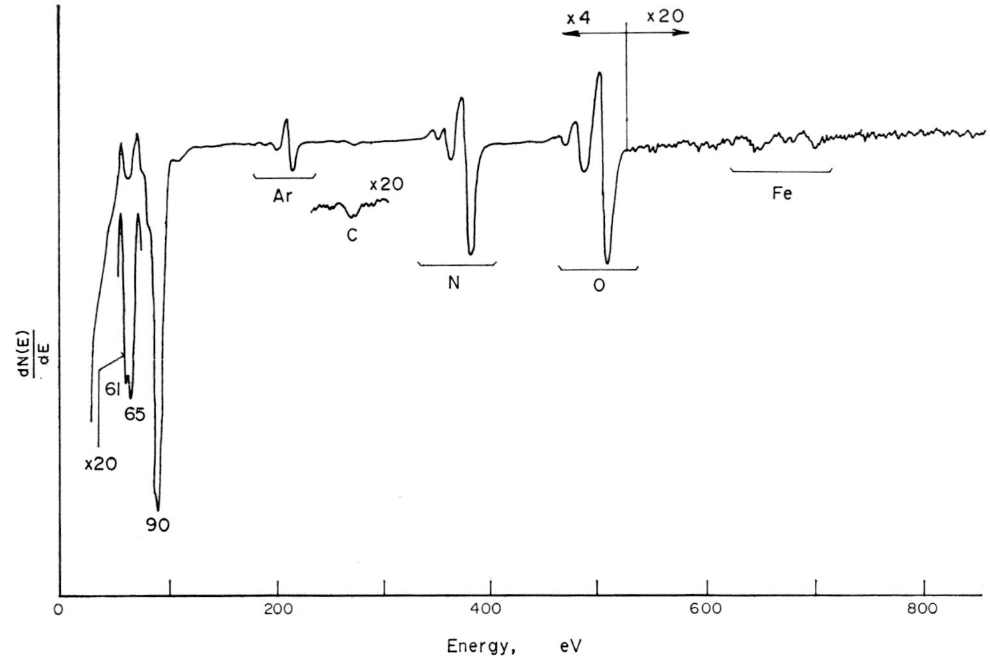

Figure 12. An Auger electron spectrum from the Si_3N_4/SiO_2 interface; depth ~ 1800 Å.

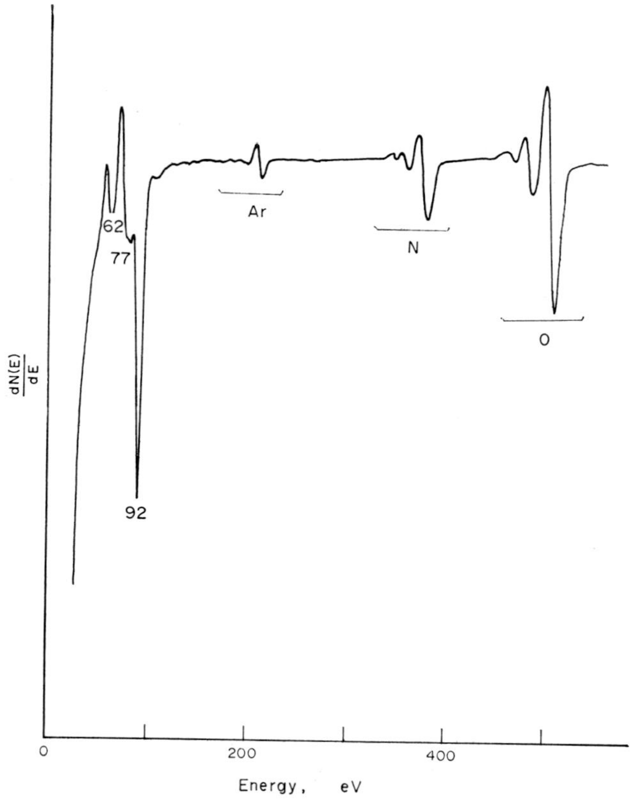

Figure 13. An Auger electron spectrum near the Si_3N_4/SiO_2 interface; depth ~ 1980 Å.

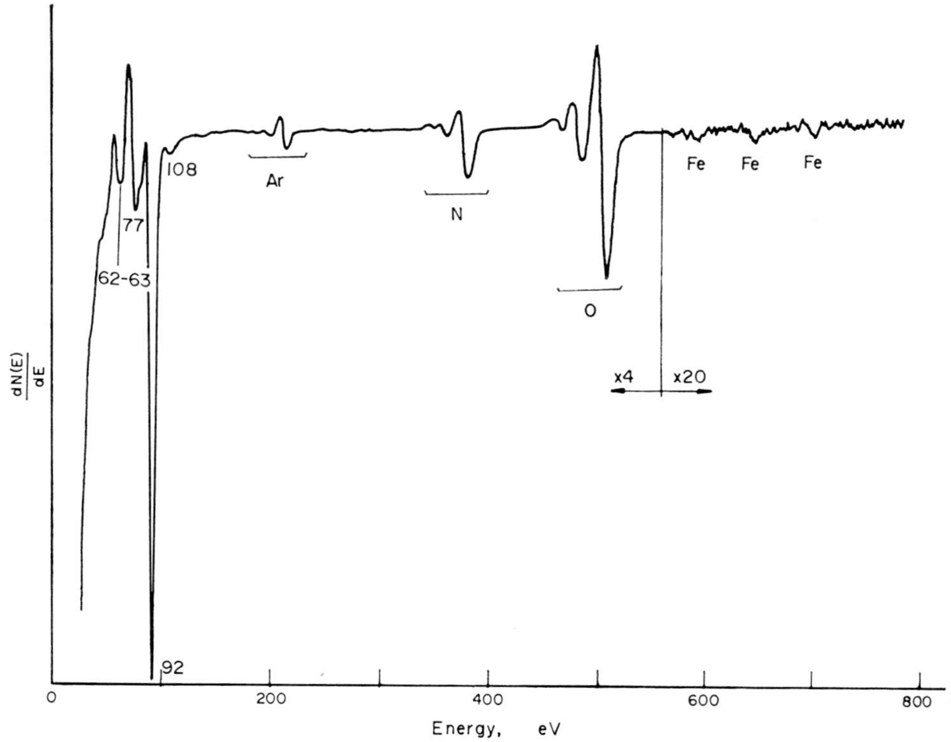

Figure 14. An Auger electron spectrum from the SiO_2 layer; depth ~ 2030 Å.

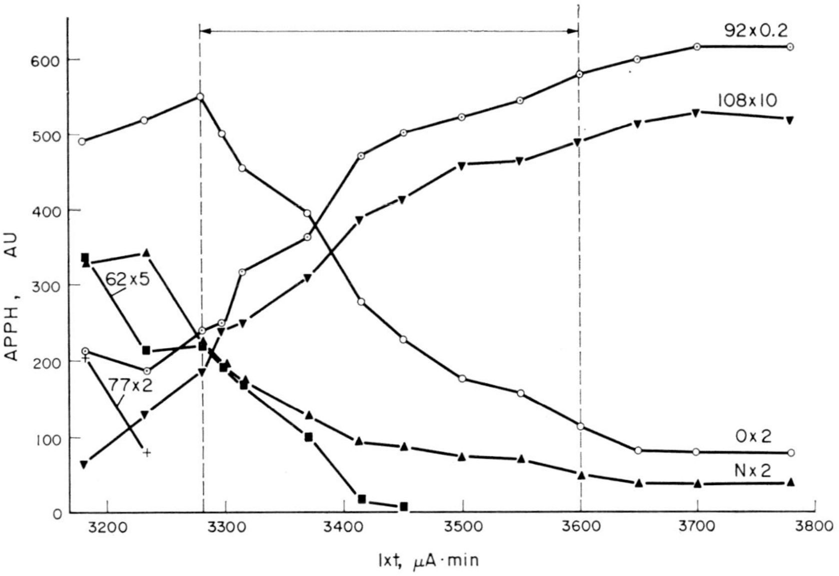

Figure 15. An Auger electron peaks depth profile of SiO_2/Si structure in MNOS sample.

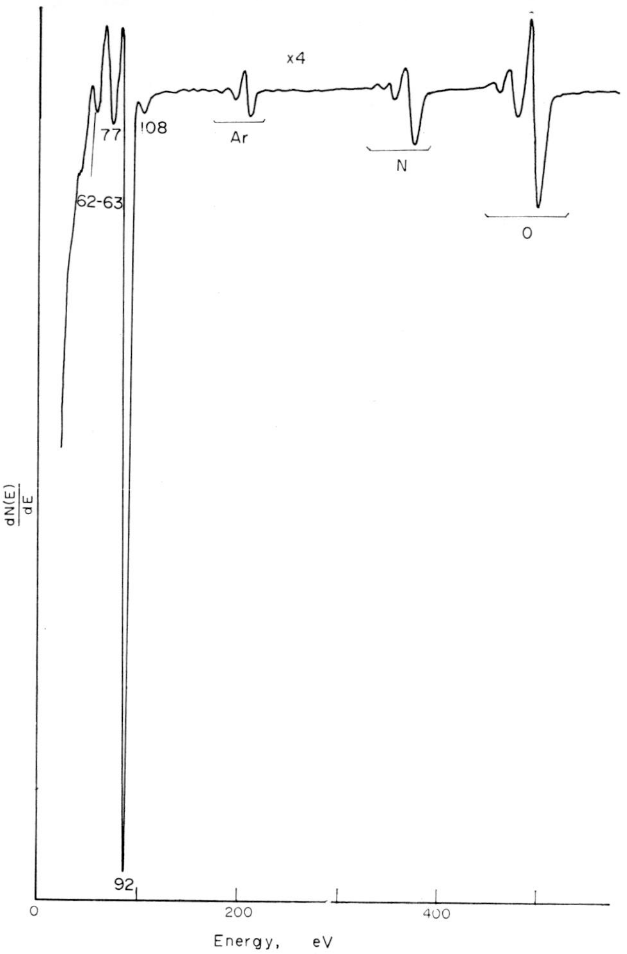

Figure 16. An Auger electron spectrum from the SiO_2/Si interface.

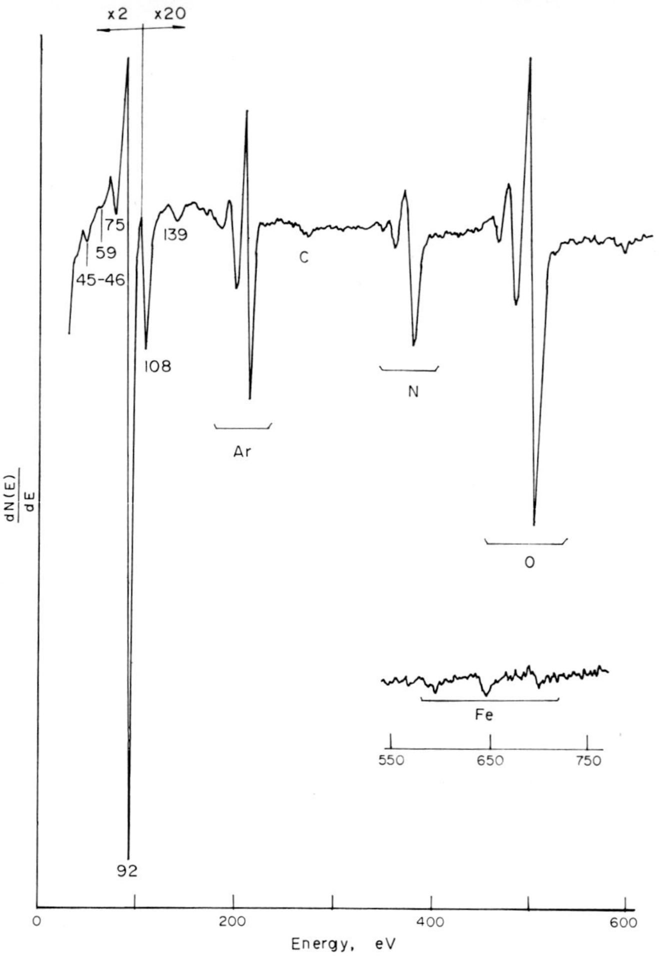

Figure 17. An Auger electron spectrum from the SiO_2/Si interface.

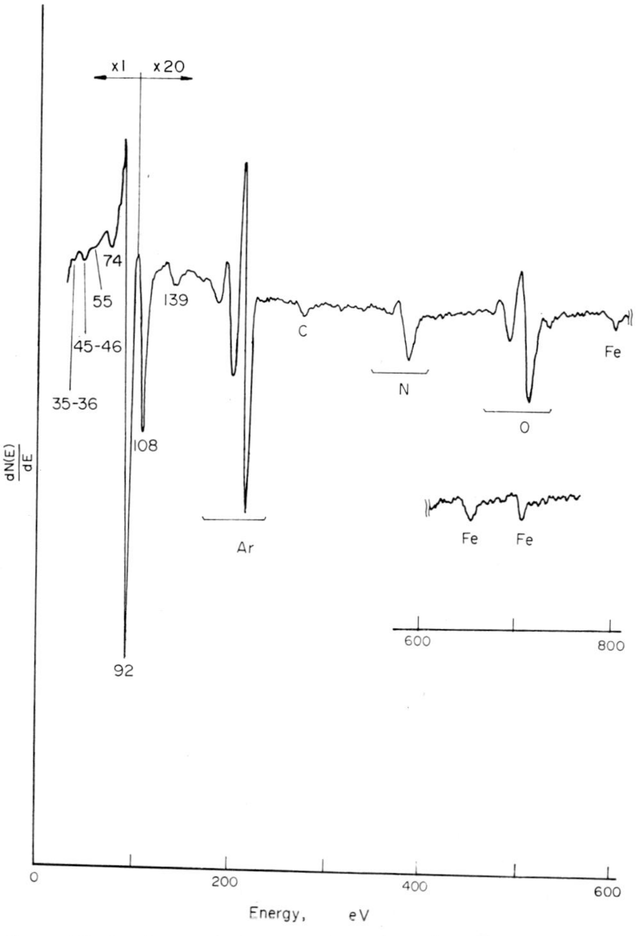

Figure 18. An Auger electron spectrum from the silicon.

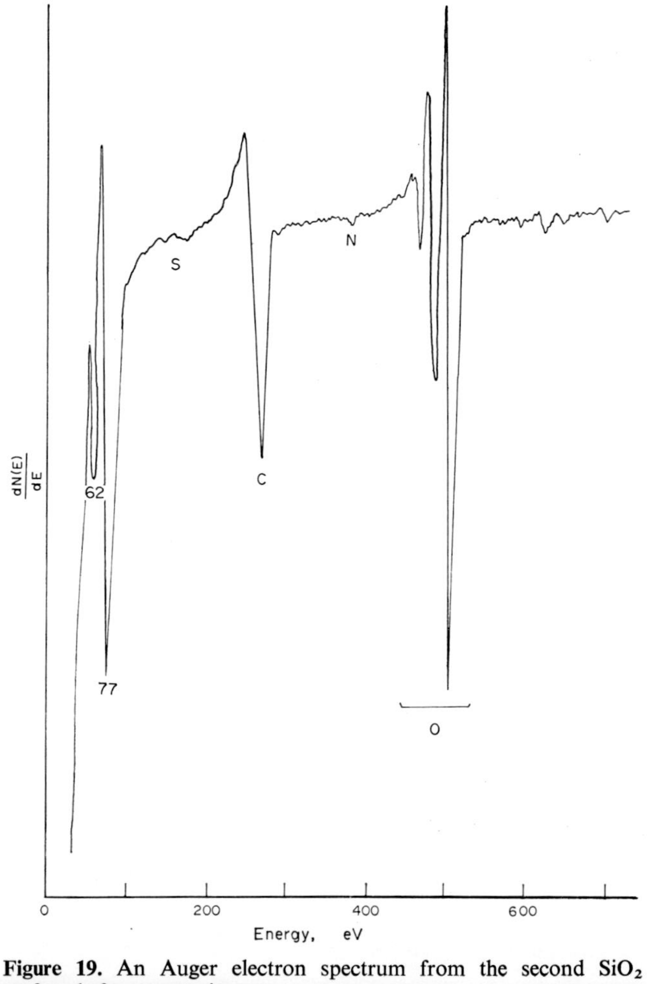

Figure 19. An Auger electron spectrum from the second SiO_2 surface before sputtering.

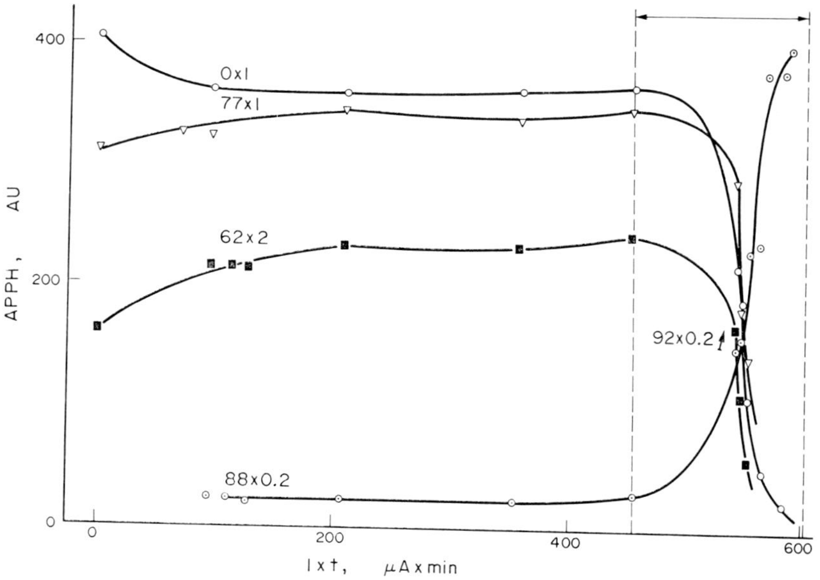

Figure 20. An Auger electron peaks depth profile of the second SiO_2/Si interface.

because we detected an Al signal through almost the whole Si_3N_4 film thickness till 600–700 Å from the Al/Si_3N_4 interface.

(h) In our CVD deposition of Si_3N_4 small amounts of silicon oxides also grew and these represent the O peaks in the whole layer of nitride.

(i) The presence of N Auger signals in SiO_2 and Si indicates the possibilities of nitrogen diffusing into SiO_2 during the beginning of Si_3N_4 deposition at $T = 830°C$, and on the next further nitrogen diffusion into Si. We do not exclude the motion of N from the interface Si_3N_4/SiO_2 as a result of argon ion beam interaction with the surface.

(j) The interface width of Si/SiO_2 in semiconductor structures creating in Si oxidation can be 100–200 Å or less, according to our MNOS and SiO_2/Si Auger analysis.

(k) The interface width between oxidized Si and CVD deposited Si_3N_4 can be 100–150 Å or less.

(l) The interface width between CVD deposited Si_3N_4 and evaporated Al can be 100–120 Å or less.

(m) The experiments showed that the depth resolution abliity of AES is decreasing with the depth of sputtered crater. For the sputtering it is more suitable to use less energetic ions.

(n) In our experiments, the Al_2O_3 was not dissociated during the ion bombardment.

References

[1] R E Weber, *AES for Thin Film Analysis*, PEI, (1973).
[2] C C Chang, A C Adams, G Quintana and T T Sheng, *J Appl Phys*, **45**, 1974, 252.
[3] T Inoue, S Horiuchi, S Yoshii and J Tanuna, *Japan J Appl Phys* Suppl, **44**, 1975, 329.
[4] N J Chou, Y J van der Meulen, R Hammer and J Cahill, *Appl Phys Lett*, **24**, 1974, 200.
[5] J S Johannessen, W E Spicer and Y E Strausser, *J Appl Phys*, **44**, 1974, 3028.
[6] Y E Strausser, J S Johannessen and W E Spicer, *NBS Spec Publ*, 400–23, (1976).
[7] J S Johanessen, W E Spicer and Y E Strausser, *Appl Phys Lett*, 1975.
[8] B D Powell and D P Woodruff, *Surf. Sci*, **33**, 1972, 437.
[9] J S Johanessen, W E Spicer and Y E Strausser, Varian publ, VR–108, 1976.
[10] G Dufour, H Guennou and C Bonelle, *Surf. Sci*, **32**, 1972, 731.
[11] D J Quinto and W D Robertson, *Surf Sci*, **27**, 1971, 645.
[12] P W Palmberg, G E Riach, R E Weber and N C MacDonald, *Handbook of Auger Electron Spectroscopy*, PEI.
[13] H M Naguib and R Kelly, *Radiat Effects*, **25**, 1975, 1.
[14] R Heckingbottom, P R Wood, *Surf Sci*, **36**, 1973, 594.
[15] P H Holloway.

1.10 Growth and composition of Al clusters on NaCl

S R Veljković and P Bibović, *Institute of Physical Chemistry, Faculty of Science, Beograd, Semiconductor Factories, Niš, Yugoslavia*

In most theories on the formation and growth of thin solid films diffusion of deposited monoatoms upon substrate surface is suggested to be the deciding process.[1,2] It has been shown recently that the motion of smaller nuclei may also take place and influence the kinetics of deposition and growth of thin films.[3,4] In the present work the formation of thin aluminium film on NaCl monocrystal was studied with the aim to discover phenomena related to the motion of small nuclei. The influence of this motion on the kinetics of thin film formation has also been discussed.

Specimen preparation

Deposition of aluminium on NaCl was carried out in an all stainless-steel apparatus, in high vacuum (with deep-cooling traps) at the pressure of 10^{-7} torr. Substrate samples were mounted on the holder with controlled temperature (accuracy $\pm 5°C$) which could be manoeuvred from outside. In this work, substrate temperature was varied from 100° to 300°C. Substrate samples, $1 \times 1 \times 0.1$ cm in size, were prepared by cleavage of NaCl monocrystal. Before the beginning of the experiments, mounted samples were annealed in high vacuum at 350°C for 1 h. The crystal orientation of substrate samples was (100).

High purity, 99.999% Al was used in all experiments. Before use, it was first chemically polished by mineral acids mixture (H_3PO_4, CH_3COOH and HNO_3), washed with deionized water and then dried at 150°C in vacuum. The evaporation of aluminium was performed from tungsten evaporator, heated at 920°C. The evaporator was located approximately 20 cm from the sample holder. The rate of deposition and the mass deposited were controlled by a quartz crystal monitor (Sloan Angstrometer). The thickness of aluminium deposits was controlled optically (by an interferometer) and mechanically (by the 'Talystep' apparatus). Calculated deposition rate was 3 Å s^{-1}, and the flux of impinging atoms was estimated to be about 4×10^{14} atoms cm^{-2} s^{-1}.

Measurements

The structure of aluminium films deposited on (100) NaCl was inspected by a Phillips EM300 G transmission electron microscope with the resolution power of 10 Å. After the deposition, the deposits were fixed by a carbon layer (approximately 200 Å). NaCl carrier was then dissolved by floating on distilled water and the specimens were mounted on grids in the electron microscope. Several photographs were taken from different parts of the specimens, at the standard magnification of $\times 35,000$. Photographic plates with the fine grain emulsion were used in this work. In Figure 1 is presented the micrograph of aluminium film, deposited 3 s on NaCl at 200°C.

The micrographs were analysed as such, or were further enlarged in order to get a more detailed picture. The substructure of clusters was discovered when the negatives ($\times 35,000$) were inspected in an optical microscope (E. Leitz—Ortolux), at the magnification $\times 30$ (total magnification is about 10^6). Practically, most large clusters were seen to consist of two or three smaller nuclei held together. The number of such complex clusters, and the distance of small nuclei inside them, varies with time and temperature of the film deposition. An example from the micrograph of film deposited 3 s on NaCl at 200°C is given in Figure 2.

Temperature dependence of cluster size and composition

Our enlarged micrographs demonstrate clearly that some clusters have composite structure. This is surprising, having in mind that smaller nuclei should coalesce in cases when they approach each other sufficiently closely. For this specific behaviour, which is evidently imposed by our working procedure, we can only say to result probably from the mobility of smaller nuclei, facilitated by high temperatures of the substrate. Facilitated mobility is physically and thermodynamically justified, but the preservation of uncoalesced nuclei in clusters is a problem.

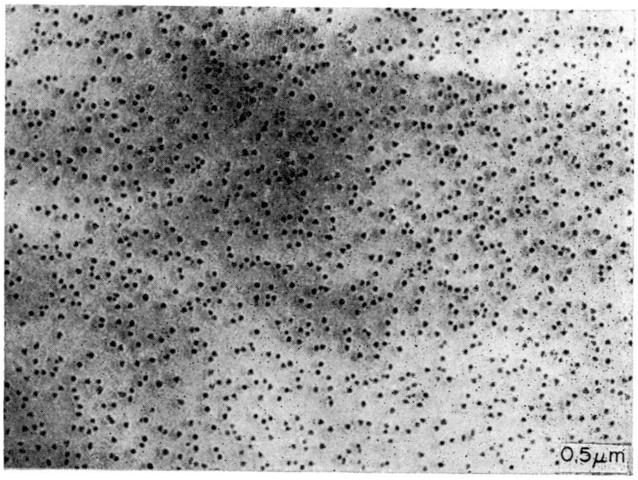

Figure 1. The electron micrograph of Al film deposited on NaCl for 3 s at 200°C. The primary magnification on the microscope: $\times 35,000$.

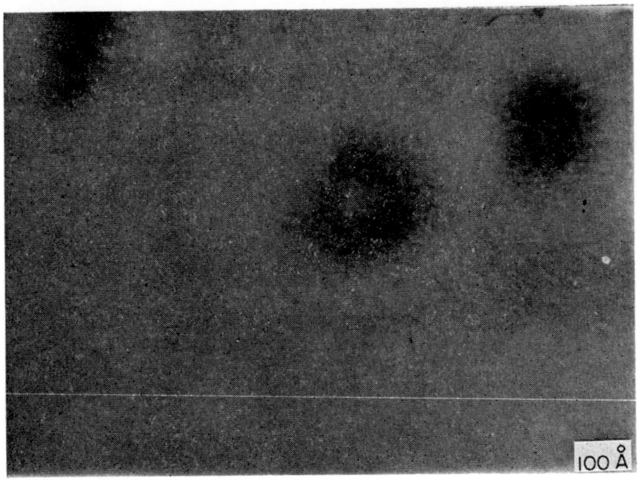

Figure 2. The enlarged micrograph of Al film deposited on NaCl for 3 s at 200°C. The magnification on the optical microscope: ×30. The total magnification: $\times 10^6$.

In order to see if the temperature of substrate is the main cause of above phenomena, we will compare data on the number and on the size of clusters at three different temperatures. Histograms on cluster size distribution (at the magnification of ×35,000) at 100, 200 and 250°C are shown in Figures 3–5.

These histograms indicate that in most cases there is an increase of the size of clusters with time and temperature, and the maxima of distributions shift towards larger sizes. This behaviour is formally analogous to the behaviour imposed by Ostwald ripening of nuclei. The surface tension and the surface Gibbs free energy of small nuclei is high and surface atoms have sufficient energy to evaporate and diffuse towards larger nuclei whose surface tension is lower. The ensuing asymmetrical size distributions (mathematically defined by Lifshitz, Slyozov and Wagner) are similar to distributions in Figures 3–5.

A special phenomenon is observed in the histogram for 3 s at 250°C (and for most samples at 300°C and higher temperatures). Namely, the distribution becomes practically symmetrical, as both smaller and larger sizes are more frequent.

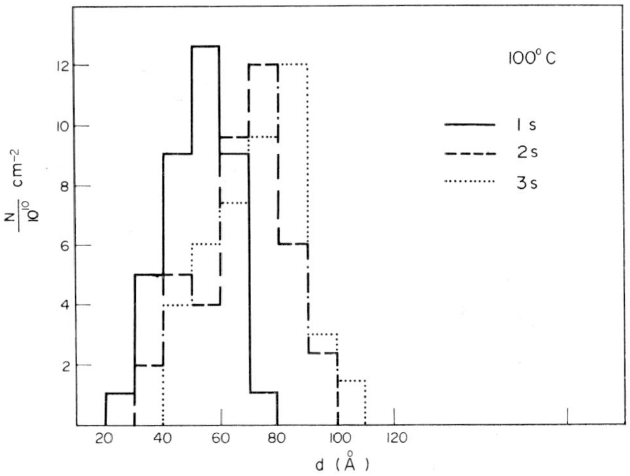

Figure 3. Histograms of the number of Al nuclei vs diameter interval in micrographs. The magnification: ×35,000. Temperature of samples: 100°C.

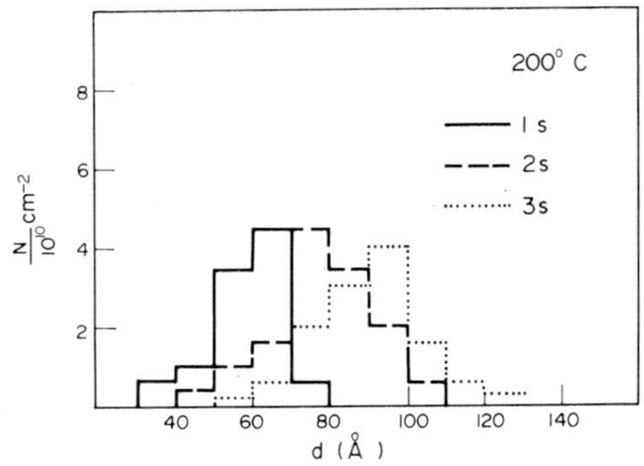

Figure 4. Histograms of the number of Al nuclei vs diameter interval in micrographs. The magnification: ×35,000. Temperature of samples: 200°C.

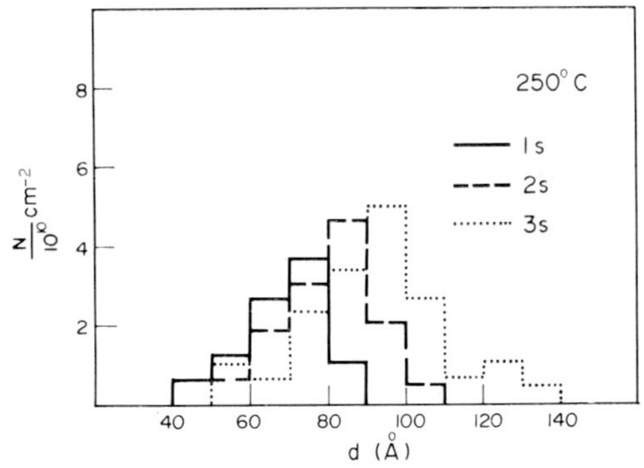

Figure 5. Histograms of the number of Al nuclei vs diameter interval in micrograph. The magnification: ×35,000. Temperature of samples: 250°C.

Figures 3–5, demonstrate an essentially 'natural' behaviour of aluminium nuclei in thin films. But some correction should be introduced on the basis of the data from enlarged micrographs. First, in our samples the Ostwald ripening is not the only explanation of the appearance of larger clusters, since they should have a compact structure produced by the gradual capture of incoming atoms. However, the photos shown in Figure 2, as well as in Figures 6 and 7 do not show compact clusters, but the groups of smaller nuclei inside the clusters.

Figure 6. The enlarged micrograph of Al film deposited on NaCl for 2 s at 250°C. The total magnification: $\times 10^6$.

Figure 7. The enlarged micrograph of Al film deposited on NaCl for 2 s at 200°C. The total magnification: $\times 10^6$.

On account of the above data we may say that small nuclei in aluminium film do not undergo only Ostwald ripening if the temperature of the substrate is sufficiently high, but move as a whole either towards each other, or towards some surface sites (defects in substrate lattice, etc.). They are incorporated in a sort of complex agglomerate.

If we suppose that complex agglomerates have sufficiently long lifetimes they may then decompose by the same process, the diffusion of nuclei. At the highest temperature, the small nuclei may reappear, as in fact, we see in the symmetry of the distribution in Figure 5.

Why do small nuclei not coalesce? High temperatures at substrate surfaces and the unstable nature of the thin film which is produced by arrival of monoatoms of aluminium having large kinetic energy (evaporated at 920°C) is the main reason. Namely, it is necessary to remove excess energy of products at each step of the formation and growth of nuclei and clusters. NaCl has modest heat conductivity (as most ionic crystals) and the film remains 'excited' as a whole. Only at higher coverage, when heat is dissipated faster through a denser metallic layer, the stabilization, coalescence, etc. become immediate. Our smaller nuclei are, thus, sufficiently mobile to oscillate in complexes and to fly apart. Even Venables suggests a similar phenomenon by saying that at the highest temperatures stable pairs (initial nuclei) may evaporate directly into the vapour.[2] These facts should be kept in mind, since in practical productions of thin films, 'excited' deposits may produce uneven agglomerates and locally depleted regions in spite of correct working arrangements.

Many authors have suggested that small nuclei may be largely mobile in the course of the thin film formation, but there are still many doubts about their eventual encounter and the subsequent coalescence.[2,5,6] Our enlarged micrographs indicate that under certain conditions their encounter may not be ultimately followed by their coalescence. Since their behaviour in the course of the formation—and eventual decomposition—of large clusters bears some characteristics of reversible processes, we may introduce some kinetic treatments based on reversible reactions. This will be first done in connection with the evaluation of reaction enthalpy for the complex formation. On this basis we may describe quantitatively the motion of small nuclei.

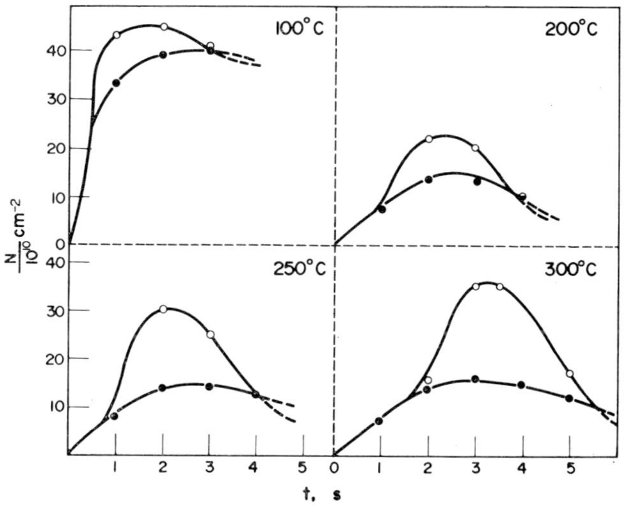

Figure 8. The plot of nucleus density, N vs time, t, at different temperatures. The upper curves (—○—): data from enlarged micrographs, at $\times 10^6$ magnification. The lower curves (—●—): data from micrographs at the magnification of $\times 35{,}000$.

Rate data for nucleus density

We may construct diagrams for time dependence of the density of nuclei in both: the original micrographs (at the magnification of $\times 35{,}000$) and also enlarged pictures of parts of micrographs (at the magnification of about $\times 10^6$). Data for 100, 200, 250 and 300°C are shown in Figure 8. The lower curves, at each temperature, refer to the magnification of $\times 35{,}000$, and the upper ones represent the data for the magnification of $\times 10^6$).

It is observed in Figure 8 that the difference between upper and lower curves increases with temperature. This means that the occurrence of complex clusters is favoured by heat. If we take the ratio of maxima in upper and lower curves at each

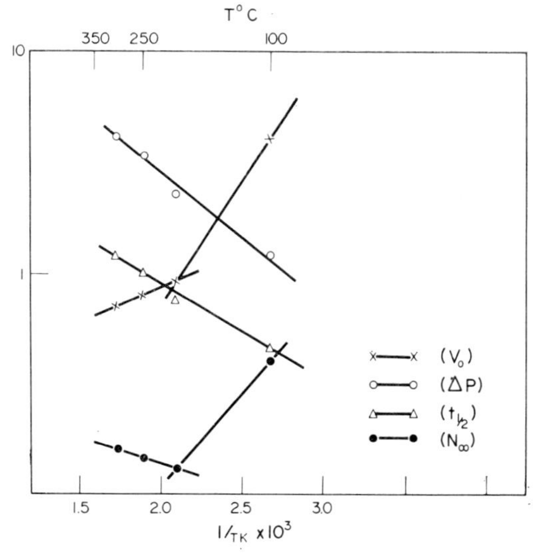

Figure 9. Plots of parameters from Figure 8 $1/T$ K. $V_0 = dN/dt$. Data from micrographs at the magnification of $\times 35{,}000$. ΔP $P_{(○)} - P_{(●)}$; P(○) = the surfaces limited by rate curves for 5 s (data at the magnification of $\times 10^6$); P(●) = the surfaces limited by rate curves for 5 s (data at the magnification of $\times 35{,}000$); $N_{00} = N_{00}^{(●)}$ in Figure 8. $t_{1/2}$ = the half-life of the accumulation of $N_{00}^{(●)}$ in Figure 8.

temperature, it changes from 1.13 at 100°C to 2.24 at 300°C. The same tendency is observed in results on ΔP, obtained when we integrate surfaces under the upper and the lower curves, at each temperature, and take their difference. The integration is performed over surfaces corresponding to the deposition in the course of 5 s, since the difference between upper and lower curves vanishes at that time. Practically, integrated surfaces represent total yields of respective nuclei (or clusters), and ΔP measures the excess amount of smaller nuclei discovered at the magnification of $\times 10^6$, compared to the amount of clusters in the original micrograph (at the magnification of $\times 35{,}000$).

It is most convenient procedure to plot data for log ΔP vs $1/T$ with the aim to evaluate the enthalpy of the cluster formation from the slope of the plot. This plot is given in Figure 9. From the slope we have calculated the enthalpy of the association–dissociation process for small nuclei to be 2.4 kcal/mole. This value is very near to the value of the enthalpy of melting of aluminium metal: $\Delta H_m = 2.6$ kcal/mole. On this basis we may suggest that the state of thin film of aluminium in our samples is very similar to the state of that metal in the course of the melting, and that the behaviour of small nuclei is determined by enhanced mass transport inside the film. This is justified on the basis of the relatively low melting point of aluminium: 660.1°C, which renders possible massive diffusion at about 200°C (the Tamman temperature for aluminium metal). If we take into account that a great number of our samples was kept at 200°C and above that temperature, it is obvious, that the cooling of most arriving aluminium atoms onto substrate surface was poor, and that the aluminium thin film was 'excited' in reality. This means that our samples may serve as reliable examples of the behaviour of excited thin solid films.

When other parameters of curves given in Figure 8 are plotted in the same way as ΔP, we obtain from their slopes the values of the enthalpy of reaction satisfactorily close to those given above. These parameters include: N_{00}, which represents the nucleus density at the maximum of curves based on data at $\times 35{,}000$, $t_{1/2}$ which is the half-life of the accumulation process of N_{00} nuclei, and V_0, the initial change of nucleus density, dN/dt, before the onset of the upper curves. Because of an increased tendency to dissociate, the composite nuclei attain equilibrium early and their recombination is slow, especially at higher temperatures. The values of N_{00} may rise slowly by some secondary processes, but all changes are slow at higher temperatures, since small nuclei are preserved in greater quantities reflected in upper curves in Figure 8.

The enthalpy calculated from the slope of the plot of log $t_{1/2}$ versus $1/T$ is 1.9 kcal/mole. Numerical values of the above parameters are given in Table 1.

Table 1. Essential parameters of nucleus density N, of Al deposited on NaCl enlargements: $\times 35{,}000$ (●) and 10^6 (○). Data from Figure 8.

Temperature of substrate (°C)	$N_\infty^{(\circ)} / N_\infty^{(\bullet)}$	ΔP ($P^{(\circ)} - P^{(\bullet)}$) (ms^2)	$P^{(\circ)}$	$t_{1/2}$ (for $N_\infty^{(\bullet)}$) s
100	1.13	120	1145	0.4
200	1.54	226	683*	0.75
250	1.98	355	810	1.05
300	2.24	495	970	1.25

* The decrease of the surface is the result of changes in diameters and volumes of nuclei (ref 2, 3).

P = the surface under the plot: N vs t.

Conclusion

Our work demonstrates the complex behaviour of Al thin films deposited on NaCl monocrystal at high temperature. In this work we introduce the picture of 'excited' thin films. It is based on insufficient removal of excess energy from nuclei and clusters. which then take part in the mass transfer on a large scale, eventually facilitated by gliding phenomena due to traces of impurities and surface irregularities.

Acknowledgements

We wish to thank Dipl Fiz R Milinković for valuable help in the preparation of micrographs.

References

[1] G Zinsmeister, *Thin Solid Films*, **2**, 1968, 497.
[2] J A Venables, *Philos Mag*, **27**, 1973, 697.
[3] J G Skofronick and W B Phillips, *J Appl Phys*, **38**, 1967, 4791.
[4] K Heinemann and H Poppa, *Thin Solid Films*, **33**, 1976, 237.
[5] J J Métois, M Gauch and A Masson, R Kern, *Surf Sci*, **30**, 1972, 43.
[6] H Schmeisser, *Thin Solid Films*, **22**, 1974, 83.

1. Vacuum and thin films—coating processes and pumps

1.11 Silicon films growth in vacuum by pyrolysis of silane

L N Alexandrov, F L Edelman and V V Voskoboinikov, *Institute of Semi-conductor Physics of the Academy of Sciences, Novosibirsk, 630090, USSR.*

The characteristics of growth of polysilicon films on silicon oxide or nitride substrates are discussed. Operative mechanisms of growth are studied, and the epitaxial formation of polysilicon film is given for comparison. The comparison of the observed rates of grain growth with the calculated ones and distribution of grains over sizes and on a surface indicate the action of diffusion processes and processes of building-in of atoms into a lattice. It is shown that recrystallisation growth of grains in the film influence considerably their finite sizes.

1. Introduction

Polysilicon films are promising material for electrical engineering. These are used as separating insulation in large integrated circuits and as resistive elements. There are attempts to produce Polysil MIS-transistors, solar batteries, piezoelectric transducers. However, now and in the nearest future Polysil will be used on a large scale as a conductive gate electrode of MIS-transistor with MNOS-and MOS-structure as well as in devices with charge coupling.

There are several preconditions:

(a) the possibility of deposition of Polysil in a united technological cycle with silicon surface decontamination and growth of silicon (or without it), and with deposition of a dielectric;

(b) possibility to decrease the stray capacitance and to increase the density of elements of the large integrated circuit and the circuit speed;

(c) reduction of threshold voltage of transistors.

Figure 1 shows the scheme of the effective three-phase charge coupling device (CCD) on silicon using three-layer metallization with Polysil. The use of Polysil for integrated circuits, especially in the case of multilayer metallization, inevitably requires the control and reproducibility of a shallow developed film texture. In this connection it is necessary to observe a gas phase composition and Polysil grain sizes in a wide temperature range and after repeated heating of films. It is clear that deposition of Polysil on dielectric in the pyrolysis of silane has features similar to those of epitaxial growth in the case of homogeneous origin and different in the case of heterogeneous origin,

The purposes of this work are as follows:

(1) to find out microscopic phenomena at Polysil formation and after film annealing;

(2) estimation of such an important parameter as thickness of a continuous Polysil;

(3) comparison of origin of Polysil on an amorphous substrate with that of monosilicon epitaxial film.

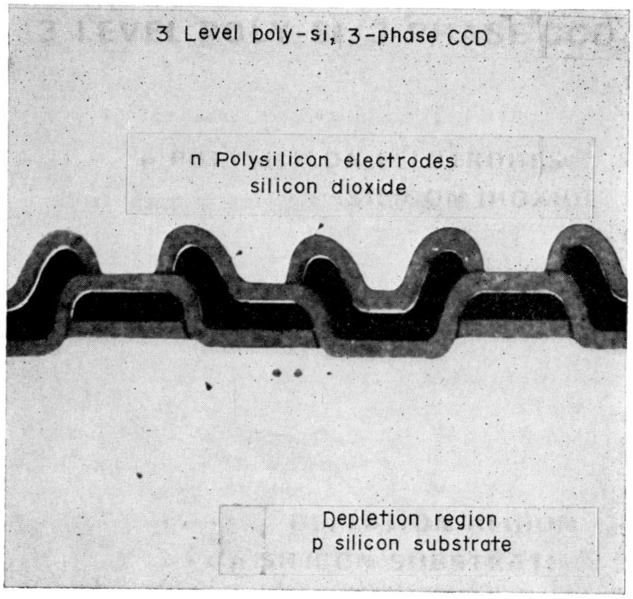

Figure 1. Schematic diagram of the three-phase charge coupling device on silicon with three-layer metallization on Polysil.

2. Method of obtaining of Polysil

The present paper offers mainly (below we shall make reservations illustrating exceptions) the results of deposition of Polysil in a low pressure (10^{-1}-10^{-2} torr) reactor which is heated from outside.

For substrates we used silicon plates of 0.2 mm thickness and 35–40 mm diameter which were first chemically treated in a standard way and coated with silicon nitride of 0.1 μm thickness, obtained by ammonolysis of silane at 850–900°C (in particular cases with a silicon oxide layer).

The schematic diagram of the reactor and some details of the processes were given before.[1–4]

The deposition of Polysil on silicon nitride is one of the most widely used operations in microelectronics. It is the process that is carried out in obtaining MNOS-transistors with self-adjusting gate. In CCD IC Polysil is more often deposited on thermal silicon dioxide. We have studied this process in detail and later we shall publish the results.

Pyrolysis of silane seems to be a simple reaction. Its equation with no intermediate stages is

$$SiH_4 \rightleftarrows 2H_2 + Si\ (\text{Polysil}) \qquad (1)$$

However a quantitative description of the results of pyrolysis is difficult due to imperfect knowledge of a number of secondary factors. For example, in many cases an elementary mechanism of decay of silane is unclear (homogeneous or heterogeneous, with formation of intermediate products, such as Si_2H_6, or without it, and so on). On that ground, in some works the activation energy of decay of silane, which determines the temperature dependence of the rate of Polysil thickness increase (V_p), varies from 10 to 37 and even up to 52 kcal mol^{-1} (0.5–2.5 eV). It means that the spread in values can reach almost eight orders of magnitude at various, sometimes badly controllable, conditions of silane decomposition. Of course, these are extreme cases. In reports on heterogeneous silane decomposition the activation energy for $V_p(t)$ is often given at about 0.9–1.6 eV, and in some not numerous investigations on homogeneous decomposition 2–2.4 eV.

Some new features of silane decomposition reaction and of silicon growth have been noted in one of the last works on pyrolytic deposition of silicon on a silicon substrate, cleaned from oxygen and carbon at the moment of deposition of a layer in high vacuum. Thus, as it turned out $V_p \sim P_{SiH_4}$. It proved that Si atoms impinging on the substrate re-evaporate considerably, and what is more, a greater part of the atoms re-evaporate. As with epitaxy, the film growth is carried out in high vacuum with the aid of a known sequence of steps, the temperature dependence of the motion speed is exponential with an activation energy, which is characteristic of surface diffusion of silicon on silicon, $F_{sd} = 1.5$ eV. As can be seen from Farrow,[5] according to the data of different authors, F_{sd} is measured from 0.2 to 2.5 eV in this case. It is the motion of steps but not the silane decomposition, which determines the rate of growth in the case described (activation energy of the silane decomposition is estimated as 0.7 eV). Ref 6 offers the analysis of the pyrolysis of silane in argon during the deposition of Polysil on the oxidized silicon surface. Of course, under these conditions the silane decomposition rate will be greater and, perhaps, the activation energy in the temperature function of the growth rate will differ from the results obtained at pyrolysis of silane in a mixture with hydrogen. According to the data of Seto[6] the stage determining Polysil growth is the decomposition of silane chemisorbed on the substrate. The activation energy of the Polysil deposition rate is 1 eV.

In general, according to a great number of publications the values of the maximum growth rate of about 1–10 μm min^{-1} are characteristic of both epitaxial silicon growth and Polysil deposition on an insulating substrate. These values are achieved at high silane concentration and in the range of deposition temperatures of 1000–1200°C.

3. Initial stage of film deposition

At the deposition of silicon films on silicon in vacuum Joyce *et al*[7] observed the formation of three-dimensional nuclei at a contamination of the substrate surface by molecules of O or C in the amount of more than 0.01 of a monolayer. The heating temperature of substrates increased up to 1200°C, the nuclei width-to-height ratio grew, and in 10 min the nuclei became two-dimensional. The use of a getter[8] and the increase of annealing temperature up to 1300°C[9] permitted to clean the silicon surface from carbide particles, to facilitate the motion of steps at the film growth. The system of steps on the substrate surface added Si atoms, moved, reconstructed and straightened them (Figure 2). The step height was 10–30 Å,

Figure 2. A non-stationary surface relief of the homoepitaxial silicon film at 800°C; an effective film thickness is 50 Å, growth time is 10 s.

the distance between steps was 0.1 μm. No pits or growth islands were observed on the films with an effective thickness of more than 25 Å, thus the film was continuous. The surface film relief was stabilized at a thickness of hundreds of Ångströms, but further optimization of conditions of substrate surface by cleaning and of the angle of its deviation from the (111) plane permits, in principle, to provide a nucleus-free layer growth with stable surface relief at all stages of the film growth. In this case of homoepitaxy, the substrate is fully overgrown with layer growth at a thickness of the order of the height of the initial surface steps. On the atomically clean step-by-step surfaces of silicon, the growth of film, even from another substance, e.g. germanium, can occur with no formation of three-dimensional nuclei from dihedral and trihedral angles of steps by formation of substrate-oriented pseudomorphic coherently conjugate regions.[10] At large supersaturations, the work of formation of critical nuclei decreases, the role of point defects of the substrate surface as active centres of atom accumulation grows, and orienting action of the substrate becomes weak. The disorientation of islands apparently causes the formation of a polycrystalline continuous film at the coalescence of islands.[11] On amorphous and polycrystalline layers coating the silicon surface, such as silicon nitride or oxide, the polycrystal films forms at any supersaturation.[1–4] At the transition from a clean silicon surface to that coated with the oxide or nitride film and, all the more, to the silicon nitride or oxide surface, an increase of the surface energy of the silicon nucleus–substrate interface changes the conditions of formation and growth of islands-grains. Therefore of interest are sufficiently reliable estimates of kinetic and thermodynamic parameters of growth which determine the distribution of islands over sizes at the moment of their coalescence into a continuous layer as well as the kinetics of recrystallization of continuous grain growth.

For the homoepitaxial film growth in high vacuum, we calculated an activation energy of surface self-diffusion of silicon atoms, lifetime in an adsorbed state, and time of building-in, including diffusion towards the step, at 840°C to be 6×10^{-4} and 1×10^{-6} s, respectively.[8] At the silicon island growth on a non-orienting substrate, the participation of diffusion was shown by the analysis of a correlation function of distribution of islands on the substrate surface and distribution of distances between islands.[4] The distribution of distances differed from the normal one

$$\theta(D) = \frac{1}{\sigma_l\sqrt{2\pi}} \exp\left(-\frac{D_l^2}{2\sigma_l^2}\right) \tag{2}$$

where D_l determines the deviation of distances between grains l from the mean one $\bar{l}$; σ_l is the dispersion. At small distances (400–600 Å) nuclei were not always observed, the distribution shifted towards large distances. Each nucleus was surrounded by a region whose dimensions corresponded to a diffusion wave. Adatoms diffuse from these regions to islands, decreasing supersaturation and eliminating the possibility of formation of new nuclei in a depletion region. At comparable temperatures (800–840°C) the diffusion length of adatoms on the silicon nitride surface was 4–5 times smaller than that on silicon; it corresponds to a relative increase of an adsorption activation energy, as compared with adatom diffusion activation energy, by 0.2–0.4 eV.

Figure 3 shows the Polysil microstructure at the initial stage of growth at 850°C on the silicon nitride surface (a) and the histogram (b) of grain sizes after 6 s of film growth. The Polysil structure and the histograms of grain sizes, which form at the same conditions of deposition on the silicon dioxide surface, are given in Figure 4. It is seen that the film texture on silicon dioxide is rougher, the grain height is higher.

4. Formation of continuous Polysil layer

At the deposition of silicon on the silicon oxide or nitride surface, the grain-nuclei have a quasispherical form, the rate of their formation (J) varies from 10^7 to 10^9 cm^{-2} s^{-1}. The rate of growth of grains within the deposition temperature interval of 750–1100°C varies from 1 to 4 nm s^{-1}. The formation of a continuous layer lasts 40–1 s at a thickness of 40–4 nm, respectively. One can use two approaches when estimating the continuous layer thickness. If the rate of growth of island-nuclei is limited by adatom diffusion on the island surface and decreases with an increase of the size according to the parabolic law, then the critical thickness of the continuous film is determined from the dependence

$$d_m = K_d(D_s/J)^{1/4} \tag{3}$$

where D_s is the surface diffusion coefficient of adatoms, the kinetic coefficient being $K_d \simeq 1$.

(a) (b)

Figure 3. The Polysil microstructure at the initial stage of growth at 850°C on the surface of silicon nitride (a), and the histogram of grain sizes (b). Time of grain growth is 6 s.

(a)

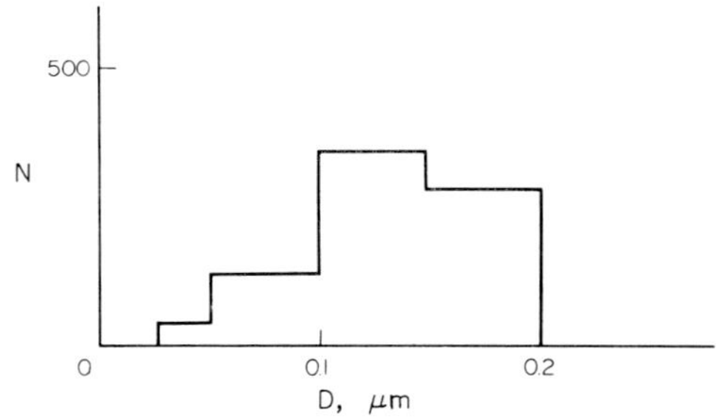

(b)

Figure 4. The Polysil microstructure at the initial stage of growth at 850°C on the surface of thermal silicon dioxide (a), and the histogram of grain sizes (b). Time of grain growth is 6 s.

If the rate of growth of islands is determined by the time of building-in of adatoms, it is independent of island sizes and does not change in the course of growth. The rate of island growth can be considered, in the first approximation, to be constant at the small interval of its change. In this case, allowing for the difference between tangential V_t and normal V_n components of the island growth rate, we have for d_m

$$d_m = K_k\left(\frac{V_n^2}{JV_t}\right)^{1/3} \qquad (4)$$

where $K_b \simeq 0.8$. A good agreement of the values calculated from (4) with the experimental ones is achieved at $V_t/V_n \simeq 2$–4. The similar connection was observed with epitaxial growth, and this relation increased depending on the surface cleanliness. As for the dependence (3), the estimation of the diffusion coefficient D_s over this range under the assumption of a limiting role of diffusion in the nuclei growth ($D_s = d_m^4\ J\ K_d^{-4}$) yields evidently too low values; with an increase of the deposition temperature the calculated value decreases, which shows that it is impossible to use this scheme of calculation. Thus, one should consider the building-in of silicon atoms into islands to be a limiting stage of growth. If 10^{-6} s is the building-in time, and assuming a maximum possible number of active places, corresponding to the product of a number of nuclei $N = Jt \simeq 10^8 - 10^9\ \mathrm{cm}^{-2}$ by a number of atoms in the island perimeter, the growth rate is determined as $D_l/N_0\ 10^6\ N\pi D/\pi D_a \simeq 10$–100 nm s^{-1} for islands of 10 nm diameter. The rate realized in the experiments is somewhat less, which is associated with a smaller number of active growth points of islands. This difference increases for large grains, but the conclusion about a determinative action of the atom building-in under the growth conditions studied is confirmed, in essence. The delay in supply of atoms caused by diffusion processes appears to tell something about less perfect parts of the growth surface.

5. Recrystallization changes in grain sizes and finite Polysil structure

Enlargement of grains in the polysilicon films occurs due to accumulative recrystallization in the grown continuous film during the time of further growth. Investigations of the temperature dependence of the rate of recrystallization growth of grains[4] yielded the values for an activation energy of 1.2–1.7 eV, which corresponds to both the activation energy of the surface (interface) self-diffusion in silicon and the bond energy per atom. The growth of individual grains at the expense of neighbouring ones occurs via the transition of atoms from one grain to another, overcoming the interface potential barrier, however the supply of silicon atoms is facilitated by their diffusion on the grain interfaces. The limiting grain sizes corresponded to the film thickness.[11] Special prolonged annealings (30 min and 2 h at 1350°C) have shown that further growth of individual grains proved to be possible due to film decomposition. Grains reach 8–15 μm in length and film free regions appear on the substrate surface. The change of free energy of grain interfaces is a moving force of the recrystallization growth of grains. This change is far less than that of the free energy during deposition,[4]

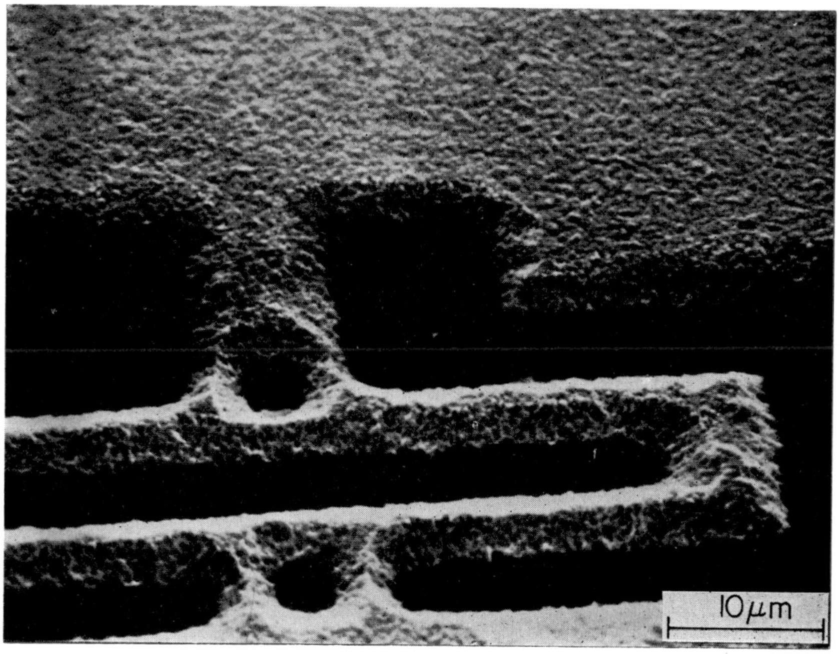

Figure 5. The Polysil microstructure on silicon nitride used as a MIS-transistor gate.

however the rates of grain growth at recrystallization exceed these during deposition. At 1000°C the grain size for the deposition time varied from 0.2 to 1.6 μm; it corresponded to the growth rate of 0.5 μm min^{-1}, but at deposition this growth was only 0.16 μm min^{-1}. Here a considerably high density of an atomic flow at recrystallization, as compared with a flow of adsorbed atoms, takes place on the grain surfaces. The characteristic microstructure of Polysil on silicon nitride used as a MIS-transistor gate is shown in Figure 5.

The results of electron-diffraction examinations show that polysilicon films are textured as a rule, The texture axis is 110. However the causes of texturing are unclear, and the formation of texture is not associated with recrystallization, since often textures arose at the early stage of film growth before coalescence of individual grains into a continuous layer. At the deposition of films in a outside reactor heated from outside, textures appear more often than in the case of induction heating of the substrate on a graphite pedestal.

The most important factor, which complicates the reproducibility of obtaining and the successful use of Polysil in MIS gates and metallization, is imperfect knowledge of microscopic processes following film doping. The grain interfaces gives a large contribution to a specific resistance, reaching 10^6 to 10^8 ohm cm (at Polysil thickness of 1 μm it corresponds to $10^7 . 10^4 = 10^{11}$ ohm/□). In order to lower the resistance down to a value acceptable for metallization of 10^2 ohm/□ at deposition the Polysil is doped from the gas phase (or other ways) by up to a concentration of 10^{19}–10^{21} atoms cm^{-3}, i.e. at the limit of solubility. In this connection, there is a danger of precipitation of phase-compounds with silicon in the doped Polysil. The most undesirable are boron–silicon phases insoluble in usual etches which remain on the plate surface after photolithography on Polysil and are detrimental to further treatment (diffusion, metallization by aluminium and so on). The investigation of the phase composition of these insoluble deposits[2] has showed the presence of SiB_4, SiB_6 and elementary boron (Figure 6).

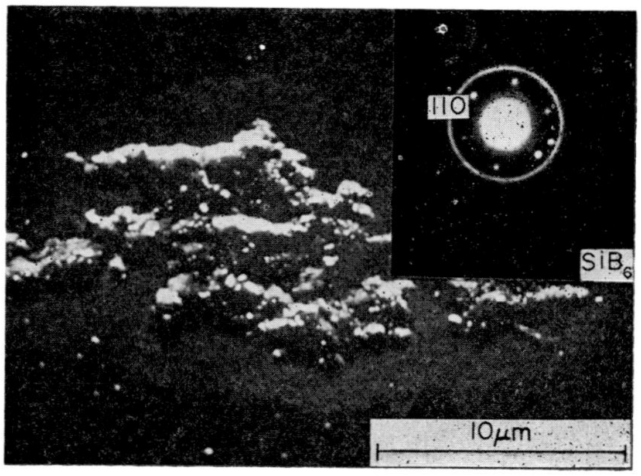

Figure 6. The deposit microstructure after etching of Polysil.

Conclusion

The investigations of growth and recrystallization of silicon films on surfaces of silicon nitride, silicon dioxide and silicon have shown that the rate of film growth is sensitive to an initial state of the substrate surface and growth conditions in vacuum. In particular, we have found some peculiarities of surface diffusion and of building-in of adatoms during the growth of grain-nuclei. Phenomena of accumulative recrystallization are essential at the initial stage of growth and after formation of a continuous layer. Of special interest are investigations of peculiarities of Polysil origin and mechanisms of growth on the surface of silicon dioxide (pyrolytic and thermal), mainly in the connection with problems of multilevel metallization and of CCD-structures.

References

[1] F L Edelman, V V Voskoboinikov and V E Latuta, *Mikroelektronika* **3**, 1974, 418.
[2] F L Edelman, *Elektron tekhnika Ser Materialy*, **8**, 1974, 102.

[3] F L Edelman, V V Voskoboinikov, V V Smirnov and V V Pavlov, *Mikroelektronika*, **3**, 1974, 554.
[4] L N Alexandrov, F L Edelman and V V Voskoboinikov, *Thin Solid Films*, **32**, 1975, 241.
[5] R F C Farrow, *J Electrochem Soc*, **121**, 1974, 899.
[6] J Seto, *J Electrochem Soc*, **122**, 1975, 701.
[7] B A Joyce, J N Neave and B E Watts, *Surf Sci*, **15**, 1969, 1.
[8] L N Alexandrov and R N Lovyagin, *Thin Solid Films*, **20**, 1974, 1.
[9] L N Alexandrov and R N Lovyagin, *Japan J Appl Phys*, Suppl 2, 1974, 609.
[10] L N Alexandrov, R N Lovyagin, O P Pchelaykov and S I Stenin, *J Cryst Growth*, **24/25**, 1974, 298.
[11] L N Alexandrov, *Kinetika Obrazovaniya i Struktura Tvoyrdykh Sloyov*, Nauka, Novosibirsk (1972).

1.12 Inline production magnetron sputtering

A Aronson and S Weinig, *Materials Research Corporation, Orangeburg, NY, USA*

Considerable interest has been shown in the use of planar magnetron sputtering to obtain higher deposition rates than those obtained by conventional diode sputtering. However, there are other valuable benefits of magnetron sputtering that greatly extend the range of use. These include reductions in substrate heating, in radiation damage and in film contamination as well as being favorable for operation in modes that result in lower coating costs. The deposition of multilayer films on to moving substrates using magnetron cathodes simplifies the problem of attaining uniformity and at the same time, provides high substrate throughput. A microprocessor controlled inline sputtering system has been designed that is capable of providing a high order of process repeatability.

Introduction

Within the past decade the sputtering process has become a widely accepted manufacturing method for the production of thin films for both electronic and non-electronic applications. The recent development of the planar magnetron sputtering cathode has provided an important tool for the thin film industry and has resulted in the development of a new family of sputtering equipment. These are inline systems that operate in the moving deposition mode. This mode has proven to be advantageous when used with conventional rf diode cathodes, both independently and in conjunction with magnetron cathodes.

Magnetron

The mechanics and the advantages of magnetron sputtering have been known for many years, but only in the form of equipment that possessed limited capabilities for thin film deposition. Over the past four decades, the design has evolved to that of the planar magnetron cathode. It is this cathode that has justified the development of the inline systems discussed below, which are providing a major contribution to the coating field.

The use of magnetron cathodes provide three major benefits which can result in substantial reductions in coating costs due to high throughputs and improved yields. These are high deposition rates, lower substrate temperatures and improved thickness uniformity.

The important difference between magnetron and normal diode sputtering is in the use of the electrons in the plasma. Higher power densities are obtained at lower cathode voltages. In normal sputtering, the electrons travel in straight lines and drift out of the cathode area. Permanent magnets (Figure 1) in the magnetron cathode are located behind the target material and generate a field which exits and re-enters through the target surface and thus between these points is parallel to the target surface. The electric field at the target is normal to the target. Under the influence of the combined fields, the electrons travel in curved paths. The extended path length and its formation into a closed loop greatly increases the probability of an ionizing collision.

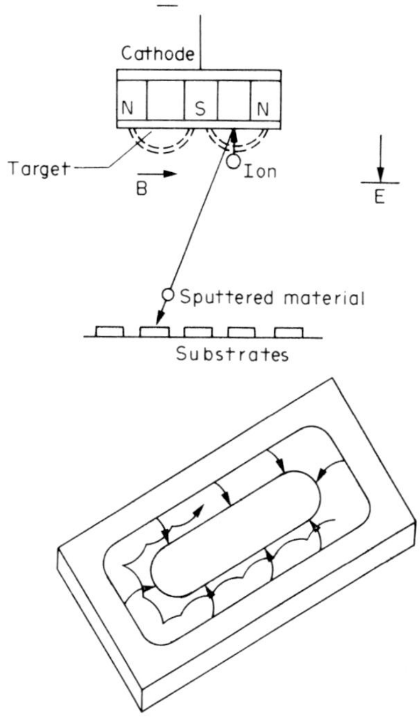

Figure 1. Basic magnetron operation.

The high density, low impedance plasma results in sputtering voltages between 200 and 550 V at high currents. These lower energies are the cause of the reduction in substrate heating. The reduced energy of the particles and the close confinement of the plasma at the target face in conjunction with a fast film growth rate result in minimal radiation damage to sensitive substrates.

Deposition rates are typically 7–10 times higher than obtained with conventional diode sputtering. Deposition of conductive materials, even those that are highly reactive is normally performed using dc excitation at pressures below 1 pascal (7.5×10^{-3} torr). rf magnetron deposition is used to obtain high rates using insulating materials. The same proportional rate increase exists, but some insulating materials are power limited due to thermal stresses resulting from the high power densities.

Moving deposition mode

The planar magnetron is best utilized in the moving deposition mode where the substrates are coated in a single pass as they move under the narrow dimensions of the target. The direct ratio of target to coating area provides maximum film growth rate which minimizes the reactive effects of system background contamination.

All of the coating material is supplied from the area of the target defined by the magnetic field and the target performs as a dual line source. The motion provides high uniformity in the direction of travel. In the transverse direction, cathode design occasionally aided by minimal aperturing provides uniformities of $\pm 5\%$ or better. This mode of operation eliminates the familiar uniformity dependence observed with conventional sputtering upon pressure, type of material, target voltage, target to substrate spacing and so forth.

It is desirable, when using the moving deposition mode, to use the term 'angstroms/cm min'. A fairly close comparison to that static deposition in angstroms/min is obtained when the target width is used as the travel distance per minute.

Film thickness will vary linearly with speed and approximately linearly with power. With removal of the magnet assembly, the same rectangular target can be operated in the conventional rf diode mode. This is sometimes used where thin layers are desired or initially when transferring a process from conventional static to moving deposition and when performing simultaneous sputtering so as to not require a second magnetron power supply. Table 1 lists the moving deposition rate for some common materials.

Table 1. Moving deposition single pass film thickness.

dc Magnetron 7 kW, 15 cm min^{-1} Material	Ångstroms	rf Diode 1.5 kW, 15 cm min^{-1} Material	Ångstroms
Ag	18,800	Cr	300
Al	7,000	NiCr	330
Au	15,500	Ti	180
Cr	7,200	Au	700
Cu	12,600	InSn	600
Mo	5,800	TiW	300
Nb	3,600		
Pd	13,300		
Pt	10,000	rf Magnetron	
Rh	8,300	1.5 kW, 15 cm min^{-1}	
Ta	3,300		
Ti	4,000	Material	Ångstroms
TiW	2,500	SiO_2	600
W	3,300	Cerment Cr/SiO	600

Equipment

The work described in this paper was performed on a Materials Research Corporation model 902 system. This unit has two 12.5 × 38 cm targets and deposits on to 31 × 31 cm pallets. Operation was by different combinations of rf diode, rf magnetron and dc magnetron. The two level loadlock on the left permits loading, unloading, venting and repumping during the period that a pallet of substrates is being coated in the main chamber. The main chamber is pumped by a 6-in. diffusion pump while the vacuum for the lock is provided by a two stage mechanical pump. Figure 2 shows the system with the top plate raised. This unit was provided with a 1.5 kW rf generator and a 10 kW dc magnetron power supply. The inline design provides enhanced reliability by greatly simplifying the mechanical requirements of the system.

Combined rf diode and magnetron

Table 2 compares the process times obtained when depositing a two layer film consisting of 150 Å of nichrome and 3000 Å of copper using different combinations of cathodes. The two pass mode is for depositing discrete layers by making a separate pass under each target. The one pass mode provides a graded interface between layers obtained by exciting the targets simultaneously and making a single pass below both. Chromium–gold films deposited in this mode do not have the problem of long term interface failure.

Table 2. Nickel–chromium copper deposition (150/3000 Å).

Passes	N/chrome cm min	Power/kW	Copper cm min	Power/kW	Cycle time minutes
Two	30	rf 1.5	2.5	rf 1.5	16.6
	30	rf 1.5	66	Magn 7.0	3.72
	66	Magn 0.5	66	Magn 7.0	2.9
One	3	rf 0.2	3	rf 1.5	24.9
	30	rf 1.5	30	Magn 3.3	3.25
	66	Magn 0.5	66	Magn 7.0	2.06

The process time includes 1 min for exchange of processed substrates for uncoated substrates and an additional ½ min in the two pass mode for repositioning of the shutter and pallet carrier. The substrate travel distance is 46 cm for each material in the two pass mode and 70 cm in the one pass mode. The use of two magnetron cathodes in the single pass mode can result in single shift throughputs in excess of 100 square meters per week.

When depositing chromium–gold films, a temperature limitation of about 200°C is necessary to avoid diffusion which can increase resistance and create bonding difficulties. Table 3 shows the drastic improvement in throughput obtained by using magnetron deposition when limited by this constraint.

Table 3. Chromium gold deposition (150/3000 Å).

Passes	Chromium cm min	Power/kW	Gold cm min	Power/kW	Cycle time minutes
Two	30	rf 1.5	1.4	rf 0.75	36
	30	rf 1.5	38	Magn 3.4	4.2
One	30	rf 1.5	30	Magn 2.8	3.25
	38	Magn 0.3	38	Magn 3.4	2.82

Additional inline applications

Transparent conductive films of indium–tin oxide (ITO) deposited reactively from an indium–tin target are metallic in appearance as deposited. They become transparent and have a low sheet resistance after annealing in air. However, if they are sequentially coated by magnetron with an insulating layer of SiO_2 they come out of the system transparent. A typical cycle to produce 1000 cm^2 of the two layer film is in 15 min.

Titanium and titanium–tungsten films followed by platinum

Figure 2. Sputtering system. (1) target 1, (2) target 2, (3) heater, (4) loadlock, (5) shutters, (6) etch platform location.

are being deposited by both magnetron–magnetron and magnetron–rf diode. The magnetron can also be used for reactive sputtering. Tantalum nitride can be deposited from a tantalum target with the same electrical qualities as obtained by the conventional techniques, but at much higher rates. Magnetron sputtering of SiO_2 functions like conventional sputtering showing similar film quality and rate variations with the addition of oxygen.

Alloys deposited by magnetron provide the same duplication of target composition as do other sputtering methods. This is of great value for depositing material such as aluminium–silicon–copper which requires sophisticated techniques in order to maintain composition when deposited by evaporation. In addition, the moving mode has proven to provide excellent step coverage that is typically above 70%.

Automation

A program was undertaken to automate a multitarget inline sputtering system using a microprocessor. There were three basic objectives set up. First, the system must operate reliably in the high electrical noise plasma environment. Second, the system must be able to handle the diversified parameters and procedures used to obtain good films. Third, no special computer or programming background should be required. The system is made to be operator interactive. This means that the machine leads the way through the program with lights and arrows indicating the parameter to be entered and requiring only the entering of the quantity by means of the keyboard. Proceeding through the program once with the machine in a 'Learn' mode completes the programming. Cassettes can be added to permit the storing of up to 40 step programs. Closed loop control of all of the sputtering parameter provides high process repeatability. Noise immunity is obtained through optical coupling and extensive shielding. A diagnostic program is built into the machine to provide rapid location of malfunctions, whether process variations or equipment problems.

Conclusion

The use of inline systems in the moving deposition mode, with magnetron and sometimes in combination with conventional sputtering is proving to offer major benefits in reducing coating costs. While many coating combinations remain to be evaluated, there does not appear to be any inherent limitation to this process. It is likely that this type of system will find use for applications that are at present being performed by other coating techniques.

1.13 Plating with ion accelerators

G Sørensen and J L Whitton, *Institute of Physics, University of Aarhus, DK-8000 Aarhus C, Denmark and Physics Laboratory II, H C Ørsted Institute, 2100 Copenhagen, Denmark*

Formation of thin films by sputtering and ion plating are well-known techniques. The present contribution describes ion beam plating which is a novel technique somewhat related to ion plating. Thin films of metal compounds on a substrate are decomposed by ionic bombardment resulting in films with good adherence to the substrate. The formation of thin palladium films on silicon surfaces will be discussed in detail. Halide compounds of palladium dissolved in organic solvents are sprayed onto the substrate. Subsequently these thin films are exposed to ion beams of inert gases in the energy range below 100 keV. After ion exposure, remains of the halide compound are removed chemically. The ion beam plated film, which appears mirror-like, is then investigated by scanning electron microscopy, proton back-scattering in order to characterise the ion beam plated film. The possible applications of ion beam plated films will be discussed.

I. Introduction

The application of ion beams in thin film technology is well known. Sputtering as well as ion plating has been used in various coating processes, and often an extremely good adhesion to the substrate surface is obtained. This is mainly due to the fact that no transfer from cleaning media to coating media is necessary in these processes. The good adhesive properties are also a result of the high kinetic energies imparted to the coating particles. Stubb[1] thus reports an improved adhesion of copper on titanium, if a thin copper layer is ion plated on titanium before electrolytic deposition of a thicker layer. Recently, Colligon *et al*[2] have used an ion accelerator to produce thin surface films. They obtained a 600-Å thick lead film, when a 60 eV lead ion beam on a surface was impinging for 6 h at a flux of 15 μA.

The present study also deals with the formation of thin films using an ion accelerator.[3] The procedure applied is somewhat similar to the photographic process, but instead of exposure to light, an exposure to keV-ion beams has been used in order to reduce a compound to metal. Thus, the penetration of fast ions through a thin film of a metal halide causes chemical changes due to the deposited energy. It is not within the scope of the present contribution to discuss the complex reactions following the ion bombardment of thin metal halide films. We will, however, show that as a result of this ion beam plating technique, it is possible to obtain essentially pure metal films.

II. Penetration of ion beams

In the ion beam plating process, the penetration of keV ions into solids is an important parameter. This parameter has been studied extensively, experimentally and theoretically, during the last 10 years. Generally, the experimentally measured ranges of fast ions in solids agree with the theory of Lindhard *et al*[4] with a confidence of approximately 10%. The theoretically projected ranges of 50–500 keV inert-gas ions in halides of the platinum metals have been estimated from the range of calculations of Schiott.[5] In Figure 1 are shown inert-gas ranges in palladium chloride and, with neon as projectile, in palladium chloride and in palladium iodide both as divalent halides. From this it is clear that by adjusting the initial ion energy of the ion beam it is feasible to deal with ion beam plated films from some tens of Ångstroms to a maximum of some hundreds of Ångstroms with the commercially available heavy-ion accelerators operating between 40 and 300 kV. However, the use of Van de Graaff accelerators with acceleration voltages of several MeV could easily result in the formation of films several microns thick.

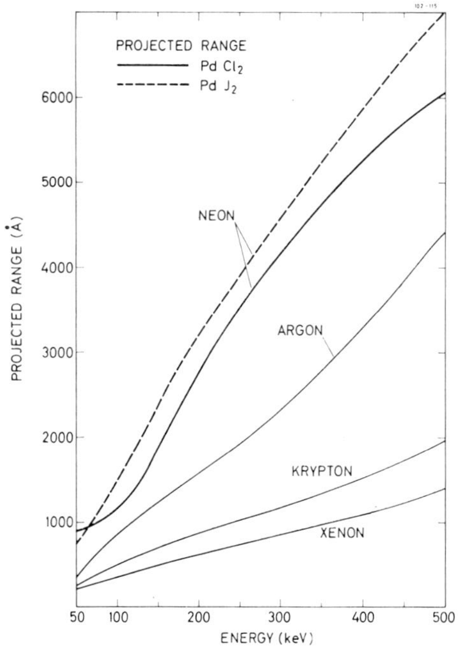

Figure 1. Projected ranges of 50–500 keV inert-gas ions in PdJ_2 (dashed curve) and $PdCl_2$ (full-drawn curves)

III. Experimental procedure

The first step in the ion-beam plating technique is to make a uniform film of a metal halide on a substrate surface. In principle many chemical compounds can be used, but the present study deals exclusively with metal halides. So far the technique has been tried with $PdCl_2$, $PdBr_2$, $PtCl_4$ and $AuBr_4$. These compounds are easily soluble in volatile organic solvents such as acetone and chloroform. It has been found that the solubility of the divalent palladium halides varies from batch to batch. The addition of minor amounts of gaseous hydrogen halides can increase the solubility considerably.

In the early work the solvent containing the metal halide was sprayed on to a substrate of silicon on glass. The coatings obtained in this way were not uniform and consisted of macroscopic crystallites. After the ion bombardment, the exposed area was mirror-like and smooth. Better results were obtained in the following way: A drop of 15–50 μl of a concentrated metal halide solution is placed at the centre of a rotating disc of the substrate. When the liquid is floating due to the centrifugal force, the evaporation of the solvent is leaving a uniform metal halide film. An alternative technique is the so-called molecular plating technique discussed by Parker.[6] It was originally developed for depositing radioactive materials uniformly, but it is also well suited to coat surfaces with films of halides, nitrates and other chemical compounds.

The metal halide film is subsequently exposed to an ion beam with currents of the order of 1 μA to doses which correspond to $\sim 10^{16}$ ions cm^{-2}. For this purpose the 60-kV isotope separator or the 600-kV heavy ion accelerator at the Institute of Physics, University of Aarhus, was utilized.

Some metal halides such as $AuBr_4$ and $PtCl_4$ interact with the atmospheric moisture. In such cases it is advantageous to work in an atmosphere of dry nitrogen.

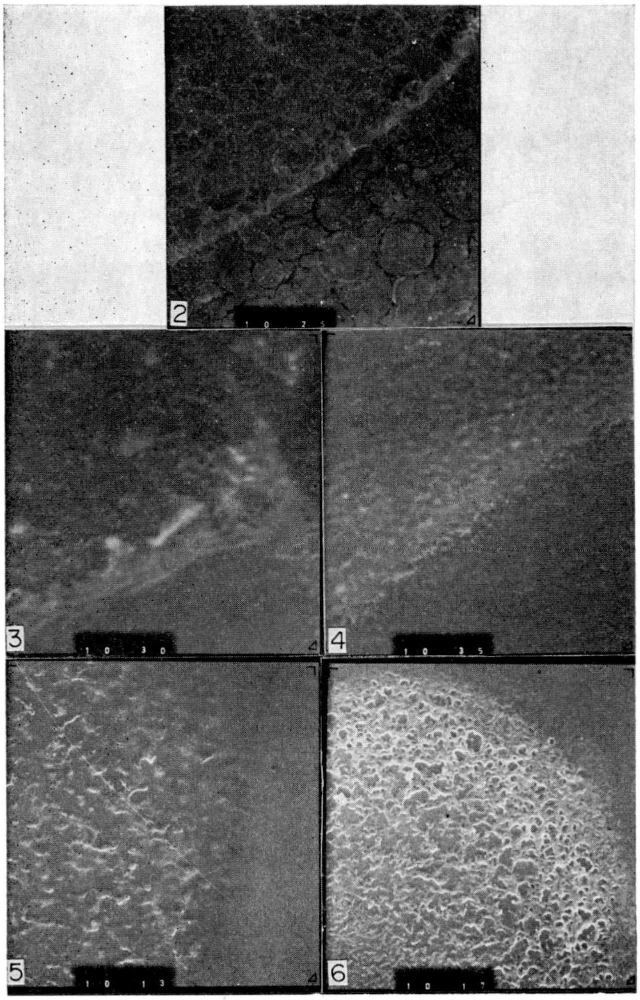

Figure 2. Film of $PdCl_2$ exposed to 10^{16} ions cm^{-2} of 50 keV Ne. Borderline between exposed (upper part) and non-exposed (lower part) is clearly visible (magnification $\times 60$).

Figure 3. Film of $PdCl_2$ exposed to 10^{16} ions cm^{-2} of 50 keV neon. Excess of $PdCl_2$ removed with acetone. Magnification $\times 10{,}000$.

Figure 4. Film of $PdDl_2$ exposed to 2×10^{16} ions cm^{-2} of 50 keV Ne. Excess of $PdCl_2$ removed with acetone. Magnification $\times 10{,}000$.

Figure 5. Film of $PdCl_2$ exposed to 6×10^{16} ions cm^{-2} of 50 keV Ne. Excess of $PdCl_2$ removed with acetone. Magnification $\times 10{,}000$.

Figure 6. Film of $PdCl_2$ exposed to 6×10^{16} ions cm^{-2} of 50 keV He. Excess of $PdCl_2$ removed with acetone.

IV. Experimental results

Ion-beam plated films appear homogeneous and their adhesion to the substrate surface seems satisfactory. Preliminary investigations by means of scanning electron microscopy (SEM) and nuclear backscattering have been made in order to characterize the film in structure and composition.

Figure 2 shows a SEM picture (magnification $\times 60$) of ion-beam plated palladium on a silicon surface (dose 10^{16} ions cm^{-2} of 50-keV Ne). The border line between the exposed area (upper part) and non exposed $PdCl_2$ film (lower part) is clearly visible, but no apparent changes in structure are observed. In Figure 3, some irregularities in the film are observed, while in Figure 4, which shows the situation when the dose of 50-keV neon ions has been increased from 10^{16} to 2×10^{16}, a very homogeneous film is observed.

Since the most likely mechanism of dissociation is that of energy transfer from the incoming ion beam to the halide molecules, we tried the effect of lighter particles than Ne. In Figure 5 is shown an exposure to 6×10^{16} ions cm^{-2} of neon with a magnification of $\times 10{,}000$ and Figure 6 shows an exposure of 6×10^{16} ions cm^{-2} of helium ions at 50 keV with the same magnification. There appears to be a slight difference between the two films. However, it is important in this connection to take into consideration the different sputtering properties of helium and neon. Control experiments in which a silicon surface without any metal halide film was exposed to identical inert-gas doses have been performed. In these cases no deposit on the surface was observed. We have investigated the composition of the metal halide film both before and after the ion bombardment by nuclear backscattering using the 2.5-MeV Van de Graaffs both at University of Aarhus and at Salford University, UK. The principle in the nuclear backscattering technique as well as some of the experimental results are shown in Figure 7. Nuclear backscattering or ion backscattering is a technique dating back from the beginning of the century.

In the present case we have used 2-MeV He^+ ions and, to improve the resolution, we have used the modified glancing angle backscattering technique.[7] This gives a resolution of ~ 25 Å compared to the ~ 200 Å obtained normally.

The position of the peaks in Figure 7 is dictated by the elastic energy loss, therefore the ions backscattered from high-mass target atoms appear at higher energy than those from low-mass

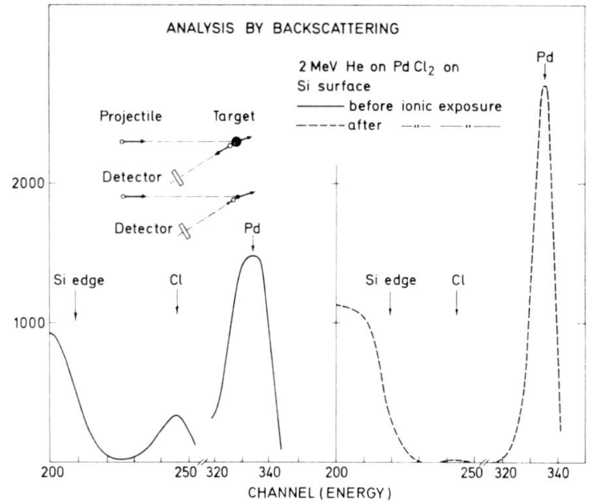

Figure 7. Backscattering analysis for 2-MeV He^+ on $PdCl_2$ on a silicon surface. Left part (full-drawn curve) show the situation prior to the ion bombardment. Right part (dashed curve) shows the ion-beam plated film.

atoms. In the left-hand part of Figure 7 peaks can be seen corresponding to Cl and Pd as well as a Si edge due to scattering from the substrate surface. This spectrum comes from an unbombarded area of film. Measurement of the areas under the peaks, with due allowance for difference in scattering cross-section, shows the film to have the composition $PdCl_2$.

On the right-hand part of Figure 7 is seen the spectrum from the film after bombardment with Ne ions. Here the Cl peak is essentially absent. The width of the Pd peak is narrower than formerly, in fact the ratio of widths corresponds to the ratio of the respective densities of the metal halide and the metal.

From the spectra it is clear that in this ion beam plating process, a thin film of $PdCl_2$ can be converted to a Pd film of a thickness of approximately $\frac{1}{3}$ of the metal halide. Other properties of the ion beam plated films, as for example, electrical conductivity, will be reported elsewhere.

V. Applications

The technique described in this communication can be utilized to produce metal films of thickness from tens of Ångstroms to some hundreds of Ångstroms. These films could find applications as substrates for evaporation or electrolytic deposition where good adhesion is a requirement. Additionally small amounts of impurity atoms may be implanted by 'knock-on' into a sample with the technique. Since a large amount of energy is deposited in a very small surface area, the normal properties of the metal may be modified. These modifications could well affect the catalytic and corrosion properties of such films. Dearnaley,[8] Williams[7] and Grant and Williams[9] have recently discussed changes of surface properties by ion implantation into metals. Since the range of ions in metals is similar to the thickness of the ion beam plated metal films described here, the present technique could equally well be applied to the modification of target surfaces with respect to corrosion rates, catalytic properties, wear and hardness.

Finally we would like to emphasize the simplicity of this process. A wide range of film thicknesses may be selected simply by adjusting the concentration of the solution. Degrees of adhesion and mixing of metal atoms with the substrate may be controlled by the incident ion energy and the positioning of a mask during ion bombardment allows the manufacture of any configuration of metal on the surface, which might be important in semiconductor technology.

Acknowledgement

The co-operation with the University of Salford with respect to the nuclear backscattering analysis is gratefully acknowledged.

References

[1] B C Stubb, *Plating*, **4**, 1974, 1070.
[2] J S Colligon, W A Grant, J S Williams and R P Lawson, *Inst Phys Conf Ser No* 28, 1976, 357.
[3] G Sørensen, *Proc 1st Intern Conf Phys in Industry*, Dublin, 10–13 March 1977, Plenum Press, New York, in press.
[4] J Lindhard, M Scharff and H E Schiøtt, Kgl Danske Videnskab Selskab, *Mat-Fys Medd*, **33**, 1963, No 14.
[5] H E Schiøtt, private communication, 1976.
[6] W Parker, Methods in preparation of radioactive materials, thesis, Chalmers University of Technology, Gothenburg (1965).
[7] J S Williams, *Nucl Instr Meth*, **126**, 1975, 205.
[8] G Dearnaley, *New Uses of Low-Energy Accelerators*, p 283, Plenum Press, New York (1975).
[9] W A Grant and J S Williams, *Sci Prog Oxford*, **63**, 1976, 27.

1.14 Energy distribution of the ions produced by saddle field ion sources

M Khorossany* and R K Fitch, *Department of Physics, University of Aston in Birmingham B47ET, England*

A retarding field energy analyser has been used to measure the energy distribution of the ions produced by two forms of the saddle field ion source for argon, nitrogen and helium. It has been shown that with the cylindrical ion source the distribution is broad but always contains two distinct peaks at energies equivalent to about 35% and 75% of the anode potential. However, with the spherical source the ion energy spread is much narrower and most of the ions occur at an energy equivalent to about 75% of the anode potential with a peak half width of about 10% of the anode potential. It has also been found that the proportion of high energy ions in the beam decreases towards the edge of the beam for both sources.

1. Introduction

During the last few years there have been substantial developments in two forms of the saddle field ion source which originated from the idea of McIlraith[1] for an electrostatic charged particle oscillator. One of these is of cylindrical geometry and was first described by Fitch *et al*[2] and subsequently in improved forms by Fitch and Rushton,[3] Rushton *et al*[4] and Fitch *et al.*[5] The other version was developed later by Franks[6] and is of spherical geometry and the characteristics of this source are given in a paper by Franks and Ghander.[7]

These sources have now been used in a variety of applications. For example, preparation of specimens for electron microscopy by ion etching,[8] preparation of field electron/ion emitters,[9] and as a vacuum ultra violet source of line radiation of the rare gases for photoelectron spectroscopy.[10] However, there has been little work reported on the energy distribution of the ions produced by these sources. Franks[11] used an electrostatic analyser with the cylindrical source and found that the energy spectra contains one main peak and several subsidiary peaks which varied with source geometry, nature and pressure of the gas and anode voltage. Ghander[12] used a simple retarding field energy analyser and found that the energy spectra were rather complex but Swaminathan and Venkatasubramanian[13] found that there were two main peaks in the energy spectra of the ions produced by a cylindrical source incorporating thermionic electron injection.

The purpose of this paper is to report the results of the measurements of the ion energy spectra for both the cylindrical and spherical ion sources using an energy analyser developed from the one used by Ghander.[12]

2. Description of the ion source and the energy analyser

The cylindrical ion source consists essentially of a stainless steel cathode, 75 mm in length and 25 mm in diameter, surrounding two 1.5 mm diameter tungsten anodes symmetrically disposed about the axis of the cylinder at a separation of 5 mm. The anodes are supported in stainless steel end-caps and insulated from them with ceramic bushes. Gas is admitted into the source through a stainless steel tube in the cathode wall. The diameter of the ion exit aperture could be varied from 1 to 5 mm and it was also possible to rotate the aperture from −10° to +40° with respect to the central position. In all cases the depth of the aperture was about the same as its diameter in order to minimize the field distortion.

The spherical source was purchased from Ion Tech Ltd and is normally provided with a cathode aluminium insert containing a 1.5 mm diameter aperture. This source was modified so that the size and position of the aperture could be varied in a similar manner to that described for the cylindrical source.

A schematic diagram of the cylindrical source and the ion energy analyser is given in Figure 1. It was necessary to design the Faraday collector to have a relatively large entrance aperture to admit the total ion beam at a distance of about 60 mm from the aperture yet still be efficient at retaining the secondary electrons. A number of designs were constructed and tested and the collector shown in Figure 1 was found to be the most satisfactory. The conical collector is made from nickel sheet, entrance aperture 27 mm and has an overall length of 125 mm and base diameter of 75 mm. Several nickel sheets, 40 mm in length are also attached to the base of the collector, as shown in the

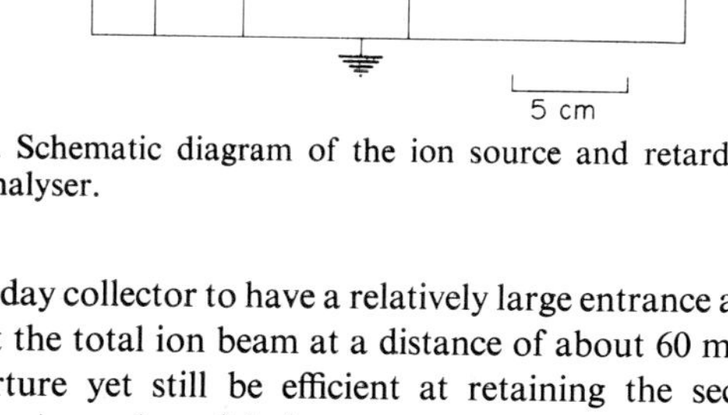

Figure 1. Schematic diagram of the ion source and retarding field energy analyser.

* On leave from National University of Iran, Tehran.

figure, to improve its efficiency at retaining secondary electrons. The figure shows that the collector is also completely surrounded by an earthed copper screen to avoid collection of any reflected ions. Double grids were used to give a large transparency yet still produce a reasonably uniform field.[14] They were made by winding 0.1 mm diameter tungsten wire round an aluminium frame such that the grids were 40×40 mm and with the wire spaced at a distance of 5 mm this should give a transparency of 98%. In practice due to some misalignment of the grid wires the transparency was found to be just greater than 90%. Grid R is the retarding field grid and grid S is the shield grid to shield the collector from the positive retarding field at R. The collector current, i_c, was recorded with a digital meter as a function of the retarding voltage V_R and the corresponding energy spectrum was obtained by differentiating this curve.

It is known that these sources produce a significant proportion of energetic neutrals which are, of course, unaffected by the retarding field but can still produce secondary electrons at the collector. Thus the efficiency of the collector at retaining the secondary electrons could be assessed by recording the collector current with V_R greater than the anode voltage, V_A. It was found that under these conditions the escape of secondary electrons contributed to always less than 3% of the collector current depending upon the source anode voltage.

The source was tested in a stainless steel chamber evacuated by a conventional rotary pump/diffusion pump system using 'Santovac 5' fluid in the diffusion pump. The pressure was measured with a Penning gauge situated in the chamber.

3. Operating modes and energy spectra of the ion sources

In an earlier publication[4] it was reported that there are three principal modes of operation for the cylindrical source—namely the oscillating, transition and glow modes. In the present work the existence of these modes has been confirmed for both the cylindrical and spherical sources by monitoring the ion beam current with the Faraday collector shown in Figure 1. Figures 2a and b show how with argon, V_A varies with pressure for the cylindrical and spherical sources using central ion exit apertures of 2 and 1.5 mm at anode currents, i_A, of 3 and 4 mA, respectively. It can be seen that when the pressure is high V_A is only about 500 V which corresponds to the normal glow mode. As the pressure is reduced there is a sudden rise of anode voltage during the transition mode. At still lower pressures the rate of rise of V_A decreases at the onset of the more efficient oscillating mode. This is much more apparent with the spherical source where it can be seen that the anode voltage is almost constant at 5 kV between pressures of 3.5×10^{-5} and 1.5×10^{-4} torr. Figures 2c and d show the corresponding variation of i_c with pressure and it can be seen that the source efficiency—defined as the ratio i_c/i_a—increases to a maximum at pressure of 6×10^{-5} and 3×10^{-5} torr for the cylindrical and spherical sources, respectively.

The variation of collector current, i_c, with retarding voltage, V_R, was recorded for both sources at various pressures for argon, nitrogen and helium. Figures 3a and 4a show two such typical curves for argon for the cylindrical and spherical sources respectively. The corresponding energy spectrum of di_c/dV_R against V_R are shown in Figures 3b and 4b. In Figure 3 the

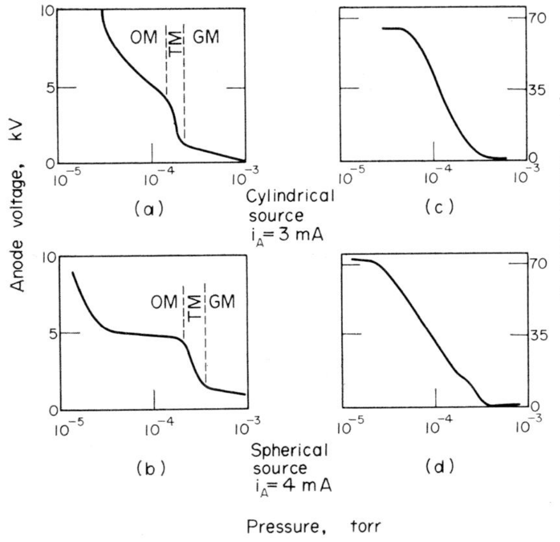

Figure 2. Modes of operation of the cylindrical ion source (a) and the spherical source (b).

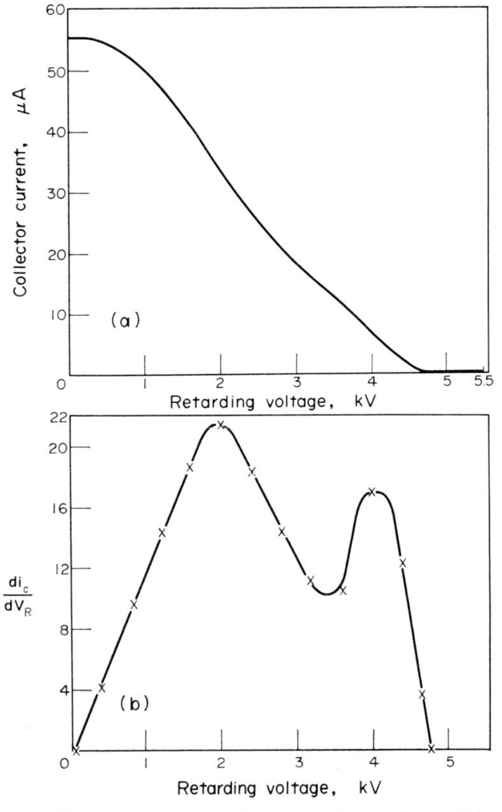

Figure 3. Collector current against retarding voltage (a) and ion energy spectrum (b) for the cylindrical source for argon.

central ion exit aperture was 5 mm and the source was operated at $V_A = 5.5$ kV, $i_A = 3$ mA at a pressure of 3.8×10^{-4} torr. The corresponding values for the spherical source in Figure 4 are 1.5 mm, 6 kV, 2 mA and 1×10^{-4} torr. With the above conditions both sources were operating in the oscillating mode.

It is clear from Figures 3 and 4 that the ion energy spectra are very different for the two sources. In the case of the cylindrical source two distinct peaks occur at 2 and 4.1 keV which is equivalent to 36 and 75% of the anode potential. In fact it was found that for all pressures with argon, nitrogen and helium two similar peaks were always present and the low and high

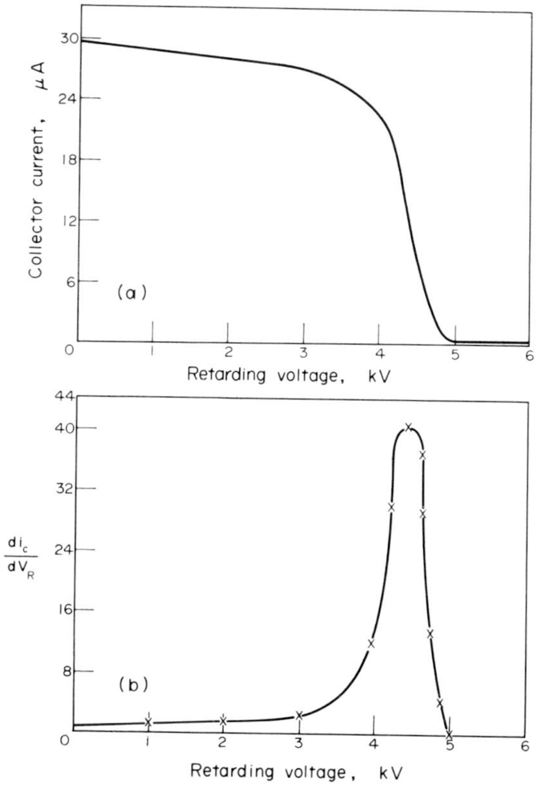

Figure 4. Collector current against retarding voltage (a) and ion energy spectrum (b) for the spherical source for argon.

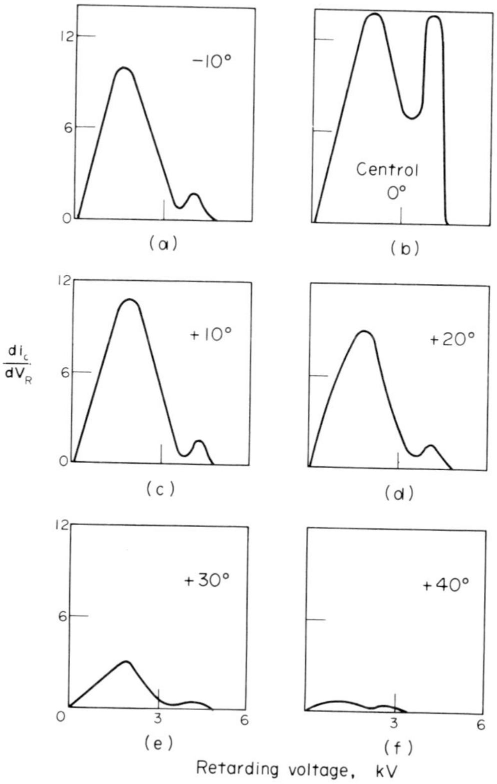

Figure 5. Ion energy spectrum for argon with the cylindrical source using different angular positions of the ion exit aperture.

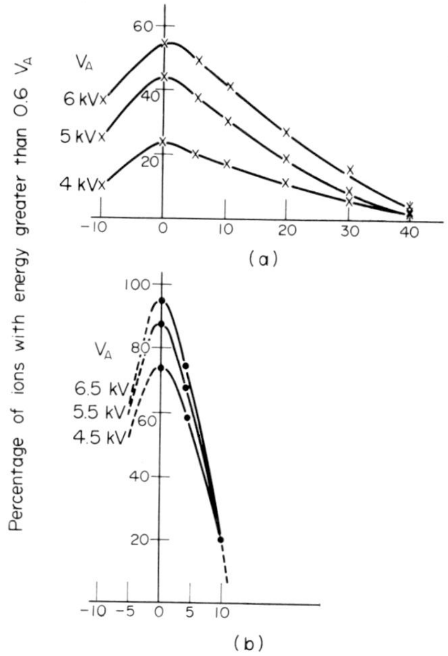

Figure 6. Percentage of ions having energy greater than 0.6 V_A for different angular positions of the ion exit aperture using the cylindrical source (a) and the spherical source (b) at various values of V_A.

energy peaks always occurred in the range 35–40% and 70–80% of the anode voltage. This spread is no more than the uncertainty of ±5% in the determination of the position of the peaks. The mean value of 0.75 V_A for the high energy peak corresponds to the saddle point potential between the two anodes. For the spherical source the energy spectra shown in Figure 4 shows only one peak occurring at 0.73 V_A and is superimposed on a broad low energy background. Once again the position of this peak varied from only 0.7 V_A to 0.8 V_A for all pressures with argon, nitrogen and helium.

Figures 5a–5f show the ion energy spectra for argon with the cylindrical source using a 5 m aperture at angular positions of −10°, 0°, +10°, +20°, +30° and +40° with respect to the central position at 0°. It is clear from the figures that the proportion of high energy ions is much greater at the central position—shown in Figure 5b—and decreases when the aperture is moved either side of the central position. The similarity of Figures 5a (−10°) and 5c (+10°) demonstrate that the aperture was correctly in the central position in Figure 5b.

This effect is also shown in Figure 6a in which the percentage of ions having energy greater than 0.6 V_A-estimated from the appropriate areas under the $i_c \rightarrow V_R$ curves—is plotted against the angular position of the aperture for $V_A = 4$, 5 and 6 kV. Similar observations were taken with the spherical source and are shown in Figure 6b but it can be seen the high energy ions are much more confined to the central portion of the ion beam than with the cylindrical source.

4. Discussion of results

The form of the energy spectrum for the cylindrical source has been found to be rather similar to that given by Swaminathan and Venkatatasubramanian[13] but comparison is only partly justified as their measurements were made on a cylindrical source using a thermionic source of electrons whereas the present measurements were taken with the normal cold cathode source. The energy spectrum for the spherical source is similar to that reported by Franks and Ghander[7] but does not show any evidence of any additional low energy peaks. Furthermore in the present work the high energy peak always occurred at about 0.75 V_A with a peak half-width of 500 V whereas in the work of Franks and Ghander the corresponding values were 0.85 V_A and 250 V. It is thought that these differences are due to the more satisfactory design of the grids and Faraday collector used in the present sork.

The difference in the energy spectra between the cylindrical and spherical source was not unexpected. This is due to the axial symmetry of the spherical source in which the end effects of the cylindrical design have been eliminated. It is thus proposed that the low energy peak at the edge of the ion beam produced by the cylindrical source arises from a potential gradient across the plasma inside the source which is in a direction transverse to the emerging ion beam. In future experiments it is hoped to be able to confirm this by measuring the potential inside the source using probe techniques.

Both designs of the saddle field ion source should continue to find new applications in various aspects of vacuum technology. In general the cylindrical source should be used if a wide ion beam with a broad energy spread is required but if a fine ion beam of much lower energy spread is necessary, then the spherical source is more suitable.

Acknowledgements

One of us (M K) would like to thank the National University of Iran for granting him study leave and both of us would like to acknowledge support from the Iran Ministry of Science (M K) and the University of Aston in Birmingham (R K F) to attend the symposium.

References

[1] A H McIlraith, *Nature, Lond*, **212**, 1966, 1422.
[2] R K Fitch, T Mulvey, W J Thatcher and A H McIlraith, *J Phys E*, **4**, 1971, 553.
[3] R K Fitch and G J Rushton, *J Vac Sci Technol*, **9**, 1972, 379.
[4] G J Rushton, K R O'Shea and R K Fitch, *J Phys D*, **6**, 1973, 1167.
[5] R K Fitch, A M Ghander, G J Rushton and R Sing, *Jap J App Phys* Suppl 2, Pt 1, 1974, 411.
[6] J Franks, British Patent No 44718/73 (1973).
[7] J Franks and A M Ghander, *Vacuum*, **24**, 1974, 489.
[8] R K Fitch, K R O'Shea and G J Rushton, *Proc Vth European Congress on Electron Microscopy*, 1972, 296.
[9] J M Walls, H N Southworth and G J Rushton, *Vacuum*, **24**, 1974, 475.
[10] F Burger and J P Maier, *J Phys E*, **8**, 1975, 420.
[11] J Franks, *Proc 2nd Int Conf on Ion Sources*, 1972, 812.
[12] A M Ghander, PhD thesis, University of Aston in Birmingham (1974).
[13] S Swaminathan and V S Venkatasubramanian, Private Communication, 1974.
[14] S Stephanakis and W H Bennett, *Rev Sci Instrum*, **39**, 1968, 1714.

1.15 Deposition of thin chromium films on substrates heated between room temperature and 300°C in the uhv chamber of an annular cryopump

H Sánchez, E Posada, J Panqueva, R Paéz, I Jaramillo and G Fritsch, *Departamento de Física, Universidad Nacional de Colombia, Bogotá, Colombia*

A high vacuum evaporation apparatus is described which includes an annular cryopump. The inner cylindric uhv chamber of the annular cryopump is confined by the cylindric cooling surface, the rotary substrate holder and the shutter and evaporator arrangement. With liquid helium filled in the cryopump a pressure of lower than 10^{-9} torr can be attained. The substrate holder is equipped with a temperature-controlled heater for temperatures up to 400°C and with a mask-changing device which can be operated under vacuum.

Thin chromium films were deposited on glass substrates above room temperature up to 300°C. The sheet resistance was measured and the film thickness determined using Tolansky's method of multiple beam interferometry.

1. Introduction

The production of high purity films requires the evaporation and deposition process to take place in very high vacuum or in ultra-high vacuum (uhv). In principle, it is sufficient to confine the uhv chamber to the solid angle formed by all possible direct paths of the evaporated atoms between evaporator and substrates. Practically this can be achieved by a cylindrical chamber: the bottom surface can include the evaporators and the top surface may be formed by the substrate holder; the cooled opaque baffle of an annular cryopump may form part of the cylinder jacket of this system.

2. Description of the apparatus

Figure 1 shows a diagram of the complete system. Three parts can be considered separately: (a) the high vacuum chamber, (b) the uhv chamber and the cryopump, and (c) the evaporators, substrate holder with heating system and mask holder.

(a) The high vacuum chamber. The evaporation assembly and the cryopump are contained in a cylindrical stainless steel chamber (1) which has lateral flange ports for measuring instruments, viewing windows and the connection for the diffusion pump (2). The chamber is provided with an external heating belt for de-gassing. The high vacuum pumping system consists of a 35 $m^3 h^{-1}$ two-stage mechanical pump and a 900 $l. s^{-1}$ oil diffusion pump with a LN_2 cooled baffle. The pressure is measured in the chamber wall by a Bayard–Alpert gauge.

(b) The uhv system. The lower part of the uhv space (b) is separated from the high vacuum space by a stainless steel cylinder (4) placed between the evaporators and the cylindrical baffle of the cryopump. The bottom of this cylinder is provided with four rectangular openings, which can be closed by the shutter (5). The upper part of the uhv chamber is formed by the LN_2 cooled cylindrical baffle of the cryopump and closed at the top by the substrate holder disc (6).

The cryopump consists of two parts: the LN_2 cooled shield (7) and the 2.5 l. annular stainless steel helium reservoir (8). The cylindrical radiation shield and the baffle are cooled by a permanent circulation of liquid nitrogen through a soldered copper coil. Appropriate connections for liquid helium transfer (10), helium gas vent (11) and liquid helium level sensor (12) are provided. The different parts of the uhv system can be radiation heated by the substrate heater for de-gassing. The internal pressure is measured by a uhv Bayard–Alpert gauge (14). Since the uhv region is surrounded by the high vacuum space, only normal high vacuum seals and feed-throughs are required.

(c) The thin film production assembly. The system is equipped with four evaporators (15) which can be operated simultaneously. Each evaporator is placed below one of the openings of the bottom of the uhv chamber. The current for the evaporators is supplied by three transformers. The maximum power available for one evaporator is 3.8 kW.

The deposition assembly (6) consists of three copper discs. The bottom disc holds the masks. The substrates are supported by the intermediate plate, while the upper disc serves essentially to obtain a uniform temperature distribution all over the substrate surface. Eight substrates and 16 masks can be mounted simultaneously. In order to change the masks, the mask holder disc can be rotated gradually in relation to the substrate holder by a lead-in. This system is being modified in order to avoid seizing between the discs at high temperatures. Furthermore, the mask holder will be made out of stainless steel due to its lower outgassing rate as compared with copper. The three discs can be rotated as a whole by a rotary lead-in (16), with a speed between 0–30 rpm. Since the evaporators are unsymmetrically placed with respect to the substrates, the rotation is necessary in order to achieve a sufficient thickness uniformity. The discs can be radiation-heated from above by two tungsten filaments, placed inside two concentric radiation shields. The substrate temperature can be varied between 20 and 300°C with

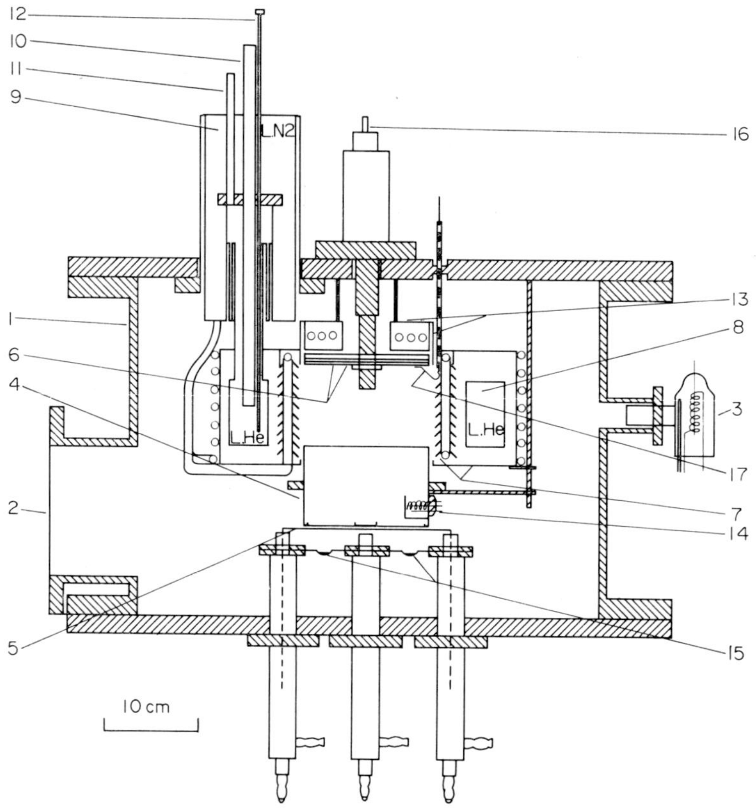

Figure 1. Diagram of the apparatus. (1) High vacuum chamber. (2) Diffusion pump connection. (3) External vacuum gauge. (4) Stainless cylinder. (5) Shutter. (6) Substrates and masks holder. (7) Cylindrical LN_2 cooled shield and baffle. (8) Annular LHe vessel. (9) LN_2 vessel. (10) LHe transfer line. (11) He gas vent. (12) LHe level indicator. (13) Substrate heater and radiation shields. (14) Internal vacuum gauge. (15) Evaporators. (16) Rotary lead-in. (17) Substrate temperature sensor.

an accuracy of ± 1°C. An electronic circuit is used to stabilize the temperature. The substrate temperature is measured by mechanical contact of an externally actuated thermocouple (17).

3. Performance of the cryopump

Assuming a specific pumping speed[1] of 2, 4 l. cm^2 s^{-1}, the theoretical pumping speed of the cryopump (considering the total surface of the helium vessel) is 4000 l. s^{-1}. However, in this annular geometry, most of the pumping is due to the inner surface thus reducing the pumping speed to about 1000 l. s^{-1}. The effective pumping speed is determined by the conductance of the chevron baffle. With the figure given by Davis *et al*[2] the final pumping speed in the cylindrical uhv chamber is about 250 l. s^{-1}. The experimental value is approximately 100 l. s^{-1}. This discrepancy can be attributed to the unknown outgassing rate of the warm surface of the mask holder. The lowest pressure measured inside the uhv chamber was less than 10^{-9} torr.

4. Deposition and electrical measurements of chromium films

Chromium films were deposited on soft glass substrates at substrate temperatures between 320 and 50°C, at a deposition rate between 1 and 2 Å s^{-1}. Before the beginning of the deposition the evaporator was heated for 5 min with the shutter closed. During this period the pressure in the uhv chamber increased to about 5.10^{-9} torr. After the shutter had been opened to start deposition, the pressure rose to 8.10^{-9} torr (the substrate holder being at 280°C). Consecutive depositions were always carried out beginning with the highest substrate temperature followed by depositions at lower temperature. Thus, the annealing effect on the first deposited films (at high substrate temperature) was reduced to a minimum.

The electrical resistance was measured by a four-point method and film thickness measurements were carried out using Tolansky's interferometric method. The calculated values of electrical resistivity were corrected for electron scattering at the surfaces and the grain boundaries of polycrystalline films, according to Singh.[3]

As the thickness of the chromium films was bigger than 0.2 λ_0 the mean free path of conduction electrons ($\lambda_0 = 500$ Å for chromium[4]), an approximate formula[5] can be used to calculate the intrinsic electrical resistivity

$$\rho_G = \rho_{\text{exp}} - \frac{3}{8}\frac{\rho_b\,\lambda_0}{d}(1-p)$$

ρ_b being the resistivity of bulk chromium (13.0 $\mu\Omega \cdot$ cm), ρ_{exp} the experimental resistivity and p the scattering coefficient. Assuming diffuse scattering of the conduction electrons, p is set equal to zero. The results of four experiments carried out at pressures between 10^{-7} and 7×10^{-9} torr, are depicted in Figure 2.

The TCR was about 1100 ppm °C^{-1}, at a total pressure of 10^{-8} torr and less, being nearly independent of substrate temperatures in the range between 280–20°C. At a total pressure of 10^{-7} torr and a substrate temperature of 325°C the TCR was 2000 ppm °C^{-1}. Deposition of chromium at room

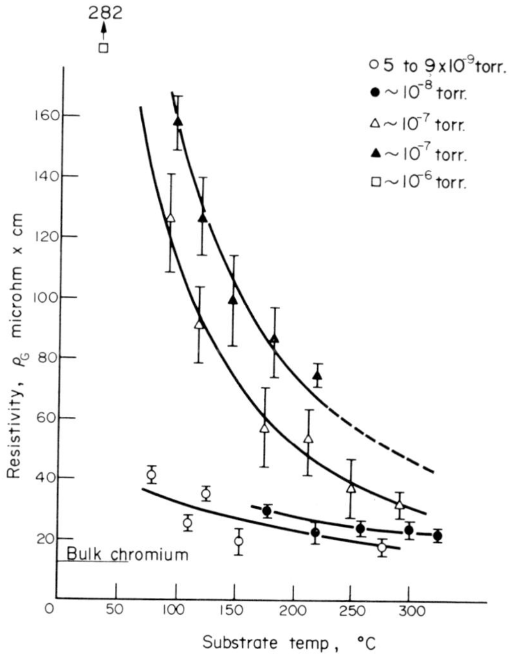

Figure 2. Electrical resistivity of chromium films as a function of substrate temperature at various pressures during deposition.

temperature at 10^{-6} torr resulted in a negative TCR of −170 ppm $°C^{-1}$ and a high resistivity of 282 $\mu\Omega \cdot$ cm, as shown in Figure 2.

These results are in good agreement with those of Hieber and Lassak[6], who studied the effect of oxygen partial pressure on the electrical properties of chromium films.

5. Acknowledgements

We wish to thank Dr K Keck for helpful suggestions. The skilful mechanical work of Mr W Schmelzer, Mr M Jiménez and Mr R Ponce is gratefully acknowledged.

References

[1] K Keck, G V Schubert and H Wiechert, *Proc 3rd Int Vacuum Congr*, **2**, 1966, 559.
[2] D H Davis, L L Levenson and N Milleron, *Rarefied Gas Dynamics*, **99** (1961).
[3] A Singh, *Thin Solid Films*, **21**, 1974, 225.
[4] R M Chapman, *Vacuum*, **13**, 1963, 213.
[5] E Mola and J M Heras, *Thin Solid Films*, **18**, 1973, 137.
[6] K Hieber and L Lassak, *Thin Solid Films*, **20**, 1974, 63.

between the source and other rim of the tube. h was, however, in most experiments, 0. The tube consisted of stainless steel and had an inner radius R; it could be heated by a nichrome coil wound around it. The thickness t as a function of A at any point in the substrate plane (x, y) due to deposited materials reflected from the walls of the tube is given by the integral[3]:

$$t(A) = \beta \frac{mR^2}{\pi^2 \rho} \int_\phi \int_z \frac{H + L + h - z}{[R^2 + (L + H + h - z)^2]^2} \cdot \frac{z(R^2 - AR \sin \phi)}{R(R^2 + A^2 - 2AR \sin \phi + z^2)^2} \, d\phi \cdot dz \tag{4}$$

The factor $\beta < 1$ takes care of the fact that not all material arriving at the substrate will stick. ρ is the density of the material; m the evaporation rate in [g s^{-1}]. Integration of equation (4) over ϕ leads to[4]:

$$t(A) = \beta \frac{mR}{\pi \rho} \int_z \frac{(H + L + h - z)z}{[R^2 + (L + H + h - z)^2]^2} \cdot \frac{2R^2 a^2 - b^4}{(a^4 - b^4)\sqrt{a^4 - b^4}} \, dz \tag{5}$$

where

$$a^2 = z^2 + R^2 + A^2$$

and

$$b^2 = 2AR$$

Equation (5) was evaluated[4] by using the Gauss–Legendre method.[5] Two examples are plotted in Figures 3 and 4. In Figure 3 a tube source was used that had a cylinder height $L = 17.2$ cm, radius $R = 7.95$ cm, and substrate to tube distance $H = 6$ cm which makes the total distance of substrate to source $h' = 23.2$ cm.

In Figure 4 a lower and wider tube source was applied with a cylinder height $L = 10.4$ cm, a radius $R = 10.2$ cm and substrate to tube distance $H = 10$ cm which results in a substrate to source distance of $h' = 20.4$. In each figure dotted curves indicate the thickness variation as it would occur without the 'tube source' at the same substrate–source distance h' as calculated from equation (2).

The theoretical curves for the thickness distribution of films produced with the tube source have the shape of an M when extended over the diameter; they are similar to the curves obtained when using rotating substrates.[6] One can see that with a tube source it is possible to deposit films with a uniformity better than 1% over a substrate length of 10 cm. For example, Figure 3 shows that the thickness of a film obtained with a tube source of $L = 17.2$ and $R = 7.95$ cm at a distance $H = 6$ cm from the top of the tube does not vary more than $\pm 0.4\%$ over a substrate length of 10 cm assuming $\beta = 0.10$.

The curves in Figures 3 and 4 are obtained by assuming $h = 0$, which means that the vapor source plane falls within the lower base of the cylindrical tube. Only one half of the curves are depicted, since the thickness distribution is symmetrical.

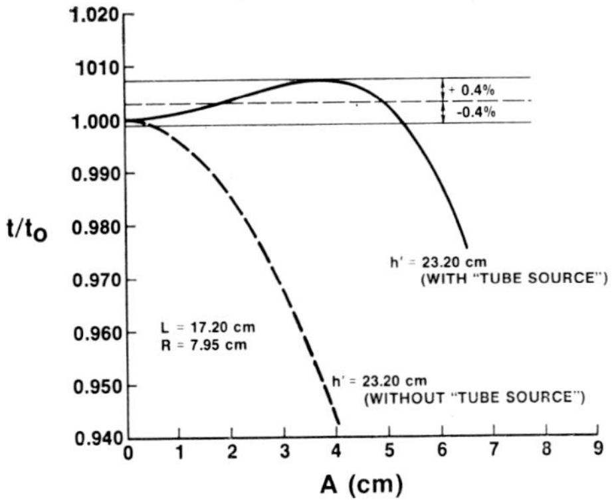

Figure 3. Analytical thickness variation curves. Thickness variation t/t_0 as a function of distance A from centre of substrate. Solid curve for tube source of cylinder height $L = 17.2$ cm and cylinder radius $R = 7.95$ cm; substrate–source distance $h' = 23.2$ cm; $h = 0$; $H = 6$ cm. Dotted curve without tube source at same h'.

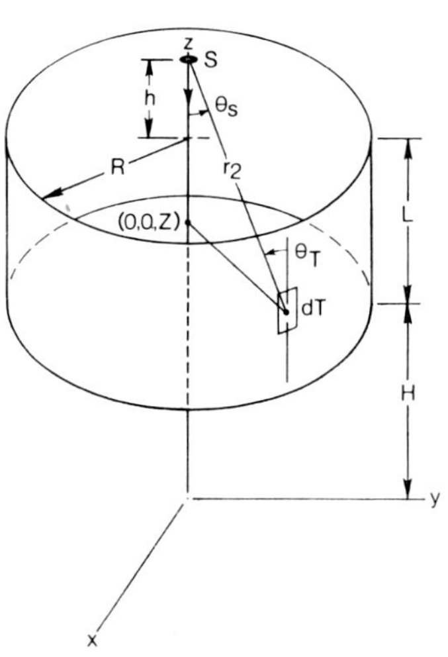

Figure 2. Schematics of tube source for analytical calculation of thickness distribution. Tube suspended at height H above substrate. Substrate in x, y-plane. Distance substrate to source S: $h' = h + L + H$.

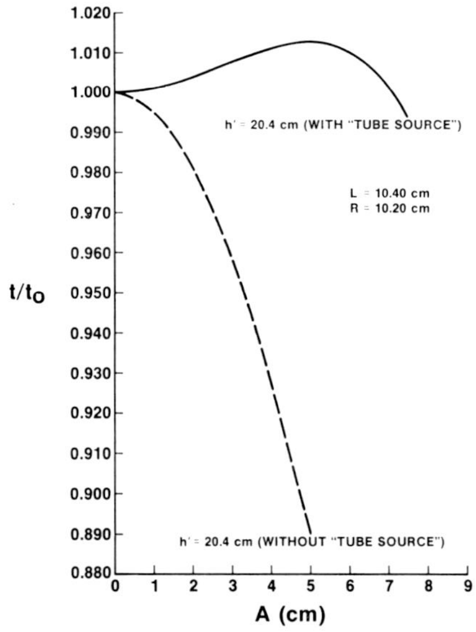

Figure 4. Same as Figure 3, but different cylinder height and radius; $h' = 20.4$ cm. Dotted line at same h'.

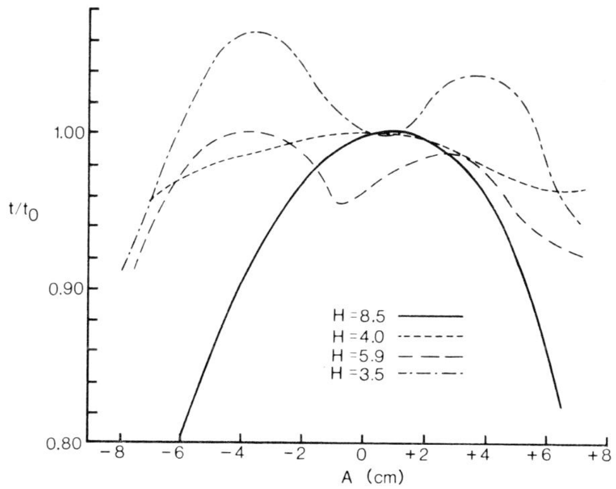

Figure 5. Thickness distribution of four copper films deposited with a tube source of $R = 10.4$ cm; $L = 10.2$ cm at different substrate heights H. $h = 0$.

Monte Carlo method. A Monte Carlo method has been developed to simulate the process of vacuum deposition in the tube source.

The following assumptions were made:

(1) The particles move inside the tube independently, i.e., there will be no collisions between them since the pressure is below 10^{-4} torr. The number of collisions between the particles can be neglected in relation to the number of particles that collide with the walls and the disc.

(2) The surfaces are assumed to be molecularly rough, which means that particles colliding with the walls are reflected randomly, i.e., the angle of reflection is independent of the angle of incidence.

(3) The sticking probability, i.e., the number of particles sticking at the solid surfaces per number of particles arriving, is varied, but mostly assumed to be 0.3.

The program* was performed with a Burrough 6700 computer and it took about 9 h starting with 5 million particles. The thickness distribution curves have the same M-shape as those obtained analytically. However, even 5 million particles were not enough to result in curves smooth enough to verify whether the variation was better than 0.4% but the fact that the curves showed the same shape as the ones calculated from equation (5) and obtained experimentally[4] confirms the capability of the new tube source to produce films with a thickness uniformity comparable with that of conventional vacuum deposition equipment using complicated rotational substrate holders.

Figure 5 depicts some experimental results achieved with a tube source of $L = 10.2$ cm and $R = 10.4$ cm for different substrate-tube distances H. The curves represent actual x–y recordings whereby the x-axis indicates the distance on the substrate and the y-axis the relative thickness (normalized to the center of the substrate) as obtained from light transmission using glass substrates.[1]

References

[1] H Schwarz, R C McCord and A Nerken, *Jap J Appl Phys*, Suppl 2, Pt 1, 1974, 471.
[2] L Holland, *Vacuum Deposition of Thin Films*, Chapman and Hall, London (1970).
[3] Equation (4) corresponds to equation (8) in ref 1; there, however, the square in the second denominator is missing.
[4] U L de Castro, Master Thesis, Departamento de Física, Universidade de Brasília, 70,000 Brasília, DF, Brazil, August 1976.
[5] B Carnahan and J O Wilkes, *Applied Numerical Methods*, Wiley, New York (1969).
[6] K H Behrndt, *Trans 10th AVS Nat Vac Symp*, p 379, MacMillan, New York (1963).

* For details of the computer program see ref 4.

1.17 Turbomolecular pumps with corrosion-resistant oil for the lubrication and cooling of the rotor bearings

J Henning and H Lotz, *Arthur Pfeiffer Vakuumtechnik Wetzlar GmbH, D-6334 Asslar, Emmeliusstr. 33, W. Germany*

When producing vacuum with turbomolecular pumps in the presence of strongly corroding gases and vapours (e.g. HF in the UF_6-technology) the problem of the disintegration of hydrocarbon chains of the standard oil used for the lubrication and cooling of the rotor bearings occurs. This disintegration of the oil results very quickly (100–500 h) in a destruction of the bearings. The problem can be solved by using fluorinated liquids (e.g. Fomblin, Montedison, I). The effects, however, appearing due to the other physical properties of these fluids, make it necessary to check each type of turbomolecular pump separately for their properties when operating with this fluid. Some results of the tests are described.

1. General

The widespread application of turbomolecular pumps in every field of high and ultra-high vacuum techniques, due to the clean vacuum which they create, the easy operation and the high degree of operating reliability, also involves numerous problems in which the turbomolecular pumps come into contact with strong corrosive media (e.g. in UF_6 technology or on ion sources).

The turbomolecular pump is a vacuum pump with rotors which run at a very high speed of 60,000 rpm (1000 Hz) maximum for the time being. The rotors are normally equipped with ball bearings which are lubricated and cooled by oil. The oils used have a mineral or a synthetic basis consisting of hydrocarbon chains with a high content of hydrogen.

The influence of strong corrosive gases and vapours leads to destruction of these hydrocarbon chains and the oil looses quickly its most important capacities. This provokes a premature wear in the rotor bearings so that the average life time of the bearings, depending on the contents of corrosive media, may be reduced to a range of 100–500 operating hours, whilst normal operating may permit a working life between 20,000 and 30,000 operating hours.

2. Fluorinated fluids

To solve this problem the use of noncorrosible fluids is recommended. A group of such fluids are polyfluorinated polyethers.[1] They are already known as a pump fluid for vapour pumps, to avoid a contamination, because it was found that backstreaming of oil molecules of the standard oils in vapour stream pumps could be polymerized on surfaces in vacuum systems where electron beams or ionizing radiation impinged.[2–4] These fluorinated fluids are polymers with the structure of perfluorinated polyethers. These consist of carbon, oxygen and fluorine atoms only. Elemental analysis confirms the structure of Figure 1.

The crude synthetic product is obtained in a range of molecular weights. It is fractionated by distillation into standard cuts

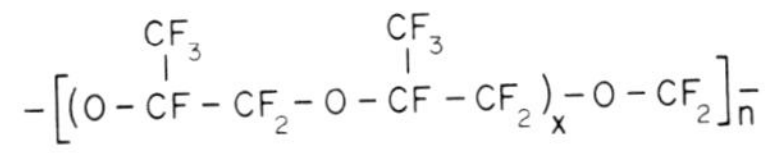

Figure 1. Structure of Fomblin[R].

of different viscosity. The different cuts do not present any residues. These fluids do not react with liquid oxygen, elemental fluorine, bromine and other oxidants (permanganate, nitric acid, hydrogen peroxide) as such or in diluted and concentrated aqueous solution at high temperatures, and chlorine and trifluoride at room temperature. Beside this these fluids are compatible with metals and elastomers at temperatures below 100°C. These capacities make these fluids highly suitable to solve the above mentioned problems.

3. Applications in turbomolecular pumps

3.1. Selection. Before the use of such fluorinated fluids in turbomolecular pumps the following points shall be considered:

the vapour pressure shall be so low that the liquid can be used at the fore vacuum side of the turbomolecular pump,

the viscosity shall have a value which permits cooling of the bearings and prevents a build-up of friction heat in the bearings,

the lubricating capacity shall correspond to the values of standard oils.

In Table 1 are compared the physical capacities of a standard mineral oil (Type F12) for turbomolecular pumps with the values of two fluorinated fluids (Types F3 and F5). The values shown indicate that the fluid F3 presents approximate similar capacities as the standard oil F12 and that it may be used in turbomolecular pumps.

3.2. Application. During far-reaching tests, it could be proved that the oil F3 complies also with the requirements of lubricating

Table 1. Data of oils for turbomolecular pumps and fore vacuum pumps.

	Conditions	Units	Type of oil F12 (mineral)	F3 (Fomblin)	F5 (Fomblin)
Average molecular weight			430	2000	3000
Density	20°C	g cm^{-3}	0.92	1.88	1.90
Viscosity	20°C	c St	25	65	250
	40°C		13	26	78
	100°C		3.4	4.2	9.4
Vapour pressure	20°C	mbar	6.7×10^{-6}	3.8×10^{-6}	8.0×10^{-8}
	40°C		6.7×10^{-5}	2.7×10^{-5}	1.2×10^{-6}

capacity. The tests have been carried out with pumps of the old and today programme.

In Table 2 the most important physical and mechanical features of these pumps are shown. The marked (asterisks) types of pumps are no longer manufactured. Accordingly three groups of turbomolecular pumps are to be distinguished with regard to the use of F3:

turbomolecular pumps with which the use of fluid F3 causes problems, e.g. nos. 4 and 6.

Due to the higher density of fluid F3, the relatively high weight of the rotors, great diameter of the bearings and the relatively high rotation speed of the rotor a heat build-up occurs in the bearings and consequently in the rotor shaft, even after reducing of the circulating oil quantity, so that the rotor reaches a temperature of about 90°C. This may cause problems in some applications.

Turbomolecular pumps in which fluid F3 can be used after some alteration of the pump, e.g. pumps number 1, 2, 3, 5, 7, 8. Because of the higher density the quantity of circulating oil shall be reduced in the pumps number 1, 2, 3 and 5. Difficulties arise in the pumps number 7 and 8 with wick lubrication as the fluid F3 is not sufficiently fed through the wick due to its bigger molecules. Therefore, a small electronic controlled fluid pump has been developed which conveys the fluid F3 in a sufficient quantity to the bearings.

Turbomolecular pumps in which fluid can be used without alteration of the production model (nos. 9–12).

All these pumps have been tested in the laboratory up to about 15,000 operating hours and no problems on bearings appeared.

Some pumps of types number 1, 3, 8 and 10 run for a longer period (up to date 25,000 h maximum) with several customers in the nuclear field to their entire satisfaction.

4. Backing pumps

As turbomolecular pumps, due to their working principles, are always used together with backing pumps, normally of one or two stages, the question for a corresponding oil for the backing pumps arose. Tests on our two stage backing pumps with a pumping speed of 4, 8, 12 and 30 m^3 h resulted that the fluid F5 (indicated in Table 2) is suitable for these pumps after some minor alterations have been carried out on them.

5. Residual gas analysis

During the tests residual gas analysis have been made with a quadrupole mass spectrometer. It resulted that the ultimate pressure of the turbomolecular pumps is not improved by the use of fluid F3. When fluid F3 is used distinct fluorine peaks were noted in the residual gas. The greatest part by far in the residual gas is, as before, hydrogen. It could be concluded, therefore, that this part of hydrogen is neither created by the standard pump oil F12 nor by way of a dissociation in the bearings.

Further tests have shown that the hydrogen part of the

Table 2. Data of Pfeiffer turbomolecular pumps.

Pfeiffer-turbo No.	Type	Pumping Speed N_2 l. s^{-1}	Rotor frequency Hz	Rotor weight kg Total	Per bearing	Bearing diameter mm Inner	Outer	Lubrication
1	TVP 250*	70	500	2.0	1.0	8	22	Circulation
2	TVP 500*	140	270	12.5	6.25	25	47	Circulation
3	900*	260	270	12.5	6.25	25	47	Circulation
4	1000*	280	300	12.4	6.20	25	47	Circulation
5	THP 900*	260	270	12.5	6.25	25	47	Circulation
6	1000*	280	300	12.4	6.20	25	47	Circulation
7	TPH 200	200	600	2.0	1.0	8	22	Wick
8	TPU 200	200	600	2.0	1.0	8	22	Wick
9	TPU 400	400	1000	2.60	1.30	8	22	Circulation
10	TPH 450	450	1000	2.50	1.25	8	22	Circulation
11	TPH 500	500	1000	2.50	1.25	8	22	Circulation
12	TPU 500	500	1000	2.50	1.25	8	22	Circulation

residual gas of a turbomolecular pump is mainly created by the hydrogen which diffuses through the seals of the fore vacuum side into the fore vacuum channel and is found again, corresponding to the compression ratio of the turbomolecular pump, on the high vacuum side.

6. Summary

The studies have proved that a fluid has been found with the fluorinated fluid F3[1] which can be used as a lubricating and cooling means for the bearings of the tested turbomolecular pumps and which meets the requirements for a 'noncorrosible' oil. A corresponding fluid F5 can be used in the tested backing pumps.

References

[1] Registered trade-mark of Montecatini Edison SpA Italy.
[2] L Holland, L Laurenson, P N Baker, *Vacuum*, **22**, 1972, 315.
[3] B K Ambrose, L Holland, L Laurenson, *J Micros*, **96**, 1972, 389–391.
[4] L Holland, L Laurenson, M J Fulker, *Jap J Appl Phys*, **12**, 1973, 1468.

1.18 A versatile cryopump for industrial vacuum systems

J Visser, B Symersky and A J M Geraerts, *Philips Research Laboratories, Eindhoven, The Netherlands*

The advantages of cryopumping, e.g. high pumping speed, flexibility, cleanliness and wide pressure range of operation, have long been recognised. Nevertheless, industrial application of this method has been rejected so far. This paper describes the vacuum performance of a cryopump equipped with a recently introduced cryogenerator. The machine offers a cold production of approximately 50 W at 80 K and 10 W at 20 K simultaneously, thus eliminating the need for a liquid helium and (or) nitrogen supply. Data presented show that industrial applications of cryopumping at 20 K impose special requirements on the design of the cryopanels and the cold capacity of the cryogenerator. It is shown that a 10 W cold capacity will offer long term operation of the pump, particularly in industrial thin film coaters, if (a) hydrogen gas is distributed uniformly over an adsorbent glued on a 20 K panel, (b) the grain size of the adsorbent is about one millimetre, and (c) contamination of the adsorbent by gases more easily condensable than hydrogen is avoided.

Introduction

Small mechanical refrigerators with a cold capacity of about one watt at 20 K were introduced in the late sixties. One watt cold production should be sufficient to create a nitrogen pumping speed in excess of 10,000 l. s^{-1} [1,2] in the conditions under which the cryopump operates at high or ultrahigh vacuum. Generally pumping speeds of this magnitude will be adequate for application in thin-film coaters. However, an industrial vacuum system will not run continuously at high vacuum but will frequently be cycled down from one atmosphere. Moreover, in many cases considerable quantities of hydrogen must be removed. These conditions impose new requirements on the design and the cold capacity of a cryopump. The study we made was focused on the performance of a 20 K cryopump under the special conditions met in an industrial vacuum system.

The vacuum system

The actual pumping performance of a cryopump cannot generally be characterized sufficiently by a measurement of the intrinsic pumping speed.[3] Unlike that of turbomolecular pumps or diffusion pumps, the pumping behaviour of a cryopump at high vacuum is influenced by the way in which the transition region at the start of the pump-down cycle is passed. If cryosorption pumping is involved, pumping performance changes in time. These considerations prompted us to carry out pumping speed measurements with the cryopump fitted to a commercial thin-film coater, frequently cycled down from the 10^{-1} torr pressure region. For this purpose we used the Veeco model VE-747 25 in. diameter bell jar deposition system. The pumping speeds for water and nitrogen do not appear to full advantage as they are conductance-limited (e.g. N_2 5000 l. s^{-1}). However, the problems to overcome in industrial cryosorption pumping at 20 K are mainly connected with the pumping of hydrogen. For this gas the molecular flow conductance from the pump opening to the vacuum chamber is about 18,000 l. s^{-1}, sufficiently high for the present study. The behaviour of the cryopump in electron beam evaporation plants was investigated using a Balzers BAK550 coater in which the flow conductance for easily condensable gases is less restricted.

The refrigerator

Several types of refrigerators capable of producing cold at a temperature of about 20 K are currently available.[4,5] The refrigerator we used was the Philips K 20 machine, which works on the Stirling principle. The K 20 can offer a cold capacity at two stages simultaneously. The first-stage capacity is about 10 W at a temperature of approximately 20 K, the second-stage capacity is about 50 W at 80 K (Figure 1). There is no need for an additional liquid nitrogen supply as the cold capacity at 80 K can be used for cooling the surfaces shielding the 20 K cryopanel. An incidental advantage for the present study is that the cold production at the 20 K stage can be varied from about 1–10 W. Consequently the influence of the cold capacity on the pumping performance can easily be checked.

The 20 K cryopanel

Cryotrapping at 20 K does not offer the potential to remove the hydrogen quantities evolved during evaporation.[6] For this reason our attention was drawn to the cryosorption of hydrogen on solid absorbents.

(a) Influence of gas distribution. Experiments on cryosorption pumping by a column of adsorbent very clearly demonstrated that the gas will be taken up mainly by the adsorbent first seen by the gas flow, if the capture probability and the mean time of sojourn on the adsorbent surface are sufficiently high.[7] For hydrogen on an adsorbent at 20 K these conditions will be fulfilled. The practical consequences of a non-uniform distribution over the adsorbent are a loss in effective pumping speed, particularly at the low pressure end of the pump-down, and a loss in effective sorption capacity. Hence the design of the

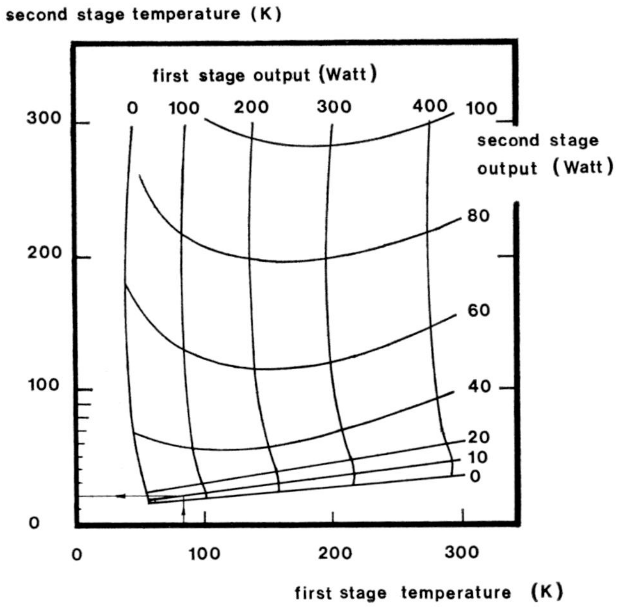

Figure 1. Energy diagram for the Philips Stirling cryogenerator K20. Cold production at the first and second stages as a function of the temperature levels.

cryopanel must be such that the gas is distributed fairly uniformly over the adsorbent. As a compromise between adsorbent surface and gas distribution we selected a finned construction (Figure 2). Scattering surfaces[8] serve to improve the hydrogen gas distribution over the panels. A design in which some discs are placed behind each other and the charcoal is glued on the surfaces facing each other was rejected because the hydrogen gas flow is first offered to the outer regions of the discs. The results obtained with a column of adsorbent also indicate that even across a few millimetres of adsorbent the influence of the hydrogen gas distribution over the adsorbent can be significant. The smaller the grains the better the gas distribution over the adsorbent can be. Moreover the mean temperature of the adsorbent attained with a smaller grain size will be lower. On the other hand we need a certain amount of adsorbent for the required sorption capacity. The optimum grain size was derived from the results of measurements shown in Figure 3. The data do not represent true equilibrium conditions but were obtained 10 min after switching off the hydrogen gas flow. The charcoal we used was activated carbon SC-116/L2 supplied by Chemviron Ltd. The test panel surface coated with charcoal was 15 cm^2. The charcoal grains were glued to the panel by means of an epoxy resin (Hysol base resin A9-4297; hardener H2-3477). Activation of the charcoal by heating above room temperature was not possible. Warming up to room temperature and removing the released gases by a rotary pump (until 10^{-1} torr was achieved) proved to be sufficient for the present situation. Apparently the effective sorption capacity of the smaller grains is better at low pressures. The sorption capacity of the 1 mm grains closely approaches the literature data[9] which, in contrast to our data, were obtained under optimum equilibrium and activation conditions. Further reduction in grain size will not be advantageous because of the loss in adsorbent weight per square centimetre. The experiments discussed later on were carried out with charcoal grains of about 1 mm. In spite of the 'open' cryopanel design and small adsorbent grain size the influence of the gas distribution could not be eliminated completely. This became clear from the following experiment. Two hundred and twenty-five torr-l. of hydrogen were sorbed on the finned 20 K cryopanel surface (1000 cm^2) covered with charcoal having a grain size of about 1 mm. The hydrogen gas was introduced while the pressure was kept constant at 2.5×10^{-5} torr. After closure of the gas inlet the pump-down performance is as represented by curve one of Figure 4. Switching off the refrigerator for a short time causes an increase in the temperature and consequently in the pressure. A mass spectrometer indicates that the pressure rise is completely due to hydrogen. The machine is switched on again at

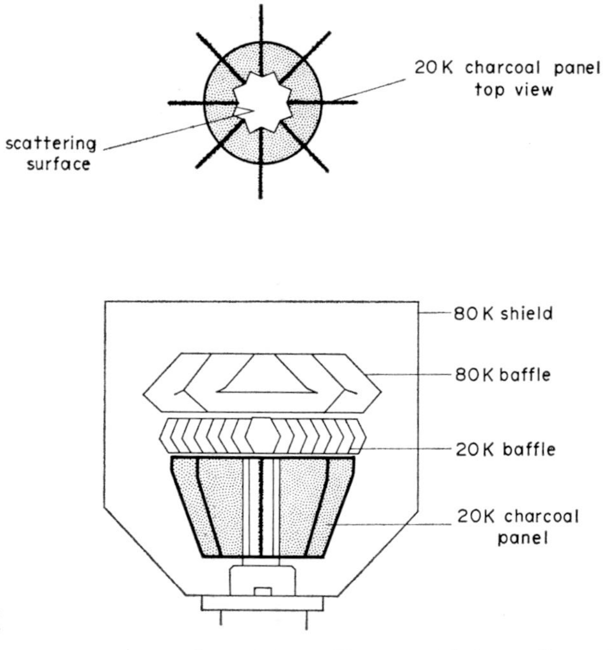

Figure 2. The cryopanel arrangement. The second design as discussed in text. In the first design the 20 K baffle was omitted. Opening in 80 K shield 33 cm in diameter.

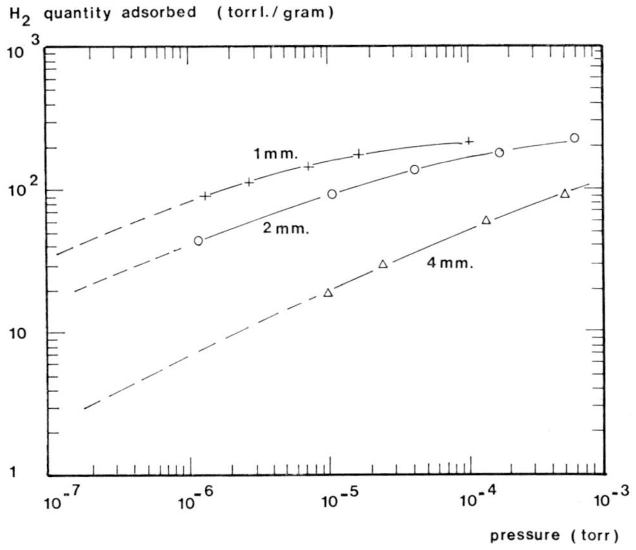

Figure 3. Effective adsorption capacity (obtained within 10 min) as a function of pressure for various absorbent grain sizes.

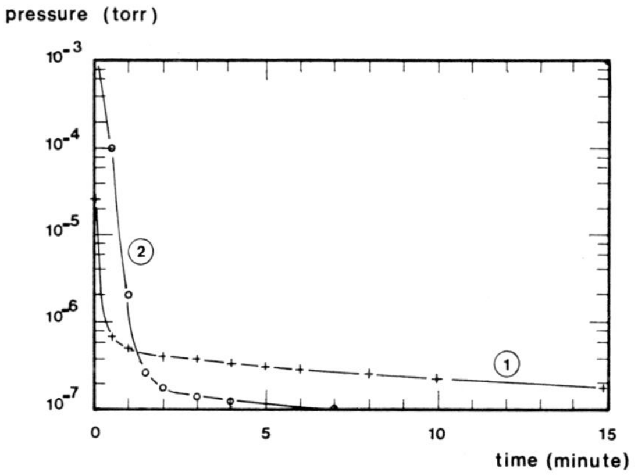

Figure 4. Curve 1 represents the pump-down process after a quantity of hydrogen has been sorbed at 2.4 × 10^{-5} torr. Curve 2 represents the pump-down process immediately after switching the refrigerator off for a short while without changing the sorbed amount of hydrogen.

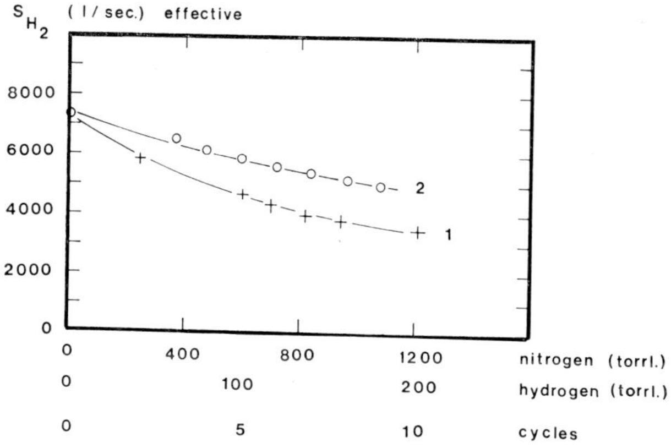

Figure 5. Effective hydrogen pumping speed as a function of the number of pump-down cycles from 0.4 torr together with hydrogen gas load. Curve 1 relates to the first design. Curve 2 relates to the second design equipped with an additional 20 K baffle.

10^{-3} torr. As a result the pressure decreases but now towards lower hydrogen pressures in a shorter time (curve 2). Obviously this procedure causes rearrangement of the hydrogen over the adsorbent in favour of a better pump-down performance.

(b) Influence of gas composition. In practice cryopumping will involve more than one gas. As the pump-down process is started at relatively high pressures (0.1–0.4 torr) large amounts of nitrogen and oxygen are to be removed. This may influence the pumping behaviour in various ways.

A frozen deposit of the easily condensable nitrogen and oxygen gas will be built up on the external surface of the adsorbent grains. This changes the accessibility of the charcoal for the hydrogen.[10] The initial gas throughput will be high and as a consequence the temperature of the charcoal, particularly at the grain surface, will increase temporarily. This condition works in favour of a transport of nitrogen and oxygen to adsorption sites at the inner surfaces of the adsorbent grains. Consequently, less sorption capacity for hydrogen may become available and the capture coefficient for hydrogen gas on the charcoal may change.

In order to get an impression of the influence of these effects on the hydrogen speed some experiments were carried out with two different 20 K cryopanels. First a cryopump consisting of a 1000 cm^2 finned panel covered with charcoal and surrounded by a 80 K chevron baffle and 80 K panels was tested. In the second design an extra 20 K baffle was installed between 80 K chevron baffle and charcoal panel (Figure 2). This baffle was not coated with charcoal, but partly shielded the charcoal from the easily condensable gases. Due to the small dimensions of the present pump chamber more efficient baffling could not be provided because of the unwanted reduction in molecular flow conductance associated with this.

A 300 l. chamber backfilled with nitrogen was repeatedly pumped down from 0.4 torr. About 20 torr-l. of hydrogen at a flow of 4.10^{-2} torr l. s^{-1} were cryosorbed during every pump-down interval. This amount fairly represents the hydrogen quantity produced during electron beam evaporation of aluminium (see Appendix). At the end of each pump-down cycle the hydrogen speed was measured.

Figure 5, curve 1 shows that in 10 cycles the net pumping speed for hydrogen decreased gradually from about 7000 l. s^{-1} to about 3500 l. s^{-1} (i.e. from about 12,000 to 4300 l. s^{-1} intrinsic pumping speed) if the charcoal panel is not shielded from the nitrogen.

It can be derived from a comparison with curve 2 of Figure 5 that even the partial shielding of the charcoal from the nitrogen flow improves the pumping performance for hydrogen. We have seen that in the case of pure hydrogen the pumping performance will be improved if the sorbed gas is allowed to rearrange itself over the adsorbent, e.g. by means of switching the refrigerator off and on. In order to find out what happens when hydrogen was pumped together with nitrogen the refrigerator was switched off for a while after 10 cycles.

The refrigerator was switched on again when the cryopanel temperature was about 30 K and the total pressure reached 10^{-3} torr (~10% H_2; ~90% N_2). The hydrogen speed we measured afterwards only came to about 2500 l. s^{-1} instead of 3500 l. s^{-1} (achieved before switching off-on) for the first cryopanel design and to about 4000 l. s^{-1} instead of 4900 l. s^{-1} for the second design. Obviously transport of the strongly bound gases to the charcoal enhanced by a temporary increase in temperature causes a decrease in hydrogen pumping speed.

The system was also cycled without admitting hydrogen continuously. Only small amounts of hydrogen (about one-tenth of the amounts at continuous flow) were bled in temporarily to permit pumping speed measurements. Hence the pumping performance could not be limited by the sorption capacity. Nevertheless the change in pumping speed observed from the first to the tenth cycle (Figure 6, curve 1) differs only slightly from the change we found with higher H_2 gas load (curve 2 Figures 5 and 6).

The hydrogen pressure achieved 3 min after valving off the hydrogen was 6 × 10^{-8} torr. This also shows that the reduced pumping speed measured in the 10^{-6} torr pressure region cannot be due to saturation of the adsorbent as a whole. We thus obtained evidence that the decrease in pumping speed for hydrogen when pumped together with nitrogen is mainly caused by a change in the capture probability for hydrogen.

Cold capacity and its influence on hydrogen pumping

The pumping performance for hydrogen was observed in successive series of pump-down cycles, each series at a different cold capacity. Four levels of cold capacity were applied 9, 6, 3

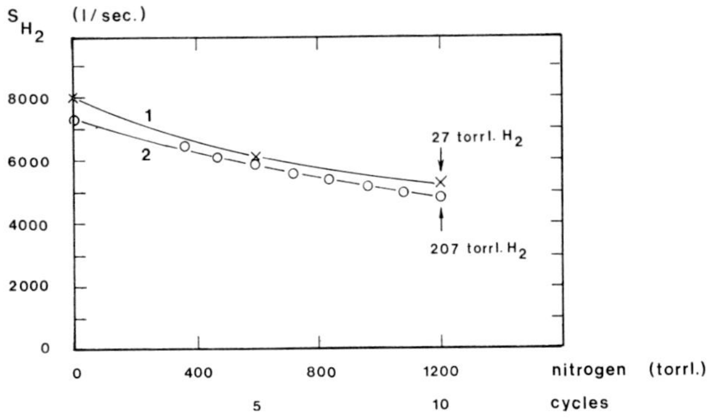

Figure 6. Effective hydrogen pumping speed as a function of a number of pump-down cycles from 0.4 torr with a small and large hydrogen gas load.

and about 1 W at 20 K respectively. The data in Table 1 illustrate that the cold capacity of the refrigerator becomes important if long term operation of a cryopump of the present design in an industrial vacuum system is required. At a cold capacity of about 1 W at 20 K 30 pump-down cycles in which 360 torr-l. of hydrogen and 3240 torr-l. of nitrogen were removed make the hydrogen pumping speed measured at a pressure of 10^{-6} torr zero. When at that time the cold capacity was raised to about 9 W a hydrogen pumping speed of 4250 l. s^{-1} at 10^{-6} torr was realized. The hydrogen pumping speed measured at 4.10^{-6} torr was about 3350 l. s^{-1} for 1 W cold capacity and about 4500 l. s^{-1} for 9 W.

Application in vacuum deposition

The K20 cryosorption pump was fitted to a Balzers BAK 550 coater. This vacuum system together with an Airco Temesal *e*-gun was used for the production of aluminium thin films. Deposition at a power of 6 kW was carried out during 10 min. As mentioned in the Appendix the hydrogen production during deposition is strongly determined by the partial pressure of water. The experiments showed that the pressure during deposition easily could be maintained in the 10^{-7} torr pressure region if the process was started at about 10^{-7} torr. In order to get a realistic impression of the possibilities of the cryopump various deposition runs initiated at a total pressure of about 2.10^{-6} torr were carried out. Between each run the system was vented to ambient atmosphere and the vacuum chamber door opened for loading during 5 min. Roughing was accomplished by a standard rotary pump until 0.35 torr was achieved. This high change-over pressure guarantees a completely hydrocarbon free vacuum system. During each run the total pressure was recorded continuously. The residual gas composition was checked frequently. The hydrogen pumping speed was measured at intervals.

Figures 7, 8 and 9 illustrate that even over 94 runs a reproducible pumping performance is obtained.

Figure 10 shows the total pressure curve and some of the mass spectral data of run no 38. Different from the other runs the deposition is started at a low partial pressure of water.

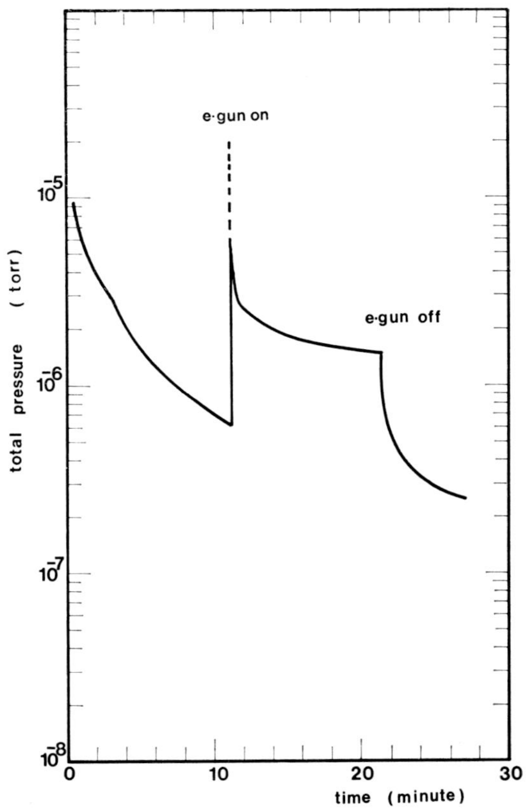

Figure 7. The total pressure recorded during pump-down (from 0.35 torr), during *e*-beam deposition (6 kW) and after deposition. Second run.

Conclusions

Cryopumping at 20 K realized by means of a mechanical refrigerator is shown to be an attractive alternative for diffusion and turbomolecular pumping even in industrial vacuum systems where relatively large amounts of hydrogen have to be removed.

Table 1. The influence of the cold capacity on the hydrogen pumping speed when a 300 l. chamber is frequently cycled from 0.4 torr and large quantities of hydrogen must be removed. For each cold capacity the pumping speeds are given at the highest cycle number.

H_2 pumping speed (l. s^{-1}) $4\,10^{-6}$ torr	10^{-6} torr	Cold capacity (W at 20 K)	Quantity H_2 sorbed (torr l.)	Quantity N_2 sorbed (torr l.)	Cycle number
4600	4600	9	195	1560	1–13
4400	4200	6	245	1920	14–17
4200	2500	3	360	3240	18–29
3350	0	1	445	3720	30–34
4490	4250	9	512	4080	35–37

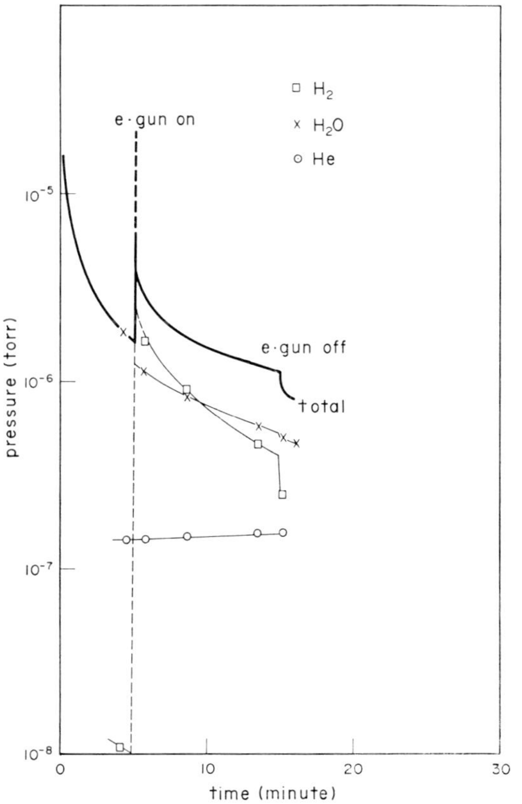

Figure 8. As Figure 7. Run number 56. The measured partial pressures (in nitrogen equivalents) are added.

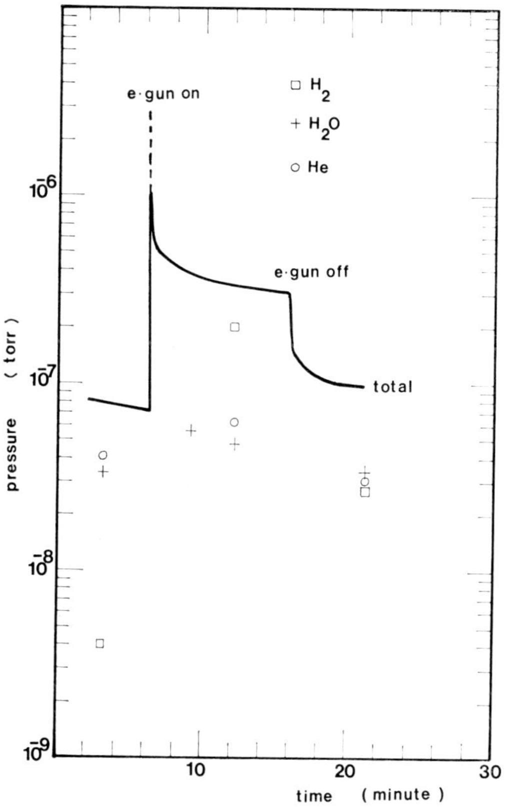

Figure 10. As Figure 8. Run number 38. Deposition takes place at a low partial pressure of water.

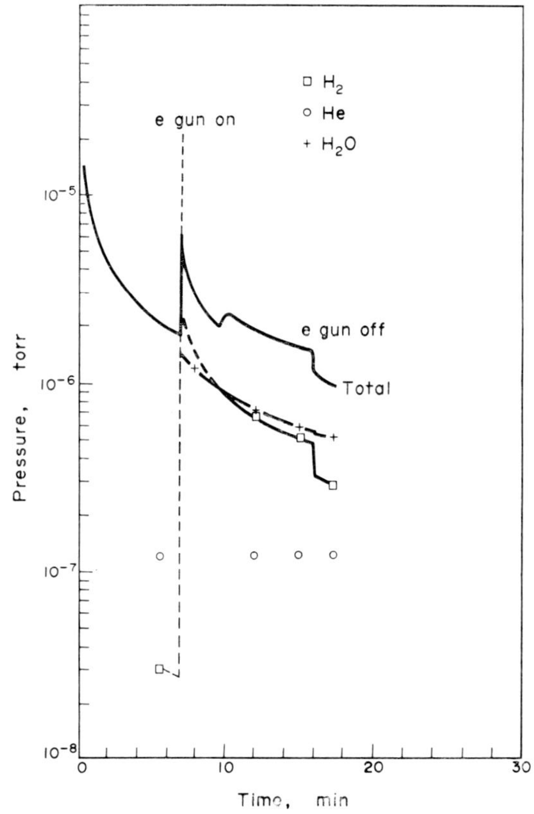

Figure 9. As Figure 8. Run number 95.

The pumping performance for gases with low vapour pressures at 20 K is extremely good and stable. Simple criteria already mentioned in the literature are fulfilled.[11, 12] For hydrogen the situation is completely different. High pumping speeds can be realized easily, but only for small amounts of pure hydrogen. If prolonged hydrogen pumping or rapid cycling of a vacuum system is required, determining factors are the cold capacity of the refrigerator, the distribution of the gas over the adsorbent, the grain size of the adsorbent, and the contamination of the absorbent gases more easily condensable than hydrogen. We found that 10 W of cold capacity at 20 K, 1 mm grain size of the adsorbent, and shielding of the charcoal panel by an additional 20 K baffle are conditions that offer the opportunity of creating a versatile cryopump for industrial applications.

References

[1] F T Turner and W H Hogan, *J Vac Sci Technol*, **3**, 1966, 252.
[2] O Winkler, H. Woessner and J Ruhe, Supplement à la Revue *Le Vide*, **169**, 1974, 190.
[3] W W Stickney and B B Dayton, *Trans 10th AVS Vac Symp*, Macmillan, New York (1963).
[4] F X Eder and H Woessner, *Ergebnisse der Hochvakuumtechnik und der Physik dünner Schichten Band II*, p 348, *Wissensch*, Verlags-Gesel, mbH, Stuttgart (1971).
[5] G J Haarhuis, *Le Vide*, **171**, 1974, 351.
[6] J Hengevoss and E A Trendelenburg, *Vacuum*, **14**, 1967, 495.
[7] J Visser and J J Scheer, *Ned. Tijdschrift Vacuumtech*, **11**, 1973, 17.
[8] B A Hands, *Cryogenics*, 1973, 699.
[9] S A Stern, J T Mulhaupt, R A Hernstreet and F S Di Paolo, *J Vac Sci Technol*, **2**, 1965, 165.

[10] K Becher, G Klipping, W D Schoenherr, W Schulze and V Toelle, *Proc Fourth Int Cryogen Eng Conf*, Eindhoven, May 1972, p 319–323. IPC Science & Technol Press Ltd, Guildford, Surrey (1972).
[11] J P Dawson and J D Haygood, *Cryogenics*, **5**, 1965, 57.
[12] J Roussel, J J Thibault and A Nanoboff, *Le Vide*, **118**, 1965, 249.

Appendix

Hydrogen evolution in thin film coaters. In thin film coating processes large quantities of hydrogen may be released from various sources. During *e*-beam evaporation hydrogen will be produced mainly

(a) by desorption of surface contamination due to electron bombardment,
(b) by degassing of the bulk evaporant,
(c) by dissociation of water at hot surfaces, e.g. *e*-gun filament and evaporant material,
(d) in some cases by reaction of water vapour with the growing thin film.

(a) and (b) will give rise to a pressure burst of hydrogen at the start of the process.

(c) and (d) both contribute to a nearly steady hydrogen production during evaporation.

(a) will always be present. In most cases it only contributes noticeably during the initial stage of the deposition process. (b) applies mainly for getter materials like Mo, Cr, Ti, etc. and is strongly dependent on the purity of the evaporant. Fortunately it can be reduced to an appreciable low level by using high purity material.

Both (c) and (d) give rise to a continuous production of hydrogen. It can be reduced to a low level by starting the deposition process at an appropriate low partial pressure of water. However in practice when rapid cycling is required the vacuum chamber will be pumped down to about 10^{-6} torr (mainly H_2O). For the case of aluminium (a) + (d) then become the dominating mechanisms in the production of hydrogen. For example 5–10 torr-l. of hydrogen will be produced within 10 min of aluminium evaporation at 6 kW. It appears that in the long run the deposition of aluminium creates one of the worst conditions with respect to hydrogen pumping. For this reason experiments were carried out at a hydrogen gas load which fairly represents the hydrogen production during aluminium deposition whereas the final applicability test was performed in the production of aluminium thin films.

2. Electrical properties of thin films—metal–semiconductor contacts

2.1 Metal–semiconductor junctions—related to deposition of thin films

K Eisele and M Schulz, *Institut für Angewandte Festkörperphysik, Freiburg, Germany*

The electrical properties of metal-semiconductor junctions are defined by the barrier height, by interface states, by interfacial layers and by the doping and carrier lifetime of the semiconductor. Schottky contacts have a high barrier at the junction which the carriers have to surmount for current transport. Ohmic contacts require a low barrier or a low tunnelling resistance through the barrier.

The surface preparation, the choice of the metal, the deposition technique and the post deposition treatment have an influence on the junction properties. The influence, however, is limited by surface states and is modified by the degree to which an interfacial layer is desired or to be eliminated.

The heating of the semiconductor by the deposition process and the radiation damage when e-gun evaporation, sputtering or ion implantation is used, must be examined as to their influence on the junction. With most of the semiconductors a thermal treatment is part of the deposition process.

1. Introduction

Semiconductor device fabrication incorporates always the deposition of metal films for interconnections, for ohmic contacts and for metal semiconductor junctions of non-ohmic behaviour, blocking or Schottky contacts. Whereas the first of these three applications occurs on insulating layers and is therefore independent on semiconducting properties, the latter two depend on the metal used, the semiconducting material, the surface preparation, the method of deposition and the post-deposition (thermal) treatment. It is understood that the manufacture of the metal-contact should not degrade other properties necessary for the functioning of the device. Metallizations on semiconductor devices which are manufactured on a large industrial scale have been widely studied, also under the aspect of reliability and economy. As the application of more exotic semiconductors for electronic devices appears advantageous new semiconductor–metal-systems and the methods to realize them have been investigated. Excellent reviews of the theory underlying the behaviour of these junctions have been written[1–3] and numerous experimental results presented.[4]

2. The metal–semiconductor junction

A metal–semiconductor junction is characterized by its current–voltage relationship. If it shows rectifying properties and therefore the current widely differs for the two possible bias directions, we call it a Schottky-contact. If it is close to linear, in both bias directions and the resistance is low, we call it an ohmic contact. Both types of contacts must be available on *n*-type as well as on *p*-type semiconductor material. A fundamental understanding of the conduction mechanism can be gained from the energy band diagram (Figure 1) where a semiconductor is brought into intimate contact with a metal. Under equilibrium conditions the Fermi levels of both partners must coincide. Because of the difference in work function of the metal and the semiconductor, there exists a contact potential V_c which appears as a difference in the vacuum potential level in both materials. This contact potential is formed by a space charge dipole layer at the interface. Such a space charge layer does hardly penetrate into the metal because of the large carrier concentration. However, it extends into the bulk of the semiconductor, where the doping is less than the number of free electrons in the metal. The barrier height Φ_B at the ideal interface is determined by the difference between the metal work function Φ_M and the semiconductor electron affinity χ_S. For an *n*-type semiconductor we obtain

$$\Phi_B = \Phi_M - \chi_S \tag{1}$$

In a *p*-type semiconductor the energy bands bend down and create a barrier for holes

$$\Phi_B = E_G - (\Phi_M - \chi_S) \tag{2}$$

With these equations one can confirm the ideal behaviour of evaporated *M–S* contacts on *n*- and *p*-type lead telluride as found by Nill *et al*[5] shown in Table 1. Depending on the metal work function and the electron affinity one can correctly predict,

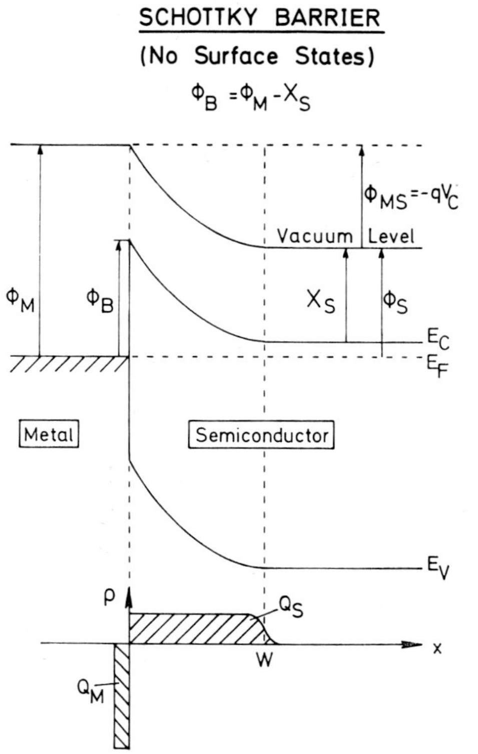

Figure 1. Energy band structure of a metal-*n*-type semiconductor contact without surface states. Space charge distribution is shown in the lower diagram.

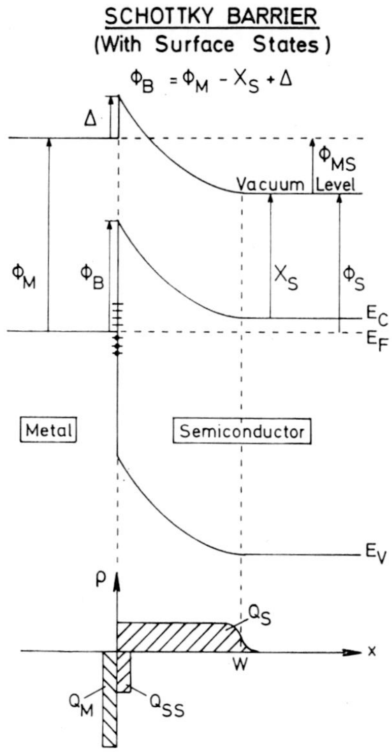

Figure 2. Energy band structure of a metal-*n*-type semiconductor contact having a large interface state density. Space charge distribution in the lower diagram.

Table 1. Evaporated metal contacts on *n*- and *p*-type PbTe[5].

Work function (ϕ_m)		Metal	Experimental Observation
$\chi < \phi_m$	5.4	Pt	Barriers on *n*-type Ohmic to *p*-type
$\chi \approx \phi_m$	4.6	Au	Barriers on *n*- and *p*-type
$\chi > \phi_m$	4.4	Sn	Barriers on *p*-type Ohmic to *n*-type
	4.2	Zn	
	4.0	Pb	
	3.8	In	

$\chi = 4.6$ eV = electron affinity of PbTe (measured from the conduction band to the vacuum level).

whether an ohmic contact ($\Phi_S <$ O) or a barrier ($\Phi_S >$ O) is obtained on either *n*- or *p*-material.

Unfortunately these predictions fail on the more common semiconductors, e.g. Si, Ge, GaAs. With these the presence of a large surface state density pins the Fermi level to an almost fixed energy at the surface. An additional dipole layer from the surface states superimposed on the already existing one increases or decreases the barrier height by an amount Δ (Figure 2) so that instead of equation (1) the actual surface barriers observed become

$$\Phi_B = \Phi_M - \chi_S + \Delta \tag{3}$$

and instead of equation (2)

$$\Phi_B^* = E_G - (\Phi_M^* - \chi_S + \Delta)$$

E_G being the energy gap of the semiconductor.

This stabilization of the Fermi level by the surface states is demonstrated in Figure 3, where the barrier energies of various metals obtained on three semiconductors are plotted against the electronegativity of the metals.[6] The slope $S = d\Phi_b/d\chi_m$ for many semiconductors, plotted against their electronegativity is shown in Figure 4. It now becomes evident that the more ionic a semiconductor lattice is bound (NaCl structure) the less the *MS* junction is affected by surface states. On the other hand,

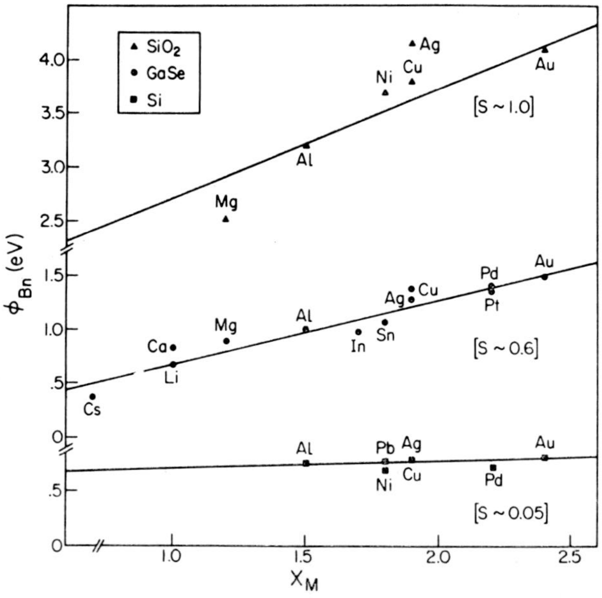

Figure 3. Barrier energies of various metals on *n*-type semiconductors plotted vs electronegativity.[6]

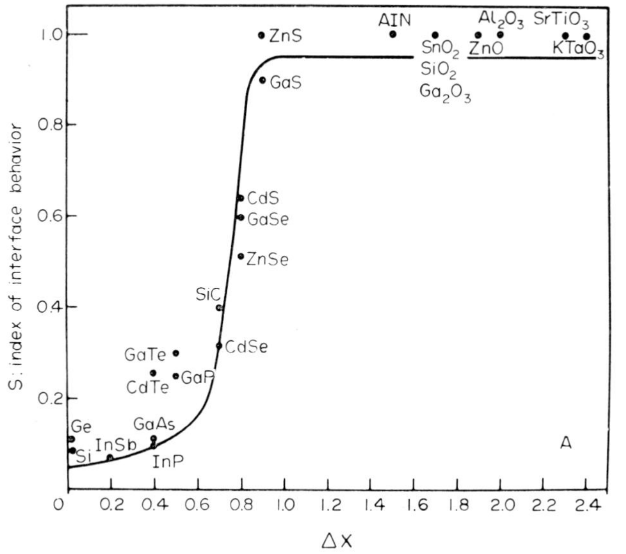

Figure 4. The slope $S = \mathrm{d}\,O_{b/dxm}$ for various semiconductors demonstrating the covalent–ionic transitions.[6]

covalent semiconductors (zinc blend structure) show little dependence of barrier height on the metal used. In fact it has been shown[7] that with covalent semiconductors the barrier energy is often close to 2/3 E_G. In spite of this empirical rule some significant variations of barrier height have been measured, mostly with silicon because of its widespread use in device technology. A useful selection of measured barrier heights has been compiled and discussed by Andrews.[8] This result is shown in Figure 5. Discrepancies in published values

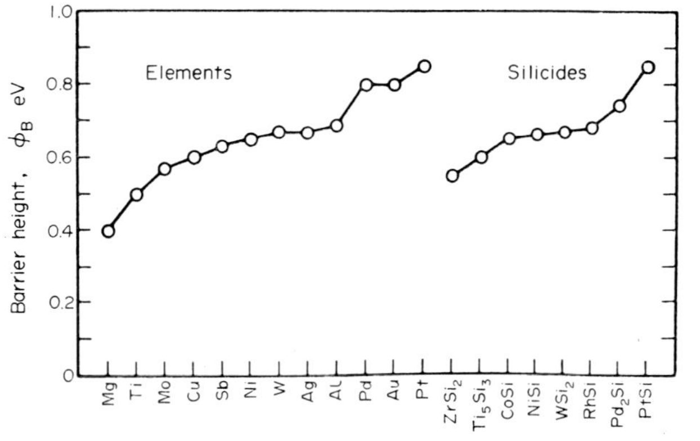

Figure 5. Potential barrier heights for elements and silicides on *n*-type silicon.[8]

may be due to differences of surface preparation, which will be elaborated on in the following sections. Contacts that have been the result of a metallurgical reaction, e.g. PtSi, WSi_2 show considerably more consistency in the quality of the *M–S* junction, because surface layers may have been lifted, or interfacial oxides may have been removed.

The current flow through a Schottky barrier entirely depends on the barrier height and is controlled by thermionic-emission. In published data the character of a Schottky contact is often expressed by a so called '*n*-parameter'. It is dimensionless and equal to

$$n = 1 + \delta V_B/\delta V$$

It expresses the change in barrier height under forward bias and is obtained by fitting the current to the Richardson-thermionic emission law

$$I = I_S \cdot [\exp qV/nkT - 1]$$

The value of n is close to 1, between 1.04–1.06 at room temperature. It is a measure of the quality of the junction and an indicator for deviation from ideal conditions as caused by an interfacial layer. The barrier width is related to the barrier height by

$$W = \left[\frac{2\varepsilon_S}{q^2 N_D}(\phi_B - \phi_{SF} - qV)\right]^{\frac{1}{2}}$$

where Φ_{SF} is the distance of the semiconductor Fermi level from the conduction-band edge and V the applied voltage. The width decreases with increasing donor concentration N_D and reaches a point where tunnelling supplies an appreciable fraction of the current and finally takes over completely at levels of 10^{19} cm^{-3}. At this point we have an ohmic contact irrespective of the other terms in the expression. If the doping is high enough, contact with any metal will result in an ohmic contact.

Deviations from expected ohmic behaviour can be caused, first, by an interfacial layer. Second, the metal may—in the deposition process and any thermal treatment thereafter—appreciably compensate the doping of the semiconductor, it may even produce a heterojunction by forming another semiconducting compound. Third, an amorphous semiconductor surface must regrow upon thermal treatment to perfect crystallinity.

On the other hand, the doping by a contacting metal can be used to advantage when an ohmic contact to a low doped semiconductor is intended. In some metal–semiconductor combinations, e.g. Si and Al, the temperature in the annealing process does not suffice for appreciable diffusion of Al into Si, but still a metal doped layer is formed between the two.[9] In this case the silicon dissolves in the aluminium and upon cooling epitaxially regrows on the bulk with inclusion of *p*-doping aluminium.

The interfacial layer. In the previous discussion we have assumed that metal and semiconductor directly join at the junction. But most semiconductors have a natural oxide at the surface, for example Si has 25 Å of SiO_2 grown on its surface when exposed to air at room-temperature. If such an oxide layer is not destroyed or removed in the deposition process, then it will certainly affect the contact properties. Figure 6 shows the band structure diagram when a thin insulating film separates the metal from the semiconductor. First we expect the surface state density to decrease. Consequently the Fermi level will be less strongly pinned and coupled. The extreme case would be the MOS structure, where the insulating oxide exceeds 500 Å. Second, the additional oxide barrier will reduce tunnelling currents in both forward and reverse direction. Third, whereas in pure Schottky contacts we need only consider majority

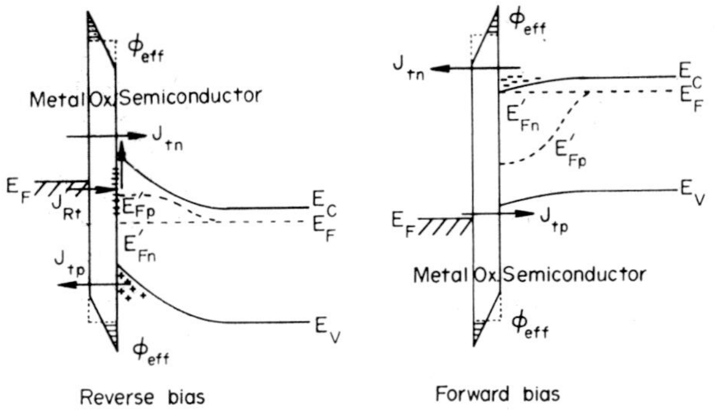

Figure 6. Band structure diagrams for Schottky barriers with a thin interfacial oxide, for reverse and forward bias.

carriers, in this case minority carrier injection will participate appreciably in the current transport mechanism. Tunnelling from the metal to surface states will also modify the current-voltage characteristic's sensitivity to image force barrier lowering. For ohmic contacts one cannot derive any advantage from the additional effects of the interfacial layer because it introduces a series impedance. But for some applications of Schottky barriers a stabilization of device properties can be obtained especially where an increased minority carrier injection is essential for the process as for instance in electroluminescence from Schottky diodes.[10]

A good demonstration of a thin SiO_2 film influencing the I–V-characteristics of a Schottky barrier is shown in Figure 7 observed by Kar[11] for SiO_2 film thicknesses between 10 and 47 Å. For thin films the slope in forward direction corresponds well to the value predicted by thermionic-emission theory with a factor $n = 1.04$ but deviates considerably as the thickness increases. Also the current in both directions decreases. The transition from a Schottky barrier to an MOS structure occurs for oxides in the region 30–40 Å. The saturation of the reverse current in the region above 30 Å is due to an interdependence between the extraction of minority carriers from an inversion layer and the then increasing effective tunnelling barrier. There are applications of Schottky barriers, where the properties of an interfacial layer are desired. In such a case the preparation of a usable interfacial layer is preferable over just using the natural oxide that forms on almost all substances. The oxides can be produced by boiling the semiconductor in water or in H_2O_2, or by dipping it in an oxidizing acid, or by applying a thin anodic oxide, or by thermal growth as used in silicon

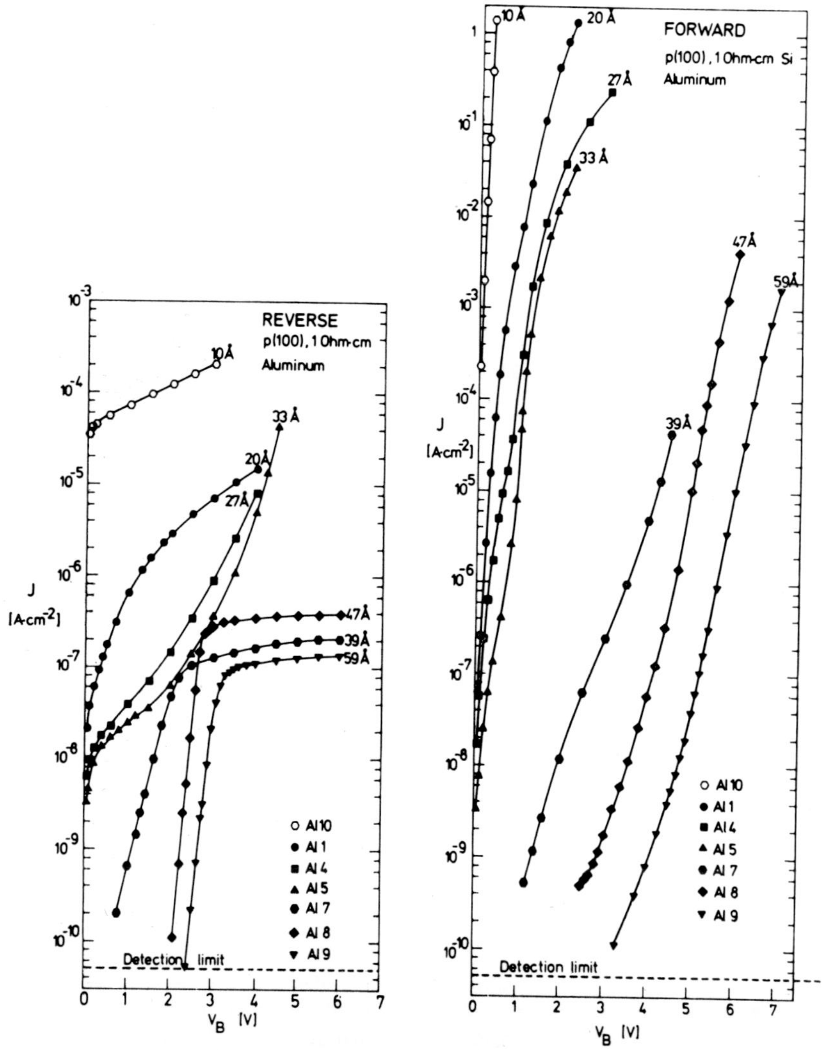

Figure 7. Current voltage-characteristics measured on Al–Si diodes with thin SiO_2 films of variable thickness on the Si.[11]

MOS technology. Other surface films than oxides have not been investigated yet. It is understood that the metal deposition on to the thin oxide layer must not damage it, nor should a reaction with the oxide occur in the post-deposition thermal treatment. An example is the application of aluminium, which at moderate temperatures (400°C) reduces the SiO_2 and forms Al_2O_3.

The metal deposition in view of the two possible requirements: Barrier or ohmic contact. In Table 2 the deposition methods, their advantages and disadvantages as well as their applications have been compiled. There are no clear-cut limits to either of the possible methods. Very often there are practical preferences of one method over another. Some of the ways applicable in a laboratory experiment are too inefficient for device production.

No reliable information is available on the effects of the crystalline structure of the deposited metal on the *M–S* barrier height. Since the choice of the metal allows only limited variations it is very unlikely that structural effects have any significance. If they are present they would surely be masked by metallurgical reactions at the interface.

(i) Evaporation from a wire, boat or crucible made of refractory metals, ceramics, graphite or compounds, like boron nitride, can be performed with the least investment in costly equipment. If purity conditions can be met, this method has the least influence on contact properties. There is little surface heating, low particle energy, and most of the common metals can be evaporated. Deposition of alloys is only possible with flash-evaporation or evaporation to dryness, it is inferior to other methods in this respect. Practical limitations exist in deposition rate, in economy, in alloying of the evaporant with the boat, and in thickness of the film obtainable with one loading. For integrated circuits in MOS technique, sodium free aluminium deposition is mandatory, which is not guaranteed by evaporation from a boat.

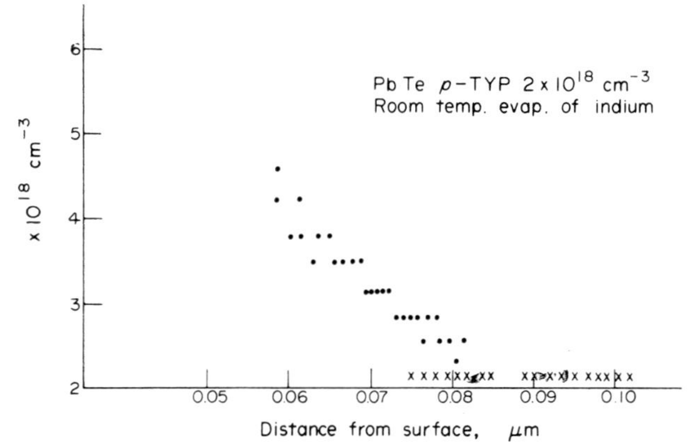

Figure 8. Influence on the doping of PbTe by an evaporated. Indium contact.

For some critical applications, even the radiative heat from an evaporation boat causes diffusion of the metal into the lattice and thereby changing the doping profile. In these cases the semiconductor material must be cooled, conveniently to liquid nitrogen temperature. An example[12] is presented in Figure 8 where indium is evaporated on to lead telluride at room-temperature where it shows a clear increase in doping concentration towards the surface. No such effect is observed when the substrate is cooled to liquid nitrogen.

(ii) Electron beam evaporation results in films of highest purity. Even refractory metals can be evaporated. For high rates the heat in the system increases less than in boat evaporation. There are no limits in thickness because of large metal reservoirs. The semiconductor substrate is exposed to back-scattered electrons[13] and to X-rays. The radiation-damage

Table 2. Deposition methods of metal films on semiconductors.

Deposition method	Metals	Advantages	Disadvantages	Application
Resistance heating (Boat, filament, crucible) Flash evaporation Inductive heating	Limited melting-point not for alloys Alloys	No influence on substrate	Low rate some alloying with boat, subject to impurities from boat, limited thickness	Exotic narrow gap semiconductors Gunn devices
Electron beam evaporation	All metals (alloys+)	High rate highest purity of all methods convenient and economical	Backscattered electrons, X-rays	Integrated circuits production
Sputtering	All metals esp. refractory metals alloys	Permits sputter cleaning, high sticking coefficient good step coverage	Amorphous layers poly cryst. layers gas inclusion	Beam lead multiple layers ohmic contacts with reactive metals Schottky contacts without interfacial layer
Plating	Ni, Pd, An, In, Cu	Simple method no heating selective deposition unlimited thickness	Purity of plating solutions may be not sufficient	Gunn devices
Chemical vapour deposition	Almost all common metals	Unlimited thickness high depos. rate	Possibly unwanted reaction products	Power devices

caused is clearly detectable in MOS devices, but is there eliminated at the usual alloying temperatures (<300°C). Any detrimental effects to the metal–semiconductor junction have not been discovered as yet. In soft materials there might be a limitation when technology in these materials is further developed.

(iii) Sputtering is most suitable for the deposition of the refractory metals Pd, Pt, W also for Ti. It is the best method for depositing alloys because it retains stoichiometry and produces homogeneous films. Sputtering permits also the cleaning of the semiconductor surface by bombarding the substrate with 500 V A^+ ions. Interfacial layers can be removed. However, it is inevitable that radiation damage already ensues if the semiconductor is struck by charged particles of relatively low energy (500 eV). In the late sixties it was generally believed that radiation as contained in a sputtering system must impair the functioning of a semiconductor device. Therefore sputtering systems have been built, which greatly reduce the exposure to charged atoms by either avoiding a field between target and substrate or by neutralizing the ion beam. Such systems also do not heat up the substrate as much as the conventional parallel plate configurations.

In many of the unorthodox sputtering geometries the ability to backsputter the semiconductor is lost. The backsputtering process is important for cleaning the surface without reforming an oxide, and immediately depositing the metal. This is common practice in metal contact deposition, expressly eliminating an interfacial layer. Backsputtering causes two changes in the surface: for once it redeposits already sputtered-off material that is reflected from the anode (the target in this case) and also deposits anode material. Second, it converts the crystalline surface into an amorphous one.

The influence of an amorphous layer intentionally produced by ion implantation has been investigated with Al on Si and Pd on Si. The presence of an amorphous silicon layer produces pronounced interdiffusion of silicon and aluminium atoms and a reordering of the amorphous layer at 350°C which is significantly below the Al–Si eutectic temperature (577°C) and the normal recrystallization temperature of Si (600°C).[14] Palladium forms a uniformly reacted layer of PdSi after annealing at 200°C regardless whether the underlying silicon is crystalline or amorphous.[15]

In Pt–Si ohmic contacts a direct relationship between electrical properties and morphology exists.[16] The micro structure is controlled by deposition parameters: PtSi-films with the lowest contact resistance could be formed by first sputter cleaning the sample and then sputter depositing platinum at 450°C. Heat treatment *in situ* at 750°C after rather than during deposition produced better electrical contacts than the same treatment external to the deposition chamber. In a comparison of samples only chemically cleaned with those also sputter etched the latter consistently had superior electrical properties. The inherently better perimeter coverage of contact holes in a sputtering process is considered an advantage over evaporation, especially on highly doped *n*-type silicon when a rectifying Schottky contact is intended. Similarly superior ohmic contacts with a W–Si system have been obtained by sputtering after sputter etching.[7] Metallization with metal-alloys, like Ti–W, where both partners react with the silicon to form complex silicides can only be deposited by sputtering. In the near past the growing importance of III–V-semiconductors for device applications has spurred work on contacting these materials.[18] Their relatively low decomposition temperature (640° for GaAs) forbids high temperatures during deposition or in the following annealing process. However, sputtering has been applied and found to be superior to evaporation in producing ohmic contacts on GaAs with a Au–Ge alloy.[19] The use of an alloy for metallization to compound-semiconductors lowers the alloying temperature and achieves higher doping levels, required for good ohmic contact. In the deposition of alloys sputtering is far superior to all other methods. Besides possible radiation damage that may not be removed in a post deposition annealing, sputtering can heat up the substrate more than is desirable in compound semiconductors. But a well designed sputter-system and measures, like potential free substrate, permit close control of the substrate temperature. It can certainly be held below 300°C without much effort. Compound semiconductors of which one component is sputtered preferentially, like the sulphur in CdS[20] exposure to a glow discharge strongly alters the electrical properties by creating an excess of one partner. A limited way to balance sputtering rates exists in choosing a different sputtering gas, for instance krypton instead of argon.

(iv) The plating of contacts in device technology is not very widespread in spite of its simplicity. The equipment needed is almost negligible in cost compared with all the other techniques mentioned. Some of the plating solutions are rather complex and for this reason a source of contamination for the process intended. There is no heating involved and the metal can selectively be deposited, which may reduce the necessary number of photolithographic—and etching processes in device manufacture, all these are advantages that come to bear with exotic semiconductors like lead–telluride, lead–tin–telluride, Cd Hg Te, etc.

The method has not only been used for ohmic contacts in rather demanding applications[21] but also for Schottky contacts.[22]

(v) Chemical vapour deposition is a well known technology in semiconductor device processing, however, mostly for insulating films or polycrystalline semiconductor layers. It is much less considered for depositing metal films for contacting purposes on semiconductors. A few characteristic parameters of this process should be pointed out. The wafer temperature has to be raised to effect the deposition. It determines the crystalline structure of the metal, the rate of deposition, the adhesion, the formation of unwanted reaction products and the purity of the metal film. With the knowledge of the maximum temperature, the semiconducting material can tolerate, it is easy to decide whether this method is applicable and which source material must be used. Among the reaction products there should be none harmful to the substrate.

The CVD-metal layers usually have a very good adhesion, they are relatively pure, and there is no thickness limit. Consequently their application is called for when films of several μm thickness are required. Almost all metals usable for contacts can be deposited, especially those of high melting point.

Those processes that employ a reducing reaction, would be suitable to remove an oxidic interfacial layer. The use of metal halides as the metal source, produces an etching effect which contributes to the good mechanical and electrical contact. Metallization of silicon power devices with tungsten has recently been announced.[23] In many instances the substrate temperature during the process will suffice to complete the alloying reaction.

Assessment of the possible radiation damage caused by the deposition of the metal. Two effects are caused by radiation damage

(a) electronic redistribution (trapping-recombination)

(b) atomic rearrangement.

The first of these subjects can be shown to occur when electron beam evaporation is applied to MOS devices. The second is the cause for an amorphous surface in a sputtering process.

Most models of radiation damage assume elastic collisions as if the striking particle is neutral. The damage consists of a vacancy-interstitial pair. But when a charged particle or a high energy photon penetrates into the lattice, the energy loss will occur in form of ionization and trapping (inelastic collision). In such a case, the size of the damaged region with respect to electrical parameters is much larger. We define a threshold for displacement T_d as the minimum energy a lattice atom must receive to be displaced to a stable interstitial position. For silicon this is in the range 20–30 eV. The values published differ considerably. Table 3 taken from compilations of Aukermann[24] gives values for T_d and the minimum electron energy in keV which can be imparted to the semiconductor material. It also lists the energies an argon ion must have to dislodge an atom from a lattice site. The electron energies necessary are far above those encountered in electron beam evaporation. Therefore no displacement can occur. In a sputter and especially in a sputter cleaning process the energies involved are above the threshold. If we assume sputtering voltages of 500 V and A^+ ions we obtain collision energies of 486 eV with Si, and similar values with other semiconductor materials. Here we must expect atomic rearrangement.

Table 3. Threshold energies for displacement of atoms from their lattice site.[24]

Substance	Si	Ge	GaAs		InSb		CdS		
Displaced atom	Si	Ge	Ga	As	In	Sb	Cd	S	
Threshold energy	15.9	14.5	8.8	10.1	6.4	8.5–9.9	7.3	8.7	eV
Minimum electr. energy	173	355	228	273	247	286	290	115	keV
Minimum argon energy	16.4	15.8	9.5	11.14	8.3	12.0	9.4	8.8	eV

A rough estimate of the maximum depth of the amorphous layer can be obtained from ion implantation tables. It is expected to be of the order of 30 Å for an argon beam at 500 V in silicon. This is still in a range where conduction electrons may tunnel through, and it can be considered as a kind of interfacial layer. In some cases of compound semiconductors, e.g. CdS, where one partner, the sulphur is, sputtered preferentially a layer of about 20 Å non-stoichiometric material with changed conducting properties (*p*-type) results. In most of the processes in which a semiconductor is contacted by a metal a thermal treatment of some sort follows the deposition. From very sensitive MOS measurements, we know that already at 300°C any radiation effects in SiO_2 from *e*-beam evaporation are eliminated. In the same way, damage to the lattice by sputtering can be removed by heat treatment. But the meaning of recovery has to be specified more closely. Figure 9 shows an isochronal annealing process in which already at 250°C the final stage to recovery is reached, 2% below the value before irradiation. The critical property in this experiment is the conductivity. But the experiment is, for example, not concerned with minority carrier lifetime restauration, which could play a role in photoluminescence applications. From ion implantation it is known, that to restore minority carrier lifetime much higher temperatures, up to 1000°C, are required. At high doses and damage the annealing may never restore the surface to its original state. LEED analysis provides a means to examine the restoration of the surface.

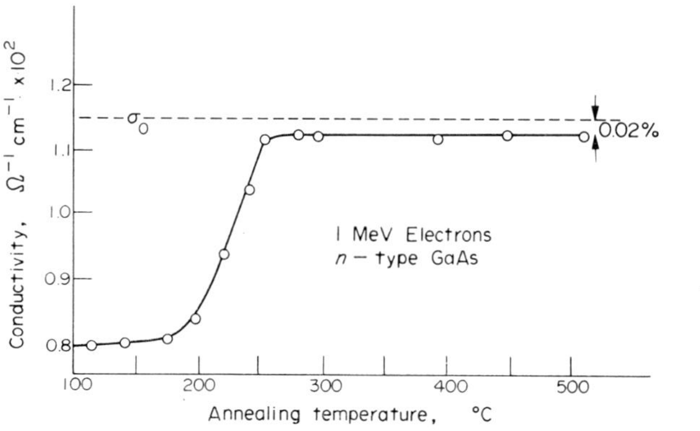

Figure 9. Isochronal annealing process of GaAs irradiated by electrons of IMe V.[24]

Metallurgical effects

In the previous sections we encountered a number of metallurgical reactions, for instance the formation of silicides, the solution of silicon in the aluminium. Another example is the application of multilayer ohmic contacts:[25] Ti–Al–Pt on CdS. The titanium is in this system called the preparative metal. It reacts with the interfacial layer and getters the adsorbed gas molecules. The aluminium as active metal forms the desired ohmic contact. The Pt serves as a protective cover.

With some semiconductor–metal combinations the alloy regrowth technique does not only result in a highly doped semiconductor layer underneath the metal, but actually forms a heterojunction. In this way excellent blocking 'contacts' have been obtained with tellurium on CdS[26] of the constituents CdTe–CdS. Copper on CdS produces a heterojunction of Cu_2S–CdS which finds application in large area solar cells.[27] Two metallurgical measures taken to prevent the deterioration of *M–S* contacts should be mentioned. The first concerns the solution of Si from the contact areas into the covering aluminium[28] which causes formation of shorts in underlying *p–n* junction.[29] An addition of 2% Si to the aluminium does effectively suppress the migration of silicon away from the contact area. Similarly a small percentage of Cu (2%) in the aluminium prevents electromigration.[30] Because of its low atomic weight aluminium is transported in direction of the electron flow in amounts proportional to the current density. At the edges of contact holes, the metallization usually thins out, so that there the current density is higher. Therefore these edges are especially subject to metal transport and, as a consequence, to electrical failure. The copper in the aluminium is preferentially positioned at the grain boundaries and blocks the A1-atoms from migrating across. The deposition of these metal alloys poses problems, because they only transfer in the intended ratio either flash evaporated or sputtered. A method that seems to evolve lately is electron beam evaporation with the beam jumping between two crucibles one with aluminium the other with silicon or copper, the dwell time of the beam in the respective troughs determines the composition of the alloy.

Summary

The conditions for obtaining ohmic and Schottky contacts have been explained using the band structure diagram. For narrow gap, and predominantly ionic semiconductors, the electronegativity of the materials, the energy gap of the semiconductor and the metal work function difference determine whether the *M–S* contact is ohmic or has a blocking character. In contrast to this behaviour, covalent semiconductors exhibit a barrier height which is essentially independent of the metal used. The high surface state density pins the surface level at the interface. This surface state density is modified if an interfacial layer, mostly an oxide, is present. This layer also introduces minority carrier injection and a decrease in the tunnelling current. Depending whether an ohmic contact, a clean Schottky barrier, or one with an interfacial layer is desired one has to choose a reactive metal for the first two and a nonreactive metal for the last type of contact and a deposition technique commensurate with the intention to destroy or to maintain the interfacial layer. For achieving an ohmic contact the semiconductor has to be highly doped to reduce the barrier width to dimensions thin enough for tunnelling of charge carriers. If the carrier density is not high enough, deposition of a suitable metal and the proper thermal treatment can lead to additional doping of the boundary zone, so that the above condition is met. If a metal deposition process produces radiation damage, the thermal treatment must anneal the damage completely. The destruction of the crystal lattice at the surface by ion implantation, by backsputtering or by the sputtering of the metal itself furthers the metallurgical reaction occurring in the thermal treatment, the forming of silicides or alloy regrowth.

It is not possible to derive a general recipe for obtaining a metal semiconductor junction of specific properties. There are too many variables and uncertainties in the system which depend on the metal, the substrate to be contacted and the treatment before and during processing. The list of possible effects mentioned is by no means complete. The intention of this paper can only be to provide guidelines along which the proper contacting of semiconductors can be attempted.

References

[1] F A Padovani, in *Semiconductors and Semimetals* Vol. 7a, p 75. Edited by R K Willardson and A C Beer, Academic Press, New York (1971).
[2] E H Rhoderick, in *Institute of Physics Conference Series No.* 22, p 3, London (1974).
[3] M Schulz, in *Proc European Summer School; Scientific Principles of Semiconductor Technology*, p 191. Edited by H Weiss, Bad Boll (1974).
[4] B Schwartz, (Editor) *Ohmic Contacts to Semiconductors*, Electron Div. Electrochem Soc, New York (1969).
[5] K W Nill, J N Walpole, A R Calawa and T C Harmain, in *The Physics of Semimetals and Narrow Gap Semiconductors*, p 383. Edited by D L Carter and R T Bates, Pergamon Press, New York (1970).
[6] S Kurtin, T C McGill and C A Mead, *Phys Rev Letters*, **22**, 1969, 1433.
[7] C A Mead, W G Spitzer, *Phys Rev A*, **134**, 1964, 713.
[8] J M Andrews, *J Vac Sci Technol*, **11**, 1974, 972.
[9] H C Card, in *Metal-semiconductor Contacts*, Conference Series Number 22, p 129. The Institute of Physics, London (1974).
[10] S Y Haeri and E H Rhoderick, in *Metal semiconductor Contacts*, Conference Series No. 22, p 84. The Institute of Physics, London, (1974).
[11] S Kar, private communication.
[12] J Baars, D Bassett and M Schulz, ESSDERC, Munich (1976).
[13] P Sommerkamp, *Z Angew Phys*, **28**, 1970, 220.
[14] D H Lee, R R Hart and O J Marsh, *Appl Phys Lett*, **20**, 1972, 73.
[15] D H Lee, R R Hart, D A Kiewit and O J Marsh, *Phys Stat Sol*, **15**, 1973, 645.
[16] R M Anderson and T M Reith, *J Electrochem Soc*, **122**, 1975, 1337.
[17] V Kumar, *J Electrochem Soc*, **123**, 1976, 262.
[18] V L Rideout, *Solid State Electron*, **18**, 1975, 541.
[19] A Y C Yu, H I Gapen and R K Watts, Technical Report, AFAL-TR-70-196, Sept, 1970.
[20] W Ruppel, in *Semiconductors and Semimetals*, Vol 6, p 328. Edited by R K Willardson and A C Beer, Academic Press, New York, (1970).
[21] J A Copeland and S Knight, in *Semiconductors and Semimetals*, p 651. Edited by R K Willardson and A C Beer, Academic Press, New York (1971).
[22] F H Dörbeck, *Solid State Electron*, **9**, 1966, 1135.
[23] P Batzies, H Haas, *BBC Nachrichten*, in press.
[24] L W Aukermann, in *Semiconductor and Semimetals*, Vol 4, p 371. Edited by R K Willardson and A C Beer, Academic Press, New York (1968).
[25] K W Böer and R B Hall, *J Appl Phys*, **37**, 1966, 4739.
[26] R W Dutton and R S Muller, *Solid State Electron*, **11**, 1968, 749.
[27] F A Shirland, *Adv Energy Conversion*, **6**, 1966, 201.
[28] J O McCaldin and H Sankur, *Appl Phys Letters*, **19**, 1971, 524.
[29] H Murrmann and L Schleicher, Metalloberfläche, *Angew Elektrochem*, **28**, 1974, 121.
[30] J R Black, *Proc IEEE*, **57**, 1969, 1587.

2.2 Substrate surface damages by rf-sputtering

S Berg, L P Andersson, H Norström and E Grusell, *Institute of Technology, University of Uppsala, Uppsala, Sweden*

Electrical properties of (111) surfaces of n-type silicon exposed to sputtered metal atoms deposited in a rf-diode sputter system were studied. The metal–silicon junction formed by this process was investigated to evaluate surface properties of the junction. It was shown that the sputtering process introduced damage to the surface, causing charged centres to appear at the surface of the crystal. A theory is presented which explains most of the experimental results of the so damaged Schottky junction. Thermal activation energies are deduced from experimental data and compared to expected values predicted by the theory presented. The energies are found to vary between 0·3 and 0·7 eV depending on sputtering conditions.

1. Introduction

It is well known that the properties of a thin film depend strongly on the sputtering or evaporation deposition parameters as well as on the original substrate conditions. However, drastic changes of surface condition of the substrate during deposition may also be important. Investigations of such phenomena have mostly been reported from ion implantation technology[1] where high energy ions are expected to damage the implanted surface region. However, much less seems to be reported of substrate damage caused by low energy particles, e.g. originating from a discharge plasma.

Sputtered metal contacts on *n*-silicon do not always show the good rectifying properties of evaporated contacts[2–4] if they are not sputtered at a low power.

Mullins and Brunnschweiler[3] suggest that during sputter deposition a high density of 'donor-like' traps is created in the surface region. The state density can be assumed to decrease exponentially into the bulk. This high density of donor centres creates a thin region ($0 < x < d$) of very high electric field. Padovani and Stratton[5] have shown that under these circumstances tunnelling can occur through the barrier. Furthermore, it is expected that the main contribution to the current comes from electrons tunnelling at an energy E_{00} smaller than the energy $q\phi_{B0}$ at the top of the barrier.[6] This situation is schematically drawn in Figure 1, for a reverse biased metal *n*-type semiconductor contact.

Variations in surface trap densities will cause a change in the position of E_{00}. A low surface trap density gives $E_{00} \simeq q\phi_{B0}$ but increasing surface trap density gives a decreasing value of E_{00}. (111) surfaces of high resistivity silicon wafers are exposed to a discharge plasma during sputter deposition of Au-atoms. By changing the sputtering parameters from one experiment to the other we can influence trap densities of the so formed metal–silicon junction. By measuring the reverse current of these junctions at different temperatures it is possible to evaluate the value E_{00}.

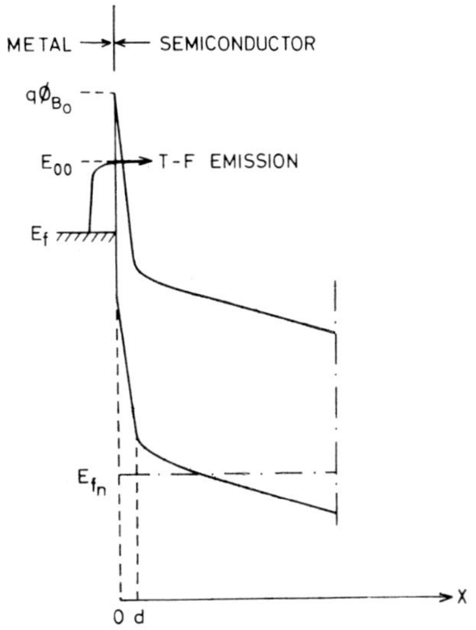

Figure 1. Schematic energy diagram of a reverse biased sputter-damaged Schottky barrier. The damaged region is $0 \leq x \leq d$. The thermionic field emitted electrons are indicated by the arrow. E_f is the Fermi level in the metal and E_{fn} is the quasi-Fermi level for electrons in the semiconductor.

2. Fabrication process

n-type silicon of resistivity $\simeq 10$ kΩ cm was used. The samples were $\simeq 0.2$ mm thick and were subjected to mechanical polishing and chemical etching in CP6 prior to the sputtering. The gold sputtering on the (111) surface was performed in an MRC rf-sputtering system model SEM-8620. After this, metal contacts were evaporated on the other side of the samples. As these contacts were forward-biased when the reverse currents of the sputtered contacts were measured, their characteristic did not influence the measurements.

3. Theory

We assume that in the sputtering process donor-like traps are created near the surface. The concentration $N_t(x)$ of these traps at a distance x from the surface is assumed to be

$$N_t(x) = N_{ts} \exp\left(-\frac{x}{L}\right) \quad (1)$$

where N_{ts} is the trap density at the surface and L is a characteristic length. L and N_{ts} depend on the sputtering voltage and on the sputtering time.

When the metal *n*-silicon contact is reverse biased, the quasi Fermi level (E_{fn} in Figure 1) for electrons is assumed to be constant throughout the depletion region. Moreover, at a bias of a few volts all traps at the surface are far above this quasi Fermi level and are thus positively charged.

The reverse current density of this device can be written

$$J_s = \int_{-\infty}^{\infty} J_{ss}(V)\mathrm{d}V \qquad (2)$$

where $J_{ss}(V)\,\mathrm{d}V$ is the current density contribution from electrons with energies in the interval qV and $q(V+\mathrm{d}V)$. Equation (2) can also be written in the form

$$J_s = \int_{-\infty}^{\infty} f(V).T(V)\mathrm{d}V \qquad (3)$$

where $f(V)\,\mathrm{d}V$ is the current density in the metal caused by electrons moving towards the semiconductor with an energy in the interval between qV and $q(V+\mathrm{d}V)$. $T(V)$ is the transparency of the barrier for electrons with energy qV.

To calculate J_s from equation (3) we need expressions for $f(V)$ and $T(V)$. Calculations based on statistical mechanics give the following expression for $f(V)$.[7]

$$f(V) = \frac{4\pi q^2 mkT}{h^3} \cdot \frac{1}{1+\exp\left(\dfrac{qV-E_f}{kT}\right)} \qquad (4)$$

where q is the electron charge

m is the electron mass

k is the Boltzmann's constant

T is the absolute temperature

h is Planck's constant

E_f is the Fermi level in the metal

qV is the electron energy. (For convenience we choose $E_f = 0$.)

$T(V)$ can be calculated by the *WKB* approximation method.[7] We then get

$$T(V) = \exp\left\{-4\pi \frac{\sqrt{2mq}}{h} \int_{x_1(V)}^{x_2(V)} [\phi(x) - V]^{\frac{1}{2}}\mathrm{d}x\right\} \qquad (5)$$

where $q\cdot\phi(x)$ is the potential energy of an electron at rest in the conduction band of the semiconductor at a distance x from the junction and $x_1(V)$ and $x_2(V)$ are found from $\phi(x_1) = \phi(x_2) = V$. The other symbols have the same meaning as above.

We assume $T(V) = 1$ for $V > \phi_B$ where ϕ_B is the maximum of $\phi(x)$. To calculate $\phi(x)$ near the junction we solve the Poisson equation with the charge density given by

$$qN_t(x) = qN_{ts}\exp(-x/L) \qquad (6)$$

When the effect of image force lowering is included the solution for $\phi(x)$ is

$$\phi(x) = \phi_{B0} - \frac{qN_{ts}}{\varepsilon\varepsilon_0}L^2[1-\exp(-x/L)] - \frac{q}{16\pi\varepsilon\varepsilon_0 x} \qquad (7)$$

where ε_0 is the dielectric constant of vacuum and ε is the relative dielectric constant of silicon. The expression (7) for the potential $\phi(x)$ holds only for small x where $N_t(x) \gg N_d$, the donor concentration of the substrate. The barrier height ϕ_{B0} is assumed to have the same value as for an evaporated contact, i.e. $\phi_{B0} = 0.8$ V.[8,9] In Figure 2 calculated values of $\phi(x)$ and $J_{ss}(V)$ are plotted for $N_{ts} = 2.55 \times 10^{25}$ m^{-3} and $L = 60$ Å. Figure 2 also illustrates that $E_{00} < q\phi_B = q(\phi_{B0} - \Delta\phi)$ where $\Delta\phi$ represents the barrier lowering.

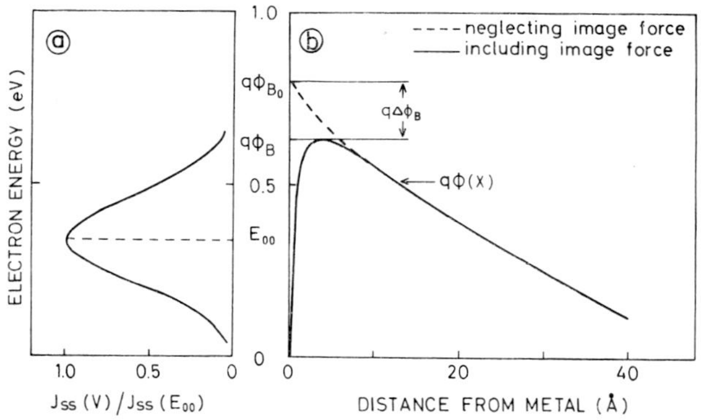

Figure 2. (a) Normalized fractional current as a function of electron energies. E_{00} denotes the most probable emission energy. (b) Electron potential energy as a function of distance from metal. ϕ_B is the maximum of $\phi(x)$.

4. Experimental results and discussion

Diodes were fabricated by sputtering in argon atmosphere of 0.6 Pa (5×10^{-3} torr) with the following sputtering voltages and times:

1. Sputtering of gold with 2500 V accelerating voltage for 30 s.
2. Sputtering of gold with 2500 V accelerating voltage for 20 s.
3. Sputtering of gold with 1500 V accelerating voltage for 90 s.

The sputtering time was chosen to give equal amount of deposited gold on samples from group 1 and 3. The reverse current–voltage characteristic at different temperatures measured on a representative sample from group 1 is shown in Figure 3.

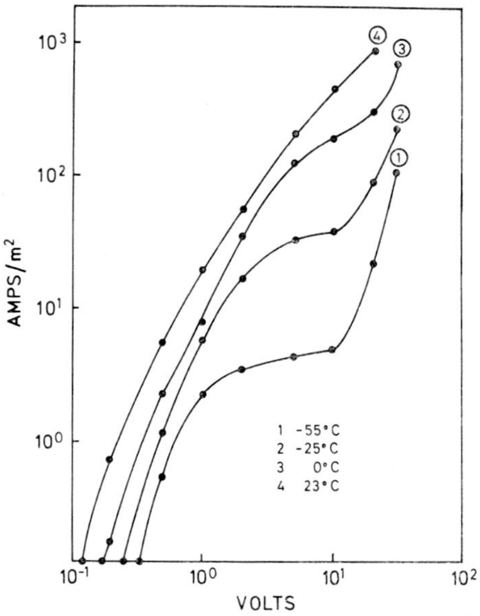

Figure 3. Reverse-current–voltage characteristics for a Schottky junction from group 1 experiment showing the effect of sputter damage. Sputtering was made by gold with 2500 V accelerating voltage in 30 s.

The reverse currents at different temperatures and at 10 V reverse bias for diodes from all three groups are shown in Figure 4. Also shown is a diode where deposition of gold was done by thermal evaporation. Calculations based on equation (3) give the theoretical curves also included in the figure. To obtain the best fit to the experimental results the following values were obtained:

Group	N_{ts} m^{-3}	L Å	E_{00} eV
1	2.55×10^{25}	60	0.34
2	2.0×10^{25}	60	0.49
3	1.0×10^{25}	60	0.70

It is thus clear that increasing sputtering voltage and/or sputtering time will cause a decrease in E_{00}.

However, a surface represents the abrupt end of a more or less perfect lattice. Surface regions will therefore always exhibit a high density of recombination centres. If these centres are assumed to be distributed according to equation (1) with $N_{ts} \leq 0.1 \times 10^{25}$ m^{-3} and $L \leq 60$ Å no significant field emission through the barrier can occur. The effective barrier for electrons penetrating into the silicon from the metal would thus appear to be $\simeq \phi_{B0}$.

Generation of carriers will however take place in this region due to the relative high density of generation centres. Thus in a surface region of a distance $\simeq 5 \times L \simeq 300$ Å a considerable generation current will be created. The behaviour of this generation current has been described by a modified Schockley–Read–Hall process.[4] In this very near surface region it has been shown that the temperature activated energy is $q\phi_g \simeq q\phi_B/2$ rather than $E_g/2$ (half the bandgap) as expected from simple Shockley–Read–Hall theory.

The above mentioned generation effect explains why results from an evaporated contact (the dotted curve) in Figure 4 exhibits a 'knee' at $T = T_c$. If $T < T_c$ generation of carriers within the surface region gives the dominating contribution to the total current. When $T > T_c$, thermionic emission over the barrier dominates.

A comparison of field emitting properties of sputtered and evaporated contacts should thus be made only in the region $T > T_c$ where such properties are expected to dominate. The activation energy 0.76 eV $\simeq q\phi_B$ found for the evaporated contact shows that the value of N_{ts} is so small that no significant field emission takes place. On the other hand, at $T < T_c$ one finds that N_{ts} is big enough to give a measurable generation current component.

Sputtered contacts have larger N_{ts}-values than evaporated contacts. This should then give rise to a larger generation current for sputtered contacts. However, N_{ts} influences the field emitting properties even stronger, thus this effect completely dominates in the whole temperature region measured. Only group 3 contacts (smallest N_{ts}-values) at low temperatures can be expected to show some minor influence of the generation current component.

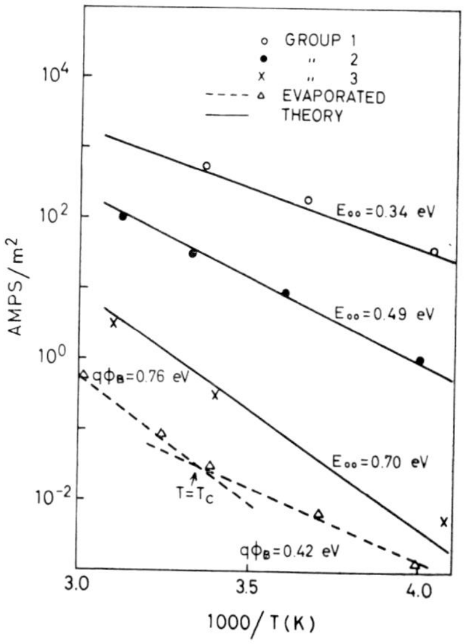

Figure 4. Current measurements at different temperatures for samples sputtered under different conditions. Also is shown a typical sample of an evaporated contact. In the figure are drawn theoretical lines for the best fit to the measured results.

Conclusion

We have shown that thermionic field emission through a thin damaged region gives the main contribution to the reverse current in Au-sputtered metal–*n*-silicon contacts. Most of the electrons are tunnelling at an energy E_{00}. The value of E_{00} is decreasing for increasing sputtering voltage and/or sputtering time. A detectable surface damage also exists on chemically etched surfaces not exposed to a plasma discharge.

References

[1] J W Mayer and O J Marsh, *Applied Solid State Science*, Vol 1, Academic Press, New York (1969).

[2] A R Wolter, *Ohmic Contacts to Semiconductors*, p 214. Electrochem Soc, New York (1969).

[3] F H Mullins and A Brunnschweiler, *Sol State Electron*, **19**, 1976.

[4] P A Tove, S Berg, L P Andersson and A Hyder, *Proc IEEE Conf Electron Device Techniques*, p 67, Sept. 1970.

[5] F A Padovani and R Stratton, *Sol State Electron*, **9**, 1966, 695.

[6] J M Shannon, *Appl Phys Lett*, **24**, 1974, 369.

[7] A van der Ziel, *Solid-State Physical Electronics*, Prentice-Hall, New York (1968).

[8] S Berg and L P Andersson, *Nucl Instr Meth*, **114**, 1974, 241.

[9] C A Mead, *Sol State Electron*, **9**, 1966, 1023.

2.3 Investigations of metal contacts to amorphous evaporated Ge films

M Hafiz*, E Mgbenu†, P A Tove, H Norde and S Petersson, *Electronics Dept., Institute of Technology, University of Uppsala, Uppsala, Sweden*

Amorphous Ge films are used as non-injecting ohmic contacts to high-resistivity n-silicon radiation detectors, but the function of this contact is not yet fully explored. One part problem is the role of the metal films used as external contacts to the amorphous film. In this paper we investigate the function of different contacting metals, such as Au, Al, Cr by measuring the I–V-characteristics of sandwich structures with two metals on both sides of the amorphous evaporated Ge film (of typical thickness 1 μm). It was found that while the symmetric Cr–Ge–Cr structure (also Au–Ge–Au) had low resistance (leading to resistivity values of $\simeq 4 \times 10^4$ Ωcm for the Ge film), the unsymmetric structures e.g. Al–Ge–Cr and Al–Ge–Au, showed higher resistance. This was again found for the Al–Ge–Al structure, which also showed some polarity dependence. Rutherford backscattering was used to investigate diffusion effects between the layers; also the microstructure of the Ge films was investigated by electron microscopy.

1. Introduction

Amorphous Ge is used as non-injecting back contact to monocrystalline silicon detectors.[1] Metal films are then employed as external contacts to the Ge. Metallic contacts to amorphous Ge have been the subject of several investigations. Many problems remain unresolved; for example, it is not clear what is the origin of the currents observed in the I–V-characteristic measurements. Walley[2] believed that bulk phenomena play a dominant role. Another school[3] ascribes a major role to the contact. This appears borne out by measurements of Morgan and Jonscher[4] which show that the pressure at which the contact is evaporated has most influence on the observed currents, rather than the pressure for the Ge evaporation.

Another source of controversy is the part played by oxygen or other impurity in the Ge film, on the I–V data. A recent view is that provided the pressure for Ge evaporation is better than 10^{-5} torr, no bulk oxidation of the amorphous Ge film occurs.[5]

Detailed comparison between measurements from different sources is difficult. This is because too many other factors influence the results. These include the rate of Ge evaporation, the substrate temperature during evaporation and the evaporation method (e.g. filament, electron gun) or even sputtering in which case the atomic velocities of the Ge atoms impinging on the substrate are higher. Another contributing parameter is the porosity of the film. This appears to influence the time stability of the film, and is dependent on the obliqueness of the evaporation.[6]

In these investigations we aim to throw some light on the role of contacts on the dc I–V-characteristics. Different contacts (e.g. Au, Al, Cr) to Ge were employed in sandwich structures and measurements were made at room temperature. The thickness of the Ge film is known not to influence the resistivity, provided it lies within the range 300–5000 Å.[7] Diffusion effects and the presence of oxygen or other impurity within the Ge film were investigated using Rutherford backscattering, while the structure of the evaporated film was obtained from electron microscope studies. The latter is essential because of doubts as to whether Ge deposited on cold substrates is truly amorphous or composed of small crystallites.[8]

2. I–V-characteristics

Glass substrates were used in the I–V-characteristic study. They were first boiled in chromic acid and then washed successively in deionized water, acetone and ethyl alcohol before final drying in a flow of dry nitrogen. They were enclosed in a vacuum chamber and metal strips (e.g. Au, Al or Cr) of thickness $\simeq$ 2000 Å were evaporated on the substrates before the Ge deposition. A second set of metal strips of identical thickness (the metal depending on whether symmetric or unsymmetric structure is desired) was then evaporated on top of the Ge film, in a direction perpendicular to the first set. The contact area was $\simeq 10^{-2}$ cm^2.

Filament evaporation was used for the metal strips (the vacuum pressure being $\simeq 10^{-5}$ torr) while the Ge film was electron gun evaporated. For the latter the pressure was $\simeq 10^{-6}$ torr, the evaporation rate was $\simeq$40 Å s^{-1} and the thickness of the layer was $\simeq$ 1 μm. The source–substrate distance was 12 cm, and evaporation was at 90° in order to reduce the Ge film porosity.

In Figure 1 we give dc I–V characteristics for the configurations Cr–Ge–Cr, Al–Ge–Al, Al–Ge–Au and Al–Ge–Cr. We get segments of straight lines with a slope of unity at low fields indicating ohmic behaviour. This occurs at nominal fields $\leq 2 \times 10^3$ V cm^{-1}, calculated from the applied voltage and the film thickness. The slope changes at higher fields, leading to a power dependence of the form

$$I \propto V^n$$

where $n \geq 2$. In addition, the Al symmetric structure exhibits

* On leave of absence from the University of Assiut, Egypt.
† On leave of absence from the University of Ibadan, Nigeria.

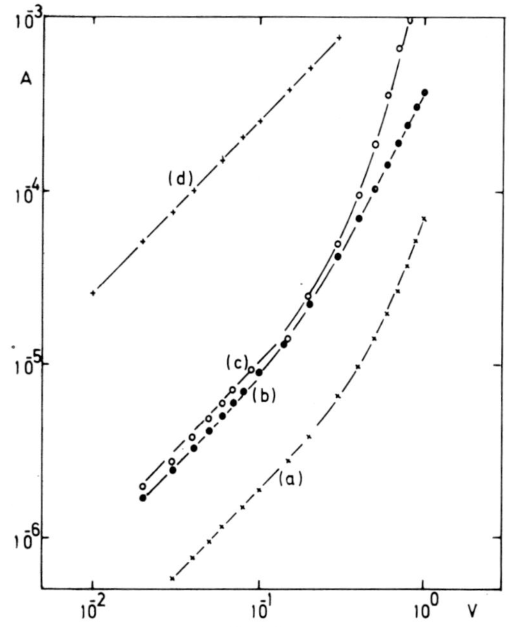

Figure 1. dc current–voltage characteristics of Ge film for different contacts, (a) Al–Ge–Au, (b) Al–Ge–Cr, (c) Al–Ge–Al, (d) Cr–Ge–Cr.

some polarity dependence (Figure 2). A possible explanation is that Al oxidizes very easily. Slight differences in the conditions for the evaporations may result in differences in the behaviour of the two contacts. On the other hand, the Cr–Ge–Cr structure shows ohmic behaviour and low resistivity even at high fields (Figure 1). Typical approximate values of resistivities obtained

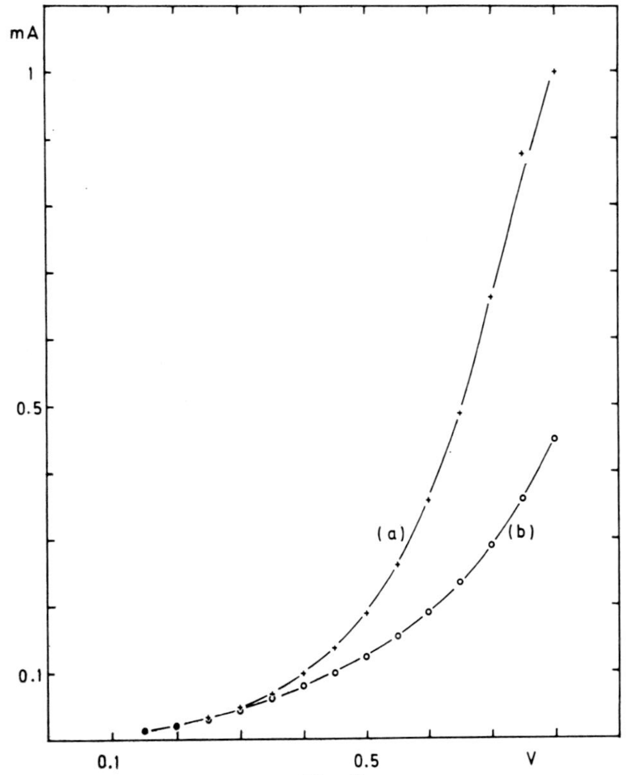

Figure 2. dc current-voltage characteristics for the structure Al–Ge–Al, (a) top electrode positive, (b) top electrode negative.

with the different configurations (using the ohmic section of the curve) are shown in Table 1. Earlier investigations by us with a different contact configuration indicated that also Au–Ge–Au structures behave ohmic, similar to the Cr–Ge–Cr system, with the same low resistance.[9]

Table 1. Typical approximate values of resistivities.

Structure	Resistivity (Ω cm)
Cr–Ge–Cr	3.8×10^4
Al–Ge–Au	5.3×10^5
Al–Ge–Al	1.2×10^6
Al–Ge-Cr	1.2×10^6

3. Rutherford backscattering

These investigations were made to give estimates of diffusion effects between contact and Ge, and to give some information about the presence, content and depth location of impurities.[10] These could influence the I–V-characteristics. We studied the interfaces Au–Ge, Al–Ge and Cr–Ge and observed no diffusion for samples deposited at room temperature.

For these measurements the substrate was a polished Al plate. A metal film was first evaporated on the substrate before the Ge deposition. To gain further insight into diffusion effects, in some cases, the substrate was held at an elevated temperature (200°C) during deposition. Spectra for the configurations Al–Al–Ge and Al–Au–Ge are shown in Figures 3 and 4. For the first structure we find no diffusion effects even at 200°C. On the other hand, for the Al–Au–Ge structure there was no diffusion for room temperature deposition but at the elevated temperature there is some evidence of diffusion between the layers. This is seen by comparing Figures 3 and 4. In Figure 4 (the heated substrate) diffusion of Au towards the surface is evident, at the high-energy side of the Au distribution.

We find no enhancement where the oxygen peak should be located in the spectra. The sensitivity, however, is only of the order of 20% (for oxygen in this measurement) but the slight drop in the Ge distribution towards the surface indicates a possible oxidization of the outer part of the layer. The small background we see in the spectra around channel 250 is due to some impurity (probably copper) in the Al substrate. This will have no effect on the I–V measurements.

4. Structure of the Ge films

Obviously it is essential to ascertain the structure of our evaporated Ge films. This we did using electron microscopy. The films were deposited on rocksalt substrates at room temperature, and at 200°C. The film was floated off by dissolving the rocksalt in water. We were able to show from the observed diffuse rings in the diffraction pattern that for room temperature deposition the Ge structure was amorphous.[9] It changes to polycrystalline for deposition on rocksalt at 200°C.

5. Discussion

Our backscattering data show that diffusion effects between the amorphous film and the contacts are not demonstrable at room temperature. *A priori* it is then not expected that such effects

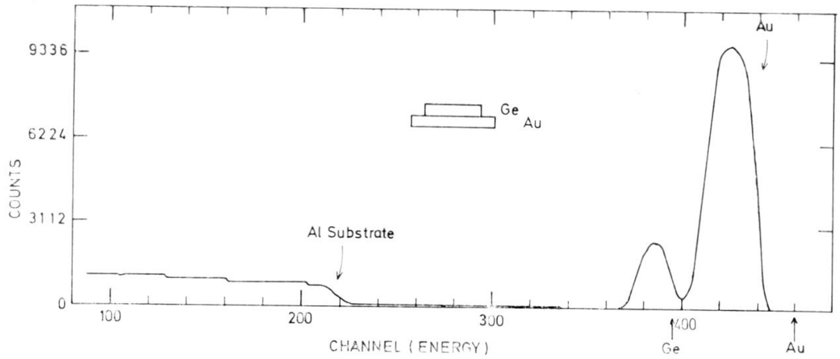

Figure 3. Backscattering spectrum for Al-Au-Ge. The α energy was 2 MeV and the backscattering angle was 170°. The substrate was at room temperature during the Ge deposition.

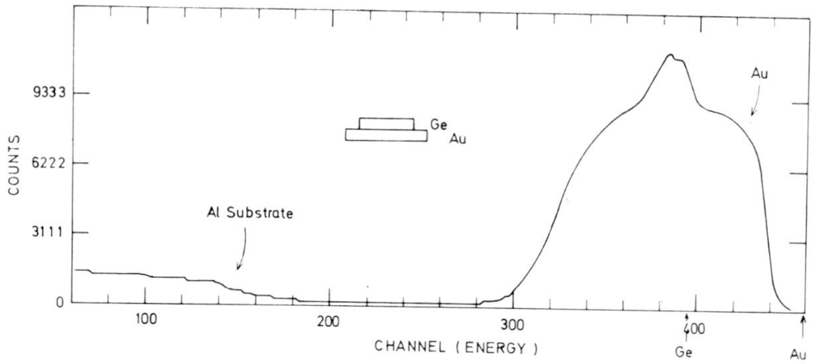

Figure 4. Backscattering spectrum for Al-Au-Ge configuration. The temperature of the substrate was 200°C during the Ge evaporation. In the spectrum the Ge peak is superimposed on the Au because the Au layer is very thick≃ 2600 Å in this case, while the Ge thickness was the same as in Figure 3. Vertical arrows indicate the calculated positions (from calibration spectra) of the indicated atom species, at the surface, while curved arrows only indicate the assignment of the distribution.

explain different I–V-behaviour. We also conclude that the oxygen content is not appreciable. Where interfacial oxide layers are critical in device operation we suggest that more accurate methods, e.g. using the *α*, *α* resonance reaction in ^{16}0 at 3.05 MeV should be used. Even monolayers of oxide can be detected in this way.[10, 11] An alternative is to use the ^{16}0 (*d*, *p*) ^{17}0 reaction.[12] However, we know from other work that a moderate oxygen content does not influence critically[5] the I–V-characteristics.

The differences we observe in apparent film resistances for the different contact structures suggest that the contacts play a role in determining the current. The symmetric Cr and Au structures present the lowest resistance while the unsymmetric structures, viz. Al–Ge–Cr and Al–Ge–Au as well as the symmetric structure Al–Ge–Al show higher resistances. For the unsymmetric structures, we can speculate whether the Al contact may be operating with a different type of carriers compared with Au and Cr. Thus, if for example Au and Cr inject electrons into the amorphous layer the situation could be that Al does not easily empty the layer of these electrons. This would lead to appreciable resistance. In this hypothesis, the high resistance of the symmetric Al structure could be attributed to a higher apparent film resistance, for hole transport, than for electrons.

This would also explain the low resistance of the symmetric Cr and Au structures. However, we appreciate that the problem is too complex to be solved with the present experimental data. Data from other types of experiments (other than I–V-characteristics) are needed. For example, capacitance and photoelectric measurements will give some information about the possible existence of barriers[13] in the structures.

Considered as information on the use of amorphous Ge contacts the work indicates that metals like Cr and Au do not introduce any extra impedance over the minimum value offered by the film itself.

References

[1] J B England and V Hammer, *Nucl Instr Meth*, **96**, 1971, 81.
[2] P A Walley, *Thin Solid Films*, **2**, 1968, 327.
[3] A H Clark, *Phys Rev*, **154**, 1967, 750.
[4] M Morgan and A K Jonscher, *Thin Solid Films*, **9**, 1971, 67.
[5] D K Pandyn, S K Barthwal and K L Chopra, *Phys Stat Sol* (a) **32**, 1975, 489.
[6] B A Orlowski, W E Spicer and A D Baer, *Thin Solid Films*, **34**, 1976, 31.
[7] P A Walley and A K Jonscher, *Thin Solid Films*, **1**, 1968, 367.
[8] J Tauc, R Grigorovici and A Vancu, *Phys Stat Sol*, **15**, 1966, 627.
[9] M Hafiz, E Mgbenu, P A Tove, H Norde, S Petersson, UPTEC 7616 R, Institute of Technology, University of Uppsala Report (1976).
[10] W K Chu, J W Mayer, M A Nicolet, T M Buck, G Amsel and F Eisen, *Thin Solid Films*, **17**, 1973, 1.
[11] B Orre, L O Norlin, K Johansson, F Falk, J E Thun, UUIP-906, Institute of Physics, University of Uppsala Report (1975).
[12] A Torreiter, Doctor Thèse, l'Université Louis Pasteur de Strasbourg, 1972.
[13] Horng-Yhi Wey and H Fritzsche, *J Non-crystalline Solids* 8–10 (1972) 236. (Cf H Frizsche in *Electronic and Structural Properties of Amorphous Semiconductors*, p 76. Edited by P G LeComber and J Mort, Academic Press, New York (1973).

2.4 Formation of various metal–semiconductor junctions by means of ion evaporation

P Glaser, Á Herman and Gy Vágó, *Industrial Research Institute for Electronics 'HIKI' H-1393 Budapest, POB 348, Hungary*

The paper deals with the result of some preliminary experiments of applying ion-evaporated metal layers to silicon and heat treatment of the samples. Ion evaporation is based on the electron beam evaporation, namely, the ions arising from the 180° deflection type e-gun are accelerated to cause implantation effects during the condensation of the neutral particles. The accelerating voltage is typically 20 kV, the ion current 5–20 mA.

Four metals have been chosen for the experiments: Al, Ni, Cu and Mo. The I–V-characteristics of the contacts deposited by ion evaporation were ohmic. The curves appreciably changed after heat treatment of the samples at 400, 500 and 580°C.

The Al–Si contacts had very high resistivity after the treatment at 400°C, but then Al alloyed diode type contacts were achieved at 500°C.

Ni and Mo formed alloyed type contacts at 580°C, and similar results were obtained in case of Cu. Very interesting results were obtained for Ni and Mo at 500°C, where we obtained I–V-characteristics of the Schottky-diode type, but with the reverse polarity.

1. Introduction

One of the fundamental processes of the semiconductor technology, especially of the IC-technology is the formation of metal contacts. There are a great deal of requirements for the metal contacts, e.g.

low resistance,
ohmic characteristic,
good adherence,
it must not adversely affect the underlying structures, etc.

The most prominent metal which meets nearly all of the above and also further requirements is aluminium. But the research of metallization processes is in progress because of the permanent development of the technologies and because of the more and more frequent special requirements. In the technology of integrated circuits there are difficulties with many materials when the deposited metal should give contacts of the same or nearly the same quality on both *p*- and *n*-type silicon of various conductivity. It is known from the literature[1] that a good ohmic contact can be expected for materials which act as dopant in the silicon or which are able to convert the conduction type of the semiconductor.

The quality of a contact depends, of course, not only on the contact metal but also on the metallization process, too. We have investigated the features of several metal–silicon junctions made by various methods.

This paper deals with the result of some preliminary experiments applying ion-evaporated layers on to silicon. This time our investigations concerned the I–V characteristics of four different metal–silicon contacts as deposited and after heat treatment.

2. Ion-evaporated films

Ion-evaporation is a hybrid vacuum deposition technology[2,3] where the material-beam, coming from an evaporation source, crosses the path of an electron beam. From this crossing-over the ionized vapour is extracted and accelerated by a high potential and directly deposited on to the substrate. Thus the condensing material consists of two components: neutral vapour and accelerated charged particles. It means—one can say—that deposition and ion implantation take place at the same time, as verified on the basis of physical tests, SIMS[3] and back-scattering.[4] Therefore the properties of these layers differ from those of the films deposited by the other usual vacuum technologies. From previous experiments we have already learned that the adherence and the solderability of the ion-evaporated films significantly exceed those of the commonly-used layers.[3] It was also tried to improve the adhesion and the ohmic behaviour of the contact by bombarding the evaporated metal films with gas-ions and/or self-ions.[5]

It can be expected that applying ion evaporation would result in contacts of special characteristics.

3. Contact metals

Four metals have been chosen for the experiments: Al, Ni, Cu and Mo. In choosing them we kept in mind that these metals should be in different column of the periodic table and should be easily applicable to vacuum technology.

3.1. Aluminium. The Al–Si contact is the best-known combination of materials in the semiconductor technology in general and in our own practice.[6]

3.2. Nickel. Data can be found also for the Ni–Si contact system in the literature, because the electroless chemical nickel deposition process is often used for obtaining contacts.[6] At the same time, nickel as a metal, located in the 8th column of the periodic system of the elements, implies the possibility of similar behaviour to the frequently investigated Pt–Si contact.[7] Moreover, nickel is a prospective material for special devices, e.g. photovoltaic cells.

3.3. Copper. Copper as contact metal does not play a role in general but it may give some interesting information because of its special influence on the minority carrier lifetime.

3.4. Molybdenum. In the last decade many researchers paid great attention to the investigation of the Mo–Si junction because of its high temperature resistance and electromigration properties that are more advantageous than those of the Al–Si system.[8]

4. Experimental procedure

The ion evaporation was carried out in a common high-vacuum equipment type BA 510/A (Balzers) using a 180-degree deflection electron beam evaporation source type GF 180 S (HIKI, Hungary) supplied with the necessary parts for ion evaporation.[2]

The metals of 5N purity were deposited on to the surface of 1 Ω cm *n*-type silicon wafers prepared with a mechanical–chemical process.[9] A metal (molybdenum) mask with holes of 0.6 mm diameter was applied for the deposition to form test 'diodes'.

Deposition parameters are summarized in Table 1. The ion current was regulated by the electron current whereby different deposition rates were obtained.

Table 1. Main parameters of the ion evaporation

Vacuum	$(2–5) \times 10^{-3}$ Pa
Electron accelerating voltage	7.5 kV
Electron current	250–500 mA
Ion accelerating voltage	20 kV (fixed)
Ion current	5–10–15–20 mA
Deposition time	180 s
Film thickness	130–270 nm
Deposition rate	0.7–1.7 nm s^{-1}

The I–V characteristics of the probes were measured after deposition and after 10 min heat treatment in dry nitrogen at 400, 500 and 580°C. A sketch of the measurement set-up can be seen in Figure 1.

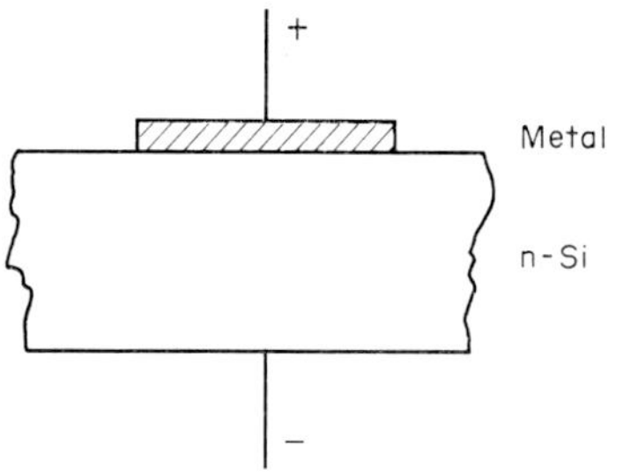

Figure 1. Measurement set up of I–V characteristics.

5. Results

Although every metal was evaporated with different ion currents, as shown in the Table 1, it did not have any effect on the I–V characteristics. Thus, the behaviour of every kind of the metal–silicon junctions can be treated summarily.

5.1. Al–Si junctions (Figure 2). The Al-films gave resistance-like contacts as deposited. After heat treatment at 400°C the forward resistance was high up to approximately 30 V, while in the reverse direction the diodes showed a soft but rather low breakdown.

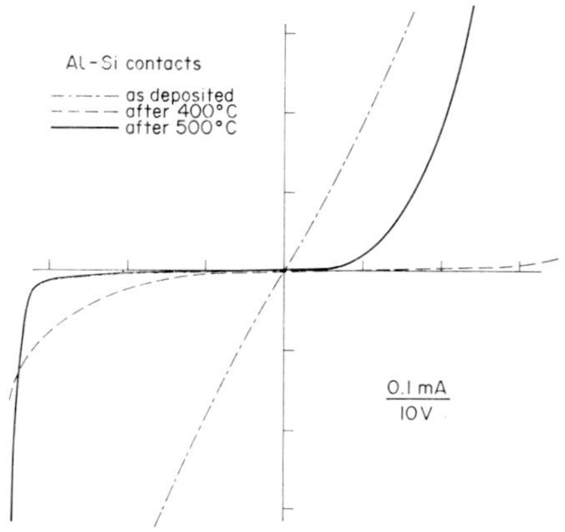

Figure 2. Typical I–V characteristics of Al–Si contacts made by ion evaporation and heat treatment.

The behaviour changed appreciably after annealing at 500°C when forward characteristics of varying quality and reverse curves with sharp breakdown at 30–35 V were obtained. These characteristics were similar to that of the so-called alloyed diodes.

The results obtained on the Al–Si contacts after the heat treatment at 580°C were not reproducible since this temperature is in the vicinity of the eutectic point.

5.2. Ni–Si junctions (Figure 3). While the deposited samples showed ohmic resistance, the heat treatment at 400°C made them somewhat similar to the alloyed diodes but bad ones, both in forward and reverse direction. The annealing at 500°C gave remarkable and very well reproducible results where the curves surprisingly resembled the characteristics of the Schottky diodes but with reverse polarity—as if the forward and reverse directions were inverted. The curves of 580°C correspond to alloyed ohmic contacts.

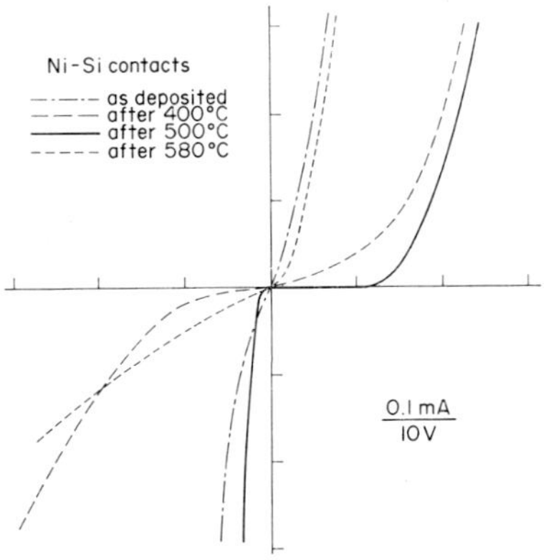

Figure 3. Typical I–V characteristics of Ni–Si contacts made by ion evaporation and heat treatment.

5.3. The Cu–Si couple (Figure 4). The samples, having ohmic resistance curves without heat treatment, were annealed at 400°C and diodes with reverse characteristics in both directions were obtained. The 'forward' direction had a sharp breakdown at about 10 V and the 'reverse' characteristics were soft.

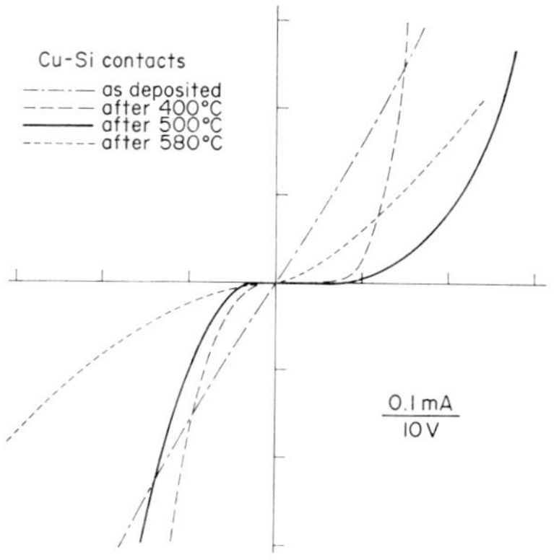

Figure 4. Typical I–V characteristics of Cu–Si contacts made by ion evaporation and heat treatment.

After 500°C annealing we obtained characteristics of alloyed diodes that changed to poor ohmic contacts after further heat treatment at 580°C.

5.4. Mo–Si contacts (Figure 5). Behaviour of the non-treated Mo–Si contacts agreed with that of the former systems. After the 400°C annealing, the ion-evaporated Mo on Si gave Schottky-barrier type characteristics with soft breakdown at a rather low (≈10 V) voltage.

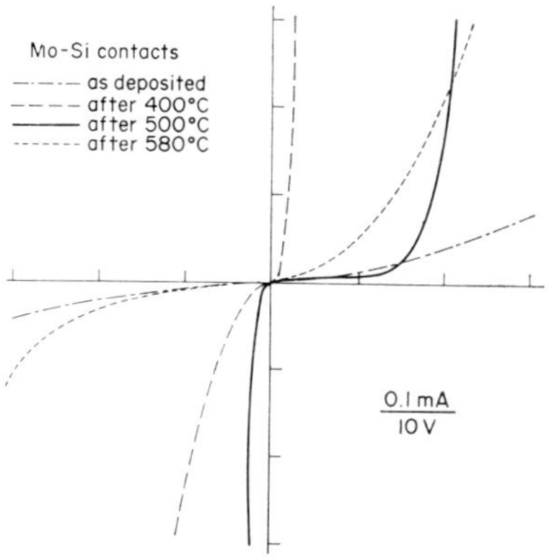

Figure 5. Typical I–V characteristics of Mo–Si contacts made by ion evaporation and heat treatment.

These characteristics turned into the opposite after the 500°C heat treatment. The picture is the same as in case of Ni, i.e. an inverted Schottky-diode characteristic with soft 'breakdown' was obtained. The annealing at 580°C resulted in the characteristics of a poor alloyed diode.

6. Conclusions

From these experimental data it is clear that there is a possibility to make metal–silicon junctions with special characteristics by means of ion evaporation and subsequent 500°C heat treatment of the Al, Ni and Mo films. The somewhat unlike behaviour of the Al–Si contact is connected with a destruction of the surface layer of the semiconductor, as it has been found also by Itoh *et al.*[10]

On the basis of these investigations we have already made ion-evaporated nickel films as solderable ohmic contact for photovoltaic cells.

7. Acknowledgements

We are grateful to Mr I Beke for measuring the I–V characteristics.

8. References

[1] B Schwartz (Editor), *Ohmic Contacts to Semiconductors*, El Chem Soc, New York (1969).
[2] Á Herman and Gy Vágó, *Finommechanika.* **11**, 1972, 233.
[3] P Glaser, Á R Herman and G Vágó, *Thin Solid Films*, **32**, 1976, 69.
[4] P Glaser, Á Herman, Gy Vágó, E Kótay and G Mezey, *'HIKI' Közlemenyei*, **XV**(4), 1975, 51.
[5] E F Krimmel, *Phys Stat Sol* (1), **7**, 1971, 141.
[6] Á Herman and P Szebeni, *'HIKI' Közleményei*, **VII**, (3), 1967, 39.
[7] M P Lepselter and J M Andrews, *Ohmic Contacts to Semiconductors*, Edited by B Schwartz, Electrochem Soc, New York, p 159 (1969). D Shinoda, *ibid*, p 200.
[8] J A Cunningham and J G Harper, *The Electronic Engineer*, **26**, 1967, 74.
[9] É Rózsa, 'HIKI' Jubileumi Évkönyve (HIKI Jubilee Year-Book), p 40, Müszaki Kónyvkiadó, Budapest (1973).
[10] T Itoh, T Inada and K Kanekawa, *Appl Phys Lett*, **12**, 1968, 244.

2.5 Behaviour of amorphous Ge contacts to monocrystalline silicon

H Norde and P A Tove, *Electronics Department, Institute of Technology, University of Uppsala, Uppsala, Sweden*

The performance of evaporated amorphous Ge films (thickness ≃500 Å) as contacts to etched n- *and* p-*type silicon crystals of different resistivities are discussed. The Ge was 3 Ωcm* n-*type (*p-*type was also used but gave no difference), and as external contact to the Ge film an Au layer was evaporated. The behaviour of the aGe–Si junction seems to be largely governed by interface effects (and thus depends on surface preparation), as is often the case with metal–Si junctions, but Ge gives more reproducible and less time-varying results.*

In the process of clarifying the function of the contact the following structures were investigated (1) aGe-^n_pSi-In(Hg), where the latter is an ohmic contact, (2) amSi-^n_pSi-metal where amSi is a surface region of the crystal which has been rendered amorphous by ion bombardment, (3) aGe-^n_pSi-metal. I–V and C–V measurements were performed. From the results we conclude that aGe–Si junctions act as low-resistance contacts when fed by electron or hole currents from the crystal. The currents (holes and electrons) that are injected into the crystal from the film are limited by barriers to small current densities, usually in the range 10^{-6} A cm^{-2}. It is suggested that the small hole currents are explained by an increase in the hole barrier, effected by positive charges at the interface or in the Ge film, which are built up when positive carriers (holes) are injected by the contact.

1. Introduction

Contacts consisting of amorphous semiconductor films to monocrystalline silicon have found use in radiation detectors[1] and the I–V behaviour of these junctions has also been the subject of theoretical investigations.[2] In the first-mentioned application the amorphous–crystalline junction is one of the two contacts to the crystal; the operation of the devices necessitates the use of another contact, usually a metal contact. The study of the I–V behaviour of two-contact structures in itself is an interesting topic and has not yet been sufficiently investigated. Some basic concepts are discussed,[3] for structures with two metal contacts on comparatively low resistivity silicon. In this work we will discuss the characteristics of structures on silicon of varying resistivity, even very high, both *p*-type and *n*-type. A common feature is that one of the contacts is an amorphous semiconductor film, either an evaporated Ge film, or an amorphized layer of the silicon crystal itself.

2. Basic considerations on two-contact structures

A. Provided the diffusion length of the carriers is much larger than the thickness of the semiconductor wafer we can separate the contributions of the two types of carriers, electrons and holes. Simple cases of a two contact structure for two different polarities of applied voltage are shown in Figure 1. In Figure 1a hole current flows from the left to the right if the barrier height ϕ_{1h} is low enough, and electron current flows from the right, if barrier ϕ_{2e} is low enough. Similarly, for the opposite polarity in Figure 1b the magnitude of ϕ_{1e} determines the magnitude of the electron current from the left and ϕ_{2e} determines the magnitude of the hole current from the right. Usually one type of carrier dominates. We see that different polarities of applied voltage give information about complimentary couples of electron and hole barriers.

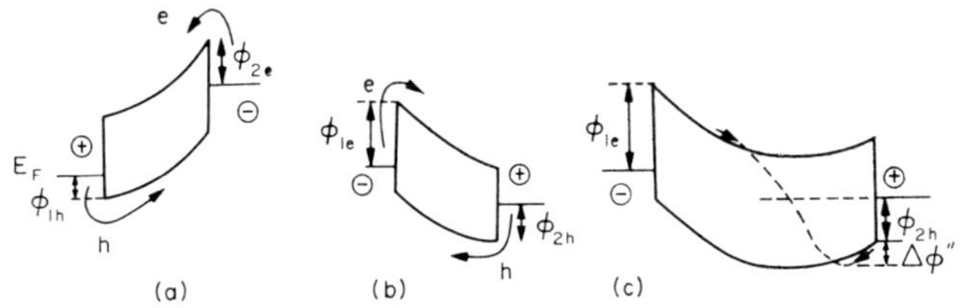

Figure 1. Band diagrams for two-contact structures. (a) and (b) are for full depletion, with different polarity of the applied bias. (c) is for an applied bias smaller than required for full depletion of the wafer, and applies to a Au–*n*Si–Al structure.

A more detailed band diagram is shown in Figure 1c. This could apply to the specific situation where contacts of Au and Al have been made to *n*-type silicon and small voltages are applied (below depletion which occurs when the depletion region extending from the Au contact approaches the Al contact). Electron current, the magnitude of which is mainly determined by ϕ_{1e} dominates. The barrier ϕ_{1e} presents a much larger obstacle to the transport than the small barrier encountered just before arriving into the Al contact, as also is seen from a comparison of the resistances of the reverse biased Au diode and the forward biased Al diode. Energetically, we infer from the band diagram in Figure 1c that the contribution of holes to the current is much smaller than that of the electrons because the total barrier $\phi_{2h} + \Delta\phi''$ is larger than ϕ_{1e}. However, as the applied reverse bias on the Au electrode increases the additional barrier $\Delta\phi''$ against hole transport vanishes (we approach the conditions of Figure 1b) and hole current may become comparable to electron current if ϕ_{2h} is comparable to ϕ_{1e}.

Direct considerations of the band diagram in itself take care of the properties of the crystal; further considerations of whether we have *n*- and *p*-type material, and its influence on the easiness with which electrons and holes are transported, are unnecessary. We assumed above that the diffusion length was longer than the thickness of the wafer. If this is not so, we will have a transfer from one type of carrier to the other as indicated by the dotted line traversing from the conduction band to the valence band in Figure 1c. In principle such transfer could also occur at an interface if the recombination rate of one carrier is high there.

B. Assuming a fixed position of E_F at one contact there is a relation between the barrier ϕ_e against electron injection and the barrier ϕ_h against hole injection of the contact, and the bandgap E_G. The smaller the value of ϕ_e the higher the value of ϕ_h is. If we find that one contact impedes both electron and hole current flow (i.e. it gives currents which are much below those given by the Richardson equation $j_{sat.} = A^{**}T^2 \exp(-q\phi/kT)$, then the conclusion is that either the proper values of A^{**} are smaller than the usual value for metals, or E_F is not fixed relative to the positions of the bands, but can shift when the polarity (field) is reversed.

The expected magnitude of the electron and hole saturation currents according to the Richardson equation are indicated in Figure 2.

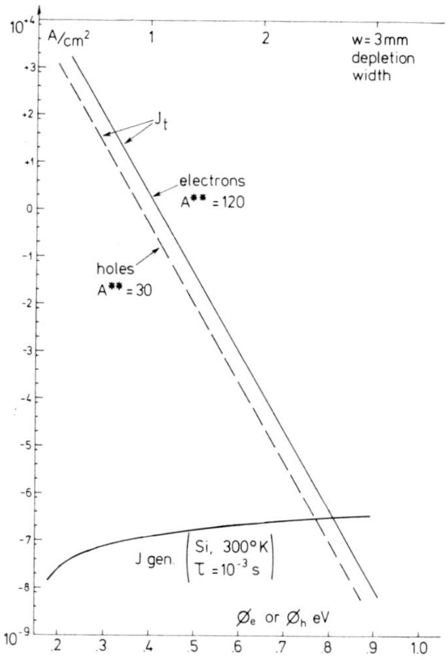

Figure 2. Saturation currents for Schottky contacts, calculated using Richardson constants appropriate to metals.[10,11]

C. We discussed above how barriers can impede current flow. There is also the possibility that current flow is impeded by space charge effects, i.e. we have one-carrier space charge limited current flow. Space charge effects appear when the moving charges influence the field in the structure. If the original field is due to fixed charges (impurity atoms) this occurs when the concentration of moving carriers becomes comparable.

D. Taking the generally encountered case where the flow of one type of carrier dominates, a depletion region extends from that contact which impedes carrier flow most effectively.

3. Structures where one contact is aGe and the other an ohmic In(Hg) contact

Fabrication process. Si wafers of 16 mm diameter and 0·5 mm thickness with lapped surfaces were thoroughly cleaned and then etched in CP4A for 2–3 min at room temperature. After etching the thickness was $\simeq$0.3 mm. Quenching was done with distilled water. When the slices were dry the aGe contact was evaporated from a tantalum boat in a vacuum of 10^{-5}, to a thickness of 200–400 Å, and covering an area with diameter of $\simeq$6 mm. Then the wafer was baked at 80°C for 12 h to remove trapped gas. This temperature was low enough to avoid crystallization of the amorphous film. On the top of this a contacting Au film was then evaporated (Figure 3.) For the back contact (also shown in Figure 3) we used an indium amalgam

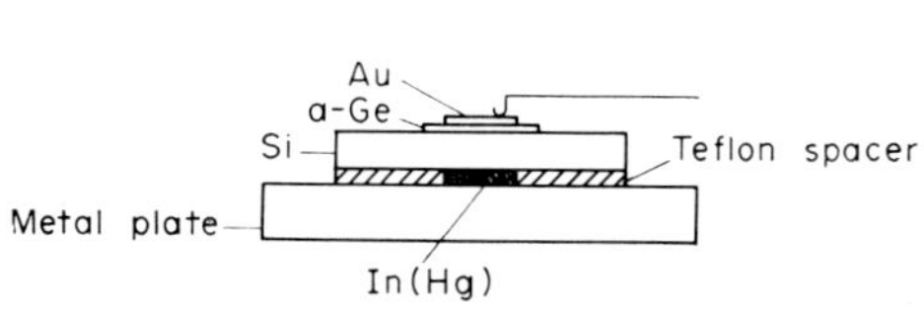

Figure 3. Contact configuration for aGe–Si–In(Hg) structures.

layer which has been reported to be ohmic for a wide temperature range.[4] The contact was not, however, quite ohmic and stable on an etched surface, but after slight rubbing of the Si surface with 75 μm diamond paste, good behaviour (ohmic but also injecting, see later) was obtained. It was also reproducible in that it could be wiped off and replaced again, without changing the behaviour.

The behaviour of this contact as compared with an ordinary metal contact is demonstrated in the following section.

I–V-behaviour. Diode characteristics for aGe-*n*Si-In(Hg) and aGe-*n*Si-Au structures made on *n*-type material are shown in Figure 4. Comparing the lower curve A in Figure 4 with the schematic band diagram in Figure 7a expected for this structure we conclude that the Ge side must have an electron barrier.

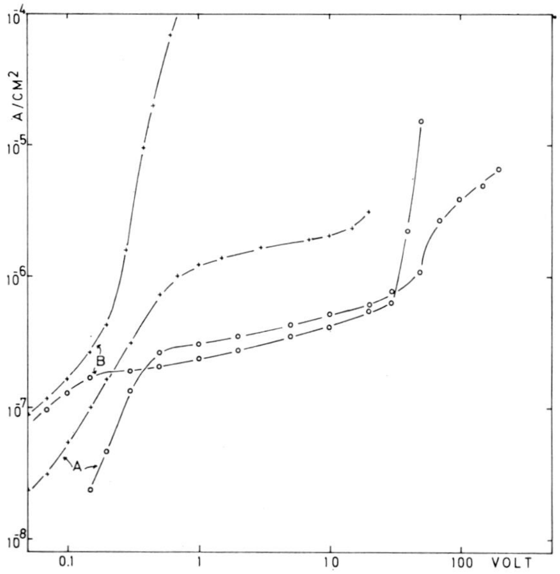

Figure 4. I–V-characteristics of aGe–nSi–In(Hg) structures (B), compared with aGe–nSi–Au structures, (A). Rings indicate the Ge contact is negative.

The Au–nSi junction is forward biased and passes electron current from Si towards the Au side but the current is limited at approximately 0.3 $\mu A\ cm^{-2}$. If the Ge contact had not presented a limit the forward current of the Au–nSi junction would have risen to several tens of mA. From the lower curve B where the ohmic contact In(Hg) has replaced the Au contact (this was wiped off) pertinent to A we get further confirmation of this. At high voltages when the depletion region which extends from the Ge contact reaches either the Au contact (lower curve A) or the In contact (lower curve B) we observe a current rise which is explained by contribution of hole current, when the field reaches these contacts. The magnitude of the hole current then supplied from the Au contact is much smaller than can apparently be supplied from the In contact. Obviously (from Figure 2) the hole barrier of the Au contact is higher than the value 1.1–0.8 = 0.3 expected from the barrier height ϕ_e of Au–nSi junctions, and we explain this as the effect of a hole blocking interface as discussed by Pulfrey[5] (cf. section 5).

For low voltages we observe a difference in the behaviour of the lower curves A and B. The voltage drop is much larger in the case where we have the Au contact, with its interface, than when we have the pure In ohmic contact.

For the upper curve B in Figure 4 the current consists of electrons from the In contact. It cannot be hole current from the Ge contact because then we should have had such current also for the upper curve A; there the current saturates at the value determined by the electron barrier on the Au side. The results for structures on nSi of different resistivities are summarized in Figure 5. GN10 has a resistivity of 70–100 Ωcm while GN20

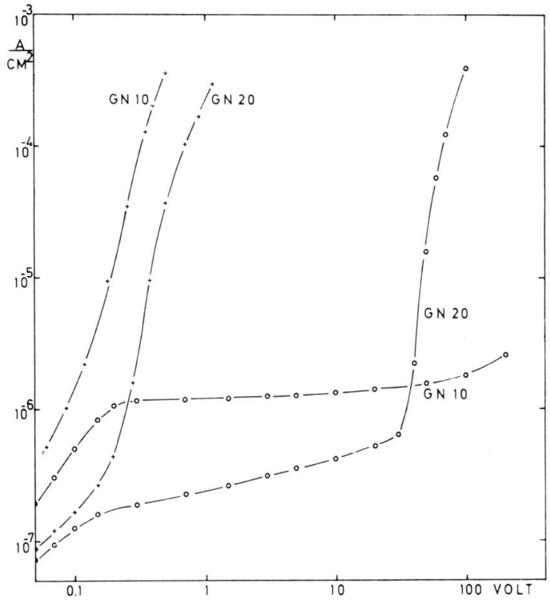

Figure 5. I–V-characteristics of aGe–nSi–In(Hg) diodes. Ring-marked points are for negative polarity on Ge. GN20 is on 28 kΩcm nSi and GN10 on 70–100 Ωcm nSi.

(already shown in Figure 4, curve B) is 28 kΩcm. There is no appreciable difference, except that full depletion can be reached only for the high resistivity material.

In Figure 6 are shown characteristics for structures on p-material. We see that for the lower GP21 curve we get no appreciable hole current from the Ge contact which can be explained by the existence of a hole barrier there (it is not possible that the saturation current below 10 V which we observe is due to the In contact because this has been shown to be ohmic and able of supplying much higher current). The current rise for $V = 20$ V is explained by electron current starting from In when the field approaches, at full depletion. In addition to GP21 which is for 40 kΩcm we show in Figure 6 data for a

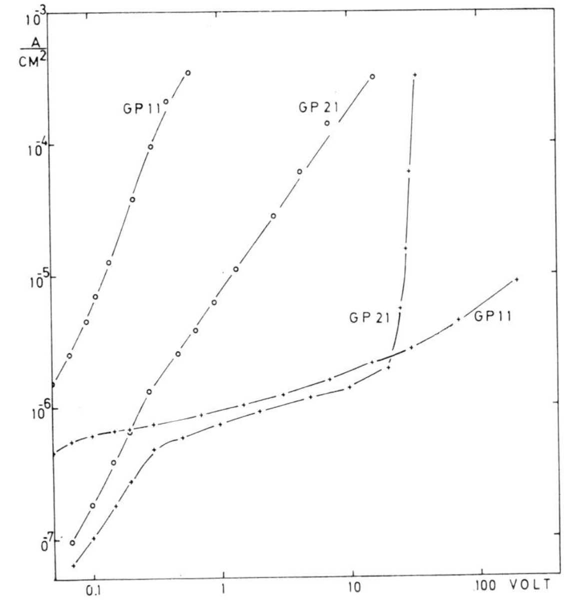

Figure 6. I–V characteristics of aGe–pSi–In(Hg) structures. Ring-marked points are for negative polarity on the Ge contact. GP21 is on 40 kΩcm pSi, GP11 on 500 Ωcm pSi.

500 Ωcm structure GP11. In this case full depletion cannot be reached. For the opposite polarity the current should mainly be holes from the In side. The effective total barrier against electron flow from the Ge contact is even higher (at least in the low resistivity material, because of the different positions of E_F in n- and p-materials) than in Figure 5 (curves for Ge negative) and this should make the electron current small (at least at low voltages).

Our conclusion from the n- and p-type structures hence is that we can obtain an electron barrier for Ge on n-silicon, and a hole barrier for Ge on p-silicon, which determine the currents in these structures. It is also possible (but we found this not to be true for all fabricated devices) to get a hole barrier at the Ge contact on nSi. It is also possible but still less sure that we also get a high electron barrier for a Ge contact on pSi. However, this was obtained for some diodes used in TCT-measurements.[6] (It was also obtained for some double Ge structures like the structure D60 discussed in a later section.)

Though the Ge contact limits injection of electrons and holes in several cases there does not seem to be any limit to the electron and the hole currents that can be fed into this contact from the semiconductor side. This is obvious from the upper curve B in Figure 4 where the Ge contact already at low voltages accepts a sizeable electron current with a low voltage drop, and the lower curve B for $V > V_d$ where large hole currents are reached. The same conclusions can be reached from the high current regions of the curves in Figure 6 from which it is seen that also in p-type material the Ge contact can accept large holes and electron currents coming from the semiconductor side, without appreciable voltage drop.

Capacitance measurements. The results of capacitance measurements at 5 kHz are shown in Figure 8 for *n*-type structures and in Figure 9 for *p*-type structures. In the latter case it is the capacitance between *p*-silicon and a positively biased Ge contact that is measured, and in the former case the capacitance between *n*-silicon and a negatively biased Ge contact. In both figures are drawn lines with the theoretically expected slope of one half ($C \propto 1/\sqrt{V}$). The position of the line depends on the doping concentration. We note for structure GN20 that the capacitance values are consistently higher (a factor 2–3) than expected from the measured resistance of the wafer. This resistance was measured by removing the Ge layer and making two identical ohmic contacts of In(Hg) and performing measurements at low injection levels. The resistance determined in this way was 27.6 kΩcm which agrees quite well with the resistance given by the manufacturer, 28 kΩcm. We attribute the excess capacitance, which is not found when one is performing measurements on ordinary reverse biased metal silicon structures,[7] as due to surface charges present in the interface and in the Ge film. As discussed by Goetzberger and Sze,[8] surface states cause an extra capacitance to be added in parallel to that due to the semiconductor junction. In an ordinary metal semiconductor diode these states would have no influence on the capacitance because they are in equilibrium with the metal,[9] when the thickness of the insulating film is small. In the case where we have a Ge film separating the metal from the crystalline semiconductor the situation could be different. We could have an intermediate case between that of a MIS capacitor where surface states are found to have a large influence on the measured capacitance[8] and the case of a metal–semiconductor diode with a very thin interface. We note that the capacitance of GN20 tends to become constant at higher voltages when depletion is approached. For the structure GN10 on lower resistivity *n*-material (nominal resistance given by the manufacturer 70–100 Ωcm), the measured capacitance is again higher than that expected from the resistivity. At small voltages we expect that the experimental points should deviate slightly from a line with slope 1/2 because of the influence of V_{diff}. The agreement with theory is quite good for GN10 and the value of V_{diff} that can be derived here is $V_{diff} \simeq 0.58$ V.

Deviations between theory and experiments are again to be found in the capacitance curves for *p*-type material. The difference compared with the *n*-type case, e.g. the smaller voltage dependence, could be due to the different sign of the net charge in the interface and film that is required because of the different polarity of the biasing voltage, in comparison with the *n*-type material. Complementary capacitance measurements, e.g. at different frequencies, are required to clarify these points.

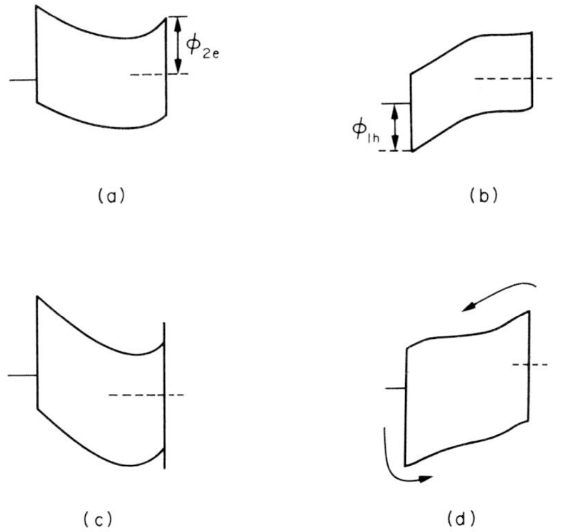

Figure 7. Schematic band diagrams for:

(a) ⊕ Au–*n*Si–aGe ⊖ structure.
(b) ⊕ Al–*p*Si–amSi–Au ⊖ structure.
(c) ⊖ Au–*n*Si–amSi–Au ⊕ structure.
(d) ⊕ Al–*n*Si–amSi–Au ⊖ structure.

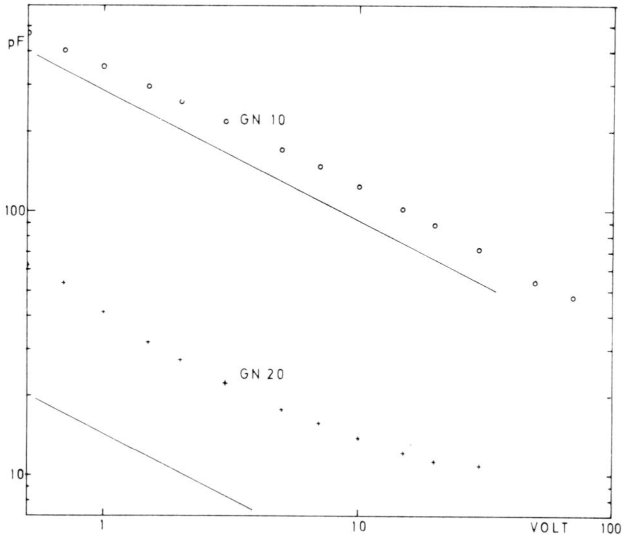

Figure 8. Capacitance curves for aGe–*n*Si–In(Hg) structures, for two resistivities.

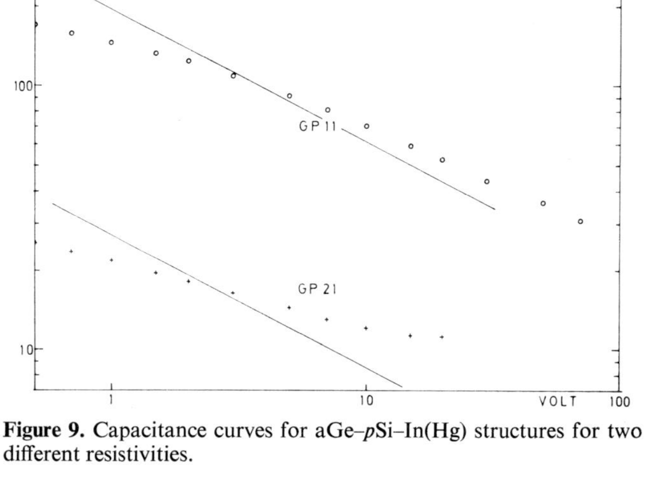

Figure 9. Capacitance curves for aGe–*p*Si–In(Hg) structures for two different resistivities.

4. Structures where one contact is amorphized silicon and the other a metal contact (Au or Al)

Fabrication process. On one side of these wafers argon ions of 50 keV energy were implanted. Before implantation the surfaces were polished but not etched. Two doses were used, 1×10^{15} atoms/cm^2 and 2×10^{15} atoms/cm^2. These data are given in the Figures 9–10 showing the I–V-characteristics. After implantation the non-implanted surface was etched and a metal layer deposited by vacuum evaporation (Al or Au). After this, wafer thicknesses were $\simeq$0.4 mm. A similar metal contact, Al or Au was also evaporated on to the implanted side of the wafer. The average implant depths of the 50 keV argon

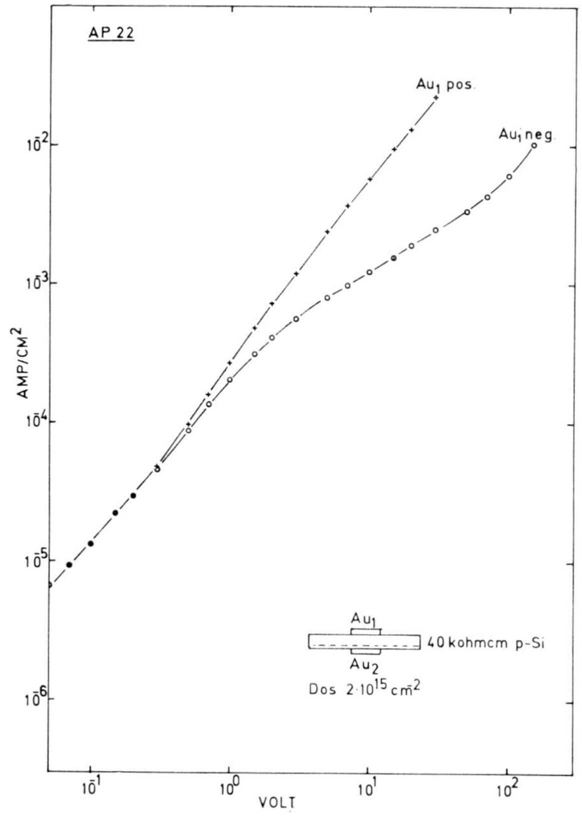

Figure 10. Characteristics of Au–*p*Si–amSi–Au structure.

ions is $\simeq$550 Å, while their depth distribution is expected to have a width of 225 Å* (standard deviation). It is also expected that the surface layer of the wafer had been damaged by the incoming ions to such an extent that the wafer becomes amorphous also in the surface region.

I–V-characteristics. We anticipate the discussion by mentioning that the I–V-characteristics have the general form expected, taking into account the type of metal used for the contact on the etched side, and the fact that the bottom contact is on a polished surface where we expect that any metal would give an ohmic contact. In spite of the amorphized region being of silicon instead of Ge, and this presumably has higher resistivity and furthermore is thicker than the Ge layers used in the other series, we did not find that the amorphous region in itself was any appreciable hindrance to current flow.

The general behaviour of the I–V-characteristics of these structures are of interest as a contrast to the results obtained for structures which on both sides had evaporated contacts on etched surfaces and where, as will be seen in section 5, interface effects are playing a large role. Also, there are a few features of the I–V data of the present structures which could merit further discussion.

Diodes on pSi. Diode AP23 (Figure 11) has an evaporated Al top contact and the expected band diagram for this is shown in Figure 7b. A depletion region extends from the Al contact when this is made positive. As seen from Figure 11 the saturation current is approximately 10^{-5} A/cm^2 which is the expected magnitude taking into account that we expect ϕ_{1h} for Al to be $\simeq$0.60–0.65 eV (a fresh prepared Al contact has a barrier height of about 0.45–0.50 eV on *n*-silicon and the quoted value of ϕ_{1h} is the bandgap complement of this value). Electron injection starts when the wafer is depleted which is at around 50 V in Figure 11. Looking at the other polarity the Al–*p*Si junction becomes forward-biased and the current increases to relatively high values, possibly also above those given by the bulk resistance of the wafer, which could then be explained by double injection of carriers, because of contribution of electrons from the bottom contact.

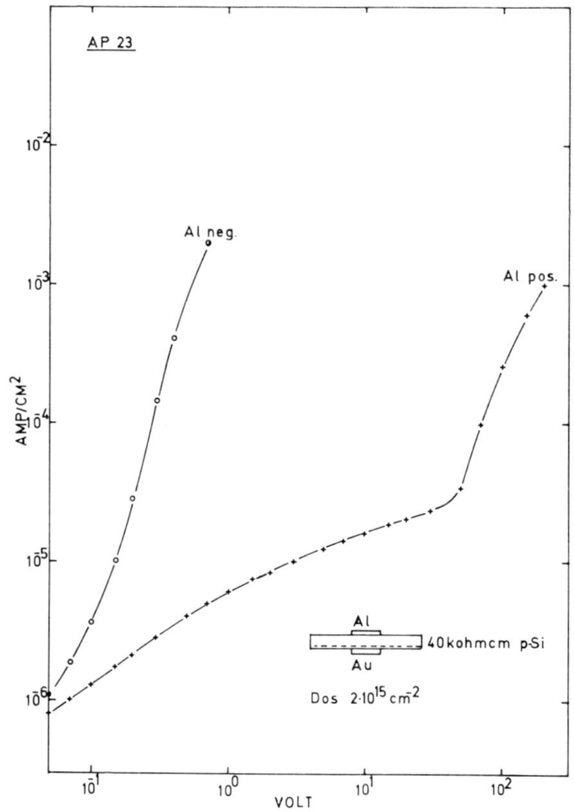

Figure 11. Characteristics of Al–*p*Si–amSi–Au structure.

Taking now the case where the top contact is Au instead of Al, for diode AP22 (Figure 10) we see that the hole barrier ϕ_{1h} which in this case is expected to be approx. 1.1–0.8 = 0.3 eV (instead of 0.65 in Figure 7b) now does not impede the flow of holes to any measurable extent but the current reached for Au_1 positive in Figure 10 increases above that due to the bulk resistance of the wafer. The corresponding band diagram is similar to that of Figure 7a. However, we note that for the other polarity, Au_1 negative, we have some tendency to saturation of the hole current that can be drawn from the bottom contact, though the current magnitude is much higher than obtained for evaporated amorphous Ge layers, in the previous section.

Diodes on nSi. The *n*-structure AN42 in Figure 12 for low voltages gives a saturation current 10^{-6} A/cm^2 for Au_1 negative. This is the expected electron saturation current from Au_1, cf Figure 7c. At full depletion Au_2 starts injecting

* T Noreland, private communication, cf D K Brice, *Ion Implantation Range and Energy Deposition Distributions*, Vol 1.

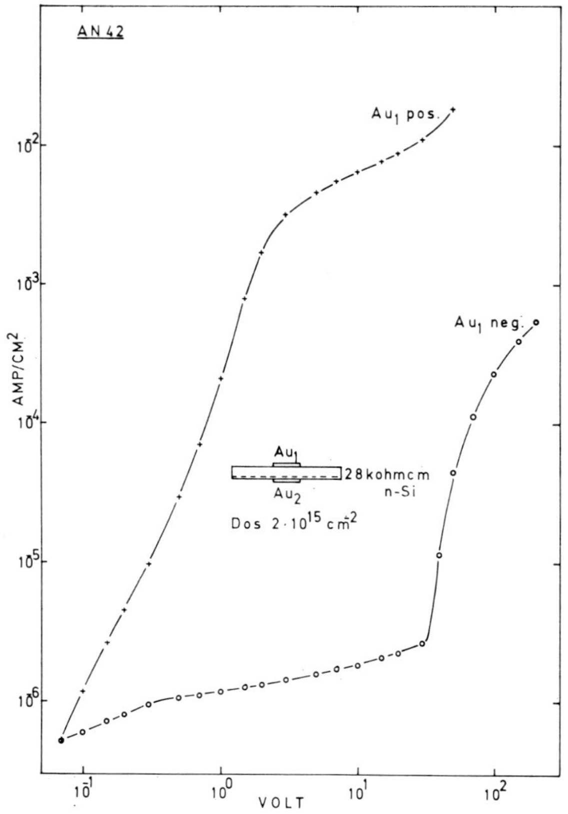

Figure 12. Characteristics of Au–*n*Si–amSi–Au structure.

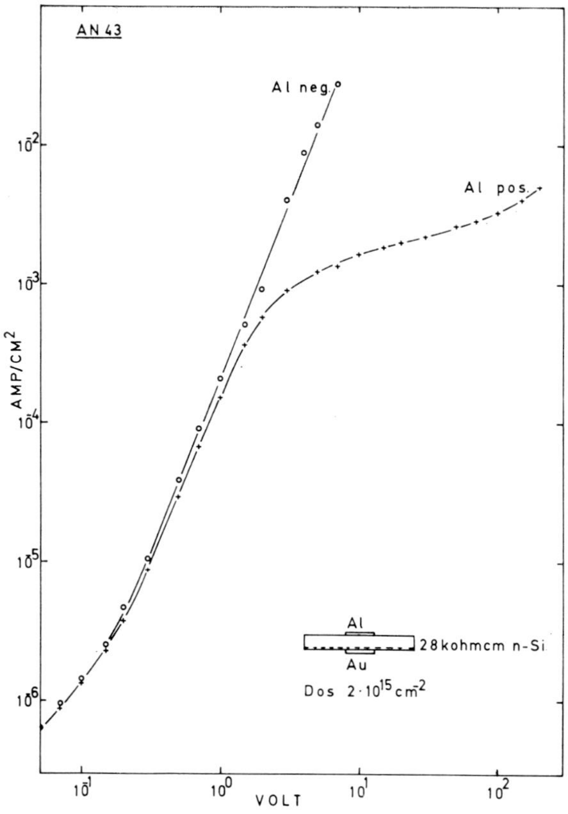

Figure 13. Characteristics of Al–*n*Si-amSi-Au structure.

holes. Because of the amorphization we expect a large number of centra throughout the band and hole generation can occur anywhere, even at the top of the valence band at the right contact in Figure 7c. However, the current does not fully reach single carrier SCLC values, possibly because of a current-limiting effect in the amorphized contact. Two carrier SCLC should not be possible because of the high barrier for electrons at Au_1.

For the other polarity we observe the expected forward characteristics of a Au–*n*Si diode. At voltages around IV this becomes modified by SCLC effects in the high resistivity material (28 kΩcm), as discussed by Sze.[10] We see no current limitation that could be traced to the amorphized contact.

For AN43 in Figure 13 the top contact is Al. For Al negative and moderate voltages the current corresponds to the wafer resistance, but at higher voltages we have a lower resistance, apparently due to two-carrier injection. For Al positive we observe a hole saturation current from this contact of 10^{-3} A/cm^2 which from Figure 2 corresponds to a barrier of $\simeq$0.55 eV which is not far from the expected value. At full depletion (cf band diagram Figure 7d) we may have contribution of electron current from the bottom contact but according to Figure 12 electron injection from this contact (for $V > V_d$) seems to be limited to values in the region of 10^{-3} A/cm^2, an addition which would hardly be visible in Figure 13.

The low current–low voltage regions in Figures 12 and 13, which are similar in shape and magnitude are apparently determined by the behaviour of the bottom Au contact, which is common for the two structures. Two normal barriers would in Figure 12 give symmetrical behaviour for the two polarities. If, however, the barrier at the bottom Au (amorphized) contact decreases successively with increased reverse bias (up to at least 1V) it would explain the shape of the upper current branch in Figure 12. At low voltages the two contacts would have the same barrier (giving the same currents) but at higher voltages only one barrier (Au_1) would remain high. Consequently the shapes of the curves Au_1 positive in Figure 12 and Al positive in Figure 13 are both determined by the bottom Au junction, the apparent resistance of which is much higher than that of the wafer. The low voltage currents are of a magnitude ($\simeq 5 \times 10^{-7}$ A/cm^2 at 5×10^{-2} V) that correspond to the expected Au–*n*Si barrier of 0.8 eV. We even observed the expected tendency that the current is smaller in Figure 12 (with 2 Au junctions) than in Figure 13.

5. Structures where both contacts are evaporated after chemical etching

Fabrication process. The contact on one side was aGe (contacted with Au), and on the other side Au, Al or aGe. The fabrication process was similar to that described in section 3. When the contact on one side was a metal it was deposited after the slight baking of the Ge contact.

I–V behaviour. A general tendency is that the currents for these structures are quite low and it seems possible to achieve saturation for both polarities on both *n*- and *p*-type material. Saturation currents are often in the region of 10^{-6} A, in some cases reaching 10^{-4} A/cm^2. Injection effects causing current rise at full depletion voltage may, or may not occur, depending on fabrication. Injection from the Ge contact is more rare than injection from the other (metal) side.

Typical characteristics for *p*Si structures are those of diodes D56 (Au–*p*Si–aGe) and D57 (Al–*p*Si–aGe) in Figure 14. We do not reach the high currents obtained for some polarities with the amorphized Si structure (section 4) or the structures with In(Hg) contacts (section 3). If we assume that the oxide interface layers formed during the chemical etching, and present both at the metal contacts and the Ge contacts, are important, we can explain the I–V data.

For positive bias on the metal (+curves in Figure 14) the current for moderate voltages must be a hole current from the metal, because of the following reasons: 1) In section 3, in connection with the discussion of 'Ge negative' curves in Figure 6 it was remarked that for this polarity the Ge contact cannot deliver large electron currents. However, contrary to in Figure 6 we must be developing a depletion region at the metal side, because the current saturates. Not until full depletion, when the field reaches the Ge contact, does current increase, as a result of electron injection from this (however, as discussed later the injection properties of the Ge contact can vary depending on the particular condition of fabrication). For low voltages we must have a significant hole barrier at the metal irrespective of whether we have Au or Al. We see in Figure 14 that the hole barrier is higher for Al (current density $\simeq 2 \times 10^{-6}$ A/cm^2) than for Au (current density $\simeq 4 \times 10^{-5}$ A/cm^2), as expected. Both these currents are, however, appreciably lower than expected from the estimated hole barriers of 0.65–0.55 eV and 0.3 eV, respectively. An increase of the hole barrier on *p*Si, because of positive charge in an interface oxide has been discussed by Pulfrey[5] and this is a possible explanation. The resulting barrier depends on the barrier ϕ_{B0} without interface, the interface thickness and the charge density in the interface.[5]

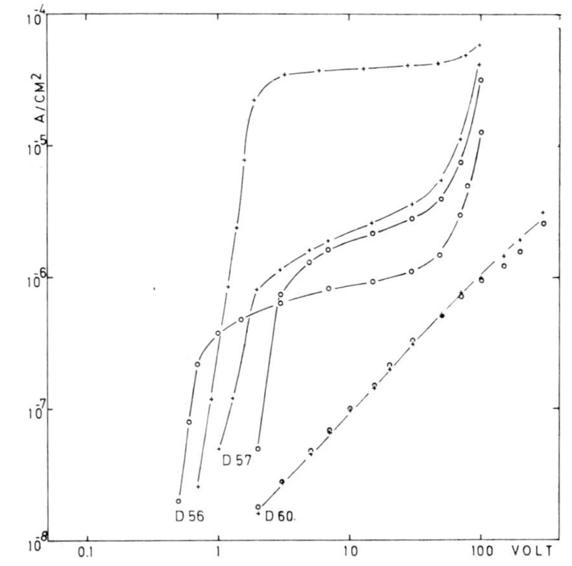

Figure 14. Characteristics of structures with two evaporated contacts. D56 is Au–*p*Si–aGe, D57 Al–*p*Si–aGe and D60 aGe–*p*Si–aGe. + points are for positive bias on the metal, ○ points for the reverse polarity.

For the opposite polarity we also observe saturation currents (though these are lower in magnitude). The currents in this case should be due to holes from the Ge contact (the effective barrier for electrons from the metal is too high for $V < V_d$) and the magnitudes are similar to those observed in Figure 6, for the same polarity on the Ge contact.

We drew the conclusion from Figure 4, lower curve A, $V > V_d$, that the hole barrier of a metal contact (Au) could be increased by charges in the interface.[5] Simultaneously we observed (for the opposite polarity, upper curve A, $V < V_d$) that the electron saturation current of the Au contact was close to normal, i.e. no significant decrease in its electron barrier was indicated. At first thought, an increase in ϕ_h could be expected to be coupled with a decrease of ϕ_e, if the change is caused by charges in the interface.

Now summing the information on saturation currents from Ge contacts with interface, we found (for $V < V_d$ and Ge positive) both in Figure 14, for D56 and D57 and in Figure 6 for GP11 and GP21 hole saturation currents of the order of 10^{-6} A/cm^2. We also found electron saturation currents in the region 2×10^{-7}–10^{-6} A/cm^2, for GN10 and GN20 ($V < V_d$) in Figure 5, and the I–V curves of the symmetric Ge structure D60 in Figure 14 do not contradict this information.

If we use the electron saturation currents for calculating the position of the Fermi level at the contact, for both metal and semiconductor contacts, using Figure 2, and then calculate the expected hole currents we thus get even larger discrepancy for semiconductor contacts than for metal contacts. One possible explanation is that the values of the Richardson constants are lower than those given in Figure 2. The theoretical model developed[2] predicts a modified Richardson type equation for amorphous–crystalline semiconductor junctions, but does not give the value of the constant. In absence of such information and considering the large current magnitude reached in several cases for forward-biased amorphous crystalline junctions in sections 3 and 4 (values expected for MS junctions are reached) we conclude that the major effect is barrier modification by charge build-up. This apparently occurs selectively, showing up as an increase in hole barrier height for positive bias on the metal or semiconductor contact. The simplest possibility is that the positive charge in the interface[5] which could explain the increase in barrier height, is due to the extracted holes. The effect appears more easily when a Ge contract is used than when a metal is used. The presence of a charge is supported by the capacitance measurements in section 3.

6. Conclusions

Junctions of evaporated Ge films on monocrystalline silicon have been shown able of accepting large electron and hole currents ($>10^{-2}$ A/cm^2) from the crystalline side, without appreciable voltage drop. The currents that can be injected into the crystal from the film are usually much smaller. Electron currents of $\simeq 10^{-6}$ A/cm^2 and hole currents of $\simeq 10^{-6}$ A/cm^2 (but sometimes up to 10^{-3} A/cm^2) are typically encountered (for the appropriate polarity in each case) and the value depends on fabrication conditions. The low value of hole current (in relation to the expected barrier height) is particularly remarkable. A similar effect is observed for metal (Au or Al) contacts, but the decrease in hole current is more pronounced for Ge contacts. As an explanation of the effect we propose that when the contact is biased for injecting holes, an increase of the effective hole barrier occurs. Such an increase has been shown to be caused by positive charges in the interface region and this could be connected with the flow of holes there.

References

[1] J B A England and V W Hammer, *Nucl Instr Meth*, **96**, 1971, 81.
[2] G H Döhler and M H Brodsky, AIP Conf Proc No 20 (Tetrahedr Bonded Amorphous Semiconductors) Amer Inst of Physics, 1974, p 351.
[3] S M Sze, D J Coleman and A Loya, *Solid State Electron*, **14**, 1971, 1209.
[4] R Hill, D Richardson and S Wilson, *J Phys D Appl Phys*, **5**, 1972, 185.
[5] D L Pulfrey, *IEEE Trans Electr Dev*, June 1976, p 588.
[6] H Norde, W Seibt, R Janney, *Solid State Electron*, **19**, 1976, 653.
[7] P A Tove and L G Andersson, *Solid State Electron*, **16**, 1973, 961.
[8] A Goetzberger, S M Sze, *Applied Solid State Science*, Vol I, p 174, Edited by R Wolfe, C J Kriessman, Academic Press, New York, (1969).
[9] E H Rhoderick, in *Metal–Semiconductor Contacts*, Conference Series No 22, The Institute of Physics, London 1974, p 17.
[10] S M Sze, *Physics of Semiconductor Devices*, p 421, Wiley, New York (1969).
[11] P A Tove, *Nucl Instr Meth*, **133**, 1976, 445.

2.6 Surface resistance measurement as an aid in controlling the fabrication of silicides

E Mgbenu*, S Petersson and P A Tove, *Institute of Technology, University of Uppsala, Sweden*

Silicides are used as ohmic contacts to silicon-based devices, and as rectifying contacts in some applications. The formation of the silicide is essential for the electrical properties as well as the long-term stability of the contact. In this paper we describe the use of surface resistance measurements to identify silicide formation. This parameter was determined as a function of temperature and duration of heat treatment, for layers of Pd, Pt and Rh deposited on Si crystals. The four-point-probe technique was used. Apart from identification of the silicide, we also found that increased annealing decreased the surface resistance of a formed silicide.

1. Introduction

Thin film metal silicides are formed by evaporating or sputtering a metal film on Si and heating the structure in vacuum or in the presence of an inert carrier gas. The formation temperatures (in K) for the silicides lie between one-third and one-half of the eutectic temperatures.[1] They are therefore formed in solid–solid rather than solid–liquid reactions. The use of silicides as ohmic or rectifying contacts is well known. The interface between this contact which has metallic character and the silicon is atomically clean. One is therefore assured more stable characteristics than obtained when an ordinary metal (e.g. Au, Al) is used for contacting Si.

In a previous paper[2] we correlate silicide surface resistivity measurements with backscattering analysis. We were able to determine resistivities of thin film silicides of Pd, Rh, Ni and Pt. Sheet resistance measurements on heat treated metal–Si systems could be valuable in other respects, when considered in detail. They could indicate precisely whether (a) silicide formation has started (b) the evaporated metal layer has been completely consumed in the solid–solid reaction (c) silicide formation has been impeded. Such information is valuable either to experimenters involved in the fabrication of new silicides or to manufacturers engaged in mass production. Some devices using silicide contacts may be sensitive to the resistivity of the silicide. This resistivity decreases with prolonged heat treatment and could easily be controlled by surface resistance monitoring.

In this paper we describe, using Pd, Rh and Pt as examples, how one can employ surface resistance measurements to quickly evaluate the state of heat treated metal–Si interface systems.

2. Treatment of data

Full details of sample preparation and measurements technique can be found.[2,3] A four point probe set-up with probe spacing 0.625 mm was used in determining the surface resistance. A current I was passed through the outer probes and the voltage V developed across the two inner probes was measured. The sheet resistance for our case, where the probe spacing is much larger than the layer thickness is given by[4]

$$R_{\text{sheet}} = \frac{\text{V}}{\text{I}} \frac{\pi}{\ln 2}$$

R_{sheet} depends on the resistivity of the layer. Where two or more layers are involved the resistivity dependence is given by[2]

$$R_{\text{sheet}} = \frac{\text{V}}{\text{I}} \frac{\pi}{\ln 2} = 1 \Big/ \int_0^{w_0} \mathrm{d}w/\rho(w)$$

where w_0 is the total thickness of the layers and $\rho(w)$ is the resistivity at depth w from the surface. It is assumed that the underlying substrate is of much higher resistivity.

The sheet resistance was measured as a function of temperature and duration of treatment. We used the percentage change in sheet resistance as a parameter. This % change K is given by

$$K = \frac{R(T, t) - R_0}{R_0} \times 100$$

where $R(T, t)$ is the sheet resistance measured for a sample annealed at T°C for t minutes, and R_0 is the sheet resistance of the as-deposited sample.

3. Results and discussion

In Figure 1 is seen a plot of K against t for Pd at $T = 300$°C. We also include experimental points for 5 min treatment at 450 and 600°C. There is an initial rise in the curve as Si atoms migrate into the transition region. We know that they are diffusing faster than Pd.[5] The resistance increases as an increasing part of the metal film is consumed for forming the silicide which has higher resistivity. For $t > 5$ min a drop occurs in K values; a similar drop is observed at the higher temperatures for 5 min treatment. This drop (after a rise) is a criterion that silicide has formed throughout the layer and is attributed to grain size increase in the silicide[2] (this decreases the resistivity). The silicide phase, e.g. for the samples corresponding to points A, B and C was identified as Pd_2Si, using backscattering analysis. The next phase, PdSi occurs at much higher temperatures ($\simeq$730°C).[6] We do not therefore expect it in our temperature range. It is interesting to compare the relative effectiveness

* On leave of absence from the Ibadan University, Nigeria.

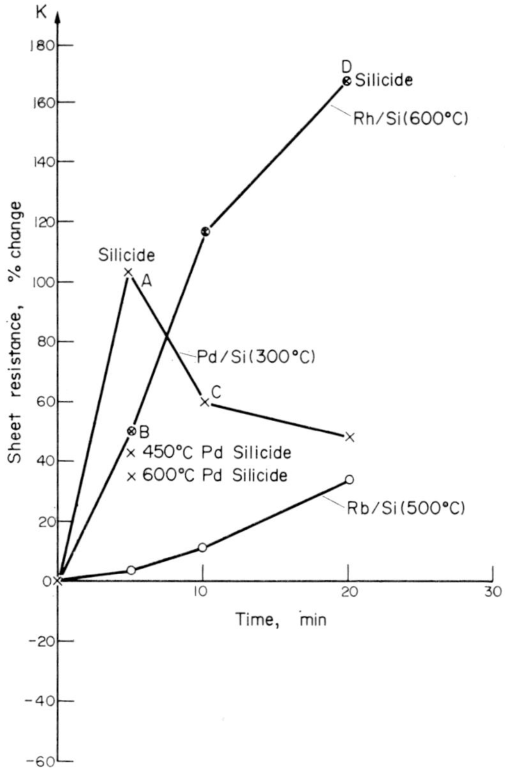

Figure 1. Percentage change in sheet resistance against duration t of temperature treatment for Pd and Rh silicides.

of temperature increase and duration of treatment on the annealing process and hence the measured resistivity. Comparing points A and B we notice that a 50% increase in temperature (for the same duration) leads to a 58% decrease in sheet resistance of the formed silicide. Comparing A and C, both for 300°C we see that a 100% increase in duration of treatment leads to a 42% decrease in sheet resistance.

The greater efficiency of temperature increase is also evident when we compare the curves for the 500 and 600°C treatments for Rh (Figure 1).

We pointed out above that if the resistance decreased after an initial increase, this indicated that silicide formation was completed (and grain size increase had set in). It can also be shown, however, that under certain conditions, a change in the slope of the plot resistance versus time occurs when formation is complete. We take as an example the case of Rh where we have a reaction-rate limited growth of the silicide and which is linear with time.[7]

In this case we have two parallel resistances. One is the formed silicide layer and one the remaining metal layer. The parallel combination of these layers gives a sheet resistance of the form

$$R_s = \frac{\rho_1 \rho_2}{\rho_2 w} \cdot \frac{1}{1 + \dfrac{kt}{w}\left(\dfrac{n\rho_1}{\rho_2} - 1\right)}$$

where ρ_1 and ρ_2 are the resistivities of the metal and the formed silicide, w is the thickness of the initial metal layer, k the silicide growth rate and n the ratio of silicide thickness to initial metal thickness (around 2). t is the duration of annealing.

If $n\rho_1 < \rho_2$ which is the case for Rh, it is easily found that the slope of the curve of R_s vs t increases with t.

This simple analysis which does not include any resistivity changes during annealing in the metal and silicide layer, shows that a change in the sign of the second derivative of the R_s vs t curve as we have in Figure 1 for the Rh/Si 600° curve can be attributed to a fully formed silicide. If the resistivity of the remaining metal layer decreases during annealing the statement above is still valid.

We see in Figure 1 that for 500°C the curve gradient increases as the silicide thickness grows with time. The silicide is not formed throughout the metal layer. This is different from the 600°C treatment, where we observe a decrease in the gradient after 10 min treatment. In this case we expect the silicide to be formed throughout the metal layer between 10–20 min. Point D corresponds to a slight annealing of the phase RhSi.

In the case of platinum (Figure 2) we identify the formation of two silicides at 200 and 350°C respectively. One case where

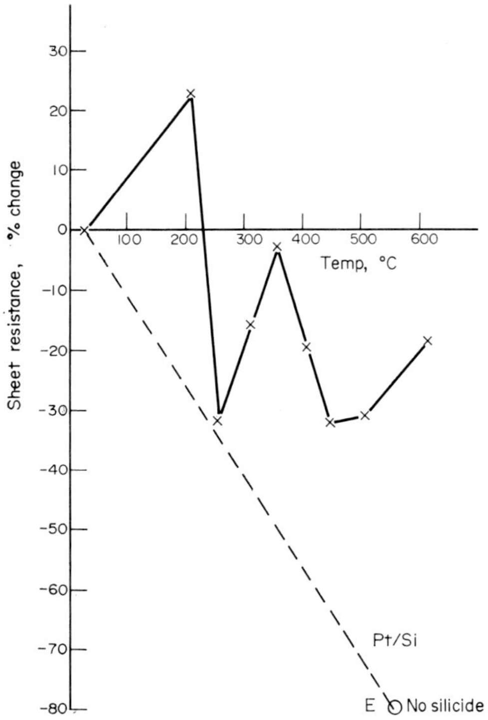

Figure 2. Percentage change in sheet resistance for Pt, against temperature of treatment. From 300°C (and on) the duration of the treatment was 30 min. Below this 60 min.

no silicide was formed despite prolonged heat treatment (550°C/30 min) is shown at point E. The platinum was electron gun evaporated on to the Si substrate. Here the drop in K is much more than average (> twice the % drop observed during normal silicide formation). It appears that the migration of Si was impeded probably by the formation of an oxide layer on the Si surface. The curve of Figure 2 in which we see normal silicide formation was obtained when platinum was sputtered on. Pre-sputter etching of the Si surface to remove any oxide layer and also produce a thin amorphous Si surface region, preceded the Pt sputter deposition. Where the pre-etch was omitted, no silicide was formed during the subsequent heat treatment. Point E would correspond to annealing of the Pt film; a decrease in resistance is obtained as vacancies and dislocations annealed out and as the grain size increases. Such an effect has been observed by Belser and Hicklin[8] for a Pt film deposited on a ceramic substrate. They observed agglomeration (island formation) leading to increased resistance at 600°C. The film we deposited on glass, at the same time as the Pt for point E was deposited on Si, agglomerated already at 400°C. This was characterized by a sharp increase in resistance and instability in the film. It appears that for deposition on glass extra stresses were introduced into the film as a result of thermal expansion of the glass.

4. Conclusion

We have discussed criterions for determining the formation of a silicide using surface resistivity measurements. This simple, inexpensive tool can be used independently of other more elaborate methods, as an aid in controlling the fabrication of silicides. It is a non-destructive method. Obviously it would not be possible to establish the silicide phase without recourse to Rutherford backscattering and/or X-ray methods. An initial increase in sheet resistance with temperature treatment would indicate that silicide formation has started. A subsequent drop would show that a phase has been formed throughout the metal layer and annealing has taken place. However, where the sheet resistance drops initially and progressively with temperature treatment, we could recognize that silicide formation has been inhibited. Invariably this is due to some oxide layer grown on the surface of the Si. It is known to increase considerably the temperature for silicide formation, by several hundred degrees; such an effect has been observed in the Mo–Si system.[9]

References

[1] J W Mayer and K N Tu, *J Vac Sci Technol*, II, 1974, 86.
[2] S Petersson, E Mgbenu and P A Tove, *Phys Stat Sol*, (a) **36**, 1976.
[3] K E Sundström, S Petersson, P A Tove, *Phys Stat Sol*, (a) **20**, 1973, 653.
[4] L Valdes, *Proc IRE*, **42**, 1954, 420.
[5] W K Chu, S S Lau, J W Mayer, H Muller and K N Tu, *Thin Solid Film*, **25**, 1975, 393.
[6] W D Buckley and S C Moss, *Solid State Electron*, **15**, 1972, 1331.
[7] D J Coe, E H Rhoderick, P H Gerzon and A W Tinsley, in *Metal Semiconductor Contacts*, Conference Report, The Institute of Physics (1974).
[8] R B Belser and W H Hicklin, *J Appl Phys*, **30**, 1959, 313.
[9] R W Bower and J W Mayer, *Appl Phys Lett*, **20**, 1972, 359.

2 Electrical properties of thin films—dielectrics

2.7 Deposition methods for dielectric films and their electrical properties

D S Campbell, *Department of Electronic and Electrical Engineering, University of Technology, Loughborough, Leicester LE11 3TU, England*

The various methods of depositing thin (< 1 μm) dielectric films are summarized. This summary is followed by a brief survey of the basic electrical properties and also of the applications of thin dielectric films with particular reference to deposition techniques. It is concluded that it is not possible to give a 'best buy' for a preparation method, as so much depends on the electrical studies to be made or the applications which are envisaged.

1. Introduction

The title of this paper is of such sweeping proportions that a book is required to give an adequate survey of all that is implied. Therefore, in the space available, all that can be attempted is to give a brief survey of the preparation methods, to look at the electrical properties which obtain, and to summarize the applications.

The reader might hope that this article would give some idea as to what is the best method of preparation for both basic studies and applications. Unfortunately, the situation is not simple enough for such a statement to be made. Methods of preparation which are of use for making films for basic studies are not necessarily of practical use in a production situation and conversely, films prepared for large-scale production processes may often be extremely impure and imprecise, so that although they are very adequate for the devices for which they are used, they are totally inadequate when it comes to identifying reproducible electrical conduction processes within them.

2. Deposition methods

2.1. Introduction. Deposition methods for dielectric films are many and varied. In the context of this particular paper it might be expected that one would concentrate on what are known as the physical methods of deposition (i.e. evaporation and sputtering), but it is felt that this would be an unfair process as it would eliminate discussion on other more chemical techniques which have major experimental or production importance. Accordingly, this section of the paper gives a brief review of deposition systems, both physical and chemical, some of which therefore, do not involve vacuum technology.

Over the past few years methods of depositing thin films have been reviewed in numerous texts, both with regard to physical techniques, (evaporation[1] and sputtering[2]) and to chemical techniques (thermal growth, anodization, vapour phase deposition and solution systems).[3] The specific application of growth methods to dielectrics has also been recently reviewed[4] and more recently still, the review of methods of deposition has been up-dated in terms of the deposition systems which have become of interest over the last few years.[5]

A summary of the techniques that can be used for preparing dielectric thin films is given in Table 1. The division into physical and chemical techniques is shown, physical techniques being those which depend on the physical evaporation or ejection of material from the source, whilst chemical techniques depend on a reaction such as the electrical separation of ions, as in electro-plating and anodization, or thermal effects on chemical reactions, as in vapour phase deposition and thermal growth. There is considerable overlap between the two basic methods, as some physical systems of deposition also involve chemical reactions as in the case of the plasma reactions shown under sputtering.

The techniques shown in Table 1 are all capable of producing films 1 μm thick or less. Other techniques are available for producing only thick films, such as glazing, electro-phoretic deposition, flame spraying, painting and screen printing. It is felt that these techniques do not come within the scope of this article. Readers interested in this area will find good summaries in 'Science and Technology of Surface Coating'[17] and sections dealing with the screen printing of dielectric layers in the 'Handbook of Thick Film Hybrid Microelectronics'.[18]

2.2. Evaporation.[1] Evaporation techniques are very widely used in the preparation of thin dielectric layers. Very large numbers of materials can be evaporated, either directly or in an oxygen atmosphere (reactive evaporation), and extremely clean conditions can be obtained in the vacuum chambers. Rates of evaporation can vary over very wide limits, dependent on the type of source used. Figure 1 shows a particular source developed by Drumheller[19] in the evaporation of silicon monoxide. The main point of the system is that no macroscopic particles are capable of being ejected from the source on to the substrate and therefore the film should not suffer from defects of this type.

A further type of source which has found considerable

Table 1. Summary of preparation methods applicable to dielectric films.

	Technique	Major references
	Evaporation (using all source types)	1, 6, 7
	Reactive	52
	Sputtering—Reactive	2, 6–8, 52
Physical	rf	2, 6–8
	Ion implantation	51
	Plasma reactions	
	Gaseous anodization	2, 3, 9, 52, 10
	Assisted thermal growth	2, 9, 52
	Polymerization	2, 9, 52
	Assisted vapour-phase	2, 9, 52
	Thermal growth	3, 7, 52
	Anodization	3, 10, 52
	Vapour base—Polymerization (EB and UV)	3, 7, 11, 52
	Oxidation	3, 7, 11, 52
Chemical	Nitriding	3
	Solution deposition oxides from colloids and	
	solutions	12
	Langmuir	13, 14
	Polymer	15
	Casting	16

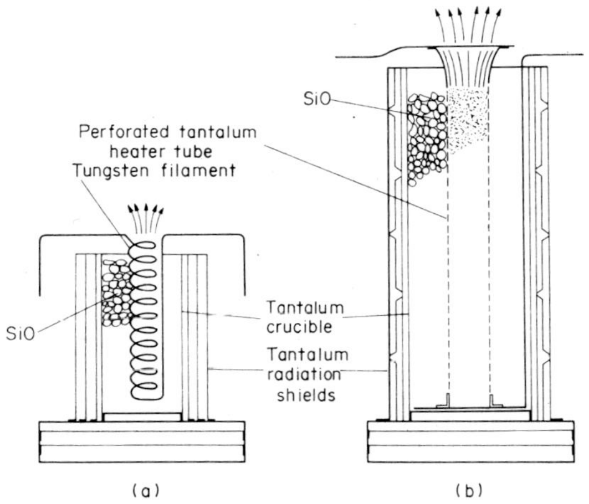

Figure 1. Two types of chimney evaporation sources for SiO.[19]

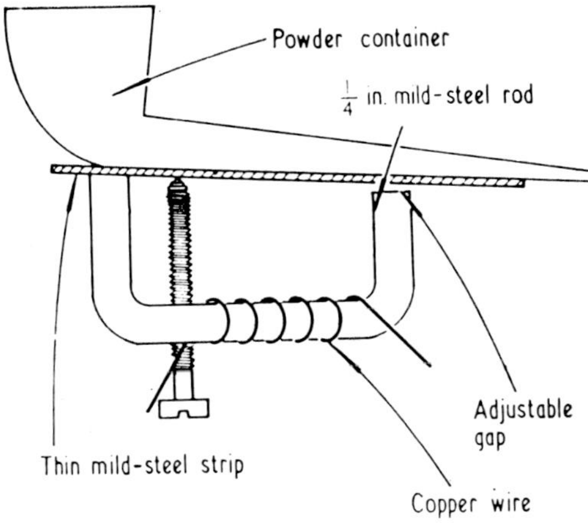

Figure 2. Flash evaporation system.[20]

applicability in the deposition of dielectric films, is that of the flash evaporation type. Figure 2 shows a diagram of a typical source developed for use in the deposition of nickel chromium alloys,[20] but which has found equal applicability in dielectrics, e.g. complex oxides such as barium titanate.[21]

The advantage of this type of source is that the composition of the final film can be the same as that of the original starting material, and effects arising from the different evaporation rates of the different constituents can be eliminated. The major disadvantage is that the material can be evaporated so quickly from the hot filament, that 'spitting' can occur and particles, although not necessarily landing on the substrate, can be ingested into the vacuum pumping apparatus unless great care is taken by the use of baffles, etc., to prevent this from happening.

Recently, laser beam evaporation has been found to be useful for the preparation of polymer dielectric films.[16] A wide variety of polymers have been prepared using this technique, including such complicated materials as polyvinylidene–difluoride and polystyrene.

Reactive evaporation,[22] as its name implies, involves the deposition of films in a reactive atmosphere such as oxygen or nitrogen. Materials such as silicon monoxide and tantalum oxide have been successfully deposited using this technique. One such system for the preparation of titanium dioxide films involves the evaporation of titanium monoxide in an oxygen atmosphere of between 10^{-3} and 10^{-4}, a source to substrate distance of 40 cm was used and a substrate temperature of 300°C. Kerner[23] has recently published studies on the kinetics of active evaporation with particular regard to alumina, silicon monoxide and chromium oxide. He has shown that in the case of silicon monoxide, one hundred times the amount of oxygen must impinge on the surface, compared with that which would be predicted if every oxygen atom was reactive. Similar figures were found for aluminium, but in the case of chromium oxide, the necessary impingement ratio was only 2.

2.3. Sputtering.[2] Glow discharge systems of sputtering have been well known for a long time and Figure 3 shows a typical apparatus which can be used. A discharge is maintained at a pressure

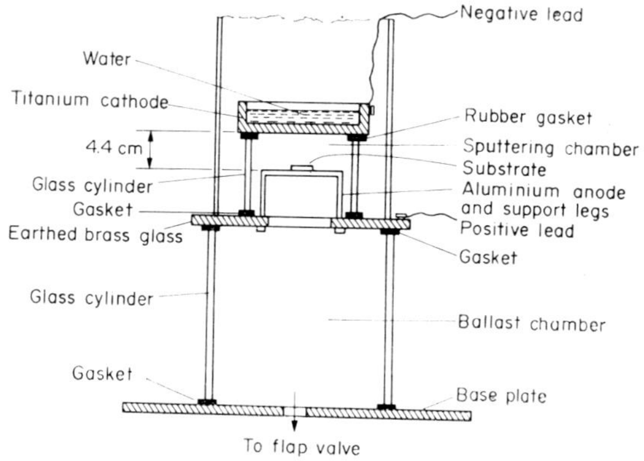

Figure 3. Glow discharge apparatus.

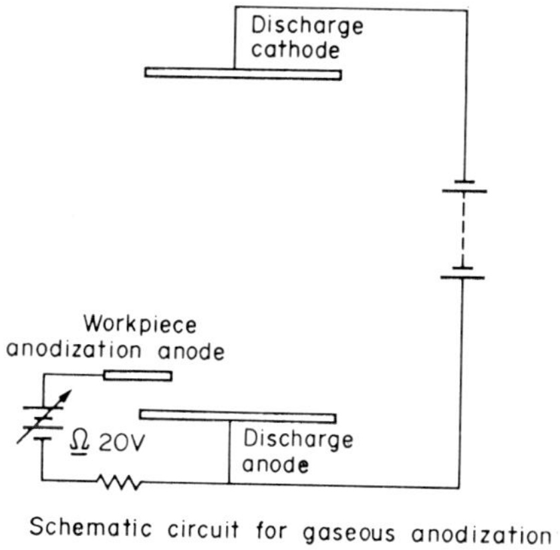

Figure 5. Gaseous anodization apparatus.

of around 10^{-2} torr between the anode and the cathode, and material is ejected from the bombarded cathode and deposited on a suitably placed substrate. The apparatus shown is that designed for the deposition of titanium, but if the argon atmosphere is replaced by an argon–oxygen mixture, then a reactive sputtering system is obtained which allows the deposition of an oxide on the substrate. The site of the oxidation process has been discussed by various workers. In the case of sputtering of tantala[24] it has been shown that there is a critical oxygen pressure (6×10^{-5} for a 1 cm^2 target) below which tantalum metal was sputtered and the oxidization reaction occurred at the substrate and above which tantalum oxide was sputtered from a tantalum surface as a result of oxidation at the target.

rf sputtering[2] using an insulator target and an rf frequency of around 50 kHz or above is an important process for preparing dielectric films. Two typical electrode systems for rf sputtering are shown in Figure 4. Not only oxides but also polymers are now being deposited using this technique.[25]

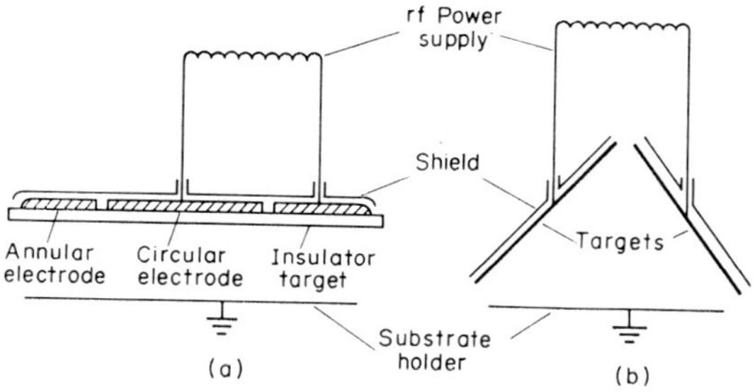

Figure 4. rf sputtering systems, (a) using concentric electrodes; (b) using adjacent electrodes at right angles.[2]

Gaseous anodization has also been used for the preparation of dielectric films.[10,26] Figure 5 shows a typical apparatus in which the material to be anodized is placed in the discharge at a slightly higher voltage than the discharge anode. Very successful films of alumina and tantala have been prepared on the parent metals and the system has the advantage of enabling films to be grown at anodizing temperatures well above the approximately 100°C limit set by the use of conventional aqueous electrolytes. However, the penetration of the plasma into any interstices in the substrate is extremely small, due to the shielding effect of the parent metal and so only flat surfaces can be successfully gaseously anodized, in contrast to the anodization of highly convoluted surfaces, which is possible with liquid systems.

Another plasma reaction that has been used for growing oxides, is that of plasma assisted thermal growth.[2,9,27] Ligenza, in his original work,[27] used an rf excited discharge and an oxygen pressure of between 0.1 and 1.0 torr, and found that silicon could be oxidized to a thickness of about 3500 Å at approximately 300°C.

Polymerization techniques have become widely used with the glow discharge acting as the polymerization medium. If a suitable monomer is introduced into the discharge chamber, it has been found that the monomer will be polymerized so that insulation films can be grown on suitably placed substrates.[9] A wide variety of monomers have been polymerized and deposited in this way.

Glow discharge conditions can also be used to effect vapour phase reactions such as the deposition of silicon nitride from a gaseous mixture of silicon hydride, ammonia and hydrogen.[28] rf excitation of the discharge has also been used and an apparatus has been described by Connel and Gregor[29] which can produce insulating films from styrene at the rate of 20 Å s^{-1}.

2.4. Thermal growth.[3] Films can be formed chemically on a large variety of metal substrates by heating them in gases of the required type. The films obtained, however, are limited in thickness because the reaction will become very slow as the film thickness increases. A typical curve for alumina on aluminium is shown in Figure 6. It can be seen that the final value of thickness is obtained only after a considerable period of time.

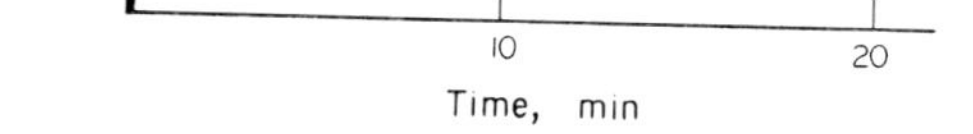

Figure 6. Thermal growth curves for Al_2O_3 on Al at 20°C.

Thermal growth can be enhanced by heating the substrate and in the case of silicon, an atmosphere of either oxygen or water vapour is used and temperatures between 700 and 1400°C.

In the case of aluminium, tantalum and silicon and other valve metals, a coherent oxide film is formed, thus preventing further oxide on the substrate. However, with a lot of other metals, a non-coherent film is obtained which will continually flake off during preparation, thus re-exposing the substrate. Such behaviour is of little use in preparing dielectric films, and explains the limited applicability of thermal growth techniques.

2.5. Anodization.[3,10] In order to avoid the necessity for heating substrates to high temperatures, it is possible to enhance oxidation by immersing parent metals in an electrolyte medium with the material to be oxidized being made the anode of the electrolytic cell. As in the case of thermal growth, only a limited number of materials form coherent oxides, but for these materials the oxide thickness obtained is now a function of the voltage applied to the electrolytic cell. Growth curves for tantalum oxide and aluminium oxide are shown in Figure 7 for an applied

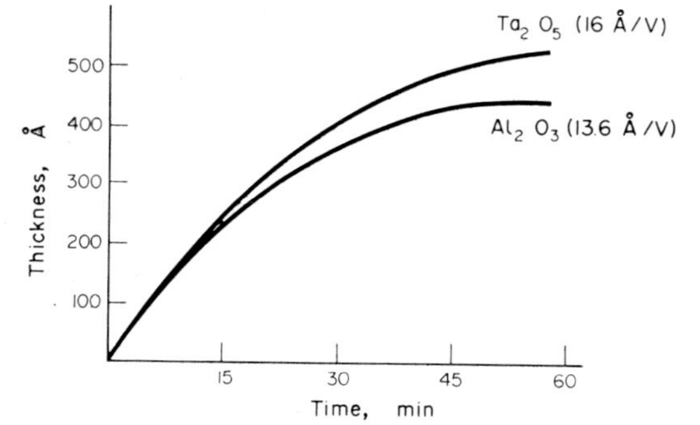

Figure 7. Anodization growth curves for Al_2O_3 on Al and Ta_2O_5 on Ta at a constant voltage of 30 V and 90°C.

voltage of 30 V. The asymptotic thickness values are again obtained, as was the case for thermal growth, and for the case of aluminium, an anodization constant of approximately 13 Å per volt and for tantalum of 16 Å per volt, is found. The disadvantage of this approach is that the initial stages of anodization require very high current densities, and one way around this difficulty is to anodize initially using constant current conditions rather than constant voltage.

The limiting condition of the thickness of the film is the breakdown under high voltages, and in the case of aluminium, this occurs at an alumina thickness of 1.5 μm and with tantala at 1.1 μm. Other factors can, however, limit the thickness to values less than those quoted, particularly the purity of the substrate and composition of the electrolyte.

Anodization is a widely used technique for obtaining amorphous, highly insulating films. In electrolytic capacitors, both those based on tantalum and aluminium, very large surface areas of very thin films are continually being prepared and used. Anodization systems for capacitors are often constructed on a massive scale and in the case of aluminium, continuous rolls of etched foil, 50 cm or so in width, are anodized to voltages ranging from 20 to 600 V. One of the features of large-scale anodization is that in order to avoid too lengthy a stay of the foil in the electrolytic bath, immersion may only be for a relatively limited time. Thus the asymptotic thickness value will not be obtained and although the foil may have been anodized with a known applied voltage, the final thickness will not be that determined from the anodization constant, but is likely to be only 60 or 70% of the expected value. Figure 8 shows a diagram of a typical anodization plant for aluminium foil for anodizing to the largest voltage normally used; washing tanks and a two-stage anodization process, are shown.

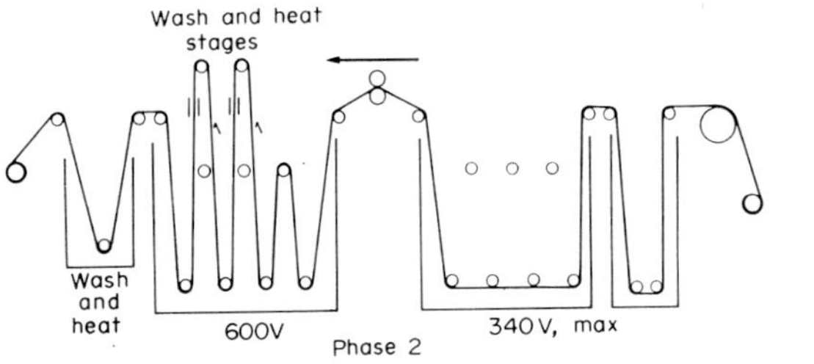

Figure 8. Schematic diagram of large-scale anodization apparatus for high voltage film.

2.6. Vapour phase systems.[3,7,11] Vapour phase growth is widely used in the preparation of single crystal silicon for semiconductor purposes. In the case of dielectric films, however, the systems of disproportionation or decomposition which is normally used for metallic or semiconducting films, are not applicable. However, dielectrics may be prepared by oxidation usually using the halide of the required metal oxide (e.g. alumina can be deposited from a vapour of aluminium trichloride mixed with steam at substrate temperatures of 600°C). However, difficulties are encountered in ensuring thorough mixing of the halide and the water vapour and also an even temperature must be maintained in the reaction vessel. One technique for effecting this is the use of a fluidized bed,[3] although in these circumstances the particles of the bed will themselves become coated with oxide.

Polymerization systems can also be used as has been indicated in the case of glow discharge vapour phase growth. Other polymerizing agents are electron beam irradiation[9] and ultra-violet irradiation.[9,25]

2.7. Solution deposition. Over recent years, it has been found possible to deposit satisfactory dielectric films by solution systems.

(a) **Deposition of oxide films from colloidal suspensions.**[12] In this technique a colloidal suspension of the material is prepared and the substrate to be coated is immersed in the suspension so that a thin layer of material is formed on both sides of the substrate. Uniformity of this liquid film can often be improved by spinning the wetted surface and after spinning, the substrate is baked, often to a temperature of between 200–500°C to convert the liquid colloidal layer to a solid, usable structure. Satisfactory silicon dioxide films have been prepared by this system, using a colloidal silicon dioxide hydrate. Film thicknesses can be as low as 100 Å and multiple layers can be built up by repeated immersion and baking, a technique which has been applied for making multi-layer dioxide/titanium dioxide structures.

Non-colloidal systems can also be used in which the baking process converts the solution into an oxide deposition. A typical example of this type of film growth is the preparation of manganese oxide layers from dilute solutions of manganese nitrate, the liquid film being pyrolized at a temperature of approximately 400°C.[30] This technique, which is widely used in tantalum electrolytic capacitors, can be used for the preparation of counter-electrodes on anodized tantalum surfaces.

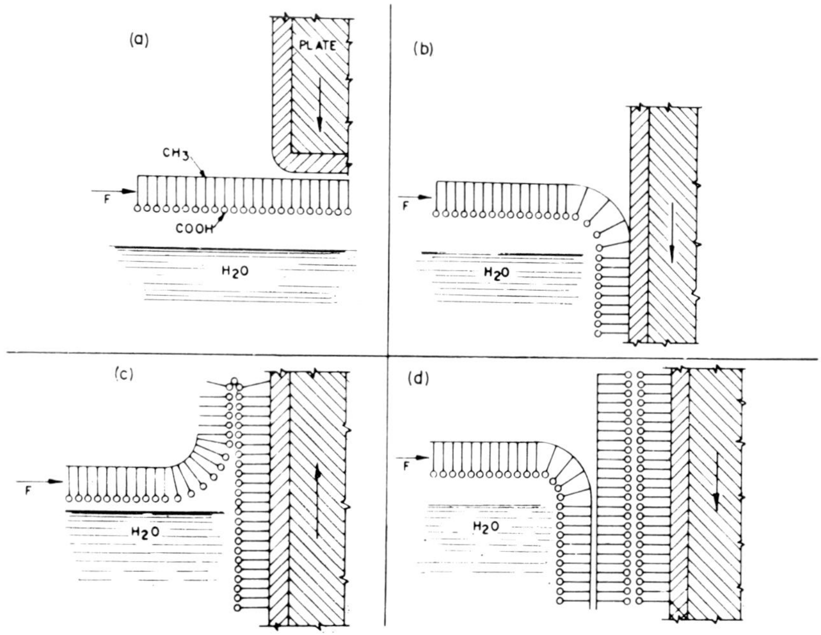

Figure 9. Diagrams showing the build up of monomolecular layers by the Langmuir process.[13] (a) to (d) is the sequence of growth.

(b) Langmuir films.[13,14] Multi-monomer layers of long-chain, fatty acids can be built up on a substrate by repeated immersion of the substrate in liquid, on the surface of which are floating long-chain, fatty acids. Barium stearate has been widely used in this connection and the layers obtained at each immersion are about 20 Å thick. Figure 9 shows a schematic diagram of the process. Successive layers can be obtained by repeated immersion and withdrawal of the substrate and successful films, up to 2 μm thick, have been prepared.

(c) Polymer deposition from solution.[15] Polymer films from materials such as polypropylene, polystyrene and PVC, have been obtained by the simple technique of direct isothermal immersion of a substrate into a suitable solution of the polymer. Material will deposit on the immersed substrate until an equilibrium is reached between deposition rate and re-solution rate. Figure 10 shows typical growth curves for solutions of PVC in cyclohexanone at 40°C.

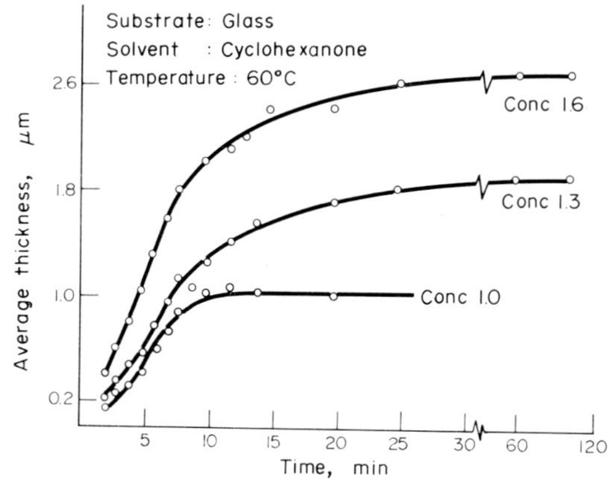

Figure 10. Growth of PVC from solution in cyclohexanone at 20°C. (Three different concentrations in g wt/100 cm^3 of solvent.)[15]

(d) Solution casting.[16] It has been found possible to prepare satisfactory film by allowing the evaporation of a polymer-containing solution placed on a substrate. Very satisfactory films of a fairly wide variety of materials have been obtained by this technique, e.g. polystyrene films from polystyrene dissolved in chloroform, but the layers obtained are usually relatively thick (greater than 1 μm).

3. Electrical properties

3.1. Introduction. Having reviewed the techniques that are available for the preparation of dielectric films, the next section of this paper is concerned with the properties of the layers obtained. As has been noted earlier, in the space available it is impossible to review extensively all the electrical properties in relationship to their preparation techniques and therefore, what will be attempted is an outline of the type of basic property which needs to be examined to determine whether or not reproducible films are being obtained, and also a summary of the applications.

For both a study of basic properties and of applications, it can be recognized that various criteria of stability must be fulfilled. These criteria can be identified in that the preparation process must produce films with the following chracteristics:

(a) Continuity. Preparation techniques which give discontinuous films, flaws, or even films with thinner areas than the main film, will lead to layers with unsatisfactory performances. Lack of continuity is often alleviated by the use of thin (less than 1000 Å) counter-electrodes, usually of aluminium, which melt above the weak point in the dielectric, where excessive current flows through the film. This melting process then isolates the weak, discontinuous area and thereby allows the rest of the film to operate satisfactorily. This type of technique is widely used in the preparation of thin film, non-polar capacitors. A second technique is to immerse the dielectric in a re-forming medium. In the case of solid tantalum capacitors,[30] a counter-electrode of manganese oxide can provide this re-forming characteristic, since any weak point and hence high current flow in the tantalum oxide dielectric which has been prepared on the tantalum metal, will cause reduction on the manganese oxide counter-electrode at that point. The tantalum oxide film will then become fully oxidized and non-conducting, whilst at the same time the manganese oxide which was initially a semiconductor, will be reduced and will become an insulator, thereby further isolating the initially weak area in the film. In the case of wet electrolytic capacitors,[30] the presence of an anodizing solution in contact with the film, will cause re-anodization of any weak area in the film during use, provided the polarity of the device is maintained such that the tantalum or aluminium base material is always an anode (i.e. positive), relative to the other electrode.

(b) Mechanical stability. It has been shown over the past few years that all films when deposited, are either in compressive or tensile stress and extensive reviews on the type of stress obtained and the mechanism involved, have been given by various authors, both with regard to thin films in general[31] and to dielectrics in particular.[32] Films in tensile stress are basically unstable and if the tensile stress is of sufficient magnitude, this can cause the film to crack and become an electrical short circuit. A wide variety of dielectric films are therefore comparatively useless because of this problem. On the other hand, films in compressive stress are much more stable provided the film stress is not sufficiently high to cause buckling of the film on the substrate. The stability of a wide variety of used dielectrics arises from this property, particularly materials such as alumina, silicon monoxide and silicon dioxide.

(c) Homogeneity. Even with continuous films in a state of slight compressive mechanical stress, satisfactory behaviour will not be obtained, particularly in basic terms, unless the films are homogeneous. The crystal structure of materials can be divided into three classes, namely, single crystal, polycrystalline and amorphous. Single crystal materials are characterized by a well-defined lattice spacing between the constituent atoms over a large volume, and dielectric films are not easily obtained in this form. Polycrystalline materials, as the name implies, consist of a collection of individual crystallites of varying sizes and shapes with grain boundaries between them. Dielectric films of this type can sometimes be satisfactory, although in the majority of cases the presence of grain boundaries affects conduction, both in terms of the accumulation of impurities in the grain boundaries or of the easy access of water vapour, etc. to the grain boundary positions. Thus, electrical conduction can occur easily along the grain boundaries themselves. Amorphous films have crystallite sizes which are only atomic in dimensions and it is not possible to describe these films in terms of the normal lattice spacings associated with single crystal materials, or the crystallites of polycrystalline materials. The most satisfactory physical description of an amorphous material can be given in terms of a radial distribution analysis in which the probability of finding an atom at a given distance from a fixed point is described. The majority of satisfactory dielectric films consist of layers of this type. They have the advantage of lack of grain boundaries and also because of this, impurities up to certain limits, can be scattered randomly through the amorphous matrix without affecting the dielectric properties to any considerable degree. To obtain reproducible basic properties, however, it should be noted that the amorphicity must itself be homogenous and any impurities present must be distributed through the matrix in a homogeneous sense, and it is this lack of homogeneity in the amorphous matrix which has bedevilled the detailed physical examination of films in the past. Uniformity of the amorphicity is very much a function of the preparation conditions used.

3.2. Basic properties. To examine a preparation technique and to determine whether reproducible, satisfactory films have been obtained, all the above criteria outlined must be fulfilled. On top of this, however, various characteristics will need to be examined and some of these can be identified. Detailed discussions of electrical characteristics of thin films have been given by various authors and the reader is particularly advised to consult the work of Campbell and Harrop,[33] and Fonasch[34] on permittivity and polarization behaviour, and Hill,[35] Jonscher and Hill[36] and Simmons[37, 38] on conduction processes.

3.2.1. Permittivity. Permittivity of the film must be a reproducible quantity and hopefully will fit into an expected value. In the case of oxides, a typical relationship which has been established is shown in Figure 11.[33] The temperature coefficient

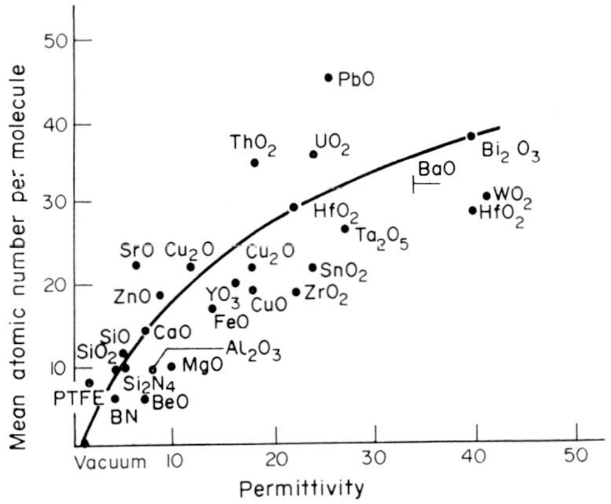

Figure 11. Mean atom number per molecule vs permittivity for thin film dielectrics.[16]

of capacitance behaviour should also be identifiable. The relationship of temperature coefficient of capacitance γ_c and temperature coefficient of permittivity γ_ε, is given by:

$$\gamma_c = \gamma_\varepsilon + \alpha_L \qquad (1)$$

where α_L is the linear expansion coefficient.

From this relationship it is possible to identify the behaviour of γ_c in terms of permittivity, polarizability and dielectric loss. For low loss materials (tan δ less than 0.1%), the relationship between the temperature coefficient of capacitance and permittivity, can be obtained as shown in Figure 12.[33] For losses greater than 0.1%, other curves can be drawn at higher values of temperature coefficient of capacitance for a given permittivity value, although these curves have not been so definitely identified.

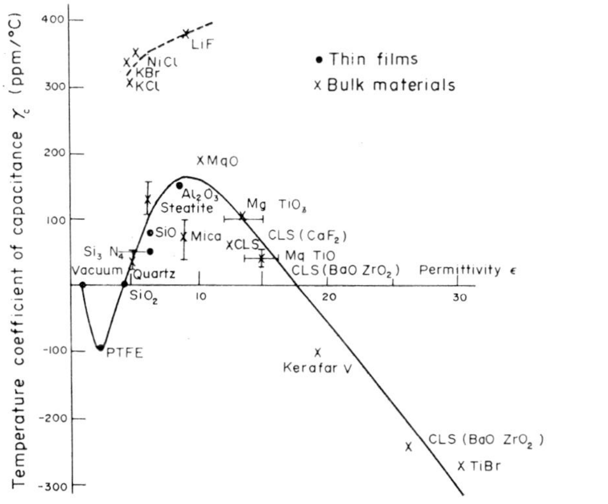

Figure 12. Temperature coefficient of capacitance, γ_c, vs permittivity for low loss thin film materials.[16]

3.2.2. dc conduction at low fields (less than 10^4 V/cm). dc conduction at low fields has been one of the most profitable areas of examination of thin dielectric films. However, often in this area, reproducibility of results has been difficult to obtain. A very valuable analysis of the present situation has been given recently by Jonscher and Hill,[36] to which the reader is referred for detailed discussion. The type of characteristic which must be looked for is in terms of a plot of current vs temperature. Various analyses have been made, given different energy band models. For example, assuming a constant density of trapping sites distributed randomly in space and centred around the Fermi-level in the dielectric, it has been shown that a relationship in which the log of the current is proportional to $1/T^{1/4}$ is obtained. This type of characteristic is most easily seen in amorphous elemental semiconductors and Figure 13 shows a typical plot for amorphous carbon, as analysed by Hill.[36]

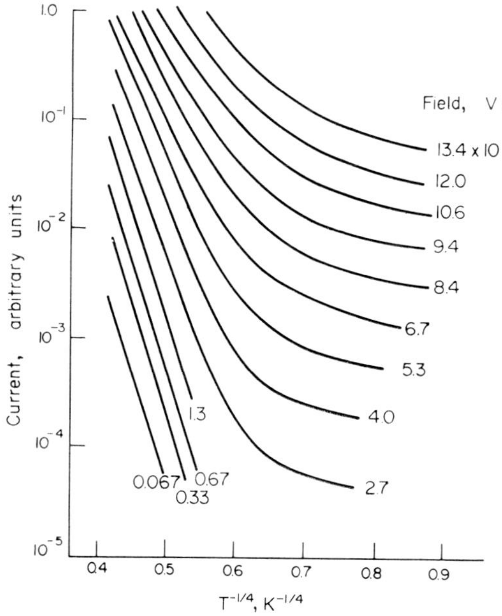

Figure 13. Current vs $1/T^{1/4}$ for amorphous Ge.[36]

3.2.3. dc conduction at high fields (greater than 10^4 V/cm). At high fields electrons may be activated out of trapping sites in the dielectric into the conduction band, giving rise to the so-called Poole–Frenkel Effect. A simple analysis of the Poole–Frenkel Effect gives the relationship that the logarithm of the current is proportional to $E^{1/2}/kT$ where E is the applied field and T the temperature. However, it has been pointed out by Jonscher and Hill[36] that this relationship is not entirely satisfactory, and a more complicated relationship should be involved. They have shown that a satisfactory proof of the existence of a Poole–Frenkel mechanism is experimental agreement of i vs V with a relationship of the form:

$$I = \gamma \cosh \gamma - \sinh \gamma \qquad (2)$$

where I is a normalized current and γ is given by:

$$\gamma = e\beta E^{\frac{1}{2}}/kT \qquad (3)$$

with e the charge of the electron and β a constant. Figure 14 shows a typical plot quoted by Jonscher and Hill[36] for results on silicon monoxide over a range of temperatures from 140 to 413 K. This type of behaviour has been found to be followed by

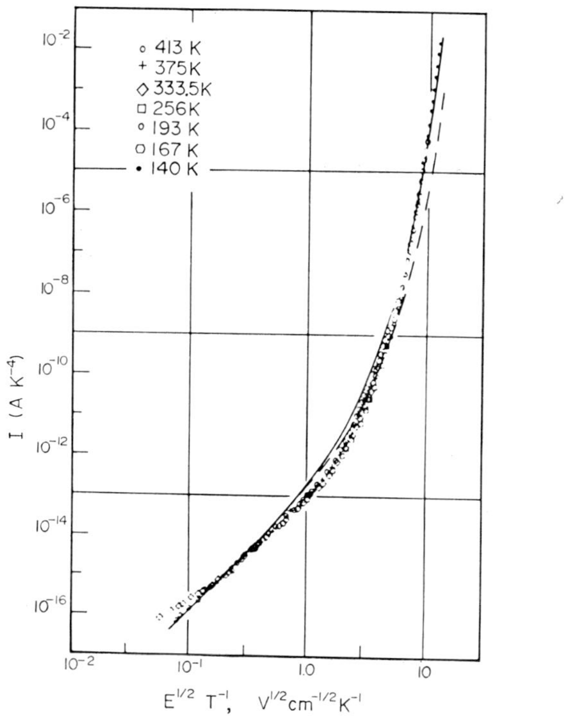

Figure 14. Normalized plot of current vs $E^{1/2}/T$ for SiO.[36] This plot shows the Poole–Frenkel Effect applies as the trap density is relatively low.

silicon monoxide, boron nitride, zirconium dioxide and chromium oxide in particular. Other modified forms of equation (2) have also been found to give good fits with experimental results, e.g. $I = \gamma^2 \sinh \gamma$ and $I = \cosh \gamma - \gamma^{-1} \sinh \gamma$.

If the impurity density and hence trapping sites density in the material is particularly high, then the Poole–Frenkel Effect behaviour will be modified, and in these circumstances log current is proportional to E/kT. For most satisfactory results, however, once again a similar plot to equation (2) can be obtained with γ being the modified function $e\beta E/kT$.

Figure 15 shows such a plot for the case of silicon monoxide and Jonscher and Hill's calculations[36] show that this behaviour corresponds to a trapping site spacing of 80 Å, a trapping site density of 2×10^{18} cm^3 and an activation energy of 0.35 eV.

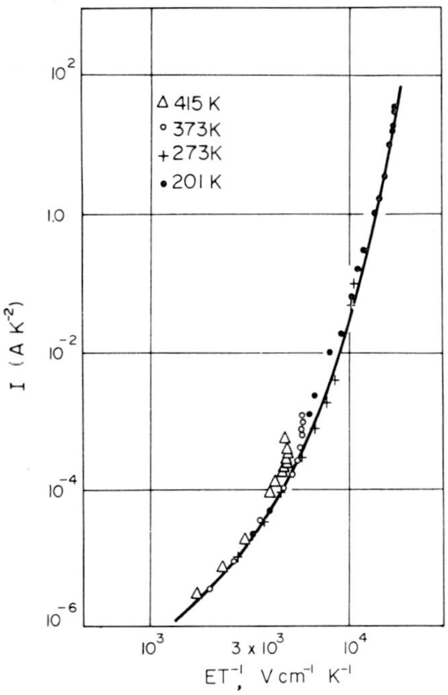

Figure 15. Normalized plot of current g vs E/T for SiO.[36] This plot shows Poole's law applies as the trap density is high.

3.2.4. Tunnelling behaviour. A further characteristic which needs to be examined in terms of identifying reproducibility in dielectric films, is that of tunnelling. For films less than 50 Å, it is possible for electrons to tunnel through the dielectric from one electrode to another, thereby giving a current that can be plotted as a function of film thickness. An analysis[37] has shown that characteristics of the type given in Figure 16 should be obtained where the conductivity is plotted as a function of applied voltage for various film thicknesses.

3.2.5. ac effects in thin dielectric films. These effects are at present receiving a great amount of theoretical and experimental attention, but as has been pointed out by Jonscher and Hill[36] there is, as yet, no way of relating in a quantifiable manner, the results to the specific amorphous characteristic of the materials, and hence to the preparation conditions. The type of characteristics that should be obtained have been discussed in the review by Simmons[37] where he has shown that a plot of ac conductivity against frequency should limit the square law relationship for a system which depends on hopping conduction processes. Figure 17 shows such a relationship for evaporated silicon oxide films. Behaviour in terms of temperature can also be analysed and Figure 18 shows a plot of ac conductivity against I/T, again for silicon monoxide.[37]

As well as ac behaviour of this type, polarizability and resonance phenomena will also be present in the film,[39] and Figure 19 shows, in a diagrammatic manner, the types of behaviour that can be obtained in terms of loss vs frequency.[6] At any given frequency a value of loss will be obtained which will be dropping proportional with frequency, since tan δ

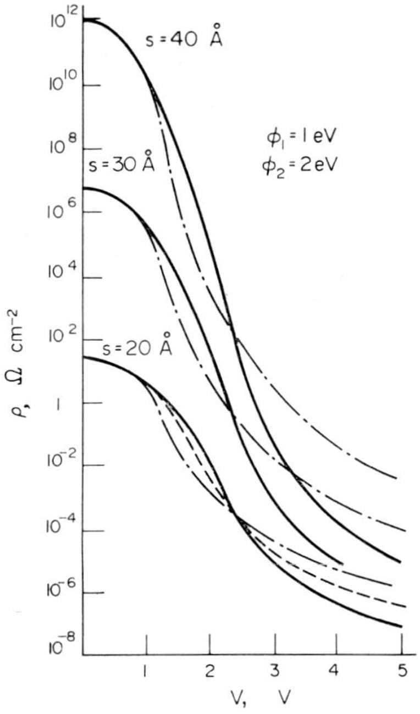

Figure 16. Resistivity vs voltage for thin dielectric layers of different thicknesses, s. Work functions of electrodes, Φ, and Φ_2 for 1 and 2 eV, respectively. Full line is the reverse and the chain line the calculated forward characteristics.[37]

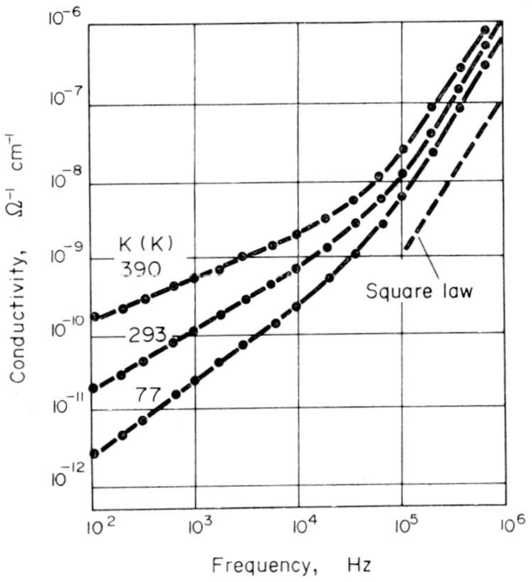

Figure 17. ac conductivity vs frequency for SiO films.[37]

$= 1/\omega RC$. However, superimposed on this will be conduction peaks due to relaxation processes, one of which is shown schematically in Figure 19. It was also found experimentally that most dielectrics have a loss value which is nearly invariant with frequency at room temperature. Frequency independent loss is found at all loss levels and this phenomenon can be considered as being due to a superposition of a wide spectrum of relaxation peaks.

3.2.6. Electrode effects. All the above properties can be modified by the electrodes themselves. The presence of the electrodes will cause band bending at the interface, the type of bending dependent on whether or not the blocking, neutral or injection contact is established at the interface. This situation

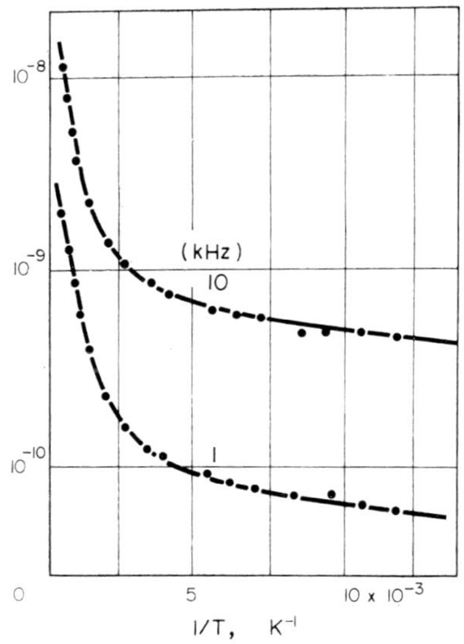

Figure 18. ac conductivity vs reciprocal temperature for SiO films.[37]

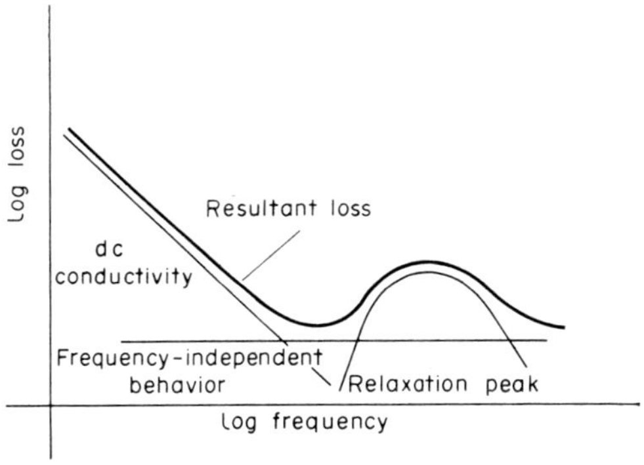

Figure 19. Schematic diagram of loss (tan δ) vs frequency showing the effect of the various ac conduction processes.[6]

has been examined in detail by Simmons,[37] who has shown that the band bending can occur over considerable distances. As a result of this, the band diagrams at one interface will be as shown in Figure 20 and for a two-electrode system, as in Figure 21. For large band bending distances which could easily be greater than the thickness of the dielectric, it can be seen from Figure 21 that the conduction processes through the dielectric are likely to be considerably modified from those expected from an ideal, isolated dielectric layer.

The band bending situation at the electrode interface can also have an effect on ac properties so that the equivalent circuit diagram for a capacitor can well be represented as shown in Figure 22. The barrier (Schottky capacitance) is shown as C_s and the interior capacitance as C_b. E_d is the activation energy of the trapping centres of the dielectric and R_o is a resistive parameter whose magnitude depends on mobility, film thickness, donor density and temperature. Results are often shown of apparent change in dielectric constant with thicknesses at a particular frequency, which can often be explained in terms of this type of equivalent circuit.

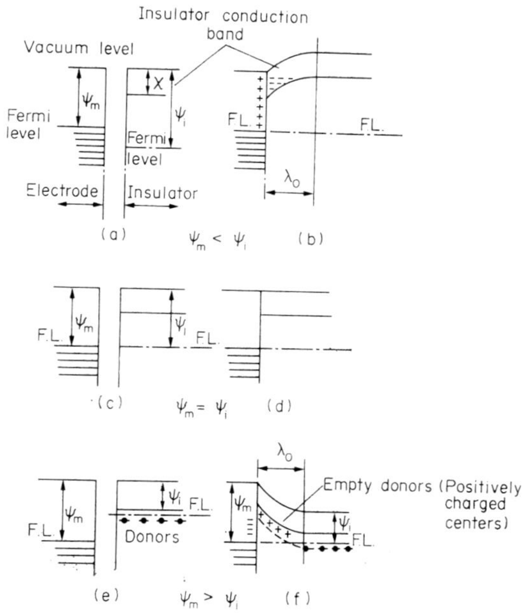

Figure 20. Energy level diagrams before and after joining a dielectric to a metal electrode. (a) and (b) injection contact; (c) and (d) neutral contact; (e) and (f) blocking contact.[37]

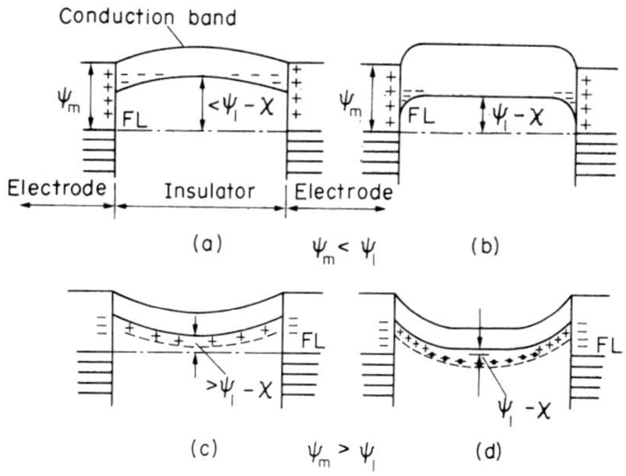

Figure 21. Energy level diagrams for two metal contacts and an insulator. (a) and (b) show the effect of different band bending distances with injection contacts and (c) and (d) similar situations with blocking contacts.[37]

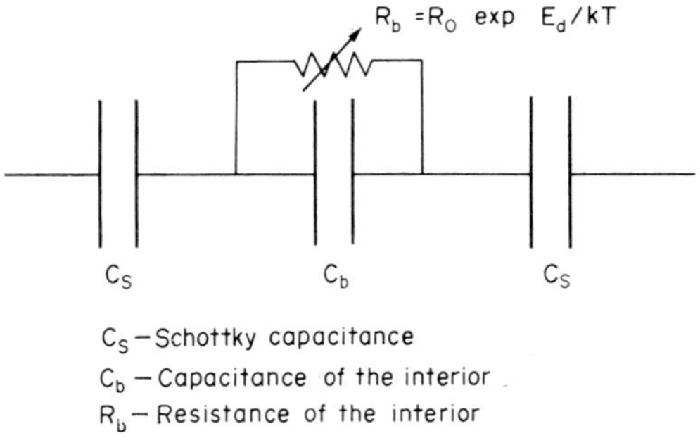

Figure 22. Equivalent circuit showing effect of interfacial barriers. (Schottky barriers.)[37]

3.2.7. Breakdown.[6,40] Electrical breakdown in thin films is far simpler than in massive materials, because of the geometry and structure which precludes a number of phenomena. The best insulating films have breakdown fields around 10 mV/cm, whereas the best massive insulators, with the exception of mica and a few plastics, have breakdowns nearer 1 mV/cm at room temperature. Three types of breakdown have been identified, and extensive studies have been made by various workers.[40]

Initial breakdowns can be identified with flaws in the dielectric and these are the type associated with the burning out of the electrode or re-forming of the dielectric, as discussed earlier.

Table 2. Typical properties of thin film dielectrics of major importance in capacitor structures.

Material	Preparation method	Typical capacitance value ($\mu F/cm^2$)	Loss (% at 1 kHz)	Permittivity	Reference
Al_2O_3	Wet anodization	0.02–0.4	0.5	10	6, 10 41, 30 42, 49
Si_3N_4	rf sputtering	0.02	0.1	9	6, 41, 43
SiO	Evaporated	0.01	0.1	6	6, 41
SiO_2	rf sputtered thermally grown	0.1	0.2	4	6, 41, 43
Ta_2O_5	Wet anodization	0.1	1.0	25	6, 41, 50

This type of breakdown does not necessarily characterize the film. At high fields, however, intrinsic breakdown occurs which depends on the basic properties of the material itself and Joule heating effects are quite often the most prominent cause.

3.3. Applications of dielectric films. These are not necessarily related to the basic properties discussed in the previous section, although reproducibility of characteristics is obviously important. Basic properties which are of relevance in device application are such characteristics as permittivity and temperature of capacitance for applications involving insulation and dc conduction at low fields, together with tunnelling, for applications concerned with conduction processes such as cermet resistors, switches and tunnelling devices.

Three areas of applications can be identified:

3.3.1. Capacitors. From a production point of view only a limited number of materials are used as capacitors, although a wide variety have been examined in experimental terms. A recent review[41] contains a useful survey of the different materials which have been used.

Table 2 summarizes the main materials which have been used for the production of capacitors. By far the most important, if only in volume terms, is that of alumina prepared by wet anodization on tantalum foil.[49] The foil is usually heavily etched into the form of long, cylindrical tunnels, often of a length of approximately 30 μm with a diameter of 1 or 2 μm. Tunnel densities are of the order of $10^8/cm^2$ and various electron microscopy studies which have been undertaken have shown the convoluted nature of the etched surface,[30,49] Figure 23(a). The films, which are prepared by the process of wet anodization on these convoluted surfaces, are usually relatively lossy ($\pm 5\%$), often due to the inclusion of impurities from the anodizing electrolyte into the film. However, this loss is usually overshadowed by the series resistance of the wet electrolyte used as the re-forming medium and the conducting medium between the dielectric and the cathode of the structure.

Capacitors based on tantalum are also an important business,[49,50] and here again, the oxide is prepared by wet anodization. The surface can either be a plain foil, an etched foil (in this case the etched surface is nowhere near as convoluted as the case of etched aluminium) or sintered tantalum powder. Sintered tantalum powder usually has a particle size of around 10 μm, and Figure 23(b) shows an example of the porous structures

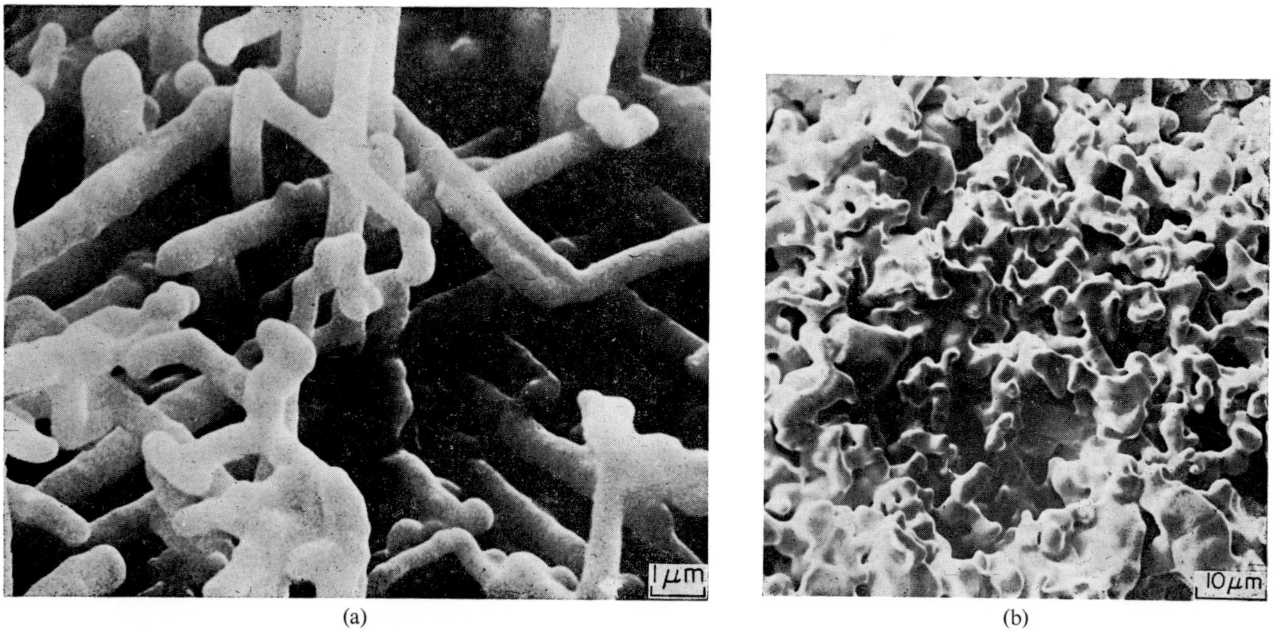

(a) (b)

Figure 23. (a) Internal structure of etched Al. (Al subsequently removed.) (By permission of the Plessey Co Ltd.)[49]
(b) Sintered Ta powder for use in Ta electrolytic capacitors. (By permission of the Plessey Co Ltd.)[49]

which are obtained. Many electron microscopy studies have been undertaken on capacitors produced on foil or sintered tantalum blocks, and Figure 24 shows on typical examination of a fractured tantalum capacitor structure.[30] This particular micrograph is of interest because not only does it show a fractured section of the tantalum metal together with the oxide film grown on by anodization, but it also shows a manganese oxide layer deposited by solution pyrolysis and acting as the second electrode. This manganese oxide layer has self-healing properties which have previously been discussed.

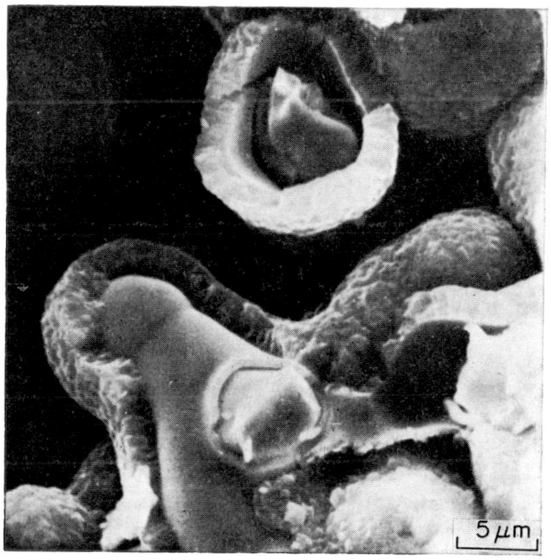

Figure 24. Internal structure of Ta electrolytic capacitor with MnO counter-electrode. (By permission of the Plessey Co Ltd.)[30]

Aluminium and tantalum foil capacitors are usually wound in spirals, together with an absorbent paper spacer in between the foils to contain the electrolyte. In the case of sintered tantalum structures, however, the electrolyte is either contained in the pores, the whole body being inside a metal case (Figure 25), or the sinter is totally impregnated with manganese oxide to form a dry structure.

Evaporated silicon monoxide capacitors have been used in thin film circuit technology for some considerable time,[41] and the compressive stresses of these films, as has already been pointed out, is of importance in determining the stability of the capacitor structure. Film thicknesses are usually greater than 3000 or less than 10,000 Å for the best stability.

Silicon dioxide capacitors have applicability in integrated circuit technology[43] and in this case, the capacitors are prepared by thermal growth of oxide on the basic silicon. However, discrete silicon dioxide capacitors have been prepared for commercial applications by the technique of rf sputtering. Provided sufficiently low loss capacitors are prepared, these structures have the advantage of a very low temperature coefficient of capacitance associated with a permittivity value of around 4 (Figure 12). Silicon nitride can also be prepared by rf sputtering for capacitor purposes in microcircuitry technology.

3.3.2. Insulators. When dielectric films are used as insulators, the loss requirements are often not as stringent as in the case of capacitors. Silicon monoxide is often widely used in this context for isolating conducting lines in thin film circuitry and silicon dioxide in integrated circuit technology.

A further important insulation application is in transistor technology as in the case of thin film devices[44] (Figure 26), and also MOS and MNOS technology[43, 55] (Figure 27(a and b)).

3.3.3. Conductors. Application of dielectric films to conductors may at first sight seem a contradiction. However, as has already been noted, an application of thin film dielectrics from a conduction point of view can be in tunnelling devices.[38, 45, 53] A further area where the conduction properties of dielectrics have been investigated is in terms of switches, using both simple materials[37, 46, 54] and complex oxides such as chalcogenides.[37, 46] Dielectrics are also important in governing the behaviour of cermet type resistors.[39, 47, 48] These resistors may be of a relatively simple type, prepared by the co-evaporation of a metal and a dielectric (Au–SiO_2) or more complicated as in the case of SiO–Cr, where an annealing treatment is involved and the conduction phase is then a silicon–chromium compound. Figure 28 shows a typical temperature coefficient of resistance α_R vs resistivity plot for an Au–SiO_2 cermet, and it can be seen that the value of α_R becomes increasingly negative as the resisitivity of the mixture increases and dielectric conduction becomes more and more important.

4. Summary

This article has surveyed the preparation of dielectric films, has noted the characteristics which must be examined to determine basic reproducibility, and has briefly discussed some of the applications. As mentioned at the beginning of the article, it is impossible to characterize dielectric films so as to identify the best method of preparation which is applicable for all situations. Stability and reproducibility are major characteristics required for studies of conduction processes and in the case of the applications of films, these requirements are coupled to those of the economics of large-scale production. Table 3 summarizes some of the major uses of thin film dielectrics, identifying the preparation technique used and the application. It should be noted that not all the applications are of commercial importance and are often concerned with device research.

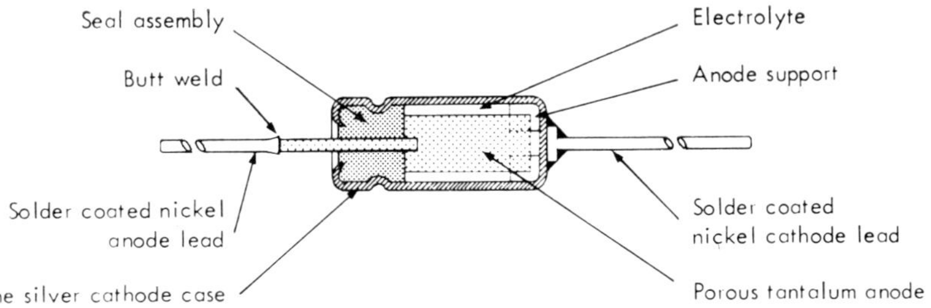

Figure 25. Diagram of Ta capacitor using sintered anodized Ta powder and a wet electrolyte.

Table 3. Use of thin dielectric films.

Material	Preparation	Application	Reference
Al_2O_3	Anodization	Capacitors	10
		Thin film transistors	49
			44
	Gaseous anodization	Tunnelling devices	38
	Thermal growth		45
$BaTiO_3$; $PbTiO_3$	Flash evaporation	High ε capacitors	41
Glasses (complex oxides)	Sputtering	Capacitors, Passivation,	7
		Threshold and Memory Switches	37
Halides (alkali)	Evaporation	Capacitors	41
Polymers	Glow discharge, UV or EB polymerization	Capacitors (research)	41
			9
			25
	Solution deposition		15
	Solvent casting		16
Stearates	Langmuir	Capacitors (research)	13
SiO	Evaporated	Capacitors, Insulators,	41
		Thin Film Transistors	44
SiO–Cr	Flash evaporation	Resistors (cermets)	47, 48
SiO–Au	Evaporation	Resistors (cermets)	47, 48
SiO_2	Reactive sputtering	Low TCC Capacitors, Insulators	41
	rf sputtering		6
	Thermal growth	Passivation, Capacitors	41, 43
		Tunnelling devices	38
	From solution	Capacitors	12
S_3N_4	Vapour phase	High ε Capacitor layers, Passivation	43
Ta_2O_5	Anodization of Ta	Capacitors; Insulators	49, 50
	Anodization of Ta_2N	Capacitors; Insulators	50

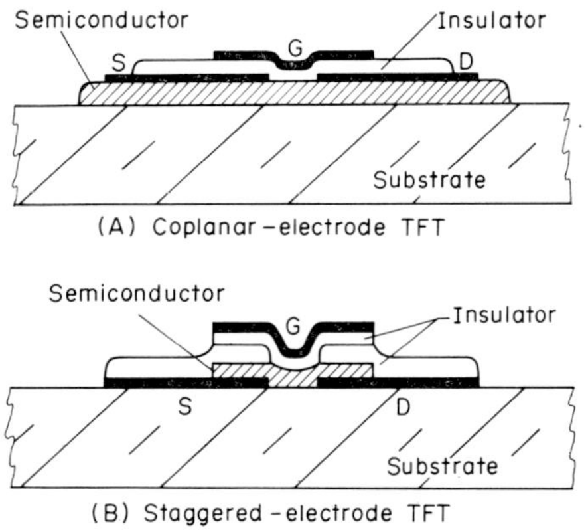

Figure 26. Cross-sectional views of two thin film transistor structures. (a) Coplanar electrode, (b) staggered electrode.

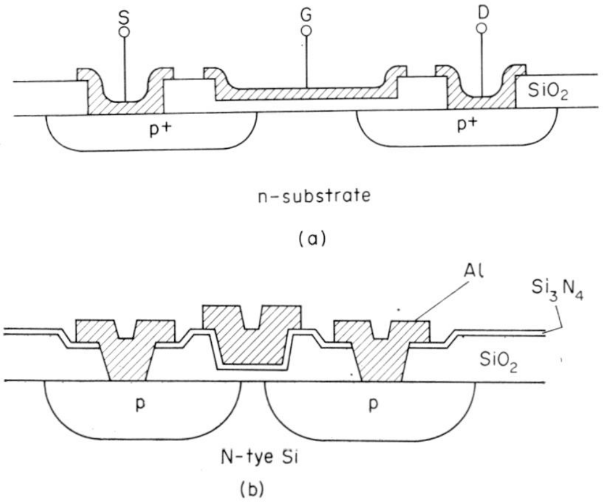

Figure 27. Transistor structures using SiO_2/Si_3N_4 insulation. (a) Integrated channel MOSFET, (b) MNOS structure.

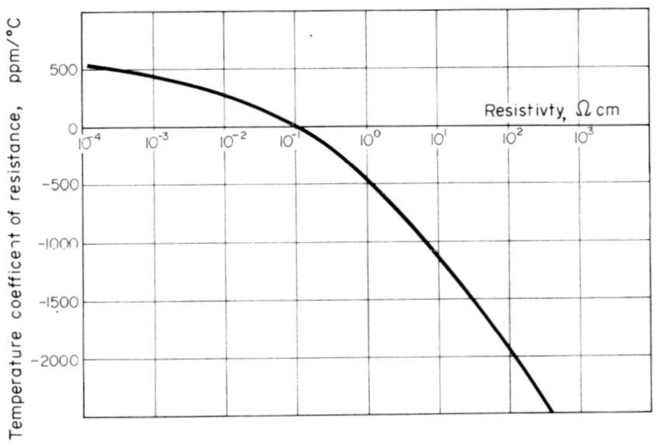

Figure 28. Temperature coefficient of resistance vs resistivity for an Au–SiO_2 cermet.[48]

References

[1] R Glang, *Handbook of Thin Film Technology*, p 1.3. (Edited by L I Maissel and R Glang), McGraw-Hill, New York (1970).
[2] L I Maissel, *Handbook of Thin Film Technology*, p 4.1. (Edited by L I Maissel and R Glang), McGraw-Hill, New York (1970).
[3] D S Campbell, *Handbook of Thin Film Technology*, p 5.1. (Edited by L I Maissel and R Glang), McGraw-Hill, New York (1970).
[4] D S Campbell, *Physics of Non-Metallic Films*, p 9. (Edited by Dupuy and Cachard), Plenum Press, London and New York (1976).
[5] D S Campbell, *Thin Solid Films*, **32**, 1976, 3.
[6] P J Harrop and D S Campbell, *Handbook of Thin Film Technology*, p 16.1. (Edited by L I Maissel and R Glang), McGraw-Hill, New York (1970).
[7] A Pliskin, D R Kerr and J A Perri, *Physics of Thin Films*, **4**, 1967, 257, Academic Press, New York.
[8] L I Maissel, *Physics of Thin Films*, **3**, 1966, 61, Academic Press, New York.

[9] L V Gregor, *Physics of Thin Films*, **3**, 1966, 131, Academic Press, New York.
[10] C J Dell'Oca, D L Pulfrey and L Young, *Physics of Thin Films*, **6**, 1971, 1, Academic Press, New York.
[11] W M Feist, S R Steele and D W Ready, *Physics of Thin Films*, **5**, 1969, 237, Academic Press, New York.
[12] H Schroeder, *Physics of Thin Films*, **5**, 1969, 87, Academic Press, New York.
[13] V K Srivastava, *Physics of Thin Films*, **7**, 1973, 311, Academic Press, New York.
[14] V K Agarwal, *Elec Comp Sci Technol*, **2**, 1975, 1.
[15] A C Rastogi and K L Chopra, *Thin Solid Films*, **18**, 1973, 187.
[16] A W Stephens, A W Levine, J Fech Jr, T J Zrebiec, A V Cafiero and A M Garofalo, *Thin Solid Films*, **24**, 1974, 361.
[17] *Science and Technology of Surface Coating*. (Edited by B N Chapman and J C Anderson), Academic Press, London (1974).
[18] *Handbook of Thick Film Hybrid Microelectronics*. (Edited by C A Harper), McGraw-Hill, New York (1975).
[19] C E Drumheller, *Trans 7th Nat Vac Symp*, pp 306–312, 1960.
[20] D S Campbell and B Hendry, *Br J Appl Phys*, **16**, 1965, 1719.
[21] A Moll, *Z Angew Phys*, **10**, 1968, 410.
[22] R F Bunshah, *Science and Technology of Surface Coating*, p 361. (Edited by B N Chapman and J C Anderson), Academic Press, London (1974).
[23] K O T Kerner, *Jap J Appl Phys* Supp 2, Part 1, 1974, 463.
[24] E Hollands and D S Campbell, *J Mater Sci*, **3**, 1968, 544.
[25] M White, *Thin Solid Films*, **18**, 1973, 157.
[26] N F Jackson, *J Mater Sci*, **2**, 1967, 12.
[27] J R Ligenza, *J Appl Phys*, **36**, 1965, 2703.
[28] H F Sterling and R C G Swann, *Solid State Electron*, **8**, 1965, 653.
[29] R A Connel and L V Gregor, *J Electrochem Soc*, **112**, 1965, 1198.
[30] D S Campbell, *Radio electron Engng*, **41**, 1971, 5.
[31] D S Campbell, In *Handbook of Thin Film Technology*, p 12.21. (Edited by L I Maissel and R Glang), McGraw-Hill, New York (1970).
[32] R W Hoffman, In *Physics of Non-Metallic Thin Films*, p 273. (Edited by Dupuy and Cachard), Plenum Press, London (1976).
[33] P J Harrop and D S Campbell, *Handbook of Thin Film Technology*, p 16.16. (Edited by L I Maissel and R Glang), McGraw-Hill, New York (1970).
[34] S J Fonash, *Physics of Non-Metallic Thin Films*, p 225. (Edited by Dupuy and Cachard), Plenum Press, London (1976).
[35] R M Hill, *Ibid*, p 189.
[36] A K Jonscher and R M Hill, *Phys Thin Films*, **8**, 1975, 170.
[37] J G Simmons, *J Phys D*, **4**, 1971, 613.
[38] J G Simmons, *Handbook of Thin Film Technology*, p 14.1. (Edited by L I Maissel and R Glang), McGraw-Hill, New York (1970).
[39] D S Campbell and A R Morley, *Rep Prog Phys*, **34**, 1971, 308.
[40] N Klein and H Gafni, *IEEE Trans Elect Devices*, **ED13**, 1966, 281.
[41] D Gerstenberg, *Handbook of Thin Film Technology*, p 19.1. (Edited by L I Maissel and R Glang), McGraw-Hill, New York (1970).
[42] D S Campbell and A R Morley, *Radio Electron Engng*, **43**, 1973, 421.
[43] D J Hamilton and W G Howard, *Basic Integrated Circuit Engineering*, McGraw-Hill, New York (1975).
[44] P K Weimer, *Handbook of Thin Film Technology*, p 20.1. (Edited by L I Maissel and R Glang), McGraw-Hill, New York (1970).
[45] C R Crowell and S M Sze, *Physics of Thin Films*, **4**, 1967, 325, Academic Press, New York.
[46] H K Henisch and C Popescu, *Physics of Non-Metallic Thin Films*, p 253. (Edited by Dupuy and Cachard), Plenum Press, London (1976).
[47] Z H Meiksin, *Phys Thin Films*, **8**, 1975, 99.
[48] L I Maissel, *Handbook of Thin Film Technology*, p 18.3. (Edited by L I Maissel and R Glang), McGraw-Hill, New York (1970).
[49] N F Jackson and D S Campbell, *Thin Solid Films*, **36**, 1976, 331.
[50] W D Westwood, N Waterhouse and P S Wilcox, *Tantalum Thin Films*, Academic Press, London (1975).
[51] P T Stroud, *Thin Solid Films*, **21**, 1972, 1.
[52] C Weissmantel, *Thin Solid Films*, **32**, 1976, 11.
[53] W T Band, *Thin Solid Films*, **34**, 1976, 225.
[54] T W Hickmott, *Thin Solid Films*, **9**, 1972, 431.
[55] Y Le Goascoz, *Physics of Non-Metallic Thin Films*, p 417. (Edited by Dupuy and Cachard), Plenum Press, London (1976).

2.8 A statistical approach to the analysis of dielectric breakdown strength of thin insulating films

K Kristiansen,* *Electronics Research Laboratory, N-7034 Trondheim NTH, Norway*

Statistical models that are widely used in the study of high voltage insulating systems are applied to thin insulating films. By use of these methods it is in many cases possible to decide whether the electrical strength of a component is bulk-limited or electrode-limited. A discussion on which type of statistical distribution function can be expected for electrical breakdown in thin films is given. By analyzing measured breakdown voltages for more than 2000 capacitors it has been shown that a description of dielectric breakdown in thin films by means of the theory of the weakest link is fruitful.

It is shown that the first selfhealing breakdown has its origin at weak spots randomly distributed over the capacitor area.

The variation in the results is too large to determine whether the breakdowns start at the interface electrode dielectric or within the bulk of the film. These techniques have also been successfully applied to previously published results, offering an explanation of results not previously understood.

Introduction

Thin insulating films are used as surface passivation, insulation between conductors and as a dielectric in capacitors. In all these applications it is mandatory that the insulating film retains its insulating properties with time and stress. Consequently, electrical breakdown strength has received great attention. The aim of this work is to further clarify some aspects of the breakdown phenomena in thin insulating films.

Although it is well recognized that electrical breakdown in insulating films has a statistical nature, only two of the 332 titles given in a recent survey[1] contain the word 'statistical'. This is a strong indication that the implications of this statistical behaviour of the breakdown strength are not fully appreciated.

In most of the previous investigations the statistical distribution of breakdown voltages is measured, and the mean values are plotted as a function of different parameters such as film-thickness, temperature, manufacturing parameters and time under stress. It is the object of this work to show that more extensive statistical treatment of experimental data will explain some previously published results that have not yet been properly understood. It is also to be shown that by using this approach it is possible to decide if the electrical strength of a component is bulk-limited or electrode-limited. Such investigations have not previously been carried out on thin films. An approach like this has been very rewarding in the study of high voltage insulating systems.[2] It is to be hoped that this approach will be equally useful in microelectronics.

First a discussion of what type of statistical distribution function that can be expected for electrical breakdown is given. Some properties of this distribution function are described. Some examples are given on the application of this distribution function to previously published results. After a description of the experimental techniques applied, results are given on the variation of the distribution of breakdown fields with film thickness and electrode area for some materials. These results are finally discussed with respect to the applicability of the previously outlined statistical model.

* Now at Norwegian Research Institute of Electricity Supply, N–7034 Trondheim, NTH, Norway.

Electrical breakdown in thin films

A thin film capacitor might in cross-section look like the sketch in Figure 1.

A metallic bottom electrode is deposited on one side of an insulating slice of glass or ceramics. On top of this a thin insulating film, usually 0.05–0.5 μm thick is deposited. This film forms the dielectric of the capacitor. A second metallic film

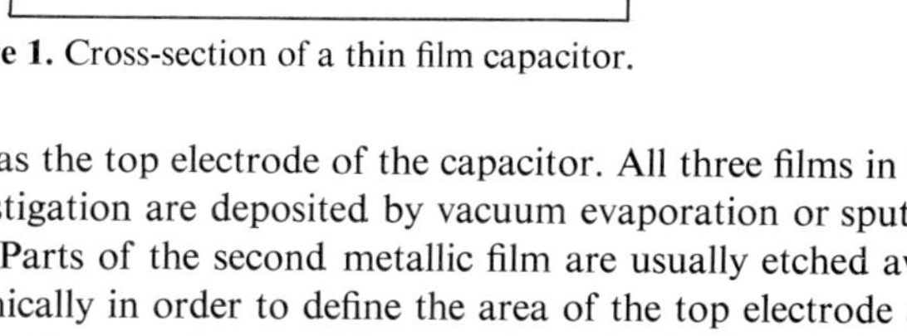

Figure 1. Cross-section of a thin film capacitor.

acts as the top electrode of the capacitor. All three films in this investigation are deposited by vacuum evaporation or sputtering. Parts of the second metallic film are usually etched away chemically in order to define the area of the top electrode and hence the capacitance of the capacitor. Further descriptions of techniques and methods are given elsewhere.[3–5]

When the applied voltage across the capacitor is increased beyond a certain level, the dielectric material is locally destroyed and the capacitor is, at least temporarily, short-circuited. This phenomena is widely discussed in the literature[6,7] and several models for the physical phenomena involved are provided.[8,9]

Statistical distribution of electrical breakdown

These models do not however take into account the recognized fact that in an actual capacitor there is a distribution of weak spots, either in the dielectric or at the electrode. These weak spots have their characteristic distribution both in critical voltage for initiating a breakdown and in location within the capacitor. From this it is clear that the sole factor determining the actual electrical strength of a single capacitor is the strength of the weakest spot within this capacitor.

An important consequence of this is that when the volume of a capacitor is increased there is always a finite chance that a weaker spot is now included than the weakest spot previously inside the capacitor. From this it is easy to see that the mean value of the electrical strength of a capacitor has to decrease as the volume of the capacitor increases. Only a short review is to be given here on some of the consequences of this statistical behaviour of electrical breakdown. More extensive treatment is given in the original report.[10] Usually, it is attempted to find an analytical expression to describe the distribution of measured values. This expression must have a few fundamental properties. First there must be a lower limit, possibly zero, for the breakdown voltage. Secondly, this distribution function is not allowed to change its shape when the size of the specimen is changed although its parameters might well change their value.

When n components from the same distribution is connected in parallel the probability that no breakdown will occur at voltages lower than U is given by

$$1 - F(n, U) = [1 - F(1, U)]^n \tag{1}$$

where $[1 - F(1, U)]$ denotes the probability that no breakdown occurs in one single component under the same test conditions. It is then assumed that the components does not influence each other.

$$F(A, U) = 1 - \exp(-\alpha \cdot A) \tag{2}$$

where α is some function of U.
The following expression of $\alpha(U)$ has been widely accepted in high voltage engineering

$$\alpha(U) = \left(\frac{U - U_a}{U_0 - U_a}\right)^b \tag{3}$$

Here U_a denote the lowest breakdown voltage possible and U_0 the 63 % value. The factor b is called shape parameter because it defines the shape of the distribution function. The final distribution function for electrical breakdown in insulating materials of unit area then becomes:

$$F(U) = 1 - \exp\left[-\left(\frac{U - U_a}{U_0 - U_a}\right)^b\right] \quad \text{for } U > U_a \tag{4}$$

$$F(U) = 0 \quad \text{for } U \leq U_a$$

This is usually referred to as the Weibull distribution. It was first applied by Weibull on the mechanical strength of materials.[11] Because the value of U_a is difficult to assess experimentally it will here be assumed to be zero. This is a common approach that might be verified.

In Figure 2 the Weibull frequency function is shown for some values of the chape parameter b. For values of b from 3 to 5 the central region is close to a Gaussian distribution. For $b = 1$ the Weibull distribution becomes an exponential distribution. A discussion of the properties of the Weibull distribution is

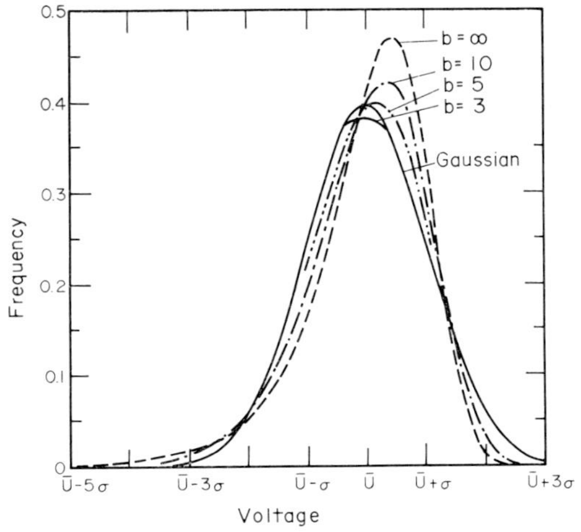

Figure 2. Frequency function of the Weibull distribution.[12]

given.[12] A combination of the Weibull distribution with the first equation for breakdown voltage of n components gives:

$$F(n, U) = 1 - \exp\left[-\left(\frac{U - U_a}{U_{0n} - U_a}\right)^b\right] \tag{5}$$

with

$$U_{0n} - U_a = \frac{U_0 - U_a}{n^{1/b}} \tag{6}$$

It is seen that this equation also describes a Weibull distribution. This means that by enlarging a sample n times, the shape of the distribution function of the breakdown voltage does not change. It can be shown that

$$\sigma_n = \frac{\sigma^1}{n^{1/b}} \tag{7}$$

The standard deviation of the breakdown voltage decreases with the scaling factor raised to the power of $1/b$.

Another important quantity is the 63 % value of the enlarged sample if U_a is assumed to be zero

$$U_{n,\,0\cdot63} = \frac{U_0}{n^{1/b}} \tag{8}$$

This means that the 63 % value decreases with the scaling factor raised to the power of $1/b$. Further properties of this distribution is given elsewhere.[12]

In a specimen with two plane, parallel electrodes an increase in electrode area can be regarded as connecting several elements in parallel. Both the electrode area and the volume of the insulating material under stress is increased simultaneously. In looking for how probability of breakdown change as a function of electrode area with constant thickness of the insulating material, one can therefore draw no conclusions whether it is the electrode area or the internal volume of the insulating material that determines the strength of the specimen. If on the other hand the distance between the electrodes is increased without increasing the area, the volume of the insulator under stress will increase. By keeping the field strength between the electrodes constant, there will be no change in probability of breakdown if this is determined by the electrode area. If the volume of the insulator is the most important factor, however, then the probability of breakdown will increase with increasing thickness. In case the volume of the insulator is the determining factor, an increase in distance between electrodes can be regarded statistically as components in parallel although they are physically placed in series. The assumption is then that partial breakdown anywhere in the insulating material leads to a complete breakdown.

One conclusion from this discussion is that if the distributions of breakdown fields are measured for several groups of samples a variation in the size of electrode area and film thickness gives a possibility to determine if the breakdown has its origin at the interface between electrode and insulator or within the insulator itself.

It is easy to see from equation (8) that by plotting the logarithm of the 63% value of breakdown field versus the logarithm of the scaling factor, a straight line with slope—$1/b$ will result. If no change in the 63% value occurs with changing film thickness and constant electrode area, whereas this relationship applies for changing electrode area and constant film thickness, the conclusion can be drawn that the breakdown has its origin at the electrode–insulator interface. If the relationship above applies both for changing electrode area and changing film thickness, the conclusion can be drawn that the breakdown starts within the insulator itself. No variation in the 63% value with a change in electrode area and film thickness, implies that the breakdown does not have its origin in randomly distributed weak spots. This is the case when thermal breakdown is the dominating mechanism.

In cases where more than one type of weak spots are present in the insulator, the situation becomes more complicated. One has then to distinguish between additively and multiplicatively mixed distributions. Additively mixed distributions are to be expected when the specimens constituting a random sample can be distinguished with regard to their breakdown behaviour due to, for instance, differences in life history or method of production. In multiplicatively mixed distributions however, at least two different cases can lead to breakdown in each test sample. The samples can be regarded as identical, but dependent on the statistical distribution of impurities, imperfections, etc, one of the two breakdown behaviours will be predominant. These cases are further treated in the literature.[12,15]

In Figures 3 and 4 examples are given on the application of the outlined statistical model to some published results. In most cases this is difficult because the initial data are not available. In a very comprehensive paper[13] a typical final breakdown distribution is given. This is redrawn on Weibull probability paper in Figure 3. It is commented in the original paper that the distribution is clearly not Gaussian. This figure shows that the Weibull distribution does give a fairly good approximation to these data.

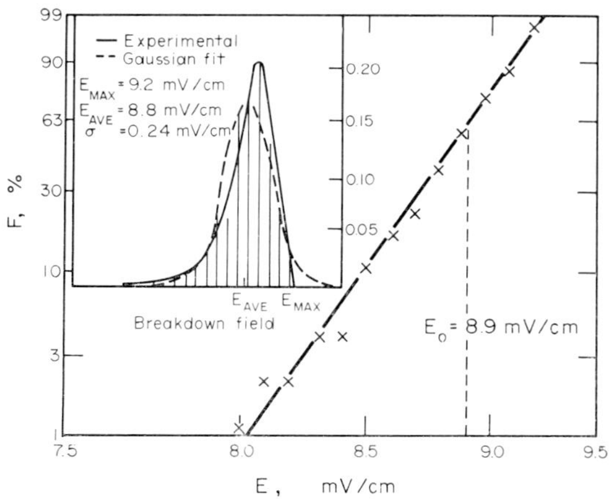

Figure 3. Final breakdown distribution according to Osburn and Ormond[13] as published and redrawn on Weibull probability paper.

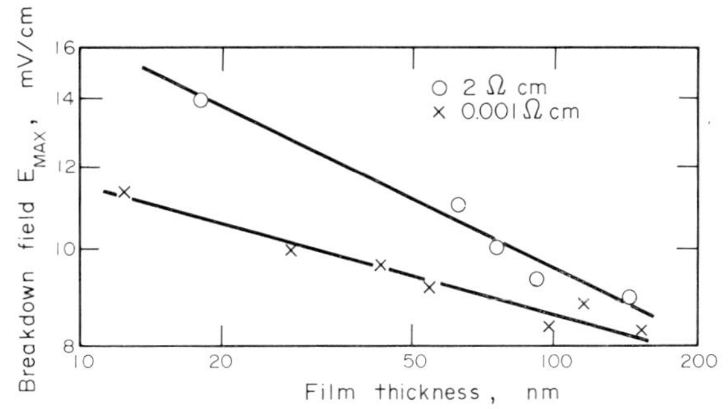

Figure 4. Maximum breakdown field vs film thickness according to Osburn and Ormond.[14] SiO_2 grown on *n*-type Si.

It has been shown that the 63% value decreases with the scaling factor raised to the power of $1/b$. This relationship also holds for other percentage values. The highest breakdown voltage observed is given by Osburn and Ormond[14] as a function of film thickness. This value might be interpreted as some high percentage value if an equal number of samples are measured for each film thickness. The logarithm of this maximum voltage is plotted against the logarithm of film thickness in Figure 4.

Since the original data and consequently the shape factor is not available, no conclusions can be drawn whether the breakdown starts within the insulator or at the interface electrode–insulator. This behaviour, however, indicates that the breakdown in this case had its origin in randomly distributed weak spots.

Experimental results

In this section experiments are described which were designed to verify if this statistical approach is useful to thin insulating films. A technique which allows simultaneous registration of both applied voltage at breakdown and examination of the voltage across the sample during breakdown has been used in this investigation. A schematic diagram is shown in Figure 5. The voltage across the sample is displayed on an oscilliscope. The trace on the oscilloscope is recorded on a video-tape recorder with video camera. By using a video-tape recorder instead of a storage scope, the applied voltage need not be reduced to zero after each breakdown. This method also allows successive breakdowns to be studied under conditions that are similar to a component in actual use.

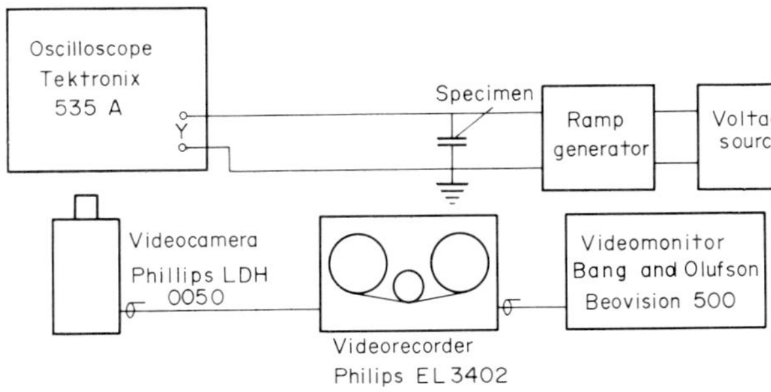

Figure 5. Experimental set-up to measure electrical breakdown.

For the samples used in this investigation, slide glass was used as substrate. The bottom electrode consisted of rf-sputtered nickel with approximate thickness 200 nm. The dielectric was then sputtered. The top electrodes were evaporated aluminium with approximate thickness 200 nm. The top electrodes were given their geometrical shape and size by chemical etching through a photo-resist mask. Capacitors with area from 0.01 to 9 mm^2 were used. Deposition parameters and dielectric properties of the samples are given in Table 1. The applied voltage at the first selfhealing breakdown was measured together with the applied voltage immediately before the capacitor was short circuited, here referred to as the final breakdown. Both voltages were measured because two different mechanisms might well be involved in these two cases. It was observed that a considerable number of breakdowns might occur before the capacitor was finally short circuited, especially on the capacitors with the largest areas. The measured values for first selfhealing breakdown, plotted on Weibull probability paper might typically look like Figure 6.

It was verified experimentally that the following parameters did not seriously affect the results:

polarity of applied voltage,
ageing without load,
rate of increasing the applied voltage,
electrode roughness,
external components in the measuring equipment.

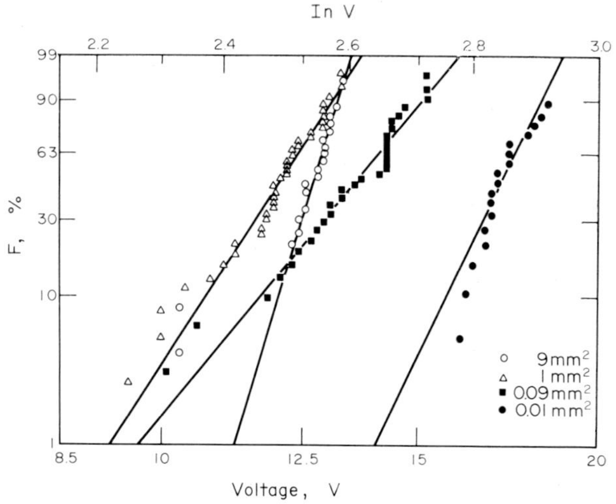

Figure 6. Distribution of first selfhealing breakdown for 50 nm thick Al_2O_3.

The reproducibility was also shown to be reasonably good.

It is shown in Figures 7–11 how the breakdown voltage varies with the capacitor area and the film thickness. The results shown are for sputtered Al_2O_3, SiO_2 and Ta_2O_5. Both the first and final breakdowns are shown.

Table 1. Deposition parameters and dielectric properties of specimens.

Dielectric	Sputter-atmosphere	Power density (W/cm^2)	Sputter time (min)	Film thickness (mm)	Dissipation factor, *tgδ* (%)
Al_2O_3	Ar + 10% O_2	3, 2	15 30 60	46 94 164	1–6
SiO_2	Ar + 10% O_2	1.9	7, 5 15 30	31 58 143	<0.2
Ta_2O_5	Ar + 10% O_2	1.9	10 20 30	65 130 194	0.2–1

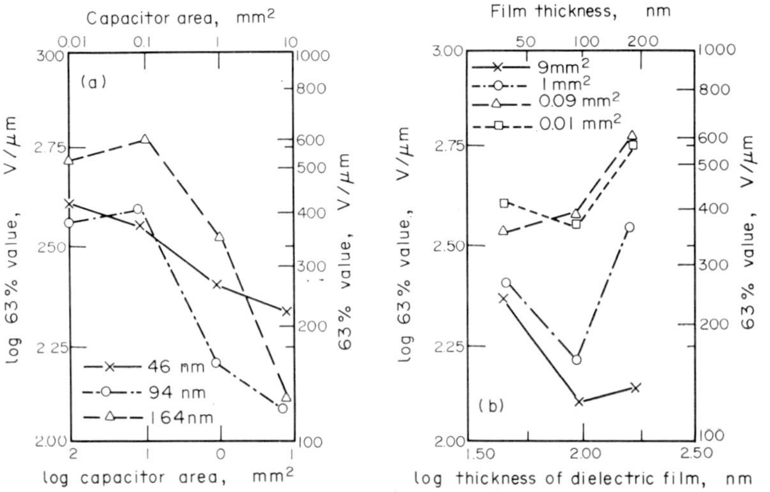

Figure 7. 63% value as a function of (a) capacitor area and (b) film thickness for first selfhealing breakdown in capacitors with Al_2O_3 dielectric.

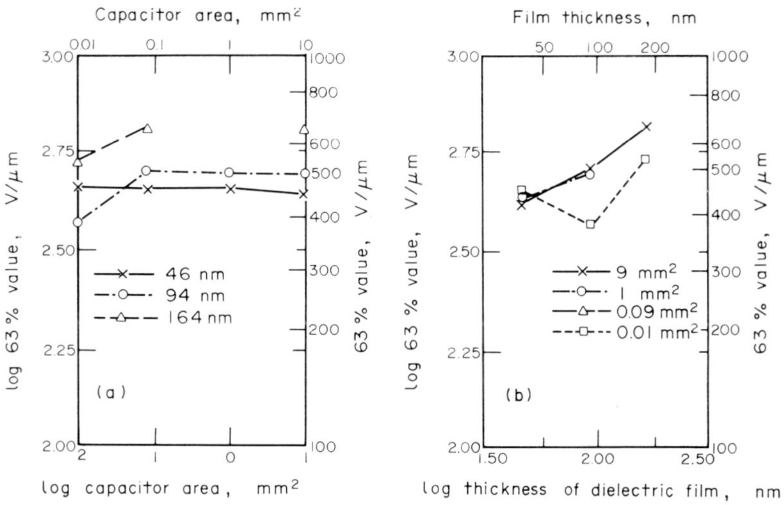

Figure 8. 63 % value as a function of (a) capacitor area and (b) film thickness for final breakdown in capacitors with Al_2O_3 dielectric.

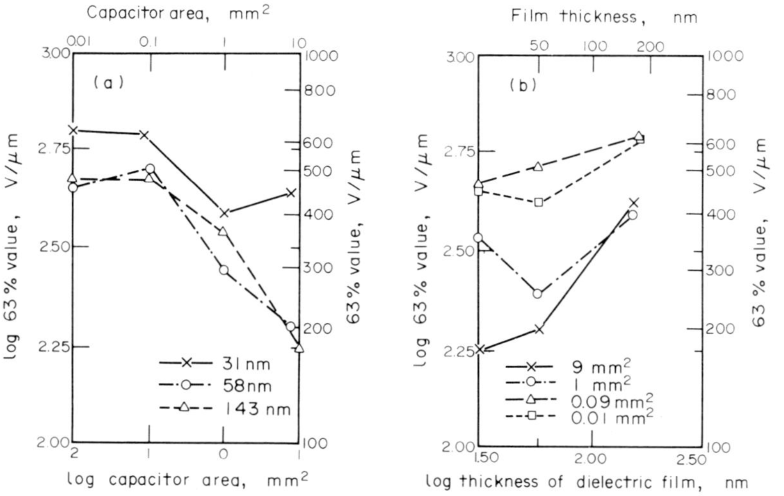

Figure 9. 63 % value as a function of (a) capacitor area and (b) film thickness for first selfhealing breakdown in capacitors with SiO_2 dielectric.

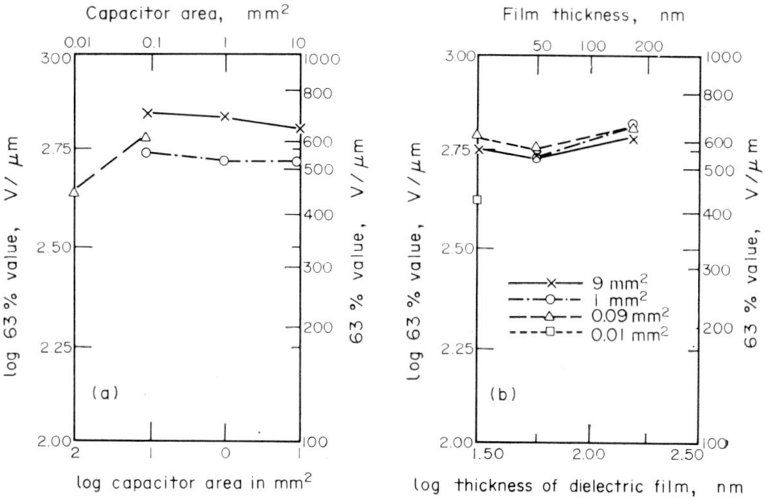

Figure 10. 63 % value as a function of (a) capacitor area and (b) film thickness for final breakdown in capacitors with SiO_2 dielectric.

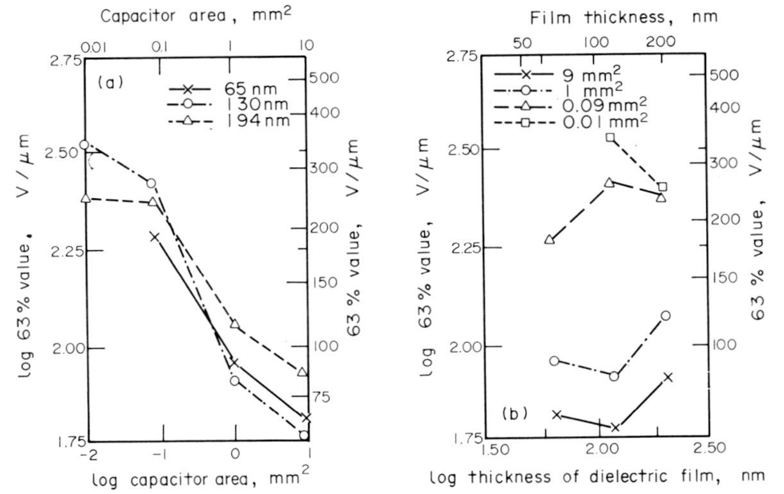

Figure 11. 63% value as a function of (a) capacitor area and (b) film thickness for first selfhealing breakdown in capacitors with Ta_2O_5 dielectric.

Discussion

Already the replotting of previously published data on electrical breakdown in thin insulating films, showed that the statistical approach introduced here is capable of explaining certain phenomena not previously understood.

As shown the measured values of electrical breakdown voltage fits well to a Weibull distribution in most cases. Even if there is, in some cases, a significant difference between two presumably identical substrates, this difference is smaller than the observed variation with capacitor area.

The figures of log 63% values as a function of film thicknesses show a somewhat erratic behaviour both for the first and the final breakdown. This work has been limited to one decade in film thickness. This covers the most interesting range in thickness for applications in microelectronic devices. The shape parameter usually lies between 5–10 for the first breakdown. This implies that the logarithm of the 63% value decreases by an amount of 0.1 to 0.2 for a ten fold increase in film thickness if the capacitor area is kept constant and the breakdown has its origin at randomly distributed weak spots within the dielectrics. However, with the large variations in breakdown voltages observed and the limited range of thicknesses investigated, no conclusion can be drawn regarding the effect of film thickness on the first selfhealing breakdown. Regarding the effect of film thickness on final breakdown strength, the situation is more clear. Since the variation in the results is much smaller in this case it can be seen that the expected reduction in breakdown strength with increasing film thickness does not take place. This means that the final breakdown does not have its origin at randomly distributed weak spots within a dielectric with otherwise constant properties. Regarding the variation in breakdown voltage with electrode area the situation is much more clear compared to the situation with varying film thickness. The range of variation in area covered is here three decades which makes any systematic behaviour more evident. The breakdown strength shows a clear tendency to increase with decreasing areas.

It is to be expected that a plot of log 63% value versus capacitor area shows a straight line with a negative slope equal to the inverse of the shape factor of the individual distributions. To determine if this is followed quantitatively, a line has been fitted to the measured values. The resulting inverse slope is given in Table 2 together with the corresponding mean value of shape parameters. From this it can be seen that the 63% value shows a dependence of capacitor area that is approximately equal to what is predictable from the statistical treatment.

Table 2. Calculated inverse slopes from dependence on capacitor area compared with mean value of shape parameters.

Material	Thickness (nm)	$\frac{1}{\text{Slope}}$	$\bar{b}$	
Al_2O_3	46	−10	11 ± 5	9, 5
	94	− 5	5 ± 2	
	164	− 5	6 ± 3	
SiO_2	31	−17	8 ± 3	7
	58	− 8	6 ± 3	
	143	− 9	5 ± 3	
Ta_2O_5	65	− 4	5 ± 3	5
	130	− 3	7 ± 5	
	194	− 5	4 ± 3	

Even if the variation in the results is so large, it seems reasonable to conclude that the first breakdown has its origin in randomly distributed weak spots. Since the variation in the results with film thickness is so large and the range in film thicknesses investigated so limited it cannot be determined whether these weak spots are distributed within the dielectric or at the interface electrode/dielectric. These experiments do not give any indications whether the cathode or the anode is the origin of the breakdown.

In contrast to the first breakdown, the final breakdown did not show any dependence on either the capacitor area or the film thickness. This means that the voltage at which the capacitor is finally short circuited does not depend on a 'weakest spot' mechanism. On the contrary the short circuiting of the capacitor seems to occur at the same field strength irrespective of electrode area. No difference in behaviour has been noted between the different materials used as dielectrics.

Acknowledgements

The investigation has been carried out at the Electronics Research Laboratory (ELAB) and is financially supported by the Royal Norwegian Council for Scientific and Industrial Research (NTNF). I wish to thank both these institutions for permission to publish this work and Professor Inge Johansen at the Electrical Engineering Department, Power Section for encouragement throughout the laboratory tests and for valuable criticism during the preparation of this report. Thanks are also due to lab ing Knut O Tangen for helpful and valuable discussions during this investigation and to ing Einar L Myhre for assistance in carrying out the comprehensive and time consuming experiments.

References

[1] V K Agarwal, *Thin Solid Films*, **24**, 1974, 55.
[2] H S Endicott and K H Weber, *AIEE Trans*, **76**, Pt III, 1957, 343.
[3] K L Chopra, *Thin Film Phenomena*, McGraw–Hill, New York (1969).
[4] L I Maissel and R Glang, *Handbook of Thin Film Technology*, McGraw–Hill, New York (1970).
[5] J L Vossen and J J O'Neill, *RCA Review*, 1968, 149.
[6] N Klein, *Thin Film Dielectrics*, The Electrochemical Society, New York, 1969, 43.
[7] J J O'Dwyer, *The Theory of Electrical Conduction and Breakdown in Solid Dielectrics*, Oxford University Press (1973).
[8] N Klein and E Burstein, *J Appl Phys*, **40**, 1969, 2728.
[9] J J O'Dwyer, *J Phys Chem Solids*, **28**, 1967, 1137.
[10] K Kristiansen, A statistical approach to the analysis of dielectric breakdown strength of thin insulating films, The University of Trondheim, The Norwegian Institute of Technology, 1975.
[11] W Weibull, A statistical Representation of Fatique Failures in Solids, *Kungl Tekniska Högskolans Handlingar*, **27**, 1949.
[12] P Dokopoulos, Thesis, Technischen Hochschule Carlo–Wilhelmina In Braunschweig, 1967.
[13] C M Osburn and D W Ormond, *J Electrochem Soc*, **119**, 1972, 591.
[14] C M Osburn and D W Ormond, *J Electrochem Soc*, **119**, 1972, 597.
[15] P Fischer and P Röhl, *Siemens Forsch Entwickl Ber*, **3**, 1974, 125.

2.9 Glow discharge polymerization of organic silicones (E-10)

J S Sandved and K Kristiansen,* *Electronics Research Laboratory, N-7034 Trondheim NTH, Norway*

This paper presents some of the experiences obtained in using the Glow Discharge Polymerization process for making thin polymer silicone films. The rate of deposition is observed to be high compared to other vacuum processes used for deposition of thin insulating films, and the polymerization takes place at low power densities in the discharge, so that heating of the substrates is negligible. The electrical permittivity of the films is in the range of 2.5 to 3.0 with losses less than 0.05%. The current–voltage characteristic shows a linear relationship even at high stress levels with volume resistivities in the range of 10^{11}–10^{13} Ωcm. The quality of the films are sufficient for use as gate insulation in field effect devices· The dielectrical strength is about 250 V for 0·25 μm thick films with capacitor areas of 1 mm^2. Thermal properties are investigated. The films are not seriously deteriorated after heat treatment in air at 400°C for 1 h. Further experiments with these films are primarily concerned with applications as surface passivation of microelectronic circuits and components.

Introduction

Thin polymer films of organic materials have in the recent years received considerable interest in microelectronics, both as a dielectric and as a surface passivation.[1–6] Certain types also find applications in electro-optical devices as thin film light-guides and as anti-reflection coatings.[7–10] Plasma polymerized thin films might also have biological tissue compatibility which rivals that of the best synthetic materials to date.[11]

A lot of methods, such as uv light-, gamma ray-, electron-beam polymerization, sputtering, vacuum evaporation and pyrolytic deposition, together with plasma polymerization, have been suggested in previous works as possible methods for making polymer films.[12] However, plasma polymerization[13–15] seems to be the process that is most easily realized and which offers the greatest flexibility both in applications and in materials to be deposited. This article will describe the plasma polymerization system used in our laboratory. System parameters to be considered are discussed and the prepared organic films are characterized in terms of electrical, chemical, thermal and some physical properties.

Glow discharge polymerization system

In the glow discharge polymerization process a monomer is introduced in its gas phase to a reaction chamber which has been evacuated to approximately 2×10^{-2} torr. In addition to the monomer gas, a carrier gas, as for instance argon, is often used. The presence of argon results in a more cross-linked polymer film.[16,17] The pumping on the chamber, which is carried out by a rotary pump, continues during the whole deposition process.

The discharge is maintained by either a dc or an rf voltage with frequencies up to 2.5 GHz.

Figure 1 shows a block diagram of a glow discharge polymerization system. When dc and frequencies below 1 MHz are used, one need electrodes in direct contact with the plasma.

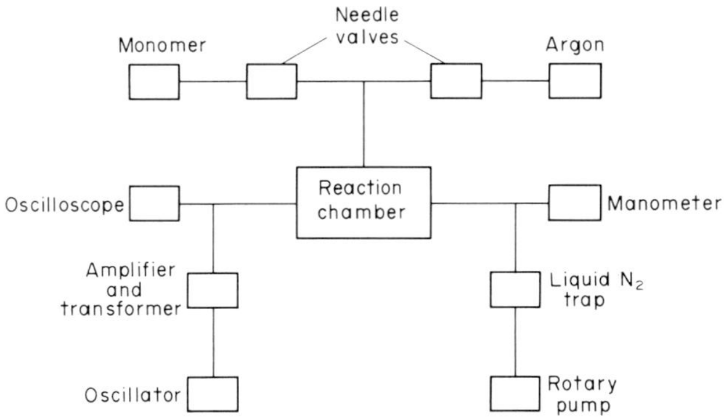

Figure 1. Block diagram of glow discharge polymerization system.

Above 1 MHz, however, the energy can be fed indirectly into the system by capacitive or inductive coupling.[18] The required voltage depends on the nature of the gas and the pressure in the chamber.

Table 1. Typical deposition parameters for the system used at our laboratory.

Restgas pressure	2×10^{-2} torr
Argon pressure	0.2 torr
Monomer pressure	0.1–0.2 torr
Discharge voltage	800–1200 V peak–peak
Frequency	190 kHz
Discharge current	0.1–0.15 A

Figures 2 and 3 show photographs of the system which has been used in preparing the films. Figure 4 shows the relation

* Now at Norwegian Research Institute of Electricity Supply, N–7034, Trondheim, NTH, Norway.

Figure 2. This photograph shows the equipment, except for the rotary pump used for the process.

Figure 3. This pictures the inside of the reaction chamber. The grids are the Al electrodes and the metal plate in front of these is the substrate holder. The metal plate to the left of the electrodes is the end plate of the chamber and contains all the necessary interconnections and the pirani gauge.

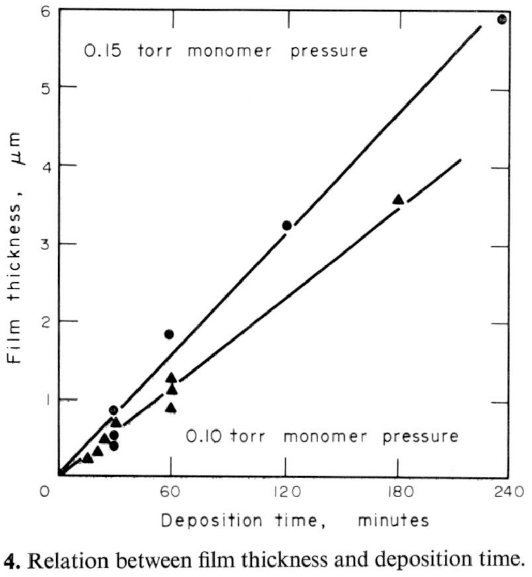

Figure 4. Relation between film thickness and deposition time.

between film thickness and deposition time for two different values of monomer pressure. The diagram shows that the deposition rate is from 0.8 to 1.8 μm h^{-1}, depending on the monomer pressure. Variations of argon pressure between 0.1 and 0.3 torr with constant monomer pressure have not shown any variations in the deposition rate. Concerning the theory of growth and chemistry involved in depositing polymer films by this process several references can be referred to.[5,6,19–22] Generally one can say that the monomer species is activated in the glow discharge, transported in the plasma by different mechanisms and then polymerized. The polymerization process may occur both in the gas-phase or on any solid surfaces of the system such as chamber walls, electrodes and substrates. Polymerization of the monomer will depend on factors as temperature, gas pressure and binding energy between the activated species and the substrate to be covered. In order to keep an equilibrium condition between the adsorbed layer and the incoming activated particles new layers are formed continuously with the result that a polymer film is obtained.[23]

Monomer source

As a monomer it was decided to use a silicone because this material has properties which make it excellent as a surface passivation. Silicones also have thermodynamic data that fit the process, and it was reported that the material formed good films using this technique.[18–20] The chief properties that render silicones particularly useful are the wide temperature range over which they are stable and their excellent properties over an extreme range of environmental conditions. The monomer used is supplied by Dow Corning under the trade name Z-6079 Silane. Typical properties for this material is shown in Table 2.

Table 2. Properties of the silicone monomer.

Chemical name	Hexamethyldisilazan, bis-trimethylsilyl-amin
Chemical formula	$C_6H_{19}NSi_2 = (CH_3)_3Si\text{–}NH\text{–}Si(CH_3)_3$
Molecular weight	161.4
Specific gravity at 25°C	0.77 g cm^{-3}
Boiling point (760 torr)	125°C
Boiling point (50 torr)	50°C
Suitable solvents	Ketones, hydrocarbons and chlorinated solvents

Preparation techniques for metal–polymer–metal capacitors

Samples are prepared by depositing polymer films on 2 × 2 in. slide glasses. Top electrodes of different areas are evaporated onto the polymer. Figure 5 shows a cross-section of such a

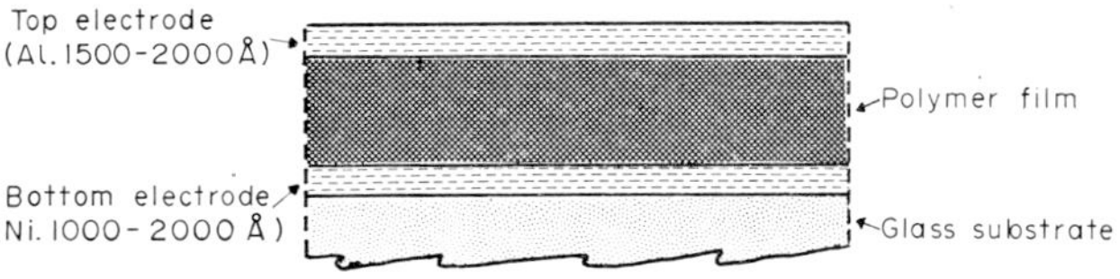

Figure 5. Cross-section of a metal–polymer–metal (MPM) capacitor.

capacitor. The deposited films are heavily cross-linked and withstand chemical attacks of both acids, alkalines and common solvents. So far no satisfactory etching procedure is found and hence a 'lift off' technique is used to remove parts of the deposited films. One covers the substrate area on which the polymer film is not wanted by photo-resist before deposition. After the required thickness is obtained the photo-resist is removed and hence the film on top of it. The 'lift off' technique was used to generate steps in the polymer films such that a Tallysurf could be used for measurements of thickness. Before using the Tallysurf the step had to be covered with an Al film of approximately 1000 Å as the polymer film itself is too soft to carry the pick up stylus.

Electrical properties

Capacitance and loss tangent were measured for different top electrode areas, and the permittivity for the dielectric film calculated. Table 3 shows mean values (approx 15 measurements) for thickness and capacitance, together with values of areas and associated permittivities. Deposition parameters are also included in this table.

Leakage currents in the film as a function of applied field were measured for temperatures between −21°–+65°C. Top electrodes of area 1 and 9 mm² are used. Figure 6 shows these

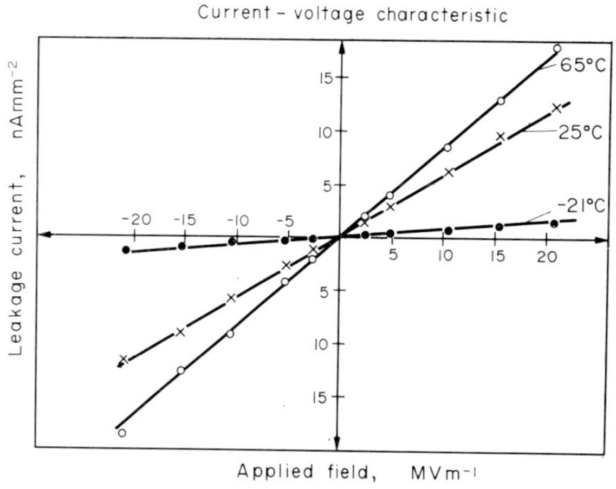

Figure 6. Current–voltage characteristics for different temperatures.

relationships. The ohmic nature of these measurements is not normal for insulators. However, similar effects are reported for PTFE.[24] Dielectric breakdown strength was measured and analysed in accordance to a method proposed by one of the authors.[25,26] Figures 7 and 8 show the breakdown voltage as a function of thickness and area, respectively. The first observed breakdown was considered. Figures 7 and 8 show that the breakdown voltage is independent of thickness of the polymer film but depend upon the area of the top electrodes. One may therefore suggest that the breakdown is an effect that occurs in weak points distributed over the total area.

Table 3. Different properties of the polymer film.

Substrate	Top electrode area (mm²)	Deposition time (min)	Pressure (torr) Argon	Pressure (torr) Monomer	Capacitance (pF)	Loss tangent (tgδ)%	Thickness (Å)	Permittivity ε
	9.0	60	0.25	0.15	159.6	0.1–0.5	14180	2.84
C33	1.0	60	0.25	0.15	18.13	0.4–0.6	14180	2.90
	0.09	60	0.25	0.15	1.72	0.3–0.9	14180	3.06
	9.0	30	0.25	0.15	300.9	0.01–0.3	7485	2.81
C34	1.0	30	0.25	0.15	33.95	0.1–0.2	7485	2.86
	0.09	30	0.25	0.15	3.31	0.9–1.2	7485	3.10
	9.0	15	0.25	0.15	683.3	0.3–0.4	3480	2.94
C35	1.0	15	0.25	0.15	76.82	0.4–0.5	3480	2.98
	0.09	15	0.25	0.15	7.18	0.9–1.3	3480	3.09

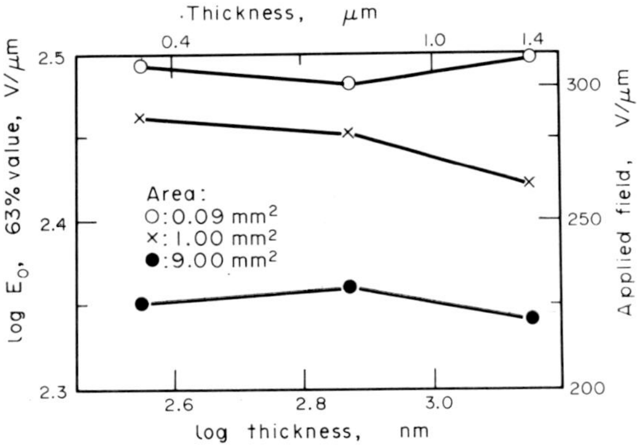

Figure 7. Breakdown voltage as a function of thickness for different top-electrode areas.

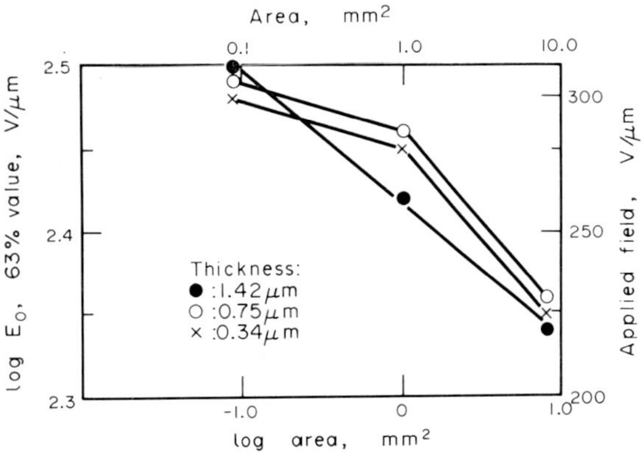

Figure 8. Breakdown voltage as a function of top-electrode area for different thicknesses.

Figure 9 shows C–V measurements measured on a metal–polymer–semiconductor (MPS) structure. Frequency was 1 MHz, polymer thickness 0.25 μm and electrode area 1 mm^2. The C–V measurements indicate a fixed charge density in the order of 10^{12} cm^{-2}, whereas the density of fast interface states was found to be less than 10^{11} cm^{-2}.

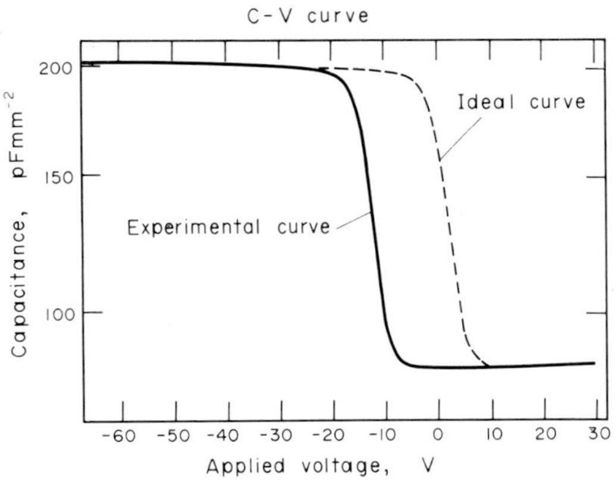

Figure 9. C–V curve for a MPS structure.

Chemical resistance

The tests were carried out by passivating, using the polymer film, Al resistors laid down in a meander pattern. By dipping these into different solutions together with an Al meander which had not been passivated, one observed the effect the solutions had on the Al resistors.

We have tried ammonia (1% of conc NH_3 in de-ionized water) hydrochloric acid (67.5 ml 37% HCl + 32.5 ml de-ionized water), Phosphoric acid (1 part of 85% H_3PO_4 + 1 part of de-ionized water at 60°C) plus iodine diluted in methanol.[27] These tests have shown that none of these solutions do actually attack the polymer film but probably due to pinholes in the polymer film some of the solutions are able to penetrate the passivation layer resulting in an etching attack on the underneath lying resistors. However, in all of the tests, the passivated resistors withstand attacks for more than twice the time that of the unpassivated.

Thermal properties

Thermal resistance was investigated by depositing polymer films on steel strips and forming capacitors as before using a metal top electrode. The steel strips were then heated by passing current through it and temperatures of up to 700°C could be obtained. Above 400°C one had to use a high melting point material, as Al started to form 'bubbles' at 400°C. Chromium on the other hand showed no such effects within the measuring range. Figures 10 and 11 show capacitance and loss tangent as a

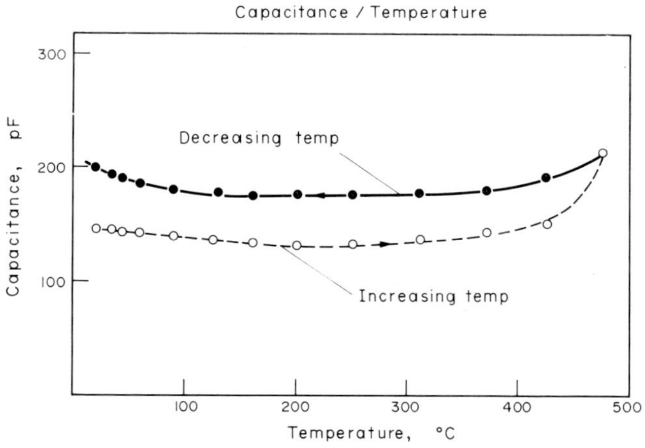

Figure 10. Capacitance vs temperature.

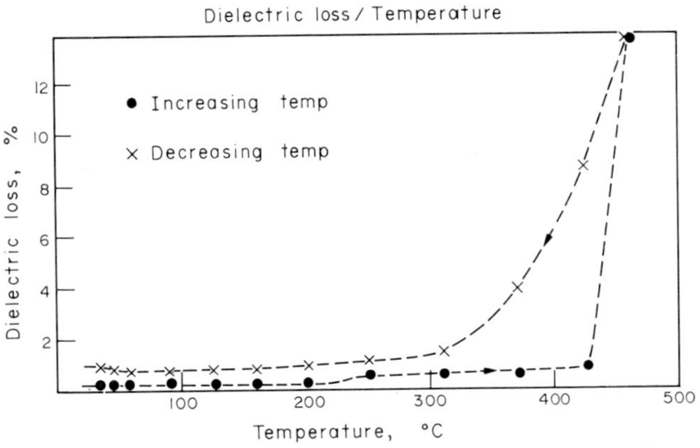

Figure 11. Loss-tangent vs temperature.

function of temperature respectively. The heating curve in Figure 10 is a typical virgin curve while repeated heating of the sample gives rise to a temperature dependence as shown by the upper curve, as long as the maximum temperature is not exceeded. The loss tangent will take any value between the two curves in Figure 11 when heating is repeated. Preliminary measurements have shown very small changes in capacitance and loss-tangent with time at elevated temperatures less 500°C.

Other physical properties

No experiments have yet been carried out in order to establish the optical properties. At this stage one can only state from inspection that the films are colourless and transparent to visual light. Adhesion to substrates is good as long as the substrates have been through an ordinary cleaning procedure before deposition. In this connection it can be mentioned that some problems have arisen in that flakes of polymer film have fallen upon the substrates under deposition from walls and other solid surfaces on which a non-adherent film has been grown. This problem has been minimized by carrying out the deposition in steps with cleaning of chamber walls, etc. in between. For this purpose we have been using an ordinary vacuum cleaner with good results. The elasticity of the polymer has not been measured but we have not observed any cracking in a film of 0.5 μm thickness supported by a steel foil of 0.1 mm which have been bent repeatedly through 90°.

From Figure 12 one may get an idea of the surface hardness, where one has made a scratch using a scalpel. To date one has not yet been in the position to carry out any chemical analysis of the deposited polymer films. The reason is that available analysing equipment require at least an isolated mass of 1 mg which is difficult to obtain with this process.

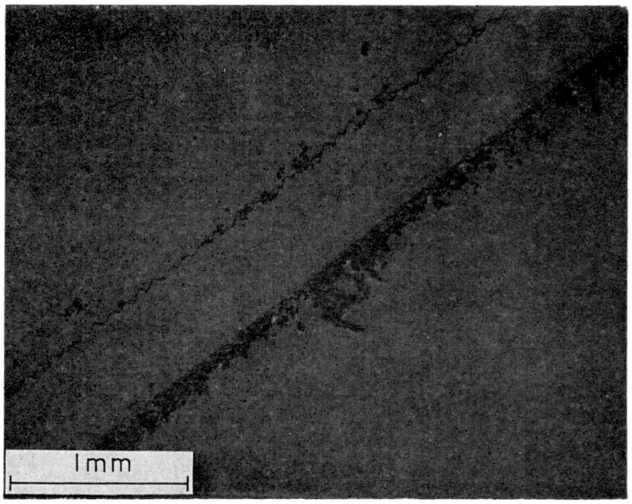

Figure 12. The photograph shows a scratch made by a scalpel on a 1 μm thick polymer film. Magnification ×32.

Conclusion

One can conclude that plasma-polymerization of silicone is a very versatile process which is easily controlled, and that the organic films which can be made by this process possesses excellent thermal, physical and electrical properties. Further experiments on these films are in progress. In the near future we will carry out environmental tests with special attention to temperature/humidity cycling after MIL STD 883 A.

In the future we will also try to deposit other materials, but at this time we feel that a thorough understanding of the properties and methods of characterizing the organic films is more essential.

Acknowledgement

Thanks are due to I Johansen for helpful discussions and to E Myhre for building of equipment and preparation of the samples.

References

[1] Electrical Design News, Oct 1960.
[2] P J Ozawa, *IEEE Trans Parts, Mat Pack*, **PMP5**, 1969, 112.
[3] J H Coleman, US pat 3069283 Dec 1962.
[4] J H Coleman, US pat 3068510 Dec 1962.
[5] G L Schnable, *J Electrochem Soc*, **122**, 1975, 1092.
[6] L V Gregor, *Solid State Technol*, 1971, 37.
[7] P K Tien *et al*, *Appl Optics*, **11**, 1972, 637.
[8] J R Hollahan, *Appl Optics*, **13**, 1974, 1844.
[9] G Smolinsky, *J Vac Sci Technol*, **11**, 1974, 167.
[10] J R Hollahan, *J Vac Sci Technol*, **11**, 1974, 167.
[11] A W Hahn and K G Mayhan, *Medical Inst*, **8**, 1974, 000.
[12] K Kristiansen, ELAB (Electronic Research Lab, NTH, Trondheim, Norway) report STF44 A76145, May 1976.
[13] A Brockers and H Konig, *Z Phys*, **152**, 1958, 75.
[14] G Smolinsky, J H Heiss, *Proc ACS 155th meeting*, **28**, 1968.
[15] A Bradley and J Hammers, *J Electrochem Soc*, 1963.
[16] B V Tkachuk and V M Kolotyrkin, *Vysolomol Soedin*, Ser B, **10**, 1968, 24.
[17] B V Tkachuk and V M Kolotyrkin, *Ukr Khim Zh*, **35**, 1969, 768.
[18] J R Hollahan and A T Bell, *Techniques and Applications of Plasma Chemistry*, Wiley & Sons, New York (1974).
[19] F W Billmeyer, *Textbook of Polymer Science*, Second Ed, Wiley, New York (1970).
[20] J J Licari, *Plastic Coatings for Electronics*, McGraw-Hill, New York (1973).
[21] J E Hanck, *1973 Materials Selector, Materials Engineering*, Reinhold, New York (1973).
[22] S M Lee, *Polymeric Films for Semiconductor Passivation*, NASA Contract, No 12–2011, 1969.
[23] S Morito, G Sawa and M Jeda, *Japan J Appl Phys*, **14**, 1975, 1459.
[24] P P Budenstein, P I Hayes, J L Smith, W B Smith, *J Vac Sci Technol*, **6**, 1968, 289.
[25] K Kristiansen, ELAB report, STF44 A76131, 1976.
[26] K Kristiansen, *Int Symp Vac Thin Film Technol*, Uppsala, Sweden, Aug/Sept 1976.
[27] R W Wilson, L E Terry, *Deposition Techniques for contacts and Interconnections on Integrated Circuits*, Balzers FL. 9496, Oct 1972.

2.10 Pulsed photovoltaic effect in thin film MIS devices

F Morin, *Centre National d'Etudes des Telecommunications, 22301 Lannion, France*

A MIS structure, illuminated by pulsed light, can generate a pulsed photocurrent. We have observed this effect in various kinds of MIS structures (Si_3N_4/Si; Polymer/Si). We have shown that the detected photovoltage pulse across a high impedance load is directly related to surface potential Φ_s ($q\Phi_s/KT$; band bending at insulator/semiconductor interface). Indeed, the photovoltage versus external bias V curves are found in good agreement with the $\Phi_s(V)$ curves obtained by capacitance–voltage measurements. For instance, photovoltage is found to reverse polarity, like Φ_s, at flatband voltage ($V = V_{FB}$).

This pulsed photovoltaic effect is then studied on thin film sandwiches: Metal/Polymer/CdSe/Metal. The CdSe film, evaporated on to a chromium contact, is coated by a polyparaxylene film (2000 A-rf plasma polymerized) and a semi-transparent counter-electrode. A pulse of light creates a 0·2 V voltage pulse across a 1 MΩ load resistor. Furthermore, after a +5 V short time bias, photo response vanishes, and reappears by reversing the bias. Hysteresis investigations of the capacitance–voltage curves is used to explain this memory effect. A short time bias is sufficient to fill or empty polymer traps near the interface. The respective semiconductor surface is then accumulated (low/ Φ_s) or depleted (high/ Φ_s). Photoresponse is in good agreement with these results.

Pulsed photovoltaic effect may be used to investigate MIS structures, and also to make up a new thin film memory device.

Introduction

A metal/insulator/semiconductor structure, illuminated by pulsed light, can generate a pulsed photo-emf. This is the pulsed photovoltaic effect. Photovoltaic effect on MIS devices has been recently studied by many authors. One may classify these studies in two ways, according the insulator thickness. Firstly, some research is performed on MIS structures with a very thin insulator; these devices generate a direct photocurrent under a continuous light and are attractive for solar cell applications.[1] Secondly, when insulator is thicker, ac or pulsed mode are studied.[2–3] The object of this work is to show that the photo-emf generated during the light pulse is related to the semiconductor–metal barrier height, just as it was supposed by the before mentioned authors.

In order to show this correlation, pulsed photo-emf and semiconductor surface potential (ψ_s) measurements were performed. M/I/Silicon samples were chosen, because on these samples ψ_s is easily deduced from capitance–voltage measurements. A good qualitative agreement is found between ψ_s and the photo-emf. Subsequently, the pulsed photovoltaic effect is studied on thin film MIS devices.

1. Surface potential and photo-emf measurements

1.1. Test-devices fabrication. Experiments on Au/Si_3N_4/Si and Au/Polymer/Si structures were performed. A silicon nitride thin film (400 Å) was deposited on to silicon by chemical vapour deposition. Dielectric constant and loss factor were 6 and 10^{-4} respectively. The counter electrode was semi-transparent (300 Å Au).

Polymer films (2000 Å) were obtained by rf plasma polymerization.[5] The apparatus is described in Figure 1. A low pressure of paraxylen vapour (10^{-2} torr) was maintained in the reactor and a rf plasma is excited. The deposition rate of

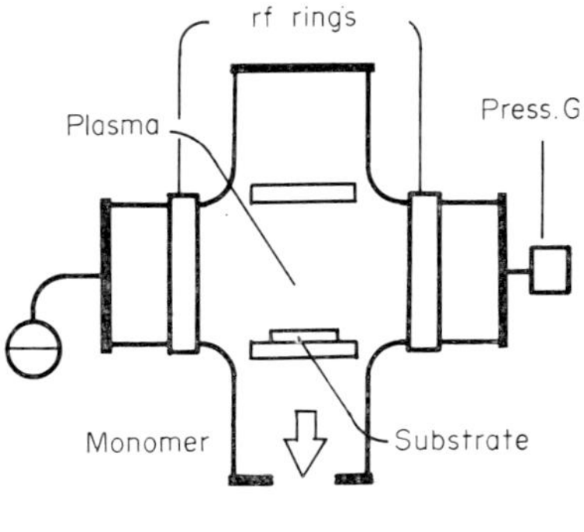

Figure 1. Polymerization reactor.

polyparaxylen film was about 100 Å min^{-1}. Dielectric constant and loss factor of these films were 2.6 and 5×10^{-3} respectively. Insulating films were deposited onto epitaxial silicon wafers (*n*-type, 5×10^{14} cm^{-3}).

1.2. Surface potential determination. To determine the semiconductor surface potential, go back to the relevant MIS capacitance theory.[6] It is well known that in a non-ideal MIS structure, the semiconductor surface is not neutral with respect to the bulk. The space charge region thickness, related to carrier concentration, may reach several thousand angströms. For an *n*-type semiconductor, the corresponding surface potential has a positive value (with respect to the semiconductor bulk) when the surface is depleted or a negative one when accumulated. Surface potential is given by:

$$\psi_s = \int_v \left(1 - \frac{C_{MIS}}{C_i}\right) dV$$

where C_i is the insulator capacitance and V the external bias voltage. Therefore, the integration of the C–V curve will give the ψ_s–V curve. Capacitance measurements were performed at 1 MHz, and also at low frequency (10^{-2} Hz) by the quasistatic method. Figure 2(a) shows the $Au/Si_3N_4/Si$ C–V curves. By

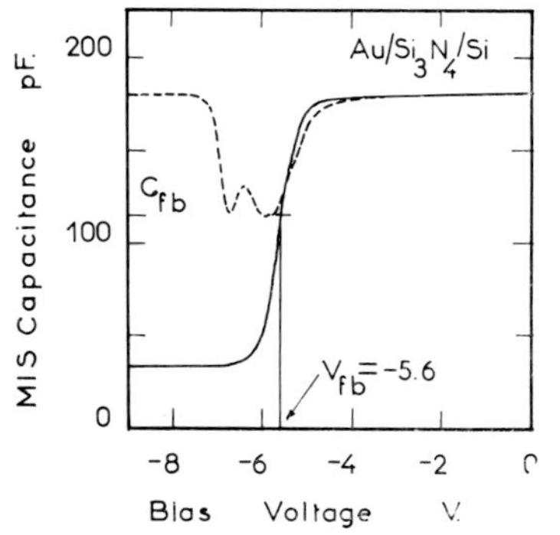

(a)

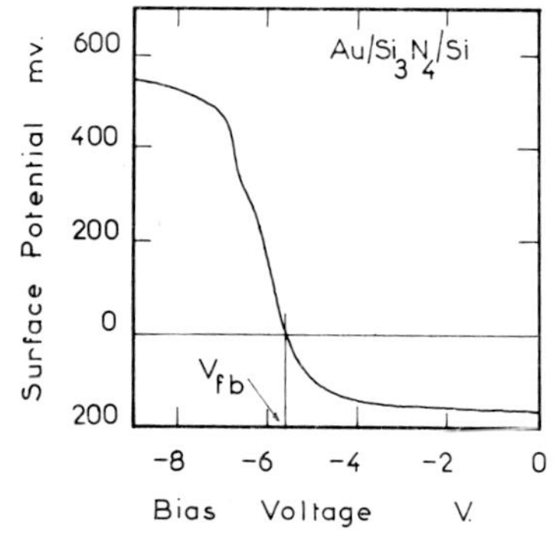

(b)

Figure 2. (a) C–V curves for $Au/Si_3N_4/Si$ samples (continuous line: HF curve, dashed line: quasistatic curve). (b) ψ_s–V curve for $Au/Si_3N_4/Si$ samples obtained by a graphical integration of the quasistatic C–V curve).

means of theoretical data,[7] flat-band capacitance and voltage are determined. In the flat-band condition, the semiconductor is neutral and then $\psi_s = 0$. Figure 2 (b) shows the ψ_s–V curve.

1.3. Pulsed photovoltaic effect. Figure 3(a) shows the basic circuit for pulsed photo-emf measurements. An MIS structure, with a semi-transparent counter electrode, was illuminated by a flash lamp (xenon, pulse duration 4 μ_s pulses rate 1 to 10^3 s^{-1}). Pulsed photo-emf was recorded on a CRT (1 MΩ input impedance).

On Figure 3(b) a bias circuit was added, in order to alter the surface potential as shown in Figure 2(b). Resistance and capacitance were used for ac and dc blocking. The photosignal and ramp voltage were displayed on the vertical and horizontal inputs of the CRT. Without external bias, $Au/Si_3N_4/Si$ samples exhibited a very low photo-emf.

Figure 4 shows the photo-emf change when the sample was biased. The pulsed photo-emf is found to reverse polarity at V = 5.6 V, just like ψ_s and the photo-emf-vs-V curve is similar

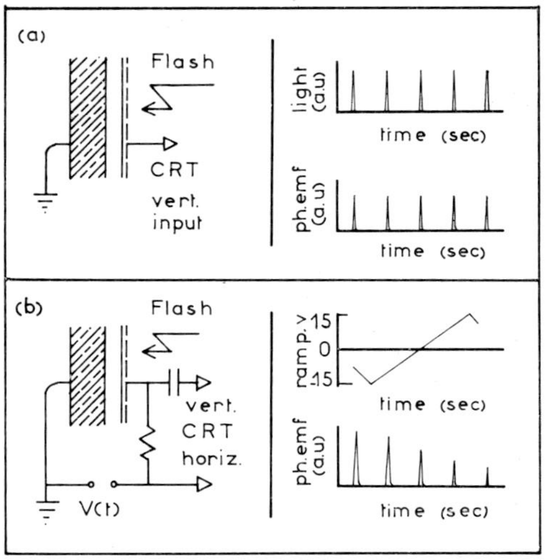

Figure 3. Basic circuits for the photo-emf measurements, (a) without and (b) with external bias voltage.

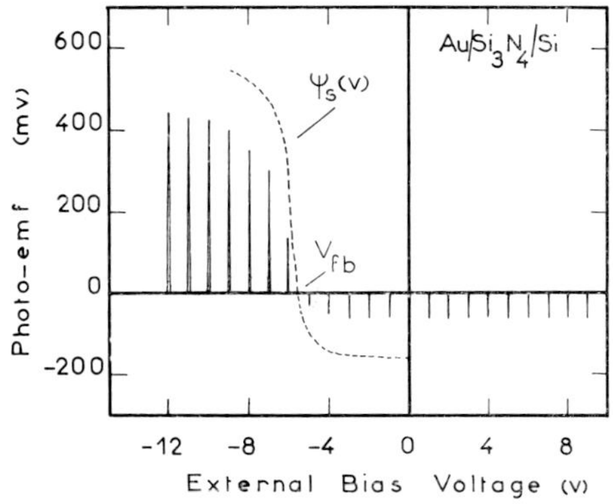

Figure 4. Pulsed photo-emf-VS-bias voltage curve, for a $Au/Si_3N_4/Si$ sample.

to that of the surface potential. Quantitative differences are discussed later. One can see that for a photovoltaic device, this is not a very good sample because at V = 0 (i.e. without external bias voltage) it generates a very low photo-emf. On the other hand, the Au/Polymer/Si Samples typically generated at 100 mV photo-emf without external bias voltage. Furthermore, after they were negatively biased for a short time, the photo-emf vanished, but reappeared when the bias voltage was reversed (Figure 5a). According to earlier results on $Au/Si_3N_4/Si$ samples, the change in photo-emf may be correlated with the change in surface potential. More exactly, a large positive photo-emf must indicate a depleted semiconductor surface, and a small negative photo-emf an accumulated one. This is well confirmed by C–V measurements (Figure 5b). Furthermore, when these samples were biased by a continuously varying external voltage, a change in photo-emf was observed, just as with the Si_3N_4/Si samples. Electron injection from the semiconductor to traps in the polymer is thought to explain the large hysteresis and, therefore, the memory effect. We shall find a similar effect in thin film MIS devices.

2. Pulsed photovoltaic effect on thin film MIS devices

2.1. Thin film MIS device. Fabrication. With regard to test-structures, the thin film devices differ only in the semiconductor used: polycrystalline CdSe thin film instead of silicon (Figure 6).

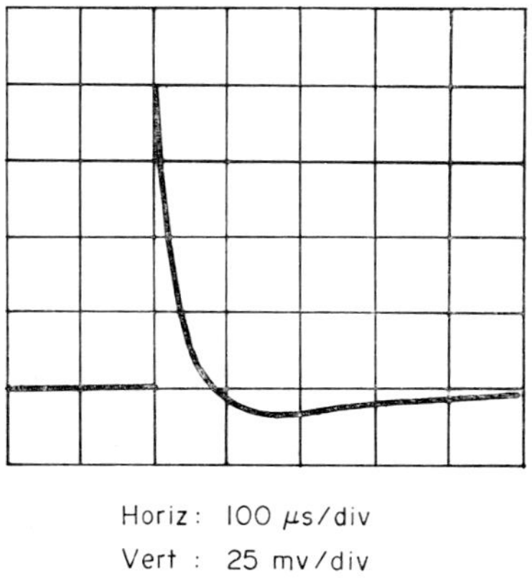

(a)

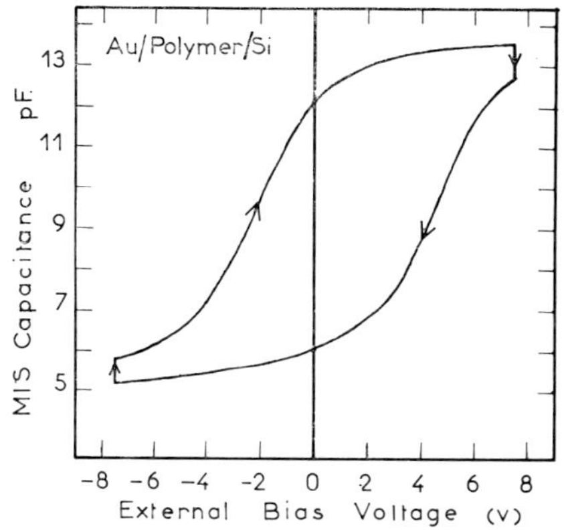

(b)

Figure 5. (a) A typical photosignal obtained with an Au/Poly/Si sample. (b) HF C–V curve for an Au/Polymer/Si sample.

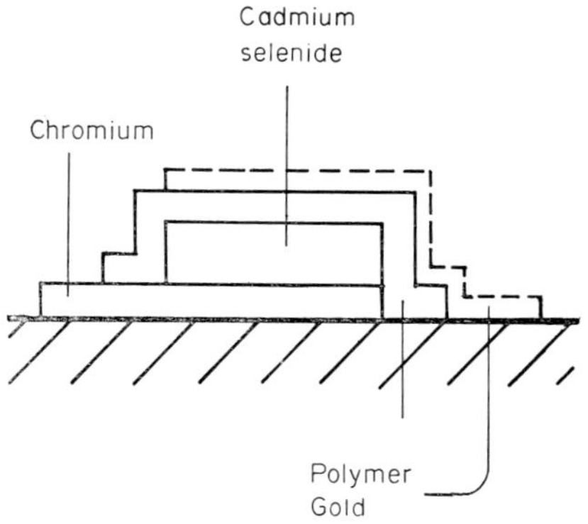

Figure 6. Thin film MIS structure.

Firstly, a chromium thin film was deposited on to the glass substrate as an ohmic contact with the semiconductor. The cadmium selenide semiconducting film was then deposited by thermal evaporation. The deposition rate was about 4 Å s^{-1} and the substrate temperature 100°C. These conditions were required to achieve a sufficiently low carrier concentration (*n*-type, 10^{17} cm^{-3}, 3000 Å). The insulator was a 2000 Å thick, rf-plasma polymerized polyparaxylen film. The counter electrode was a semitransparent Au film (300 Å).

2.2. Experimental procedure. High frequency C–V curves of these samples exhibited a large and complex hysteresis. Nevertheless a memory effect was observed like in Au/Polymer/Si Samples. When they were illuminated by pulsed light, without external polarization, these devices generated a large photo-emf. Displayed on the CRT (1 MΩ input impedance), the signal was about +200 mV high, and vanished after a brief −5 V polarization. Reversing the bias voltage, the photosignal reappeared. In order to make sure that the photocurrent was generated by electron-hole pair creation in the semiconductor, the influence of photon energy was then studied. A spectrophotometer was used to deliver a chopped light of tunable wavelength. The maximum photo response occurred at photon energies near the CdSe band gap. (Figure 8). Therefore, it is fair to conclude that the photocurrent source is the creation of electron-hole pairs in the semiconductor.

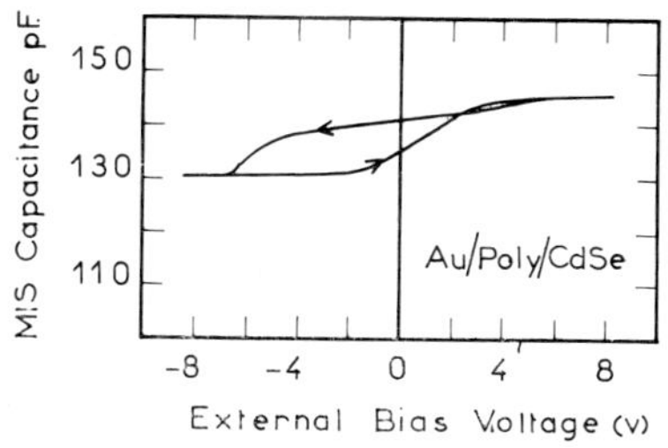

Figure 7. HF C–V curve for a thin film MIS structure.

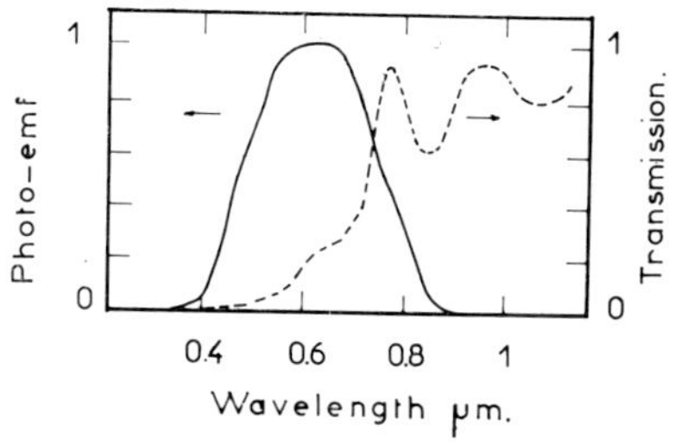

Figure 8. Photoresponse vs wavelength curve (continuous line) and optical transmission of a CdSe thin film (dashed line).

Conclusion

This experimental work shows the qualitative relation between surface potential and pulsed photo-emf in MIS devices. Nevertheless, one can see on Figure 4 that the photo-emf is lower than ψ_s. Several authors have shown (e.g. Tsarenkov[8]) that the photo-emf of surface-barrier structures reached a saturation level (close to the barrier-height) when the radiation flux density is as high as 10^{19}–10^{20} photons cm^2 s^{-1}. The radiation flux density of our xenon flash lamp probably does not reach such high values in the energies conducing to the production of electron-hole pairs in the semiconductor. Many other effects can reduce the photo-emf, for instance the trapping of carriers at the interface or inside the semiconductor. Indeed, a more complete research will be necessary for a good explanation of the pulsed photovoltaïc effect on MIS structures. Nevertheless, it appears like an attractive effect for light sensors or memory devices applications and is useful for MIS flat-band voltage measurements.

References

[1] E J Charlson and J C Lien, *J Appl Phys*, **46**, 1975, 3982.
[2] N F Kovtonyuk, V A Morozov, V G Fadin and I F Usoltsev, *Sov Phys Semicond*, **9** (2), 1975, 135.

[3] A F Plotnikov and V S Vavilov, *Sov Phys Semicond*, **7**, 1973, 599.
[4] K Okumura, *J Appl Phys*, **45**, 1974, 5317.
[5] B Vuillermoz, IUPAC Symposium on Plasma Chemistry—Rome Sept 1975.
[6] S M Sze, *Phys of Semiconducta Devices*, Wiley, New York.
[7] A Goetzberger, *The Bell Syst Techn J*, **45**, 1966, 1097.
[8] B V Tsarenkov, A Gol'dberg and T V L'vova, *Sov Phys. Semicond*, **8**, 1974, 800.

2.11 Hydrogen sensitive MOS structures

I Lundström, M S Shivaraman and C Svensson, *Research Laboratory of Electronics, Chalmers University of Technology, Gothenburg, Sweden*

It is shown that hydrogen in the ambient atmosphere changes the flatband voltage of palladium gate MOS-structures. The physical mechanisms behind this phenomenon are discussed. Experimental results on the detection of hydrogen in air, argon and nitrogen are given. It is shown that the rate constants of chemical reactions on the Pd-surface are obtained from such experiments. The detection of hydrogen in hydrogen compounds like ammonia and of hydrogen in smoke is also demonstrated.

Introduction

It is well known that a number of metals, like Pd and Pt, adsorbs, dissociates and dissolves hydrogen. We developed a new kind of hydrogen detector using Pd as the gate metal of an MOS-field effect device (transistor or capacitor).[1–3] The device works in the following way. Hydrogen is dissociated on the surface of the metal and some of the hydrogen atoms diffuse through the metal and are adsorbed at the metal oxide interface (Figure 1).

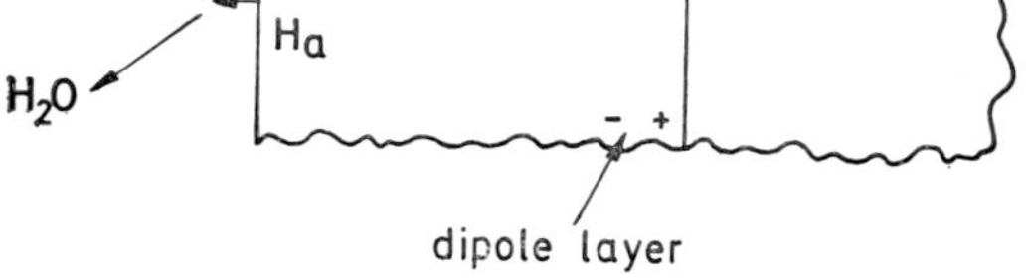

Figure 1. Schematic illustration of the process taking place in the Pd–SiO$_2$-system in the presence of hydrogen in air.

These adsorbed atoms give rise to a dipole layer at the interface, which changes the work function difference between the metal and the semiconductor or the flatband voltage of an MOS structure. (Recently the existence of dipole layer due to hydrogen atoms at interface has been shown by internal photoemission measurements).[8]

This is analogous to the change in surface potential of metal surfaces on the adsorption of gases.[4] An energy level diagram for hydrogen in and outside Pd is shown in Figure 2. The diagram is constructed from data obtained from our experiments on Pd-MOS structures and from data taken from the literature. The change in work function at the Pd–SiO$_2$ interface is determined by the hydrogen pressure outside the structure and by the chemical reactions taking place on the metal

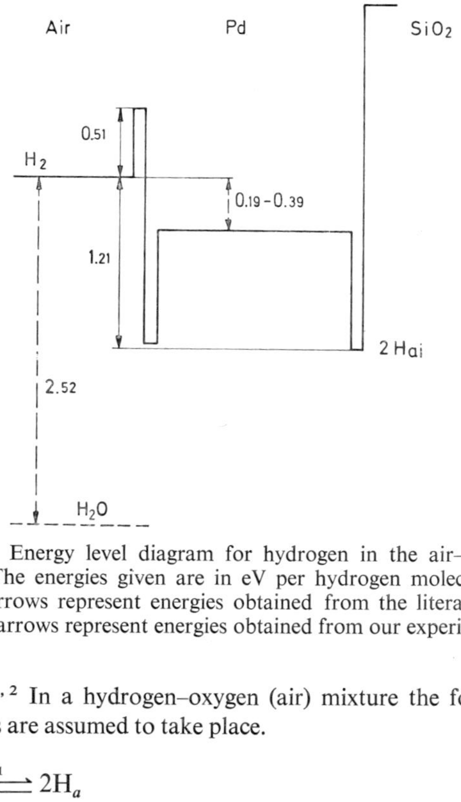

Figure 2. Energy level diagram for hydrogen in the air–Pd–SiO$_2$ system. The energies given are in eV per hydrogen molecule. The dashed arrows represent energies obtained from the literature and the solid arrows represent energies obtained from our experiments.

surface.[1,2] In a hydrogen–oxygen (air) mixture the following reactions are assumed to take place.

$$H_{2g} \underset{d_1}{\overset{c_1}{\rightleftharpoons}} 2H_a \tag{1}$$

$$O_{2g} + 2H_a \overset{c_2}{\rightarrow} 2OH_a \tag{2}$$

$$OH_a + H_a \underset{d_3}{\overset{c_3}{\rightleftharpoons}} H_2O \tag{3}$$

It is expected that the diffusion of hydrogen across a thin Pd-film is rapid compared to the time constants of the chemical reactions on the surface. This was also born out by our experiments. The reaction $H_a \rightleftharpoons H_{ai}$ is therefore always in 'equilibrium' and H_a may be replaced by H_{ai} in reactions (1–3). In analogy with adsorption on surfaces the change in work function, ΔV, is written

$$\Delta V = -\frac{\mu N}{\varepsilon_0}\theta = -\Delta V_{max} \cdot \theta \tag{4}$$

where μ is the dipole moment of an adsorbed hydrogen atom, N is the density of adsorption sites and θ their fractional coverage, all at the Pd–SiO_2 interface.

Reactions (1–3) predict the following steady state result for θ

$$\frac{\theta}{1-\theta} = \sqrt{\left(\frac{c_1\,[H_{2g}]}{2c_2\,[O_{2g}] + d_1}\right)} \qquad (5)$$

In the presence of oxygen the rate constant d_1 may be neglected. Equation (5) is in the form of a Langmuir isotherm which is often obtained for the adsorption of diatomic molecules.[5]

Measurements were performed on a number of MOS-structures with different oxide and Pd-thicknesses (10–100 nm oxide and 10–100 nm Pd). Similar results were obtained for all of these. The results presented below are mainly from measurements on *n*-channel silicon MOS-transistors with 10 nm oxide and 10 nm Pd-gate and capacitors on *p*-type Si with 50 nm oxide and 50 nm Pd-gate.

Experiments performed with hydrogen in air (20% oxygen) and argon verify equation (4). One important result is $\Delta V_{max} = 0.5$ V which was found both in studies of threshold voltage changes and from internal photoemission experiments.

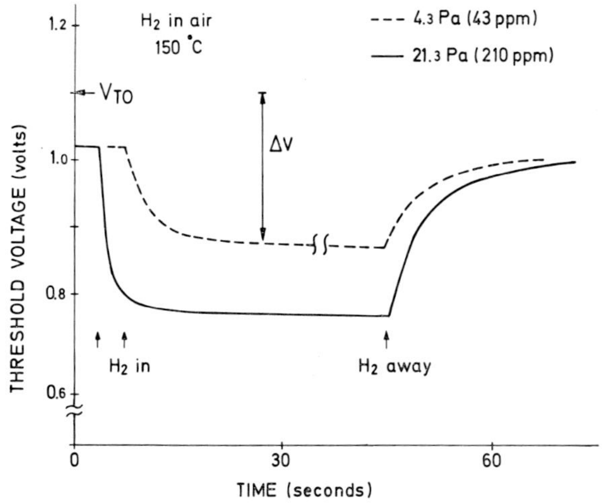

Figure 4. Typical transient response to hydrogen. Measurements were made on a Pd-gate field effect transistor (10 nm oxide, 10 nm Pd) with an extrapolated threshold voltage $V_{TO} = 1.1$ V (measured at 100 μA) in a completely hydrogen free atmosphere. The initial change of threshold voltage from V_{TO} is due to the background level of hydrogen.

Experiments

Hydrogen in air. Figure 3 shows experimental results from threshold voltage shifts on hydrogen in air (20% O_2 and 80% N_2) at different temperatures. The results show that $\sqrt{(c_1/2c_2)}$ increase with temperature. The activation energy for $\sqrt{(c_1/2c_2)}$

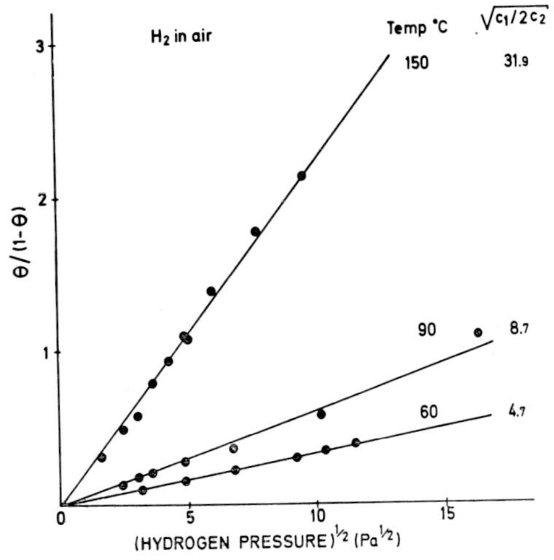

Figure 3. Experimental results for hydrogen in air at different temperatures. Results are plotted according to equation (4). $\Delta V_{max} = 0.5$V obtained from several measurements (2), was used. The measurements were performed on MOS-capacitors with 50 nm oxide and 50 nm Pd-gate metal.

within the temperature range 60–200°C is found to be 0.28 eV. c_1 and c_2 may be obtained from studies of transients. A typical transient response is shown in Figure 4. c_1 is for example determined from the initial slope of $V_T(t)$ on the introduction of hydrogen. It is found that c_1 is of the form $A\exp(-qQ_1/KT)$ where $A \approx 5 \times 10^3$ Pa^{-1} s^{-1} and $Q_1 \simeq 0.51$ eV in the temperature range 60–200°C. From these results we get $c_2 \sim \exp(qQ_2/KT)$ with $Q_2 \approx 0.05$ eV. A can be calculated from absolute rate theory[5] and is found to be 2.3×10^3 Pa^{-1} s^{-1} which is close to A above.

Reactions (2) and (3) are only tentative. Other possibilities are for example

$$O_{2g} + 2H_a \rightarrow H_2O_{2a}$$

$$2H_a + H_2O_{2a} \rightleftharpoons 2H_2O$$

Further experiments will be done to investigate these possibilities more carefully.

Hydrogen in argon. Hydrogen in argon represents a more fundamental and simple situation, only reaction (1) takes place. In this case the sensitivity to hydrogen is very large and is determined by $\sqrt{(c_1/d_1)}$, which is expected to decrease with temperature, since adsorption of hydrogen on to Pd is an exothermic process. The heat of adsorption $Q = 1.21$ eV per H_2 determined from the slope of the straight line in Figure 5 is close to the value $Q = 1.17$ eV found by other methods.[6] Furthermore the prefactor 1.64×10^{-6} Pa$^{-1/2}$ is close to the theoretical value 2.8×10^{-6} Pa$^{-1/2}$ calculated from reaction rate theory.[5]

Hydrogen in nitrogen. It was found that the hydrogen sensitivity was decreased when nitrogen was used as the 'inert' gas instead of argon. This is expected since metal surfaces are known to catalyse the formation of ammonia from H_2 and N_2.[7] The sensitivity to hydrogen at 150°C is given by a relation similar to equation (4) with $\sqrt{(c_1/c_N)} = 1.2$ Pa$^{-1/2}$. The sensitivity to nitrogen implies that the device may be sensitive to the reverse reaction, decomposition of ammonia, as well. It was also possible to detect ammonia in air, with $\theta/(1-\theta) = a\sqrt{P_{NH3}}$ where $a = 0.02$ Pa$^{-1/2}$. This indicates that $NH_3 \rightarrow NH + 2H_a$ may be one of the reactions on the Pd-surface.

Leak detection. A hydrogen leak detector capable of sensing 1 ppm H_2 in air which is comparable to that of helium leak detectors has been constructed. (The He leak detector sensitivity is limited by the normal concentration of He in air, which is about 5 ppm). The response and recovery times of the device is of the order of a few seconds. The leak detector has been successfully used for detecting leakage in vacuum systems.[3]

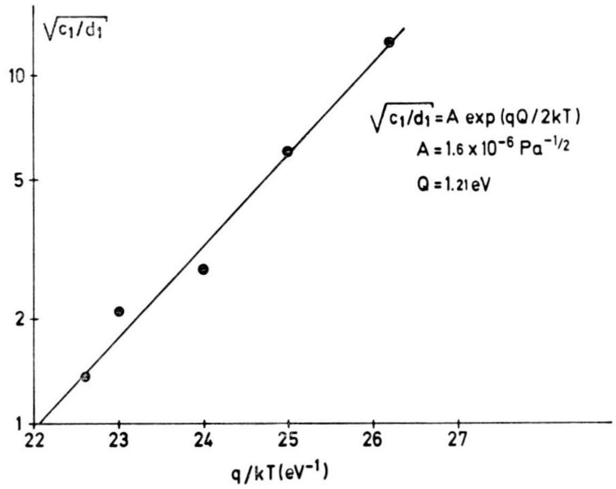

Figure 5. Temperature dependence of the sensitivity, $\sqrt{(c_1/d_1)}$, to hydrogen in argon. The experimental points were obtained from the slope of lines similar to those in Figure 3. The measurements were made on MOS-capacitors (50 nm oxide, 50 nm Pd). Temperature range covered 170–240°C.

Hydrogen in smoke. It is well known that smoke contains a certain amount of hydrogen depending on the degree of combustion. The hydrogen sensitive MOS-structure has therefore a potential application as a smoke detector or a fire alarm. Some preliminary experiments were done to investigate this possibility.[9] Figure 6 shows the hydrogen content in smoke from a

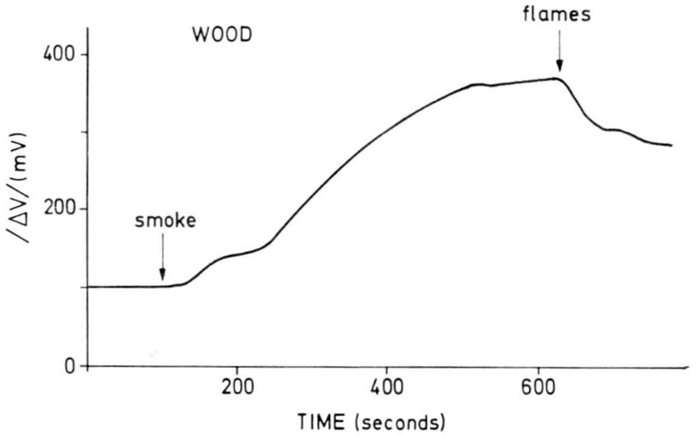

Figure 6. Change in threshold voltage of a Pd-FET showing the hydrogen content in smoke from glowing wood. The wood had a volume of about 10 cm^3 and the box it was placed in a volume of 1.44×10^5 cm^3. The wood was heated by a filament. The Pd-FET was kept under the roof of the box and held at 150°C.

piece of wood heated with a filament. The wood was kept in a metal box with a Pd-gate FET under the roof. This simple experiment shows that Pd–SiO_2–Si structures are able to detect a fire in being.

Discussion

It is seen that Pd–SiO_2–Si structures can be used as sensitive detectors for hydrogen. Hydrogen containing compounds, like ammonia and H_2S, may be detected by the Pd-gate structures in case they adsorb on the Pd-surface and leave hydrogen atoms to Pd.[10] It is also obvious that the hydrogen sensitive structures can be used to study catalytic reactions on metal surfaces. This is demonstrated by the evaluation of the rate constants for the chemical reactions on the Pd-surface and by the energy level diagram computed from our measurements.

Our investigations have so far been concentrated on palladiun and hydrogen. There are, however, other metals, like Pt and Ni, which should behave in a similar way. Furthermore it is possible that a certain metal or alloy can be found which is catalytically active for a certain reaction. To obtain a specific detector for that reaction we require that one of the reaction products is dissolved in the metal and adsorbed at the metal insulator interface where it changes the work function of the metal.

A part of this article was presented at IEDM (International Electron Devices Meeting 1975, Washington DC).

References

[1] I Lundström, M S Shivaraman, C Svensson and L Lundkvist, *Appl Phys Lett*, **26**, 1975, 55.
[2] I Lundström, M S Shivaraman and C Svensson, *J Appl Phys*, **46**, 1975, 3876.
[3] L Stiblert and C Svensson, *Rev Sci Instrum*, **46**, 1975, 1206.
[4] J Horiuti and T Toya, Chemisorbed Hydrogen, in *Solid State Surface Science*, **1**. (Edited by M Green), Dekker, New York (1969).
[5] J M Thomas and W J Thomas, *Introduction to the Principles of Heterogeneous Catalysis*, Chapter 2, Academic Press, New York (1967).
[6] G C Bond, *Catalysis by Metals*, Chapter 5, Academic Press, London (1962).
[7] G C Bond, *Catalysis by Metals*, Chapter 16, Academic Press, London (1962).
[8] I Lundström and T DiStefano, *Hydrogen Induced Interfacial Polarisation* at Pd–SiO_2 *Interfaces*, **59**, 1976, 000.
[9] I Lundström, M S Shivaraman, L Stiblert and C Svenson, *Rev Sci Instrum*, **47**, 1976.
[10] M S Shivaraman, *J Appl Phys*, in press.

2.12 Dielectric dispersion and polarization effects in thin transition metal oxide films

B Lalevic, N Annamalai,* B Leung,† G Taylor and W Slusark Jr, *Department of Electrical Engineering, College of Engineering, Rutgers University, New Brunswick, NJ 08903, U.S.A.*

Dielectric properties of thin polycrystalline and amorphous films of NiO, NiO(Li), CoO, CoO(Li), Nb_2O_5 and TiO_2 in the thickness range 0.14–15 μm were studied as a function of frequency, temperature, applied voltage and dopant concentration in the frequency range 3–10^8 Hz. These films were prepared either by anodization or thermal oxidation and in some cases by reactive sputtering. Experimental data for the real ε_1 and imaginary ε_2 parts of the dielectric constant, ac conductivity $\sigma(\omega)$ and loss tangent tan δ indicate the existence of several dispersion mechanisms. Cole–Cole plots and the modified Debye relations were used to derive the values for relaxation times at low frequencies. The frequency dependence of $\sigma(\omega)$ in the frequency region 10^3–10^6 Hz where $\sigma(\omega) \propto \omega^n$ indicates that the dispersion mechanism is due to charge carriers hopping between sites with either correlated or non-correlated potentials. The observed quadratic dependence of $\sigma(\omega)$ on frequency in the frequency region 10^6–10^8 Hz is attributed to a Debye type loss resulting from thermally activated hopping. Polarization effects at lower frequencies $<10^3$ Hz indicate the existence of Schottky barriers at the metal–insulator interfaces in agreement with Simmons theory of MIM structures.

Introduction

Thin films of transition metal oxides have potential application in various types of devices. Their transport and dielectric properties indicate their use as voltage tunable capacitors and varistors, Schottky barrier microwave detector diodes, protective switches and switching elements in logic circuits. Both memory and threshold type resistive switching have been observed in various polycrystalline and amorphous films of the transition metal oxides.[1–5] The semiconducting properties of 3-*d* transition metal oxides are due either to a lack of stoichiometry, i.e. an excess of metallic component which produces oxygen vacancies and metallic interstitials or to a suitable substitution of doubly ionized ions. Both *n*- and *p*-type conductivities have been observed.

We have performed extensive measurements on the transport, dielectric and switching properties of thin amorphous and polycrystalline films of NiO, NiO(Li), CoO, CoO(Li), Nb_2O_5 TiO_2. Two types of dc conduction mechanisms were observed. The I–V and G (conductance)–V characteristics of polycrystalline and amorphous Nb_2O_5, TiO_2 and CoO films were found to be in excellent agreement with the generalized theory of the Poole–Frenkel and Poole field assisted emission mechanism.[6] A modified theory of the small polaron hopping mechanism has been successfully applied to explain the results for NiO and NiO(Li) films.[1] The switching mechanism in the first oxide group has been fully described by the double injection theory,[7] while a small polaron delocalization process was found to be responsible for switching in NiO and NiO(Li) films.[1]

In this paper we review some of our previous results and present new complementing data on the dielectric dispersion and interface polarization effects in the above mentioned transition metal oxide films. The experimental results indicate that at higher frequencies the low field ac dispersion is due to the hopping of charge carriers while the polarization effects at lower frequencies are due to the presence of Schottky barriers at the metal–insulator interface.

Experiments

Sample preparation and measurement techniques. A detailed description of sample preparation was given in previous papers.[1,6] In general, the polycrystalline films were grown by thermal oxidation of parent metal plates in dry oxygen with an oxygen flow of 1–3 ft^3/h. The temperature and duration of oxidation were chosen for each oxide to produce an oxide film thickness in the range 1–15 μm. Oxide thickness was determined by weight gain, SEM micrographs and dielectric constant measurements. Amorphous films were obtained by anodization of parent metal plate in an appropriate solution for the given oxide. The thickness of amorphous films ranged between 1400 and 2400 Å. X-ray analysis was performed on all samples showing all the typical lines for polycrystalline films and the absence of any structure in amorphous films. The surface structure of polycrystalline films was studied with a scanning electron microscope and the observed surface roughness was approximately 0.5 μm. Counterlectrodes, 1000 Å thick of the parent metal, were either sputtered through a mask in an NRC triode sputtering module or evaporated in a vacuum of 10^{-6} torr.

Capacitance and ac conductance $\sigma(\omega)$ were measured as a function of frequency, applied voltage and temperature in the frequency region 3–2 × 10^5 Hz by using a lock-in amplifier (PAR HR-8) provided with a preamplifier (PAR 184) and signal

* Present addresses: University of West Virginia, Morgantown, W. VA USA.
† RCA, Somerville, NJ, USA.

conditioning amplifier (PAR 114). General Radio (1615A) and Wayne Kerr (B601 and B801B) impedance bridges were used in the frequency range 10^5–10^8 Hz.

Results

The results presented here are usually given for one oxide with the understanding that they are typical of the behavior of the other investigated oxides unless otherwise stated. The experimental results have shown in general the existence of three dispersion regions in the frequency range 3–10^8 Hz where $\sigma(\omega)$ exhibited an ω^n frequency dependence with n varying between 0.71–2.

Capacitance dependence on frequency has shown two distinct types of behavior. The real-part of the dielectric constant ε_1 in polycrystalline NiO and TiO_2 films had only a slight dependence on frequency as shown in Figure 1 where the plot of ε_1 vs f is given for TiO_2 5 μm thick films at 300 K. Polycrystalline and amorphous films of Nb_2O_5 and CoO showed a large increase in ε_1 at lower frequencies as seen in Figure 2 where C-f plot is presented for amorphous Nb_2O_5 films 2400 Å thick at 300 K. A more complex behavior was observed in the frequency dependence of the imaginary part of dielectric constant ε_2. At frequencies below 10^3 Hz, ε_2 varied as ω^{-n} where n varied between 1.0 (NiO(Li)) and 0.31 (amorphous Nb_2O_5), while at higher frequency ε_2 showed a less pronounced dependence on frequency. Typical behavior is shown in Figure 3 where ε_2 is plotted as a function of frequency for CoO polycrystalline films at 300 K.

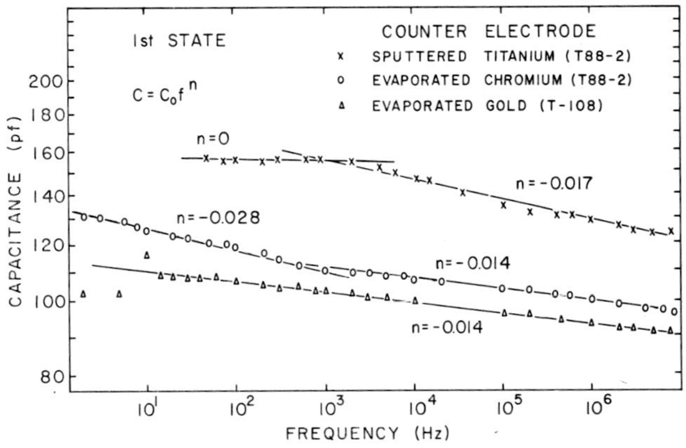

Figure 1. Frequency dependence of capacitance of polycrystalline TiO_2 4 μm thick with different counter electrodes at 300 K.

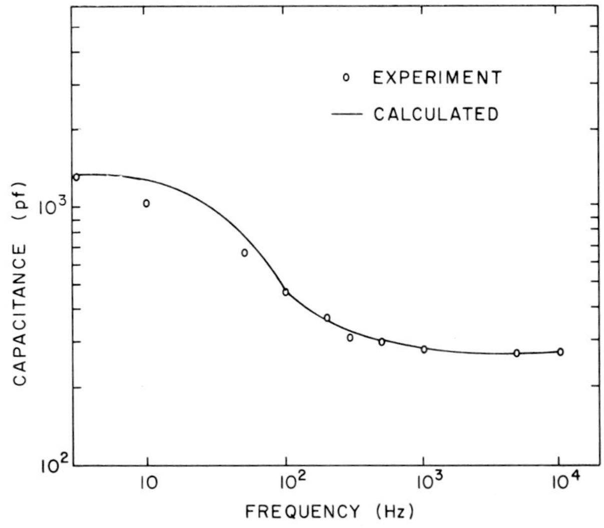

Figure 2. Plot of log C vs log f for Nb–Nb_2O_5–Nb structure with Nb_2O_5 2400 Å thick at 300 K. The theoretical C–f curve (equation 6a) is also shown.

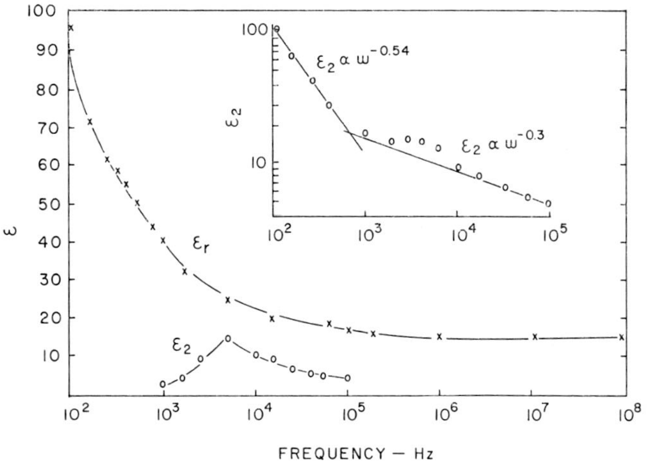

Figure 3. Plots of real ε_1 and imaginary ε_2 parts of dielectric constant as a function of frequency for CoO film of 5 μm thickness at 300 K. Plot of ε_2 vs f at the lower part of Figure 3 is obtained by subtracting low-frequency ($<10^3$ Hz) contribution $\varepsilon_2 \propto \omega^{-0.54}$.

The loss tangent tan δ has shown either a smooth decrease with increasing frequency and a leveling off at higher frequencies as is the case of NiO films, or a pronounced maximum at 200 Hz for amorphous and polycrystalline films of Nb_2O_5, CoO and TiO_2. The plot of tan δ vs frequency for amorphous 2400 Å thick Nb_2O_5 at 300 K is shown in Figure 4. The ac conductivity $\sigma(\omega)$ has exhibited a frequency dependence at 300 K of the form $\sigma(\omega) \propto \omega^n$ where n has the following values: in the frequency range 10^3–10^6 Hz $n = 0.71$ for CoO, 0.72 for NiO, 0.75 for amorphous Nb_2O_5, 0.78 for polycrystalline Nb_2O_5 and 1.0 for TiO_2. Above 10^6 Hz, $n = 2.0$ for CoO, NiO, NiO(Li) and TiO_2 and $n = 1.0$ for amorphous and

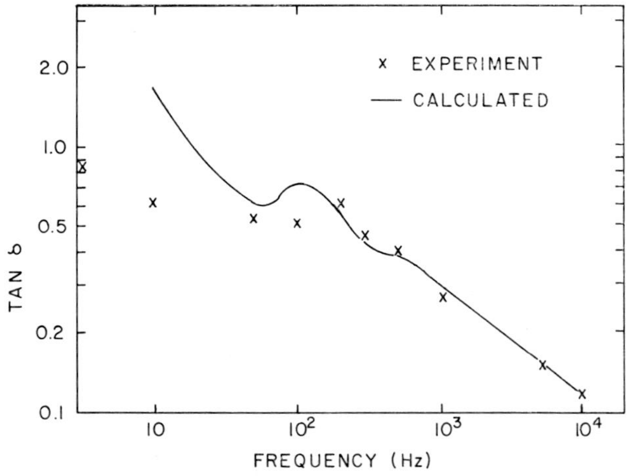

Figure 4. Plot of loss tangent tanδ vs log f for Nb–Nb_2O_5–Nb structure with amorphous Nb_2O_5 2400 Å thick at 300 K. The theoretical tanδ–f curve (equation 6d) is also presented.

polycrystalline Nb_2O_5 films. A plot of $\sigma(\omega)$ vs f for polycrystalline TiO_2 films is given in Figure 5. The capacitance of these oxides has shown a strong temperature dependence at higher temperatures and only slight dependence below 200 K. A plot of log C vs $10^3/T$ gave a set of parallel lines at different measuring frequencies. The activation energy of a dielectric oscillator calculated from these plots ranged between 0.28 eV for polycrystalline Nb_2O_5 to 0.05 eV for amorphous Nb_2O_5. A plot of log C vs $10^3/T$ for polycrystalline Nb_2O_5 film at 1 MHz is shown in Figure 6.

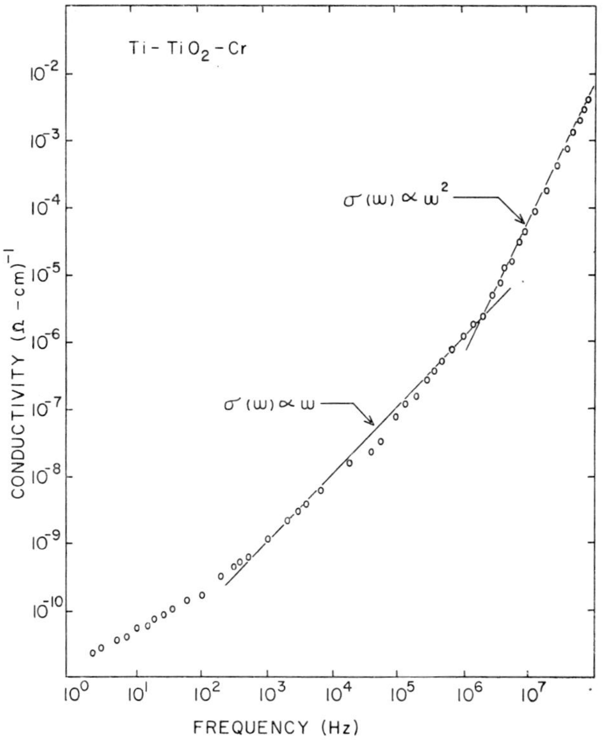

Figure 5. Low-field ac conductivity $\sigma(\omega)$ as a function of frequency for polycrystalline TiO_2 film 4 μm thick at 300 K, showing $\sigma(f) \propto f$ in the range $10^2 \leq f \leq 2 \times 10^6$ and $\sigma(f) \propto f^2$ for $f > 2 \times 10^6$.

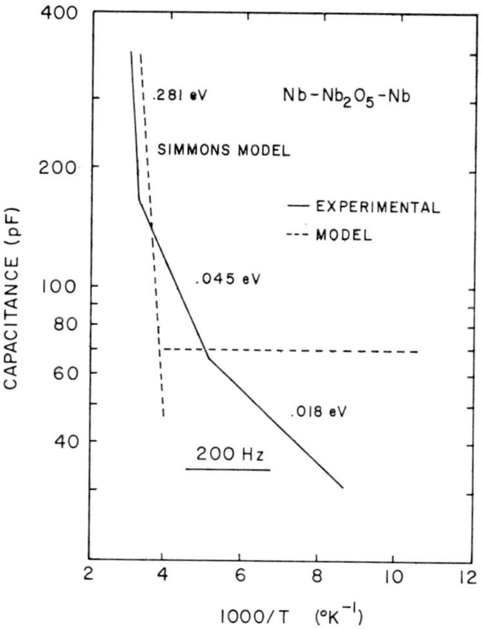

Figure 6. Plot of capacitance vs $10^3/T$ for Nb–Nb_2O_5–Nb structure with polycrystalline Nb_2O_5 5 μm thick. Measurements were performed at the frequency of 200 Hz. The theoretical curve (equation 6c) is also presented.

The voltage dependence of capacitance has also exhibited two different relationships. For polycrystalline films of NiO, capacitance was independent of the applied voltage in the temperature range 145–445 K and applied field range of 0–5 $\times$ 10^4 V/cm. For amorphous and polycrystalline films of Nb_2O_5 the capacitance has shown a symmetrical bell shape type voltage dependence. The capacitance decreased with increasing voltage and it became constant at higher voltages. This behavior is shown in Figure 7 where a C–V plot, taken at 300 K and 10 Hz is given for an amorphous Nb_2O_5 film 2400 Å thick.

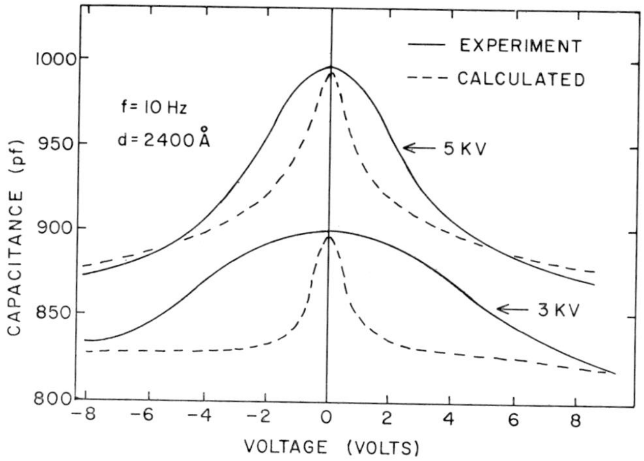

Figure 7. Plot of capacitance vs applied voltage for Nb_2O_5 film 2400 Å thick at 10 Hz and 300 K with the Nb counterelectrodes sputtered at 3 and 5 kV. The calculated C–V dependence (equation 6b) is also shown.

Discussion and Conclusions

Detailed information about the observed dispersion relations for the real ε_1 and imaginary ε_2 parts of dielectric constant can be obtained from the Cole–Cole plot. For a dielectric with a single relaxation time the Cole–Cole plot of ε_2 vs ε_1 is a semicircle defined by the Debye equations. The existence of more than one relaxation time causes a deviation from the semicircle in the ε_2 vs ε_1 plot and indicates the existence of several dispersion mechanisms. A typical plot is given in Figure 8 where ε_2 is plotted vs ε_1 for CoO films at 300 K. As shown in Figure 8 deviation from a semicircular plot occurs at the frequency of 10^3 Hz. In this case a functional relationship between ε_2–ε_1 can be found in terms of the modified Debye relations given by[8]:

$$\varepsilon^* = \varepsilon_\infty + (\varepsilon_0 - \varepsilon_\infty)/(1 + j\omega\tau_0)^\beta \quad (1)$$

with the frequency dependence of ε_1 and ε_2 given in the form

$$\varepsilon_1 - \varepsilon_\infty = (\varepsilon_0 - \varepsilon_\infty)(\cos\phi)^\beta \cos\beta\phi \quad (2)$$

$$\varepsilon_2 = (\varepsilon_0 - \varepsilon_\infty)(\cos\phi)^\beta \sin\phi$$

Here $\tan\phi = \omega\tau_0$ and $\omega\tau_0 = \tan\theta/\beta$ with $\theta = \tan^{-1}(\varepsilon_2/\varepsilon_1 - \varepsilon_\infty)$ and $\beta = 0.3$ (Figure 3). The value of the relaxation time τ_0 can be obtained from the relation $\omega\tau_0 = \tan\theta/\beta$ by plotting $\tan\theta/\beta$ vs frequency, which gives a straight line. The frequency $f_c = \omega_c/2\pi$ at which $\tan\theta/\beta = 1$ gives $\tau_0 = 1/2\pi f_c$.

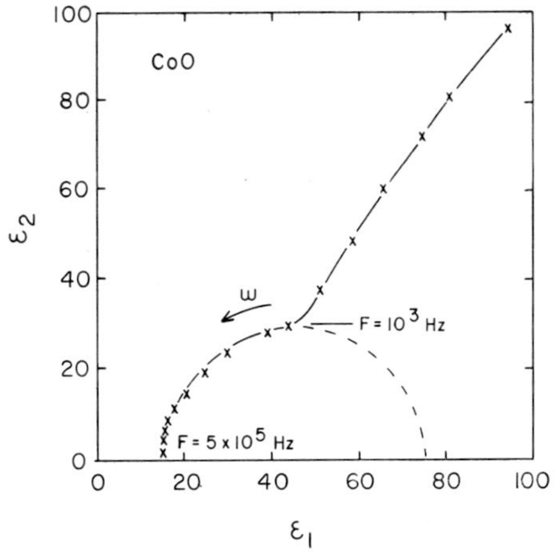

Figure 8. Plot of ε_2 vs ε_1 for CoO film of 5 μm thickness at 300 K showing the deviation from a semicircular Cole–Cole plot.

For CoO films $\tau_0 = 1.6 \times 10^{-4}$ s^{-1}, for amorphous Nb_2O_5 films $\tau_0 = 8 \times 10^{-2}$ s^{-1} and for NiO films $\tau_0 = 4 \times 10^{-4}$ s^{-1}. The Cole–Cole plots for TiO_2 films have shown several groupings of relaxation times, which were difficult to separate.

For $\omega\tau_0 > 1$ equation (2) can be written as[9]:

$$\varepsilon_1 - \varepsilon_\infty = (\varepsilon_0 - \varepsilon_\infty)(\omega\tau)^{-\beta} \sin(\beta/\omega\tau) \quad (3)$$

$$\varepsilon_2 = (\varepsilon_0 - \varepsilon_\infty)[\sin(\beta\pi/2)](\omega\tau)^{-\beta}$$

Thus for $\beta = 0.3$, $\varepsilon_2 \approx (\varepsilon_0 - \varepsilon_\infty)/2(\omega\tau)^{-0.3}$ and $\varepsilon_1 \rightarrow \varepsilon_\infty$ since $\sin(0.3/\omega\tau) \rightarrow 0$ for $\omega\tau_0 > 1$. This explains the slight dependence of ε_1 on frequency for frequencies $>10^3$ Hz as shown in Figure 3.

Further insight in the mechanism of dielectric dispersion can be obtained from the observed frequency dependence of ac conductivity $\sigma(\omega)$. It was found experimentally that $\sigma(\omega) a\ \omega^n$ where n is either ~0.8 or 1 in the frequency range 10^3–10^6 Hz and $n = 1$ or $n = 2$ for frequencies above 10^6 Hz. These relationships correspond to the conduction mechanism where charge carriers hop between sites with correlated potentials ($n = 0.79$), uncorrelated potentials ($n = 1$) and the thermally activated Debye type hopping ($n = 2$).[10,11]

The theory of hopping between sites with correlated potentials predicts a frequency dependence of $\sigma(\omega)$ in the following form[10]:

$$Re\ \sigma(\omega) \infty (14.8 - 1/2 \ln \omega)^4 \omega \quad (4)$$

A plot of the right-hand side of equation (4) vs ω shows that $\sigma(\omega)$ depends on frequency as $\sigma(\omega) \propto \omega^{0.79}$ which is in good agreement with the observed frequency dependence of $\sigma(\omega)$ for NiO, CoO and Nb_2O_5 films in the frequency range 10^3–10^6 Hz.

A linear frequency dependence of $\sigma(\omega)$ is predicted by the theory of hopping between sites with uncorrelated potentials where[11]

$$\sigma(\omega) = 1/96\ \pi^3\ N^2(E_F) kTae^2 r_\omega{}^4 \omega \quad (5)$$

Here $N(E_F)$ is the probability density of localized states at the Fermi level (cm^{-3}, eV^{-1}), r_ω is the separation distance between a pair of localized states, $r_\omega = a \ln(\nu_0/\omega)$, where $\nu_0 = 10^{12}$ s^{-1} and a is the lattice constant. Taking the appropriate values for a and ω, one obtains $r_\omega = 55$ Å for Nb_2O_5 and 49 Å for TiO_2 and $N(E_F) = 1.5 \times 10^{19}$ cm^3 eV^{-1} for Nb_2O_5 and 7×10^{19} cm^3 eV^{-1} for TiO_2. The necessary pair condition $(4/3\ \pi\ N(E_F)kT)^{-1/2} > r_\omega$ was satisfied in all cases.

The quadratic frequency dependence of $\sigma(\omega)$ in the frequency range 10^6–10^8 Hz observed in NiO, CoO and TiO_2, corresponds to ac conductivity of the Debye type and loss resulting from thermally activated hopping from one localized site to another. This predicts $\sigma(\omega) \propto \omega^2\tau/(1 + \omega^2\tau^2)$ which for $\omega \ll \tau$ yields $\sigma(\omega) \propto \omega^2$, which corresponds to the observed experimental results (Figure 5).

The large polarization effect observed at low frequencies, $f < 10^3$ Hz such as the increase in capacitance with decreasing frequency (Figure 2), capacitance dependence on voltage and temperature could be explained by assuming the existence of Schottky barriers at the metal–insulator interfaces.[12] Assuming the existence of Schottky barriers, the MIM structure can be represented by an equivalent circuit model consisting of two Schottky barrier capacitors, C_{sc}, in series with the bulk capacitance C_b and in parallel with corresponding Schottky barrier resistances R_s and bulk resistance R_b. This model[12] predicts the dependence of capacitance on frequency, applied voltage and temperature, loss tangent tan δ as a function of frequency and temperature and Cole–Cole plot for various voltage biases.

The functional relationships given by Simmons[12] for the MIM structures can be written in the approximation $R_{sc} > R_b$ as follows:

$$C(f) = C_b + \frac{C_{sc}}{1 + (\omega R_b C_{sc})^2} \quad (6a)$$

$$C(V) = C_b + \frac{C_{sco}/(1 + qV/\phi_0)}{1 + \left[\omega R_b + 1 + \dfrac{C_{sco}}{\left(\dfrac{qV}{\phi_0}\right)^{\frac{1}{2}}}\right]^2} \quad (6b)$$

with

$$C_{sco} = \frac{Aq^2 N_D}{2\phi_0{}^{\frac{1}{2}}}$$

$$C(T) = C_b + \frac{C_{sc}}{1 + (\omega C_{sc} R_0)^2 \exp(2\phi_i/kT)} \quad (6c)$$

where

$$R_b = R_0 \exp(\phi_i/kT)$$

and

$$\tan\delta(f) = \frac{1}{R(f)\omega C(f)};\ R_f = R_b + \frac{R_b}{1 + \omega^2 C_{sc}{}^2 R_b} \quad (6d)$$

Here ϕ_0 is the interfacial barrier height with the value obtained from the $1/C^2$ vs V plot; $\phi_0 \approx 1$ eV for both crystalline and amorphous Nb–Nb_2O_5–Nb structures; ϕ_i is the thermal activation energy obtained from the log σ(o) vs $10^3/T$ plot, N_D is the donor density and A is the contact area. In all oxides showing large polarization effects $\phi_0 > \phi_i$ and thus $R_{sc} > R_b$.

A C–f plot for Nb_2O_5 films as given by equation (6a) is given in Figure 2 where it is compared with the experimental data exhibiting a satisfactory agreement between the theory of MIM structure and experiments. The plot of tan δ vs f calculated for amorphous Nb_2O_5 films using equation (6d) is shown in Figure 4. The peak in the calculated value of tan δ is at 160 Hz which is somewhat displaced from the experimental peak at ≈200 Hz. The agreement between the experimental C–T data on polycrystalline Nb_2O_5 film and the behavior predicted by the equation (6c) is excellent with respect to both magnitude and shape as shown in Figure 6. Discrepancy at lower temperature is due to the temperature dependence of the bulk capacitance C_b which can be represented by the Debye relation for thermally activated dielectric oscillator in the form[13]:

$$\varepsilon_1(T) = \varepsilon_\infty + \left[(\varepsilon_0 - \varepsilon_\infty)/(\omega\tau_0)^2\right]\exp\left[-\frac{2H}{kT}\right] \quad (7)$$

where H is the activation energy for a dielectric oscillator with the value of $H = 0.018$ eV for Nb_2O_5 films. A more significant discrepancy is found between the experimental and calculated (equation (6b)) plots of C–V for amorphous Nb_2O_5 films as shown in Figure 7. This discrepancy can be attributed, however, to the observed interfacial polarization effects at frequencies <10 Hz.[5]

In conclusion, the observed dielectric dispersion in the investigated thin films of transition metal oxides can be attributed to the hopping of charge carriers. It has been shown that in the frequency region 10^3–10^6 Hz a satisfactory agreement has been obtained with the theory of ac conduction based on charge carriers hopping between sites with either correlated or uncorrelated potentials. In the frequency region $>10^6$ Hz, the dielectric dispersion was due to Debye type thermally activated hopping. At lower frequencies $<10^3$ Hz a successful explanation of the observed dielectric behavior was achieved by assuming the existence of Schottky barriers at the metal–insulator interfaces.

Acknowledgement

This work was supported by the Office of Naval Research.

References

[1] B Lalevic, N Fuschillo and B Leung, *Phys Stat Sol*, (a) **23**, 1974, 61.
[2] T Hickmott and W Hiatt, *Solid State Electron*, **13**, 1970, 1033.
[3] S Basavaiah and K Park, *IEEE Trans Electron Devices*, **ΔD-20**, 1973, 149.
[4] F Argall, *Solid State Electron*, **11**, 1968, 535.
[5] N Fuschillo and B Lalevic, *Japan J Appl Phys*, **13**, 1974, 817.
[6] B Lalevic and G Taylor, *J Appl Phys*, **46**, 1975, 3208.
[7] G Taylor and B Lalevic, *Solid State Electron*, **19**, 1976, 669.
[8] D Davidson and R Cole, *J Chem Phys*, **19**, 1952, 1484.
[9] B Lalevic, N Fuschillo, B Kuliyev and W Wang, *Appl Phys*, **5**, 1974, 127.
[10] M Pollack and T H Geballe, *Phys Rev*, **122**, 1961, 1745.
[11] M Pollack, *Phil Mag*, **23**, 1971, 519.
[12] G Nadcarni and J Simmons, *J Appl Phys*, **47**, 1976, 114 and references contained therein.
[13] P Debye, *Polar Molecules*, Chemical Catalog Co, New York (1929).

2.13 Behaviour of ceramic insulator materials during initiation of trigger discharge in TVG

Vidyardhi Nanduri, *Bharat Heavy Electricals Ltd, Corporate R & D, Secunderabad, India, 500003*

An effort has been made by the author to design a simple triggered vacuum gap (TVG) in which a plasma discharge over the surface of a ceramic insulator initiates an arc in high vacuum (p $\leq 10^{-5}$ torr). This design is reported elsewhere. A comprehensive study on the performance of different insulator materials is not reported so far.

Thermally stable insulator materials like barium titanate (LCC HTD), silicon carbide, lead titanate, rutile, supramica, alumina, sillimanite and fused quartz in the order of decreasing permittivities, were investigated. The plasma discharge is triggered with a positive voltage pulse to the trigger electrode under low voltage conditions ≤220 V, 100 A for the main electrodes. Samples like barium titanate, rutile, sillimanite showed very consistent performance even up to 1000 trials indicating that a low voltage switching device operating in high vacuum is feasible. During these experimental investigations, both external triggering and self triggering modes have been tried, the latter successfully solving the arc-initiation problem in a much simpler way. This paper presents the results obtained with the above samples at 1–5 × 10^{-6} torr.

1. Introduction

The insulating properties of high vacuum have been the subject of attention since early 19th century for the development of switching devices because the dielectric strength is of the order of 10 kV/mm even for small gaps. In such an environment, for the creation of an arc discharge, methods like drawing an arc, exploding a wire, injection of electrons and ions by external means, etc. have been tried.[1] The use of high voltage impulse generators of 20–30 kV is also not uncommon in order to give a trigger pulse to breakdown a small vacuum gap of few mm. Triggered vacuum gap (TVG) is a very versatile three electrode pulsed-arc switching device for the creation of high current discharges in a vacuum gap.

Three electrode gaps known as 'Trigatron gaps' operating in gaseous media are widely known earlier.[2] However TVG differs considerably in the following respects.

(i) The ratio of main gap withstand voltage to the trigger pulse initiation of breakdown may be as high as 100–1000 compared to 3 or 4 Trigatron gaps.

(ii) In the case of TVG, the conduction of the main discharge (metal vapour arc) ceases prior to zero current whereas there is no such control for the Trigatron gaps. The TVG has thus an excellent interrupting ability even at tens of kA.

Recent progress in the development of vacuum switches[3] and understanding of breakdown phenomena in the presence of insulators in vacuum[4] have increased the interest in the applications of TVGs. With the above advantages of switching in vacuum, several investigators[5–7] attempted different designs which however met with limited operating life.[5–7] Lafferty[5] described several possible designs in general and one of these include injection of gases. Farrall[6] reported the low voltage firing characteristic of TVG. However studies on low voltage trigger–plasma discharge in TVG on the behaviour of insulator specimens are not reported so far.

In the present study, thermally stable insulator materials like barium titanate (LCC HTD) ($\varepsilon_r = 1850$), silicon carbide (285), lead titanate (192), rutile (80), supramica-500 (32), alumina (8.7), sillimanite (6.2) and fused quartz (3.78) in the order of decreasing permittivities (shown in brackets) have been investigated in the creation of plasma discharge in simple TVG under low circuit voltage conditions i.e. 220 V, 100 A dc Lafferty[5] reported lower triggering voltages when the trigger potential is positive with the trigger electrode near the cathode compared to the other modes of triggering TVG. Hence, in these investigations, the plasma discharge over the surface of insulator specimens is triggered with positive voltage pulse. The trigger electrode is placed in the recess of the main cathode and the main electrodes may be connected conveniently in the laboratory to a low voltage dc source.

The design principles and consistency characteristic of the simple TVG has been reported by the author elsewhere.[8,9] The present paper covers the performance of these insulator specimens in triggering plasma discharge in TVG under low voltage circuit conditions both with external and self-trigger modes of operation at 1–5 × 10^{-6} torr with liquid nitrogen cold trap.

2. Breakdown across insulators in vacuum

When the electrodes in a vacuum gap are separated by a solid insulating material, the surface discharge over the insulator takes place at a lower voltage than a straightforward vacuum gap. Gleichauf,[10] Kofoid[4] and others reported the breakdown phenomena over solid dielectrics in vacuum. Fryszman[11] reported the mechanism of breakdown in the presence of insulator specimens of the order of 10 mm and above. At the cathode–insulator junction free electrons are emitted and these electrons strike the anode or the insulator near the anode thereby releasing secondary electrons. A positively charged area on the surface of the insulator near the anode moves towards the cathode so that the stress at the cathode eventually increases sufficiently to cause breakdown. Once the arc is formed, the insulator surface discharges with the decrease in field strength and the arc is extinguished. After a large number of such discharges sufficient metal vapour gets deposited on the surface. However data for specimen thickness less than 1 mm is scarce. The lower breakdown potentials with thermally stable ceramic insulator specimens are utilized in improving the design of TVG.[8]

3. Ceramic samples

The trigger electrode–insulator–trigger cathode is referred as trigger-gap which is in a recess of the main cathode. The ceramic materials chosen cover a wide range of relative permittivities. Samples like barium titanate (LCC HTD), lead titanate, fused quartz and supramica were procured as bulk samples and cut and machined to the proper size, 9 mm OD, 3.5 mm ID, 2 mm thick. Silicon carbide is a varistor from the laboratory whereas rutile and sillimanite were prepared by sintering at 800–850°C for 3 h. All these specimens are finally ground with 303 alumina powder on a grinding wheel of equal smoothness, cleaned and dried before use.

The trigger-cathode assembly is checked for concentric arrangement and trigger gap spacings of 0.12, 0.2, 0.3, 0.4 mm can be conveniently set prior to the experimental investigations. Trigger gap spacing of 0.2 mm is generally used unless otherwise mentioned in the paper. The test chamber, 300 × 300 mm high, is evacuated to 1–5 × 10^{-6} torr by oil diffusion pump provided with liquid nitrogen cold trap.

4. Electrical circuit and conditioning

The main electrodes are connected to 230 V, 25 kW dc variable generator source for various main gap settings. The triggering of plasma discharge in TVG is either by thryatron pulse generator with variable voltage up to 3.5 kV (external triggering)[8] or by the same dc source supplying the main arc current (self-triggering).[9] The circuit details are given elsewhere.[8,9] Figures 1 (a and b) show the shape of the external trigger pulse and breakdown of the trigger gap. Similarly in the case of the self-trigger mode, the pulse shape and subsequent breakdown are shown in Figures 2 (a and b). The external trigger circuit is used during the conditioning of trigger gap.

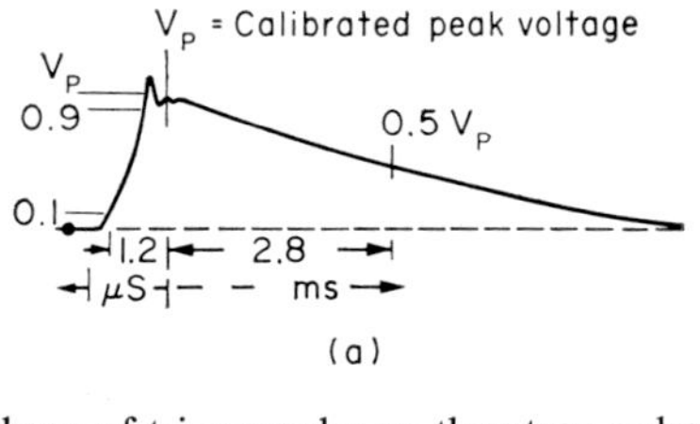

(a)

Figure 1. (a) Shape of trigger pulse on thyratron pulse generator.

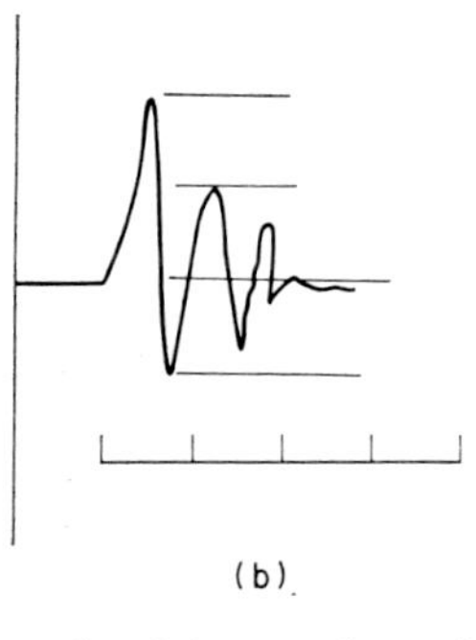
(b)

Figure 1. (b) Trigger gap breakdown and waveform, sweep 2 μs/cm.

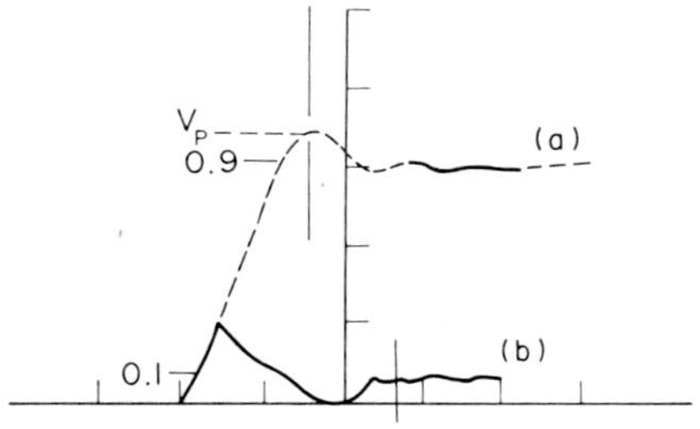

Figure 2. Self trigger mode pulse. (a) Shape dotted line, (b) breakdown trigger gap 0.2 mm specimen silicon carbide, sweep 50 ns/cm, voltage = 0.2 V/cm.

With the main dc source isolated, the trigger pulse is applied to the trigger-gap with increasing magnitude of voltage in steps of 60 to 500 V and in steps of 100 V for higher voltages. At a voltage slightly higher than breakdown voltage of trigger gap (Figure 1b), the plasma discharge across the insulator gap expands into the main gap. The firing of the trigger gap is continued until at a particular voltage (Vi) the plasma expansion was observed from the observation window for each pulse for 10 successive operations. This voltage (Vi) is near 1 kV for insulator specimens like lead titanate and barium titanate whereas for specimens like silicon carbide and sillimanite, it is near 500 V.

After the above initial firings, the main circuit voltage 220 V, dc is switched on to the electrodes and the load resistance is set to arc current of 30 A. It is obvious that at the trigger voltage Vi, the plasma discharge from the trigger gap establishes an arc in the main gap with the cathode spot moving at

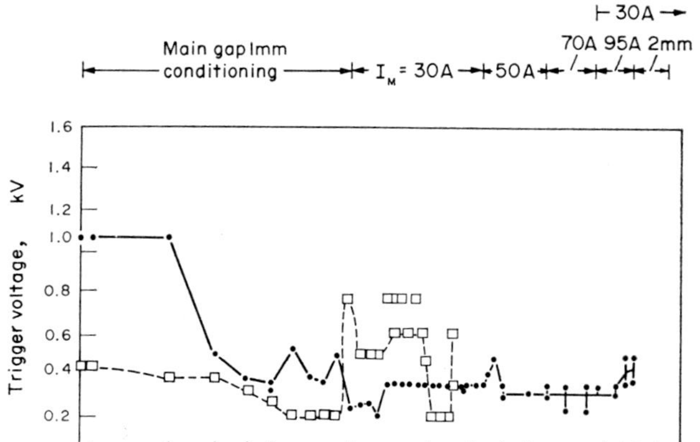

Figure 3. Variation of trigger voltage during conditioning and continued triggering of same sample in the initiation of varying main arc currents and main gap spacings ● = Barium titanate, □ = same sample after 24 h.

random. The arc is the metal vapour arc which can be switched off after 1–2 s. The thyratron pulse generator is now set to lowest voltage (146 V) and the trigger pulse magnitude is increased in steps of 25 V. At each step 2 or 3 firings are given to trigger the discharge and observations on minimum trigger voltage V_t are made to establish an arc. A number of experiments (sometimes exceeding 1000) have been repeated with the same sample and the data as a set of 25 trials is checked on an average for consistency and reproducibility.

5. Performance of insulator specimens

Figures 3 to 5 show the performance of the different insulator specimens with varying main circuit currents (30–95 A dc). In Figure 3 the performance during conditioning, continued triggering of TVG and performance after 24 h for barium titanate

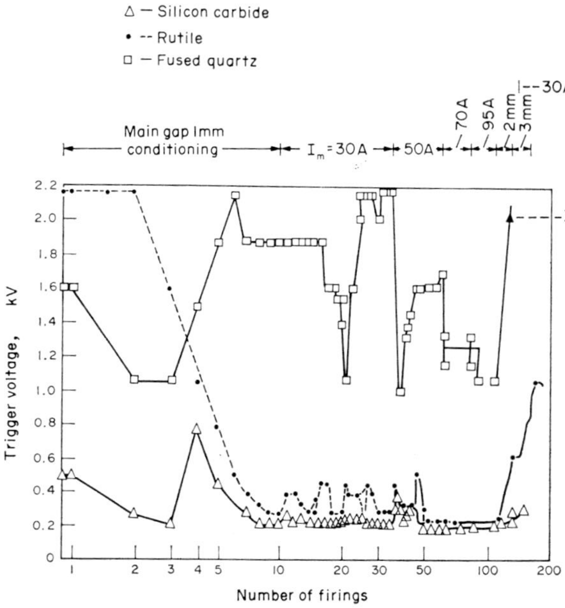

Figure 4. Variation of trigger voltage for silicon carbide rutile and fused quartz during conditioning and later 150 trials main gap 1–3 mm main circuit 220 V 30–95 A $p = 1\text{–}5 \times 10^{-6}$ torr.

specimen is shown, with varying main circuit current. The more probable trigger voltage remains within 358 V with lowest being at 236 V. The lower triggering may be due to the lower pulse rise time, with the breakdown of trigger gap occurring on the front edge of the pulse. Thus a faster trigger pulse may initiate triggering at even lower voltages. Another factor may be the source impedance of the pulse generator, lower source impedance giving rise to lower triggering. These two above observations are further confirmed by self-triggering mode.

Samples like fused quartz ($\varepsilon_r = 3.78$) alumina ($\varepsilon_r = 8.7$) and supramica ($\varepsilon_r = 32$) require trigger voltages higher than 500 V by external triggering. In comparison, the specimens like sillimanite ($\varepsilon_r = 6.2$), rutile ($\varepsilon_r = 80$), lead titanate ($\varepsilon_r = 192$), silicon carbide ($\varepsilon_r = 285$) and barium titanate ($\varepsilon_r = 1850$) require trigger voltages less than 500 V under 220 V dc main circuit conditions. Of these, specimens sillimanite, rutile, silicon carbide and barium titanate showed consistent results up to 1000 switchings.

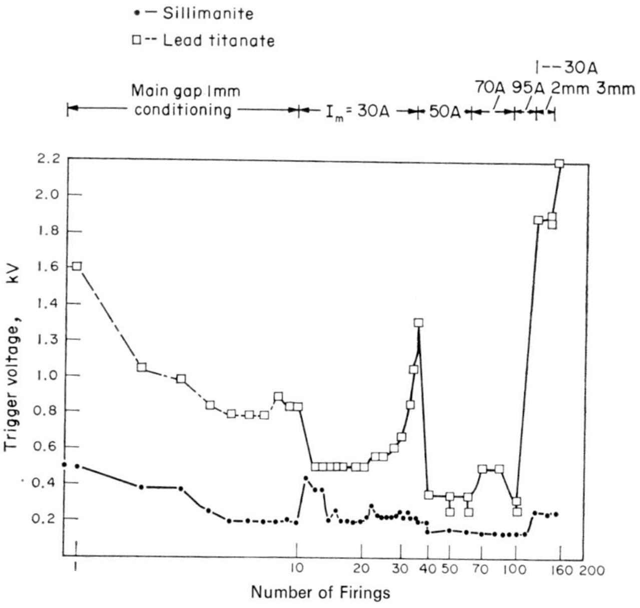

Figure 5. Performance characteristics of sillimanite and lead titanate specimens in trigger discharge main circuit 220 V $p = 1 \times 10^{-6}$ torr, trigger gap = 0.2 mm main circuit current indicated.

Table 1 shows the summary of the test results in triggering plasma discharges for initiation of arcs in TVG for varying main gap spacings of 1–5 mm. These results with trigger voltages near 200 V or less with external source triggering under 220 V dc main circuit conditions indicated that by fast switching the trigger discharge by the same source, the plasma discharge from trigger gap can be transformed to the arc in the main gap. This mode of operation is described as the self-triggered TVG.

The trigger-gap resistance is measured externally for a fresh insulator specimen and after each triggering when the trigger discharge initiates the main arc at 220 V, 30 A for the main spacing of 1 mm. Figure 6 shows the variation of the trigger

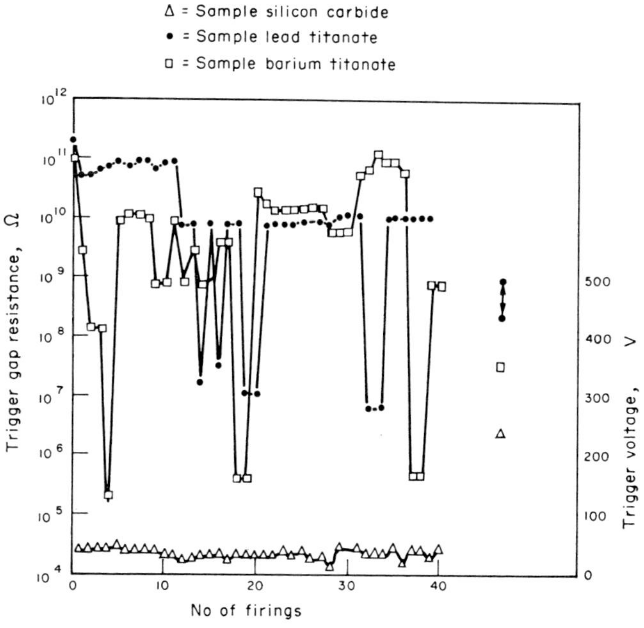

Figure 6. Variation of trigger gap resistance trigger gap 0.2 mm $p = 1 \times 10^{-6}$ torr external triggering, right end shown average probable trigger voltages.

Table 1. Performance of TVG for low voltage switching. Main circuit voltage 220 V dc, trigger gap spacing 0.2 mm, pressure 1–5 × 10^{-6} torr, external triggering.

S No	Main gap spacing mm	Insulator specimen	Triggering voltage range (V) 30 A	50 A	70 A	95 A
	1 mm	Fused quartz	2000	1600	1000	1000
		Sillimanite	200–236	146–174	146	146
		Alumina	612–780	780	612–668	612
		Supramica	380–780	1000	1050	1000
		Rutile	260	200	205	205
		Lead titanate	440–500	358	500	260–320
		Silicon carbide	205	174–205	174–205	174–205
		Barium titanate (LCC HTD)	358	236–358	358	358
		Barium titanate	380–500	397–440	397–440	397
	2 mm	Rutile	612	440–500	500	500
		Silicon carbide	200–268	200–236	236	200
		Sillimanite	268	236	236	236
		Barium titanate	500–780	612–780	612	668
	3 mm	Rutile	1050	780–888	724	724–884
		Silicon carbide	297	260	260	236–260
		Sillimanite	260	260	236	260
		Barium titanate	612–780	612–780	668	668
	4 mm	Silicon carbide	325–350	350	350	325
		Sillimanite	260	236–297	236–260	260–297
		Barium titanate	724–942	858–942	780–942	724–888
	5 mm	Silicon carbide	400	400–440	400	400–440
		Sillimanite	320	320–358	358	320–400

gap resistance for insulator specimens like barium titanate (LCC HTD), lead titanate and silicon carbide. Initial resistance of the gap is near 10^{11} ohm for the titanate samples which showed considerable change between triggerings. The resistance however remained higher than 10^{8} ohm except during 6 or 7 trials during which it is near 10^{5}–10^{6} ohm for both the titanate samples. Silicon carbide consistently showed 10^{4} ohm with trigger voltages near 200 V. In comparison the rutile specimen showed 10^{4}–10^{6} ohm with trigger voltages between 200 and 260 V. Sillimanite further confirmed the influence of trigger gap resistance of the order of 10^{4}–10^{7} on the low trigger voltages, 200–236 V for switching 30 A main arcs.

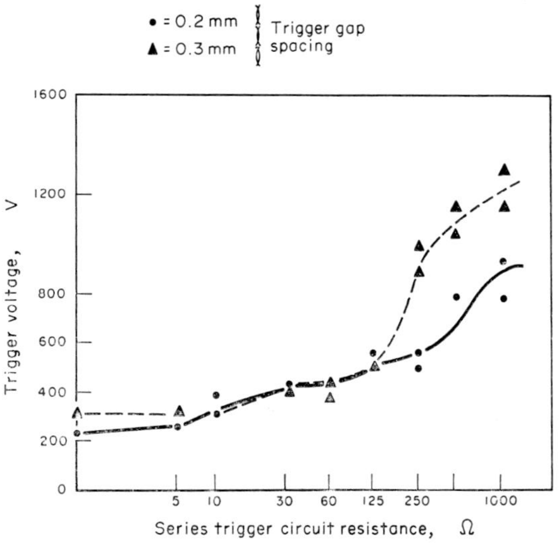

Figure 7. Influence of trigger circuit series resistance on trigger voltage by external triggering sample: silicon carbide main gap = 1 mm, p = 1–5 × 10^{-6} torr main circuit 220 V, 30 A.

On the influence of the addition of series resistance in the trigger circuit, Figure 7 shows increase in trigger voltages with higher series resistance for trigger gap spacings of 0.2–0.3 mm. These observations were checked both with increasing and decreasing values of series resistance.

6. Self trigger mode of TVG

The circuit is described elsewhere.[9] Figure 2 (a and b) shows the rise of the front edge of the voltage pulse and the fast breakdown of the trigger gap on the rising front for silicon carbide at 0.2 mm spacing. The increasing magnitude of the current immediately after breakdown helps in metal vaporization in the trigger gap which finds lower impedance path in the main circuit resulting in the establishment of an arc. In this process, in order to protect the life of the switch from damage of trigger-gap due to severe melting, vaporization and subsequent heating, a minimum 5 ohm series resistance is always included in the trigger circuit.

The self-trigger mode of operation of TVG was tried for samples like barium titanate, rutile, sillimanite and silicon carbide. Of these, barium titanate showed very limited switching at 220–240 V, 70–100 A, whereas rutile showed self-triggering for only 50–100 operations at 200–230 V dc. Such trials are not reported here.

Samples like silicon carbide and sillimanite showed best results, always triggering at voltages even between 85 and 220 V. Figure 8(a and b) show the self-trigger characteristic. In case of silicon carbide, the plasma discharge from the trigger gap is observed even at 46 V which however is insufficient to initiate main arcs consistently. It is evident that the probability, taken as 50% successful triggering from 10 trials, of initiating arcs by self-trigger-mode is higher at low circuit voltages under 220 V dc conditions compared to external triggering.

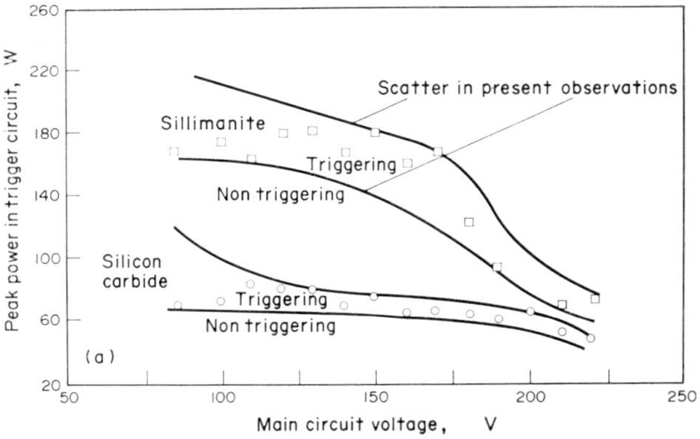

Figure 8. (a) Switching characteristics of self-excited TVG.

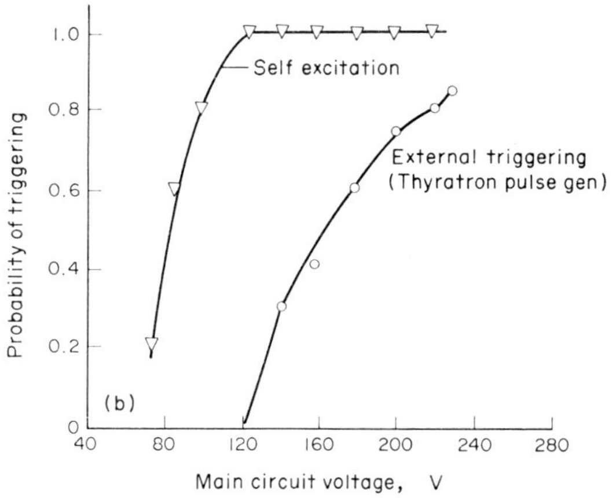

Figure 8. (b) Probability of self-sustained main arc as a function of main circuit voltage.

7. Disscussion

A. Trigger discharge initiation. When the voltage pulse is given to the trigger-gap, the field in the small gap builds up at a faster rate releasing electrons and negative ions from the negative junction. These particles impact on the anode and insulator surface releasing possibly X-rays and occluded gases. A cumulative process increases the conduction phenomena resulting in the breakdown across the insulator gap. After breakdown, the trigger current of more than 0.1 A helps in considerable evaporation of the electrode material (heating and sputtering) with the result that a metallic vapour cloud in the form of ionized plasma from the trigger-gap spreads outwards to the main gap. Figure 9 (a and b) show the trigger-gap assembly (top view) after 100 tests and the plasma discharge expansion as a bluish discharge in the TVG. A metal vapour shield may also be seen in the photograph. The trigger voltages reported in this paper are those sufficient to create such discharge conditions as in the above.

B. Physical properties of ceramic specimens. The physical properties of the insulator specimens that may likely influence the breakdown phenomena over solid insulators in vacuum are outlined by Gleichauf.[10] These are (1) density (2) resistivity, (3) vapour pressure, (4) dielectric constant and (5) thermal conductivity. Many other unsuspected factors in addition to the above may influence the initiation of trigger discharge; one among these is the vapour pressure of the electrodes in the trigger insulator gap which suffer considerable vaporization by heating and sputtering. These physical properties are shown in Table 2.

The breakdown process may be accelerated by the release of gases from the insulator surface by electron bombardment. In case of insulator specimens like PTFE and epoxy-resin samples tested earlier in the laboratory, the release of sorbed gases from the surface and vaporization of the insulator surface or decomposition are possible as these samples are subjected to softening at the surface due to heating. As a result, the TVG has limited life within 50 triggerings and inconsistency in triggering. Hence selective absorption of metallic vapour immediately after cessation of trigger discharge is likely to bridge the insulator gap reducing the trigger gap resistance almost to short circuit. The vapour pressure of these materials also influences the operating life.

Metallographic and electron microscope observations[12] on glass surfaces subjected to 100 arc discharges revealed metal

Table 2. Physical properties of insulator materials.

Material	Relative permittivity	Heat conductivity Btu/hft²/oF/ft	Vol resistivity ohm-cm	Sp Gr g/cm³	Withstand temperature °C
Mica	6.7	0.3–0.35	10^{15}–10^{16}	2.7–3.2	500–600
PTFE	2–2.2	0.14–0.15	10^{15}–10^{17}	2.1–2.3	300
Epoxy (resin)	3.3–5.1		10^{14}–10^{16}	1.1–1.3	120
Fused quartz	3.78	0.52	10^{9}	2.2	1000*
Sillimanite	6.2	1	10^{13}–10^{15}		850†
Alumina	8.7	1.6–12	10^{14}–10^{16}	3.1–3.9	1250*
Supramica (500)	32*				850*
Rutile	80†	2.9–4	10^{14}	4.1	850‡
Lead titanate	192†	1.5–3	10^{15}–10^{18}	3.5–5.5	1250*
Silicon carbide	285§	24		3.17	1250‡
Barium titanate	1850†	1.5–3	10^{15}–10^{18}	3.5–5.5	1250*

General data from Clark.[13]
* Tested in the laboratory.
† Measured at 1 kHz.
‡ Specimen prepared in the laboratory by firing for 3 h.
§ Measured from step-response.
Silicon carbide-varistor $V = KI^{b}$, $K = 190$, $b = 0.574$.

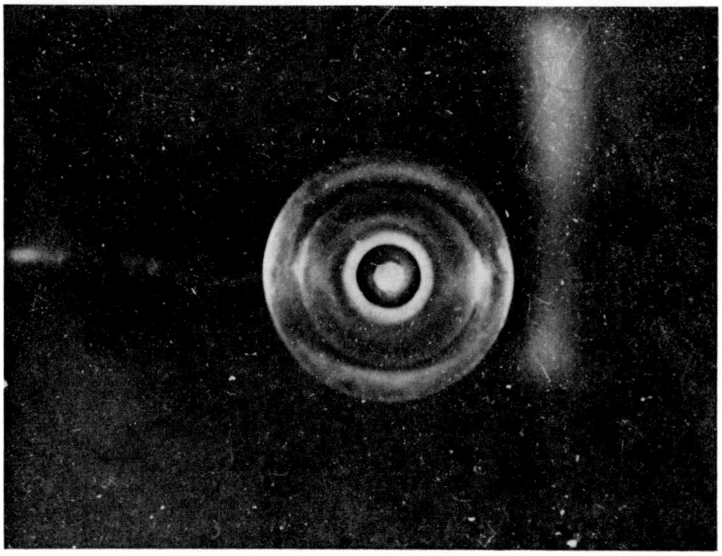

Figure 9. (a) Trigger cathode–insulator–trigger anode assembly—top view after 100 tests. Vaporization from trigger cathode and vapour condensation on trigger anode (central electrode) clearly seen in the photograph.

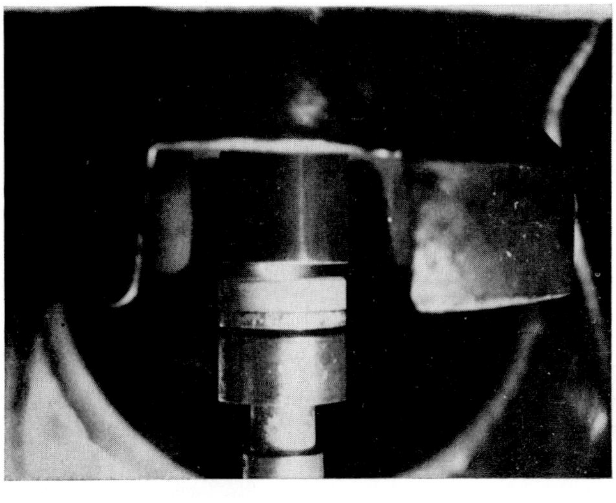

Figure 9. (b) Plasma discharge from the trigger gap expanding like a puff to the main gap. Top electrode–main anode, bottom electrode main cathode, trigger insulator gap is concentrically placed in recess to the main cathode. Metal vapour shield and the trigger electrode connection at the right end can be seen in the test chamber. Main circuit voltage 220 V, main circuit current less than 3 A, trigger voltage 500 V, pressure 1–5 $\times$ 10^{-6} torr.

deposits (copper crystals) along the surface of the insulator. Present investigations also gave the first strong clue that the metallic vapour may be adsorbed selectively, in specified regions of the insulator surface after cessation of the trigger discharge. Such selective adsorption result in the variation of trigger gap resistance. The value of the resistance is dependant upon trigger discharge currents and the equilibrium conditions between the subsequent deposition and vaporization of metallic vapour near trigger electrode–insulator surface.

The thermal conductivity of insulator materials like PTFE, epoxy-resin, mica and fused quartz is considerably lower from the other specimens listed in the Table 2. The rise in temperature from heat flux is dependent on factors like thermal conductivity, density and specific heat of the material. If the difference in the thermal conductivities of the metal and insulator specimens are considered (assuming copper electrodes) that on the basis of uniform heat flux to all surfaces, the insulator surface in the trigger gap would remain at a higher temperature than the electrodes. Thus a major portion of the metallic vapour would be redeposited or condensed on the relatively cooler electrode surfaces immediately after cessation of the trigger discharge. The ceramic insulator specimens can withstand high temperatures and thus no vaporization from the insulator surface is feasible. Hence no bridging of the auxiliary gap is observed.

In the present work, the insulator specimens were selected to cover a wide range of relative permittivities up to 1850. Some of these specimens indicated lower trigger voltage with increase in relative permittivities. However specimens like barium titanate, rutile and sillimanite showed an exception to the above conclusion.

In the case of self excitation mode, the total time delay from the application of trigger pulse to the main circuit current build up of 30 A under 220 V dc is near 100 μs. There is a tendency to lower delay time with increase in trigger current between 0.2 and 10 A. In comparison, it has been noted that the delay time is higher than 3 ms in case of external triggering. This may be due to the longer pulse duration of the thyratron pulse generator.

8. Conclusions

(i) Thermally stable ceramic insulators like barium titanate (LCC HTD), rutile, sillimanite and silicon carbide in the trigger gap of TVG showed consistent switching characteristics with less than 500 V trigger voltages because of higher thermal conductivity.

(ii) The influence of trigger-gap resistance on trigger voltage is considerable. Lower resistance showed low voltage triggering.

(iii) The trigger circuit current increases the heating and sputtering of the electrodes contributing the necessary ionized metal vapour cloud conditions for establishing the metal vapour arc in the main gap.

(iv) The probability of switching TVG is more by self-trigger mode than by external triggering due to low impedance of the source and also due to the fast rate of rise of the pulse.

(v) Switching function of TVG can be established at a preselected voltage and trigger current at any instant by the suitable choice of the insulator material in the trigger-gap with trigger voltages from 100 V to 3 kV.

The TVG with low voltage trigger characteristics finds varied applications from low voltages to high voltage switching and interrupting arc currents. The basic switching function however is achieved in a relatively small metal–insulator–metal trigger gap spacing of the order of 0.1 mm with reliability.

Acknowledgements

I thank the authorities of Indian Institute of Science for providing fellowship and the facilities for the experimental work. Thanks are also due to the authorities of M/s BHEL, Corporate R & D for encouragement.

References

[1] N Vidyardhi, PhD Thesis, Indian Institute of Science, Bangalore, India (1973).
[2] T H Lee, *Proc 4th Int Symp, Discharges and Electrical Insulation in Vacuum*, p 195, Waterloo (1970).
[3] G A Farrall, *Proc IEEE*, **61**, 1973, 1113.
[4] M J Kofoid, *Trans Am Inst Elect Engrs*, III, **79**, 1960, 999.
[5] J M Lafferty, *Proc IEEE*, **54**, 1966, 23.
[6] G A Farrall, *Proc IEEE Trans Electron Devices*, **13**, 1966, 432.
[7] A A Brish *et al*, *Intr Exp Tech*, **5**, 1958, 644.
[8] N Vidyardhi and R S N Rau, *J Phys E Sci Instrum*, **6**, 1973, 33.
[9] N Vidyardhi and R S N Rau, *Ibid*, **6**, 1973, 695.
[10] P Gleichauf, *J Appl Phys*, **22**, 1951, 766.
[11] Fryszman, *Bull Acad. Pol Sci*, **8**, 1960, 379.
[12] J W Mather and A H Williams, *Rev Sci Instrum*, **31**, 1960, 297.
[13] F M Clark, *Insulating Materials for Design and Engineering Practice*, Wiley, New York (1962).

2 Electrical properties of thin films—compounds

2.14 Polycrystalline semiconductor thin films

J C Anderson, *Materials Section, Department of Electrical Engineering, Imperial College, London, SW7 2BT, England*

1. Introduction

Because of the demands of the electronic device industry, the major effort in semiconductor thin films has been in the direction of epitaxial growth of good, single-crystal films. More recently, interest in the theory of amorphous materials has led to the preparation and study of amorphous thin films of semiconductors. The polycrystalline film has generally been considered only when its polycrystalline structure was unavoidable or where the electronic properties are unimportant.

There are now various devices whose electrical properties depend upon polycrystalline structure, for example the thin-film transistor (TFT) and photoconductive layers, both of which exploit the electronic properties of the grain boundary, as also does the silicon gate transistor in which the high resistance of a polycrystalline layer is used to support the gate field.

It is considered that a better understanding of the electronic properties and of how to control the physical properties of polycrystalline semiconductor films will lead to their more widespread use in devices.

In this paper the preparation and characterization of polycrystalline semiconductor thin films is reviewed and their electrical properties are discussed.

2. Nucleation

Any of the standard methods of thin-film deposition can lead to the formation of polycrystalline layers. In general the objective will be to control crystallite size whilst maintaining the stoichiometry and minimizing contamination. It is therefore useful to consider the factors that will influence nucleation and subsequent growth of the crystallites.

On a 'smooth' substrate, such as 7059 glass, the density of stable nucleii, n_s, after a time t, will be given by an equation of the form

$$n_s = At^x \exp\left(\frac{E}{kT}\right) \tag{1}$$

where x varies from $x = 1$ at high temperatures to $x = (i^* + 2)^{-1}$ at lower temperatures. The critical nucleus size i^* is the minimum size of nucleus that is able to grow and is a complex function of substrate temperature, the surface adsorption energy and the self-binding energy for the deposited species. For rule-of-thumb estimations in practical situations it is usually adequate to take $i^* = 1$. The energy parameter in equation (1), for the case of $i^* = 1$, depends on the difference between the adsorption energy, E_a, of the adatom on the surface and the activation energy for surface diffusion, E_d, of the adatom across the surface. These will depend on the nature both of the substrate and of the deposited material and are fixed in a given experiment. The constant A contains the rate of deposition, R, raised to the power $(i^* + 1)$.

We can therefore draw the general conclusions that, since for large crystallites it is desirable to restrict the number of nucleii, we require low rate of deposition and high substrate temperature (n_s small). It is worth noting that point defects in the substrate are preferred absorption sites for adatoms and may dominate the nucleation process. In such a case decrease in R and increase of substrate temperature will make little difference to the nucleation density. There is some evidence that this is the case in sputter deposition and ion-plating where the high energy of arrival of the deposited species creates defects in the substrate surface.

3. Coalescence

When the nucleii have grown to the point at which they touch, coalescence begins to occur.

The liquid-like rapid coalescence of nucleii, as observed in the electron microscope, is familiar to most workers in the thin-film field. For most semiconductors, with practical substrate temperatures, the nucleii will be solid and coalescence takes place by a process of fast surface diffusion. The driving energy for this process is the surface energy released by the reduction in total surface area brought about by the coalescence.

Stowell[1] has analysed the case of crystallographically-shaped islands which are in contact with an angle 2α between two straight sides as shown in Figure 1. The neck formed between the islands has a radius ρ and a length x where $x = \rho(\text{cosec}\,\alpha - 1)$. From surface energy considerations he shows that, for surface diffusion, the rate at which the neck is filled in is proportional to x^{-3} so that, for small x, the process is very fast. He finds diffusion coefficients of the order of 10^{-8} cm^2 s^{-1} for

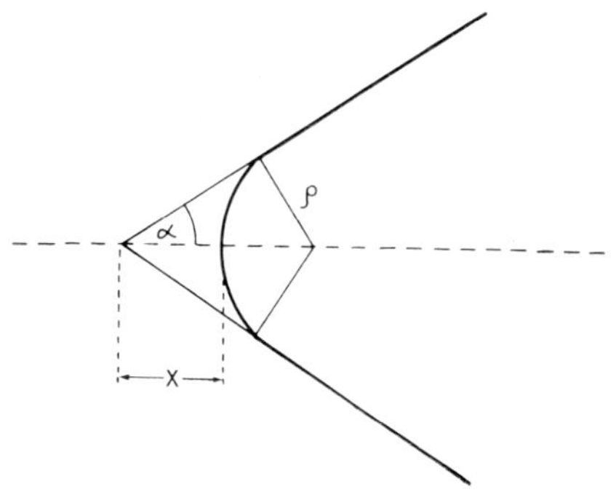

Figure 1. Schematic diagram of two islands touching at an angle α. (Courtesy M J Stowell.)

the surface diffusion of gold which is several orders of magnitude larger than typical bulk diffusion coefficients.

The shape adopted by islands which have grown from nucleii depends upon the balance between the adsorption energy of adatoms on the substrate and their self-binding energy. When E_a is very small, as is the case, for example, on a glass substrate which has not been properly degreased, the nuclei will behave as if they were condensed in free space and their behaviour, on coalescence, will be determined by minimization of surface free energy as compared with volume free energy and the islands will tend to ball up. One particular shape is that of a pentagonal bipyramid which is composed of five similar, twin-related units.[2] Allpress and Sanders[3] have shown that multiply twinned particles of icosahedral shape have greater stability than FCC clusters up to sizes of at least 2000 atoms.

When E_a is large, interfacial energy will tend to dominate the coalescence and growth behaviour. Initial nucleii become very flat and expose the surface parallel to the substrate irrespective of differing surface energies of the different crystallographic faces. On coalescence the reduction of contact area with the substrate will cost too much energy and two islands which touch will retain their shape practically unchanged. In the limit of very large E_a, such as for homoepitaxy of silicon on silicon, there is no nucleation barrier and growth proceeds by monolayer coverage.

In Figure 2(a) is shown the initial nucleation stage of CdSe evaporated on to glass at room temperature showing approximately round nucleii. At a later stage (Figure 2(b)) the islands are still separate, but show crystallographic shapes which is typical of adsorption energies in the range 0.5–1 eV with the crystallites growing as lamellae. In Figure 2(c) the coalescence stage has occurred and the film exhibits channels and holes remaining to be filled, whilst in Figure 2(d) a few narrow channels remain.

In Figures 3 and 4 is shown a sequence of CdSe films deposited on a glass substrate at 300°C by rf sputtering. Compared with the evaporated films of Figure 2 there are many more small nucleii suggesting a defect dominated nucleation process.

4. The role of misfit

When nucleii are sufficiently strongly bound to the substrate and grow as two-dimensional lamellae, each atom is positioned by the low-energy adsorption site it occupies on the substrate. If there is a misfit between the substrates and island normal lattices, a strain energy will be built into the island.[4] When the island reaches a critical size the strain may be wholly or partly relieved by the generation of a misfit dislocation at the island/substrate interface. Dislocation lines can be seen in the CdSe islands in Figure 10.

If adjacent islands are slightly misoriented before coalescence then, when they join up, dislocations will be generated at the junction unless the islands are able to rotate to match their orientations. When the islands are large, however, the energy required for rotation is larger than that needed to generate a dislocation.

Another aspect of misfit occurs on coalescence of islands having the same orientation but which have nucleated at different points on the substrate. This gives rise to displacement misfits which lead to the formation of dislocations.

Once formed, dislocations move through a crystallite in a process of energy minimization and may move out of the crystallite altogether. However, at the network stage where the film is a nearly continuous sheet filled with holes and channels, movement can only cause the dislocations to become trapped at holes as incipient dislocations. For this reason the dislocation density increases rapidly during the filling of channels and holes in the film. Stowell[5] has given the curve shown in Figure 5 for the dislocation density found in gold on molybdenite at 300°C as a function of film thickness. The dotted curve includes incipient dislocations at holes in the film whilst the solid line is the experimentally observed density of actual dislocations. In general the dislocation density in a thin film is always greater than that in a bulk polycrystalline matrix. Figure 10 has the wide range of orientation of crystallites, as shown by the diffraction contrast and exhibits many line defects within the crystallites.

5. Growth

When adjacent nucleii coalesce some substrate surface is 'swept clean' and in this area new nucleation occurs so long as there is a sufficient flux of mobile adatoms on the substrate surface to provide nucleation conditions. As growth proceeds more and more of the substrate is covered with continuous deposit and an increasing proportion of the incident atoms land directly on already-deposited material. As a result, the flux of mobile adatoms on bare substrate is reduced, eventually to a level where new nucleation is no longer possible. Increase of the area of substrate covered then proceeds by migration of adatoms to the edges of the growing islands both from bare substrate and from within a diffusion distance of the edge from the surface of the island itself.

At this stage the film exhibits gaps which are long and narrow (Figure 2(d)) due to the x^{-3} law governing the growth rate of a neck mentioned earlier. These narrow channels ultimately fill by propagation of the neck areas at a speed determined by the adatom flux. Where the coalescing islands are randomly oriented crystallites a filled channel will generally correspond to a grain boundary so that the surface will contain a groove. Stowell and Law[6] have considered the mechanism by which incoherent twin-boundaries are eliminated by thermally activated motion and show that a twin may be locked at a surface groove until a critical film thickness is reached, which depends on the thickness of the twinned region. This is in agreement with the experimental observation that deposits thicker than a few thousand ångströms are largely free of twins.

6. Experimental methods

It is assumed that, in a given experiment, the material to be deposited is fixed and that the experimenter is able to choose the nature of the substrate and the deposition conditions. If the objective is a film with large, reasonably perfect crystallites a single-crystal substrate such as mica, molybdenite or rock-salt is more likely to be successful provided that the appropriate desposition conditions can be reached. In general the adsorption energy, E_a, will be greater on single-crystals than on glass or an amorphous insulator such as fused quartz or alumina. Thus, for a given crystallite size, higher substrate temperatures and lower rates will be required with the single-crystal substrate.

For technological purposes glass substrates are frequently required so that the adsorption and surface diffusion energies on the smooth surface are fixed for the material being deposited.

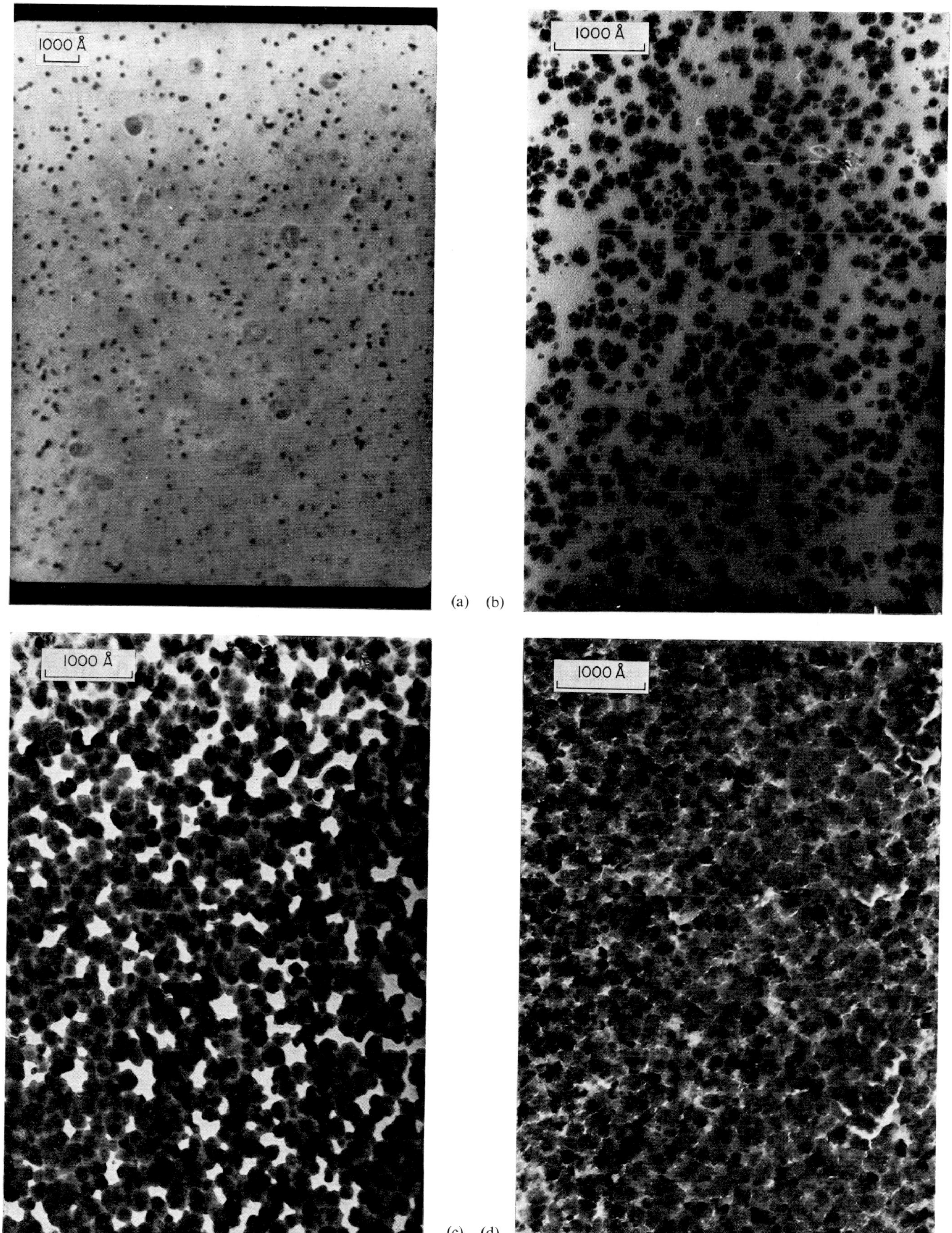

Figure 2. Electron micrographs of evaporated CdSe films; nominal thicknesses (a) 50 Å, (b) 80 Å, (c) 150 Å, (d) 300 Å, substrate temperature 300 K, rate 72 Å min^{-1}.

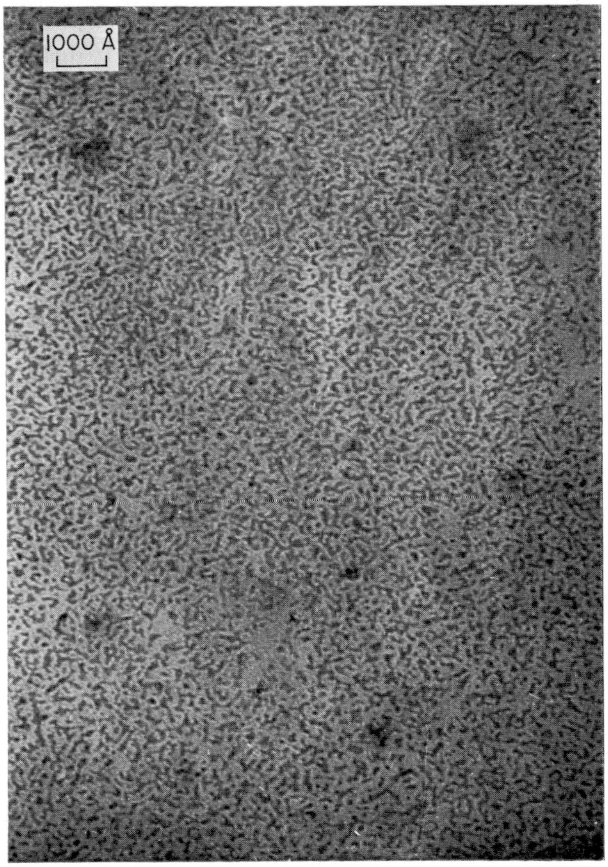

Figure 3. Electron micrograph of sputtered CdSe films; nominal thickness 70 Å, substrate temperature 300°C.

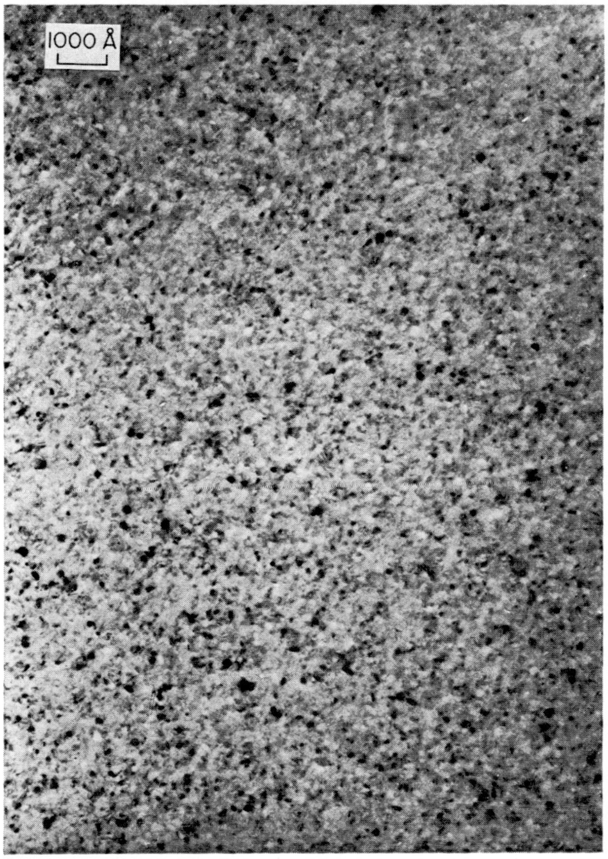

Figure 4. Electron micrograph of sputtered CdSe film; nominal thickness 300 Å, substrate temperature 300°C.

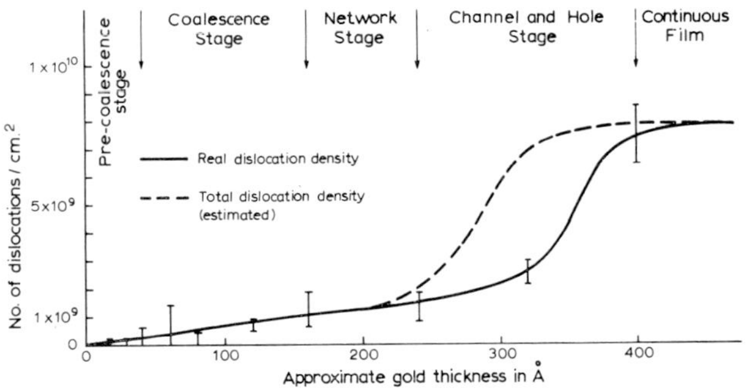

Figure 5. Number of dislocations as a function of thickness for gold films on molybdenite. (Courtesy M J Stowell.)

Control of crystallite size is then only possible through substrate temperature and rate of deposition. As a rough rule-of-thumb the rate and substrate temperature most likely to favour large crystallite size is given by $kT \ln R \simeq 0.7\, H$ where H is the heat of sublimation of the deposit material and R is rate in atoms $cm^{-2}\ s^{-1}$.[7] However the number of stable nucleii may also be reduced by treating the substrate to produce preferred nucleation sites and then depositing with the substrate raised to a temperature high enough to prevent nucleation at other than the preferred sites or else using a deposit material with a high surface mobility. This has been done[8] for Te which has a high surface mobility.[9] They used smooth glass substrates on which gold nucleii had been deposited to a surface coverage in the range 0.8–35%. Gold is electropositive whilst Te is strongly electro-negative so the adsorption energy of Te on Au will be much greater than on glass. It was found that ~5% Au coverage gave a maximum crystallite growth as shown in Figure 6.

Contrary to the theory given above, however, the greatest crystallite sizes were obtained with high rates of deposition and low substrate temperatures as shown in Figure 7. Okyuma and Kumagai[8] explain this by citing the high supersaturation of Te so produced which gives rise to a large amount of thermal energy due to the latent heat of condensation of the Te on the substrate. This provides the driving force for the crystallization.

Filby and Nielson[10] succeeded in producing large area silicon crystallites on amorphous quartz substrates. They evaporated a gold layer of ~3000 Å thickness on the quartz and then deposited Si at a rate of 2 $\mu m\ h^{-1}$ with a substrate temperature of 700–800°C. Grain sizes as large as 300 μm were achieved. Whilst it is tempting to relate this to the fact that E_a is smaller for Si on Au than for Si on quartz, Ishiwatari *et al.*[11] have interpreted the results in terms of the formation of an Au–Si alloy on the substrate. Referring to the Au–Si phase diagram of Figure 8 we take the substrate to be at the temperature T_1. As soon as Si atoms have been deposited a liquid alloy film of composition A is formed. With further deposition of Si the alloy changes to the Si- rich side until all the solid gold is consumed and reaches composition B. Further Si now causes supersaturation and silicon nucleii appear, growing larger as Si continues to be supplied. The largest crystal Ishiwatari *et al*[11] obtained was 850 μm in length and 10 μm in thickness and was grown by depositing Si at 10 Å s^{-1} on a gold film 2.2 μm thick.

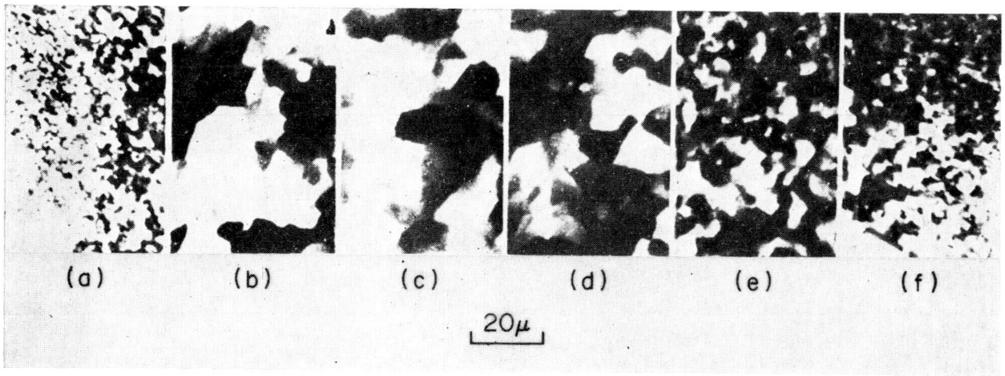

Figure 6. Te Films deposited on to glass pre-covered with nucleii of gold to percentage coverages: (a) 0.8, (b) 2.5, (c) 5, (d) 10, (e) 18 and (f) 35; thickness 1000 Å, rate 4000 Å. (Courtesy Okuyami and Kumagai.)

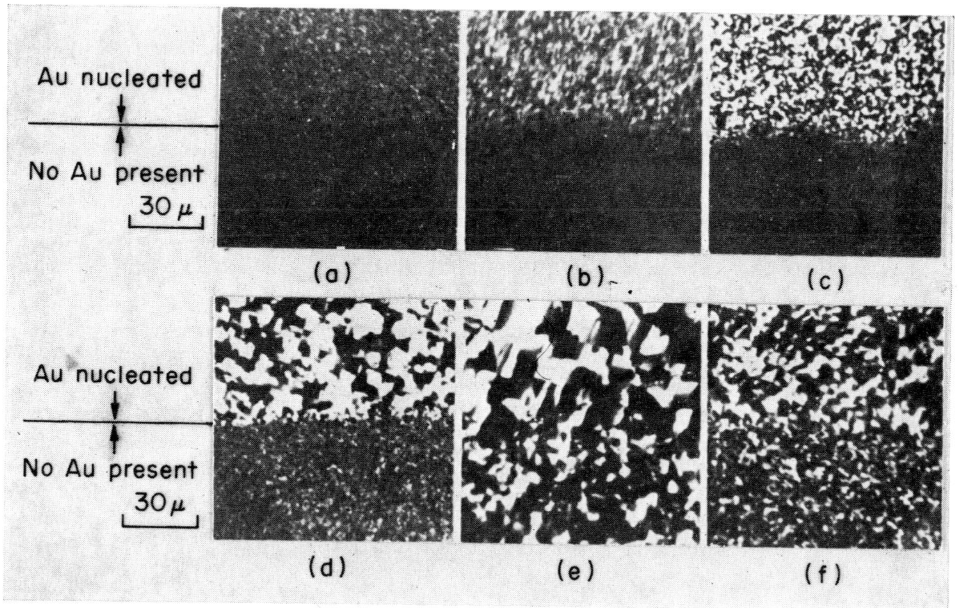

Figure 7. Te films on glass pre-covered with Au deposited at substrate temperatures: (a) 54, (b) 27, (c) −2, (d) −60, (e) −84 and (f) −134°C; thickness 1000 Å, rate 1600 Å min^{-1}. (Courtesy Okuyama and Kumagai.)

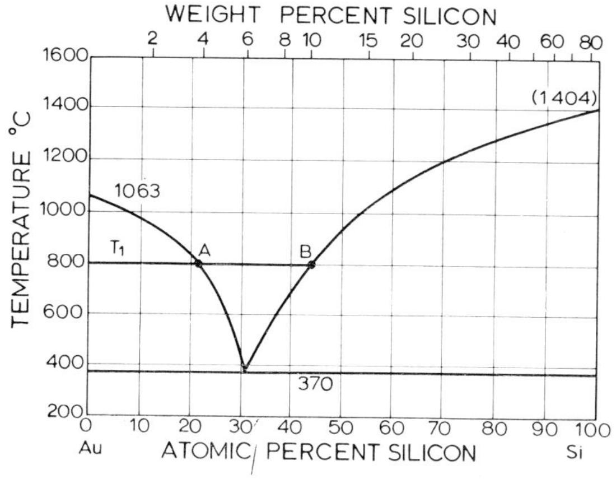

Figure 8. Phase diagram of silicon–gold. (Courtesy Ishiwatari *et al.*)

7. Post-deposition annealing

The most thorough study of crystallite growth in thin films has been carried out by[12] using *in situ* transmission electron microscopy. α-Ge amorphous films were prepared by evaporation onto silicon monoxide film substrates at 25°C at 10^{-9} to 5×10^{-9} torr. The deposition rate was 5 Å s^{-1}. Subsequent to deposition the substrate temperature was raised to 350°C and the appearance of crystallites observed. In α-Ge films of high purity in uhv (10^{-9} torr) nucleation starts on the film surface and results in spherulites of small grain size. (Figure 9). Nucleation occurs randomly and the number of nucleii grows uniformly with time. If the film surface is contaminated, for example by raising the vacuum pressure to 10^{-7} torr during deposition, surface nucleation of crystallites is inhibited and the film crystallizes by nucleation in the bulk. This leads to the formation of large, flat single-crystals as shown in Figure 9.

After surface nucleation subsequent growth of the crystallites, in the very thin films studied, may proceed by surface growth, mainly at the substrate–film interface, leading to coalescence of the spherulites. This process is very fast so that arrays of voids and even channels develop at the boundary of the coalescing crystallites. Volume crystal growth occurs when surface crystallization is hindered by gaseous contamination. This is a much slower process which produces mono-crystallites of some microns in size. Surface crystallization can be avoided by exposing the film to an oxygen atmosphere of 10^{-5} torr at a temperature above 300°C.

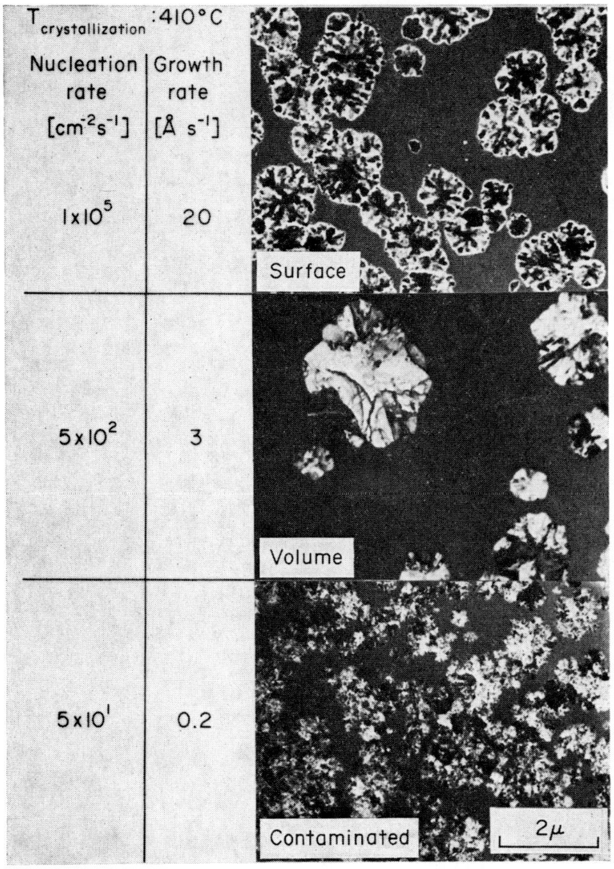

Figure 9. Recrystallization of amorphous Ge film. (Courtesy Barna *et al.*)

Whilst these results have been obtained in respect of recrystallization of amorphous films, the growth processes observed apply to small-grain polycrystalline films which are subjected to post-growth annealing. Thus in most practical cases, where the film is prepared in a vacuum of $\sim 10^{-6}$ torr, bulk recrystallization will be the dominant process. Barna *et al* confirmed that the recrystallization follows an Arrhenius relationship in which the rate of crystallite growth is proportional to $\exp(-E/kT)$ where the activation energy E was of the order of 0.13 eV in Ge films.

There is a general consideration governing crystallite growth in thin films. For small values of the surface adsorption energy E_a, the crystallite will attain a minimum free energy when it has a minimum surface area for its volume, i.e. it would like to be spherical. In terms of a planar, continuous thin film this means that the minimum free energy will be reached when the grains are cubes of side equal to the film thickness. However, for larger values of E_a, the interfacial energy between the crystallite and substrate is much lower than the surface free energy of the crystallite. In such a case the crystallite may then continue to grow so long as the reduction in total energy is greater than the energy required to remove a grain boundary and crystallites will tend to form large lamellae. The practical implication of this is that a semiconductor film on a glass substrate, for which E_a is fairly small, will show very little crystal growth beyond the point where the crystallite length is equal to the film thickness.

In Figure 10 (a–e) are shown micrographs of CdSe films of 400 Å thickness deposited on glass at room temperature. (a) is the unannealed film, (b) was annealed for $1\frac{1}{4}$ h at 350°C in dry nitrogen, (c) for 52 h, (d) for 94 h and (e) shows the diffraction pattern for the (d) film. It will be seen that the main crystallite

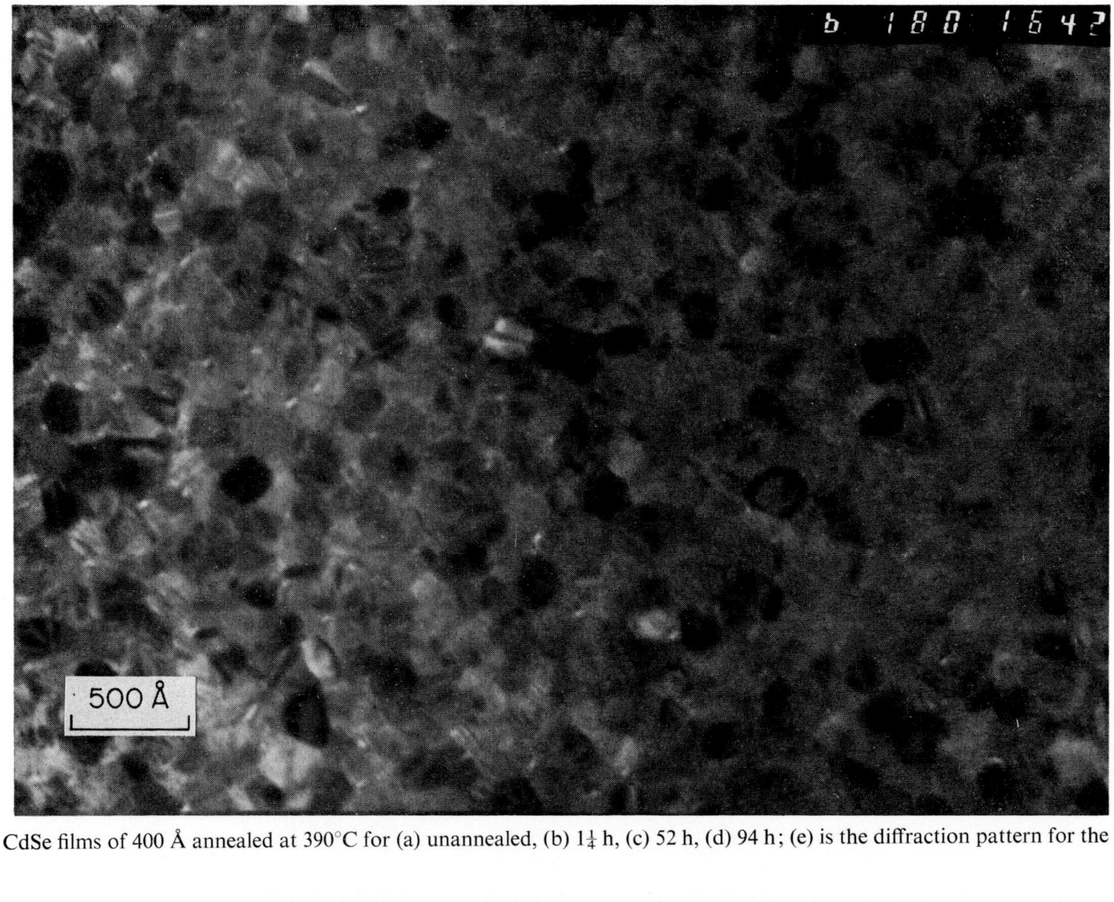

Figure 10. CdSe films of 400 Å annealed at 390°C for (a) unannealed, (b) $1\frac{1}{4}$ h, (c) 52 h, (d) 94 h; (e) is the diffraction pattern for the (d) film.

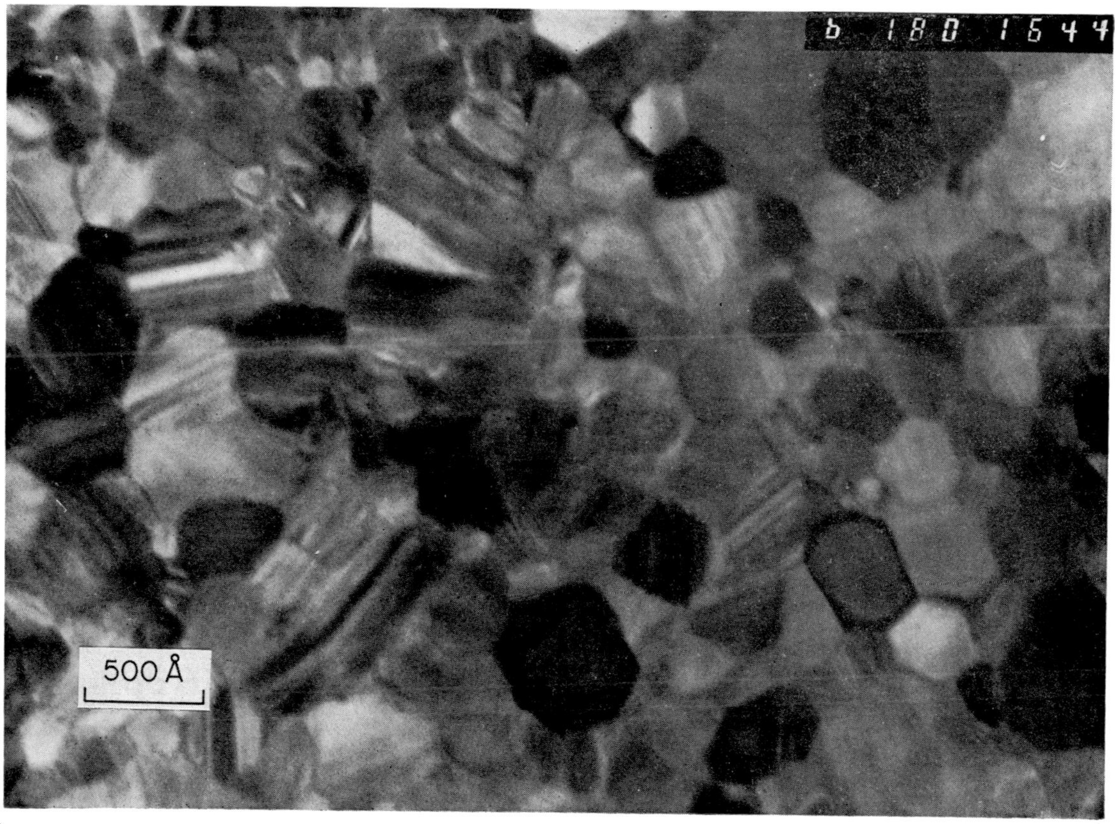

Figure 10(b).

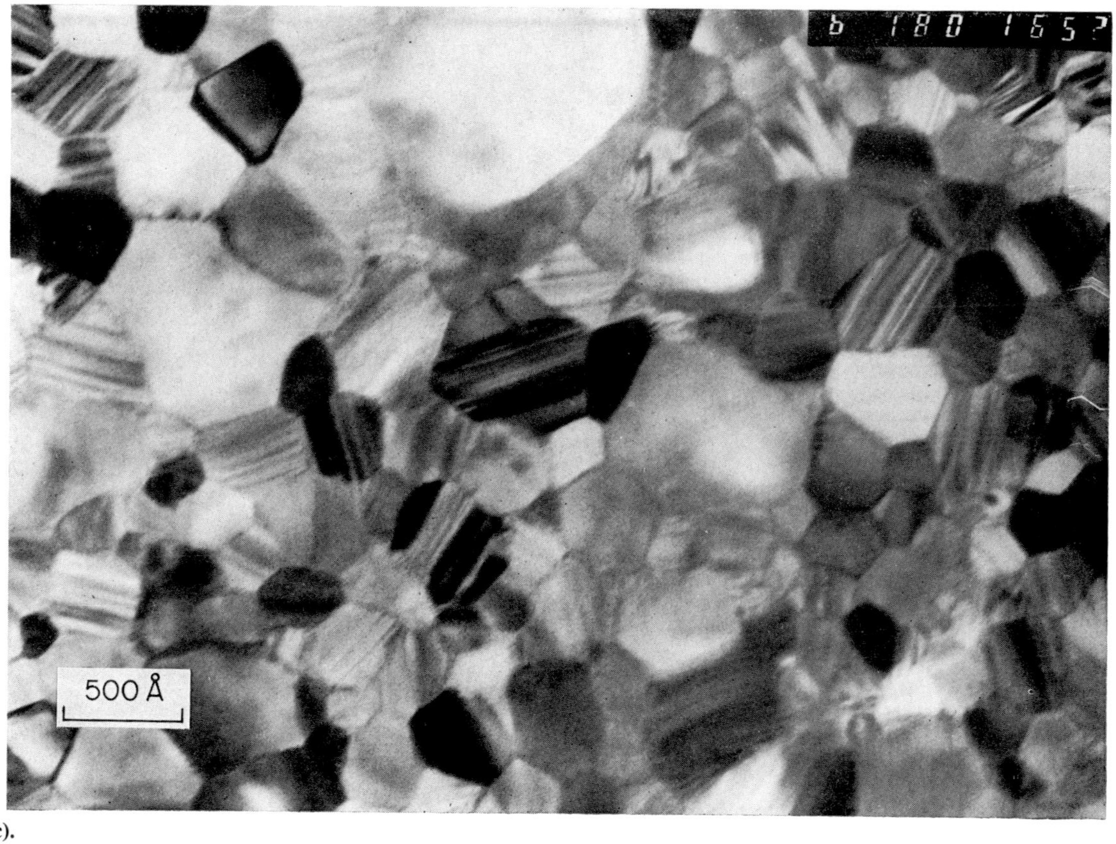

Figure 10(c).

Figure 10(d).

Figure 10(e).

growth occurs in the first $1\frac{1}{4}$ h and that between 54–92 h there is little further growth. A distribution curve of crystallite size is plotted in Figure 11 for the (c) film and shows a mean size of ~300 Å, where the film thickness is 400 ± 50 Å.

Vojdani and Hashemian[13] have suggested the use of the Peltier effect to control recrystallization of a semiconductor film. This makes use of the fact that solid and liquid forms of the same material have different resistivities. Passage of a dc current through a sample containing a molten zone will cause Peltier heating at one liquid–solid interface and cooling at the other, causing the zone to move. Thus the current provides a means of controlling the progress of the molten zone through the film. The molten zone is localized by scratching a groove in the film, heating it to just short of melting temperature and then using Joule heating by the current to melt the thinned region which has a higher resistance than the rest of the film. Whilst no complete

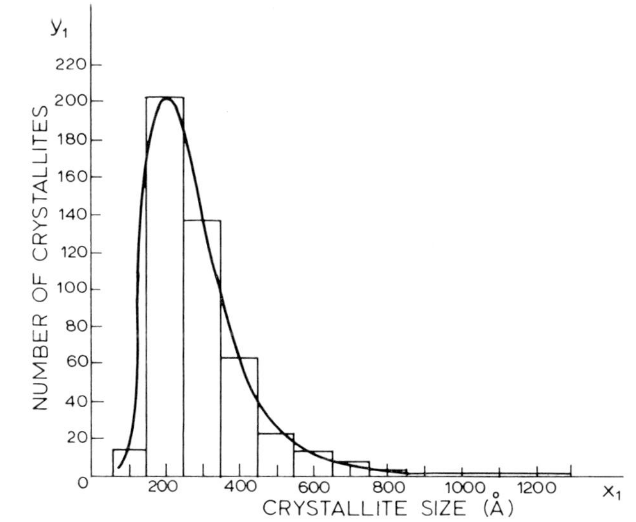

Figure 11. Distribution curve of crystallite size for the film of Figure 10(c).

recrystallization has been achieved, some results have been obtained with germanium films using dc currents of 1–2 A. It is interesting that these results indicate that it is surface recrystallization, of the type observed by Barna *et al*[21] in ultra high vacuum, that is produced.

8. Carrier mobility

The conductivity mobility of carriers in a polycrystalline thin film will be determined by the sum of the effects of the various scattering processes which may be listed as (a) lattice, (b) impurity, (c) dislocation, (d) surface, (e) grain boundary, (f) potential barrier.

These have been discussed.[14, 15] It is sufficient, for the purposes of this paper, to consider the general features and, in particular, the temperature dependences of the different mechanisms.

8.1. Lattice. This arises from scattering of electrons by lattice phonons and gives a mobility $\mu \alpha T^{-2.5}$ for low band-gap and $\mu_L \alpha T^{-1.5}$ for high band-gap semiconductors.

8.2. Impurity. The theory of scattering by ionized particles, due to Conwell and Weisskopf[16] gives $\mu \alpha T^{3/2}$ and μ is inversely proportional to the density of ionized impurities.

8.3. Dislocation. Dexter and Seitz[17] have shown that the mobility due to unionized dislocations, distributed uniformly within a semiconductor, is given by $\mu_D = a_D T^{-1}$ where a_D is proportional to the dislocation density.

8.4. Surface. For surface accumulation layers the mean mobility is given by

$$\bar{u}_a = \frac{\mu_B}{1+\lambda\left(\dfrac{1}{L_c}+\dfrac{1}{d}\right)(1+v_s)^{\frac{1}{2}}} \tag{2}$$

where λ is the mean free path of carriers in the bulk crystallite and $2d$ is the film thickness. L_c is the depth of penetration of surface space–charge into the bulk of the film and is given by

$$L_c = \frac{|v_s|}{F_s} l_D = \frac{|v_s|}{F_s}\left(\frac{\varepsilon kT}{q^2(n_B+p_B)}\right)^{\frac{1}{2}} \tag{3}$$

where v_s is the surface band bending in units of kT, n_B and p_B are bulk carrier densities and F_s is the surface function, given approximately for extrinsic material by

$$F_s \doteq \sqrt{2}\,(\exp|v_s| - |v_s| + 1)^{\frac{1}{2}} \tag{4}$$

For depletion layers

$$\bar{u}_d = \mu_B \frac{\left(1+\dfrac{\lambda}{d}(1-\exp v_s)\right)}{1+\dfrac{\lambda}{d}} \tag{5}$$

where

$$F_s \doteq \sqrt{2}\,(\exp(-|v_s|) + |v_s| - 1)^{\frac{1}{2}} \tag{6}$$

Conventionally v_s is positive for an accumulation layer and negative for a depletion layer.

The temperature dependence is determined by the variations of μ_B, L_c and $v_s = qV_s/kT$ with T. The net effect for both accumulation and depletion is dominated by the temperature dependence of μ_B, the normal bulk mobility.

8.5. Grain boundary. The grain boundaries represent a physical interruption to carrier transport which is characterized by allocating a mean-free path to the carrier which is determined by the distance between boundaries. The resulting mobility is given by

$$\mu_g = \frac{q\bar{c}}{4N_1 kT} \tag{7}$$

where N_1 is the number of grain boundaries per unit length in the direction of current flow and $\bar{c}$ is the thermal velocity of carriers. For Boltzmann statistics

$$\bar{c} = \left(\frac{2kT}{\pi m_i^*}\right)^{\frac{1}{2}} \tag{8}$$

where m_i^* is the inertial effective mass and so $\mu_g \alpha T^{-1/2}$.

8.6. Potential barriers. The usual model for barrier-limited conductivity assumes that there is a sufficient density of trapping states at the intercrystalline boundary effectively to pin the Fermi level. The trapped charge at the boundary gives rise to a potential barrier which is represented by a rise in the conduction band edge above its bulk equilibrium position with respect to the Fermi level. This is illustrated in Figure 12a and in Figure 12b is shown the effect of applying a voltage V across the barrier. Following Miles[18] and Green and Miles[19] it can be shown that the current density J is given by

$$J = J_0 \exp\frac{q\Delta V}{kT}\left(1-\exp\left(-\frac{qV}{kT}\right)\right) \tag{9}$$

where J_0, the equilibrium current, is given by

$$J_0 = \frac{1}{4} n_{0c}\, \bar{c}\, q \exp\left(-\frac{q\phi_0}{kT}\right) \tag{10}$$

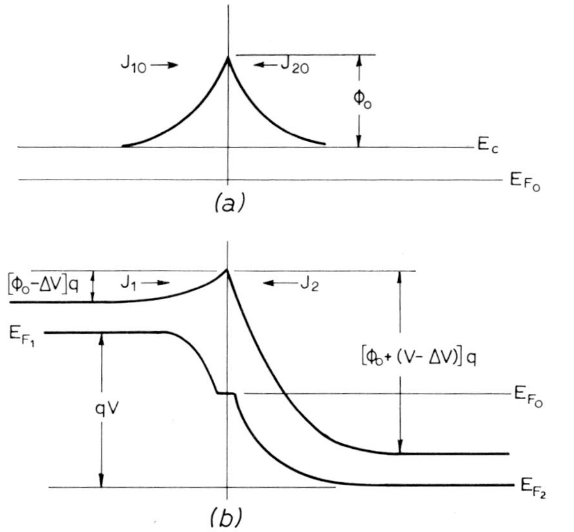

Figure 12. Potential energy diagram for an electron at grain boundary (a) without field, (b) with field.

In this equation n_{Oc} is the bulk crystallite carrier density per unit volume, $\bar{c}$ the thermal velocity of the carriers and q the electron charge; the remaining symbols are shown in Figure 12.

If we assume that $qV \ll kT$, so that $\Delta V \approx 0$, as is normally the case in most polycrystalline devices, we can write

$$J = \frac{1}{4} n_{Oc} \frac{q^2 \bar{c}}{kT} V \exp\left(-\frac{q\phi_0}{kT}\right) \tag{11}$$

where $V = V_d/N_1 L$, V_d being the source-drain voltage, N_1 the number of barriers per unit length and L the length of the film.

The above calculation is based on uniform intercrystalline boundaries stretching from top to bottom of the semiconductor film with their planes perpendicular to the flow of current, as shown schematically in Figure 13, which also shows the symbols used for TFT dimensions and voltages.

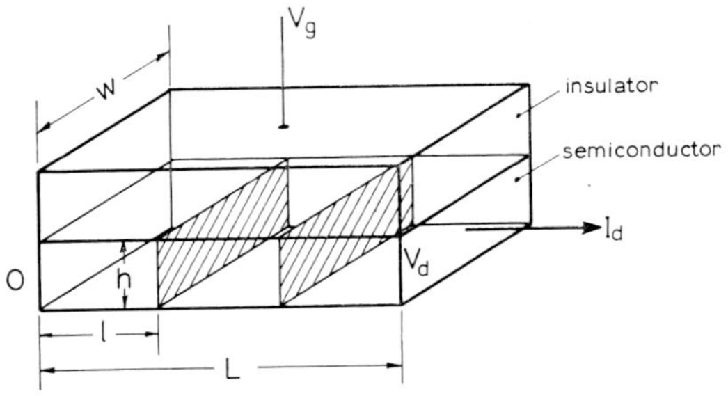

Figure 13. Schematic diagram of grain boundaries in the semiconductor film.

We now take account of the effects of trapping at the semiconductor–insulator interface. This will produce a depletion layer in the semiconductor crystallites which can be regarded as reducing the carrier density at the surface of the crystallite below its bulk value. Now the intercrystalline potential barrier height in the bulk is related to the bulk carrier density by

$$\phi_{ob} = \frac{kT}{q} \ln\left(\frac{n_{Oc}}{n_{Ob}}\right) \tag{12}$$

where n_{Ob} is the carrier density at the top of the barrier. At the surface the bulk carrier density $n_{Os} < n_{Oc}$, and whilst n_{Ob} may also be reduced in the surface region we assume that the ratio changes such that the intercrystalline barrier ϕ_{Os} at the surface is less than ϕ_{Ob}. The two-dimensional potential energy diagram up to the interface is visualized as shown in Figure 14, in which

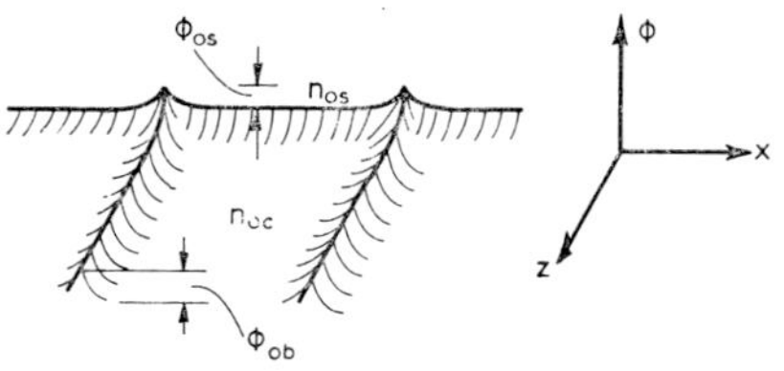

Figure 14. Two-dimensional potential barrier model.

the coordinate z is the distance below the interface and coordinate x is the direction of current flow. The third coordinate is potential energy.

A general quantitative expression has been derived by Anderson[20] from this model with $n_o(z)$ the carrier density and $\phi_o(z)$ the barrier height as linear functions of depth below the surface. With the simplifying assumptions that n_{Ob}, the carrier density at the top of the intercrystalline barrier is the same both in the bulk and at the surface of the film and that the surface depletion layer penetrates to at least half the film thickness, the current I_d for a thin (~1000 Å) film is given by

$$I_D = \frac{wh}{L} V_d \, q \, \mu_g \, n_{Oc} \exp(-\phi_{Ob}) \cdot \frac{2(\cosh \Delta\phi - 1)}{(\Delta\phi)^2} \tag{13}$$

where $\Delta\phi = (\phi_{Ob} - \phi_{Os})$ and μ_g is the grain-boundary-limited mobility given by equation (7).

In a typical CdSe film with a mean crystallite size of 1000 Å, thickness 1300 Å the barrier height is found to be 0.13 eV in the bulk and 0.035 eV at the surface. When the film becomes sufficiently thick the surface band bending is only a small fraction of the total thickness; the effect of the reduction of the intercrystalline barrier in the surface region then becomes small and the simpler expression of equation (11) applies.

Thus, the barrier-limited mobility for a thin film may be written as

$$\mu_b = \mu_g \exp(-\phi_{Ob}) \cdot \frac{2(\cosh \Delta\phi - 1)}{(\Delta\phi)^2} \tag{14}$$

For $\Delta\phi \leq 4$ the $(\cosh \Delta\phi - 1)/(\Delta\phi)^2$ term has a negligible temperature dependence and, since $\mu_g \alpha T^{-1/2}$, the temperature dependence of mobility is dominated by the exponential, i.e. $\ln \mu_b \alpha T^{-1}$.

9. Experimental results

In general the carrier mobility will be given by

$$\frac{1}{\mu} = \frac{1}{\mu_L} + \frac{1}{\mu_I} + \frac{1}{\mu_D} + \frac{1}{\bar{u}} + \frac{1}{\mu_b} \tag{15}$$

and in any given film one or more of the scattering processes may be dominant. The lattice, impurity, dislocation and surface scattering components will be present whether the film is single crystal or polycrystalline and may be conveniently lumped together in a single-crystal mobility μ_c. In a polycrystalline film grain boundaries may well introduce additional charged impurity centres so that μ_I is slightly below the single-crystal value and the effective surface space charge area is increased by addition of the grain boundaries, so that $\bar{\mu}$ is also slightly reduced. Neither of these effects will be important in most films.

Consider a chain of crystallites, each of length l with potential barriers at each grain boundary having a width $2l_D$, where l_D is the Debye length given by equation (3). The bulk crystallite region of length $(l - 2l_D)$ will have a bulk conductivity given by $\sigma_c = n_{Oc}q\mu_c$. The boundary region, of length $2l_D$, will have a conductivity σ_b derived from equation (14) and we may write the total conductivity as

$$\sigma = \frac{\sigma_b\,\sigma_c}{\sigma_b\left(1 - \frac{2l_D}{l}\right) + \sigma_c\frac{2l_D}{l}} \qquad (16)$$

In all but the most heavily doped and dislocated films $\sigma_c > \sigma_b$ and for a small-grained film in which $2l_D \lesssim l$

$$\sigma = \sigma_b \cdot \frac{l}{2l_D} \qquad (17)$$

and is independent of σ_c. The mobility is then simply given by equation (14). For very large crystallites with $2l_D \ll l$, $\sigma \lesssim \sigma_c$ and the effect of the grain boundaries is negligible.

Intermediate cases in which $2l_D/l$ is of the same order as

$$\frac{\sigma_b}{\sigma_c} \text{ gives } \sigma = \frac{\sigma_b}{\frac{\sigma_b}{\sigma_c} + \frac{2l_D}{l}} \qquad (18)$$

9.1. Germanium films. Barna *et al*[21] measured the Hall mobilities and conductivities of the films described in section (7). Arguing that the geometry of the crystallites has a random distribution, they represent μ_b by an expression of the form

$$\mu_b = \frac{\beta'}{T^{-\frac{1}{2}}\exp\left(-\frac{\gamma'}{T}\right)} \qquad (19)$$

where β' includes some sort of average value of $2l_D/l$ for the crystallites. The bulk mobility is taken to be limited by impurity and lattice scattering only so that total mobility is given by an expression of the form

$$\frac{1}{\mu} = \frac{\alpha'}{T^{3/2}} + \frac{\beta'}{T^{-\frac{1}{2}}\exp\left(-\frac{\gamma'}{T}\right)} + \frac{\delta'}{T^{-2{,}33}} \qquad (20)$$

Writing $x = 1000/T$ they resolve their experimental results into three components:

$$\begin{aligned} a &= \alpha\, x^{3/2} \\ b &= \beta\, x^{-\frac{1}{2}}\exp(\gamma x) \\ d &= \delta\, x^{-2.33} \end{aligned} \qquad (21)$$

The results are given in Figure 15(a) for a surface recrystallized film (type A) having a small grain size, in Figure 15(b) for a volume recrystallized film of large grain size (type B) and in Figure 15(c) for a contaminated volume recrystallized film of small grain size, which is believed to contain gas occlusions at grain boundaries (type C). Using the simple barrier model of equation (11) they find type A to have potential barriers of 0.01 eV with a crystallite size 500 Å, type B has barriers of 0.001 eV with crystallite size 5000 Å and type C has barriers of 0.026 eV with crystallite size 500 Å. They conclude that potential barriers at the grain boundaries represent the dominant scattering mechanism. The increase of barrier height due to contamination is of considerable practical importance.

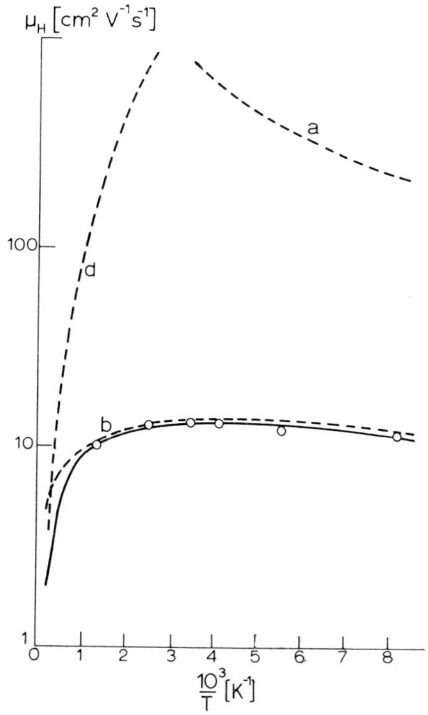

Figure 15. Mobility as a function of temperature for recrystallized Ge films of (a) type A, (b) type B and (c) type C. Dotted lines are theoretical curves (equation 2) for, (a) lattice, (b) impurity and (d) barrier scattering. (Courtesy Barna.)

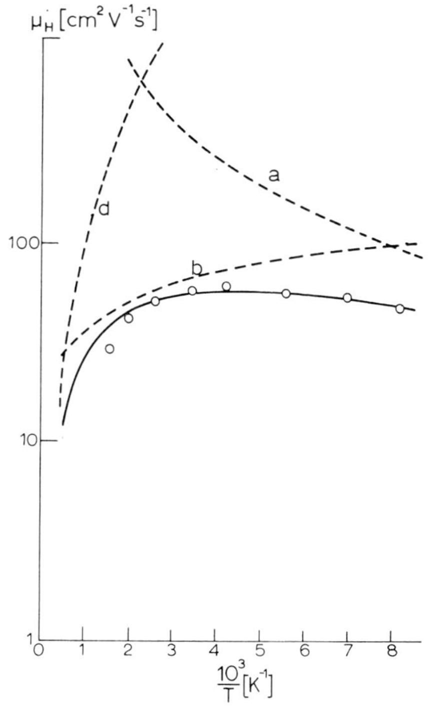

Figure 15b.

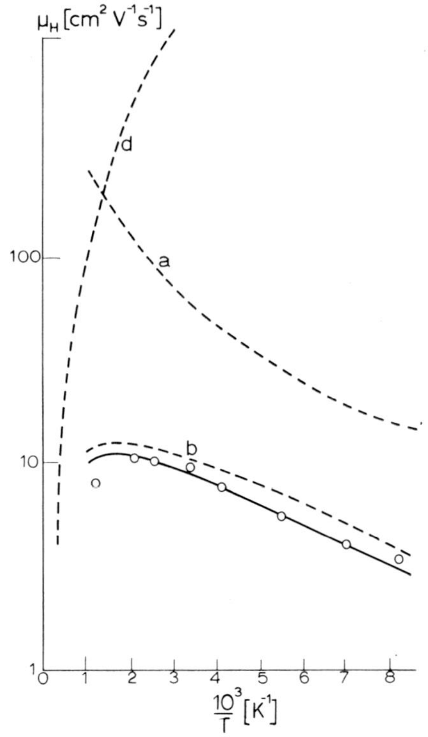

Figure 15c.

9.2. Tellurium films. Okuyama[22] has measured Hall mobilities in films of the type described in paragraph 3 above, in which gold nucleation and substrate temperature were used to control crystallite size. In Figure 16 is shown his result for the variation

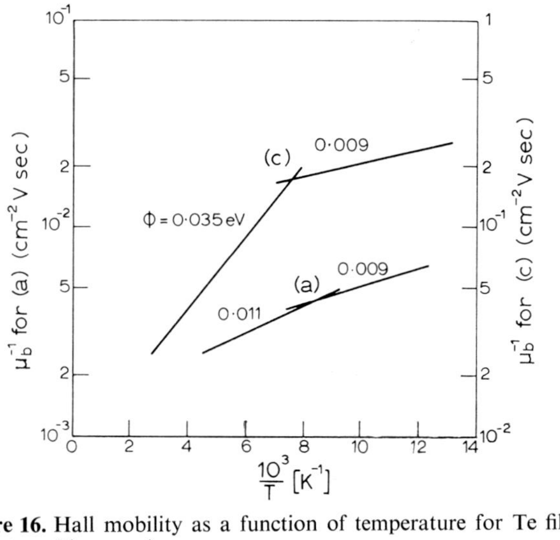

Figure 16. Hall mobility as a function of temperature for Te films. (Courtesy Okuyama.)

of Hall mobility of holes with grain size, indicating that the mobility is grain-boundary limited. Taking the total mobility to be given by

$$\mu^{-1} = \mu_L^{-1} + \mu_I^{-1} + \mu_b^{-1} \tag{22}$$

he separates μ_b^{-1} and plots it as a function of reciprocal temperature to obtain the grain boundary barrier height. For an average crystallite size of 0.1 μm he finds a barrier of 0.035 eV above 125 K and 0.009 eV below 125 K whilst for 1.5 μm average size the barrier height is 0.011 eV above and 0.009 eV below 125 K. The film thicknesses were 2000–4000 Å and the expression

$$\mu_b = \mu_o \exp\left(\frac{-\phi}{kT}\right) \tag{23}$$

is used with μ_o treated as a constant. However, correction for the temperature-dependence of μ_o makes only a small difference to the calculated barrier height. The lower activation energy at low temperature is explained in terms of tunnelling through the potential barriers when the Debye length becomes small at low temperature. The author considers it more likely that it is associated with a bulk trap in the crystallite. As the Fermi level passes through the trapping level it will fill with holes from the valence band. The ratio p_{oc}/p_{ob} is then reduced, where p_{oc} is the hole density in the crystallite and p_{ob} that at the barrier. Since barrier height is proportional to $\ln(p_{oc}/p_{ob})$ it will be reduced.

9.3. CdSe films. Several workers[20,23–25] have reported barrier limited conductivity in CdSe films, which have been widely used in thin-film transistors. Some recent results obtained by the author are given in Figure 17 in which ln *IT* is plotted against

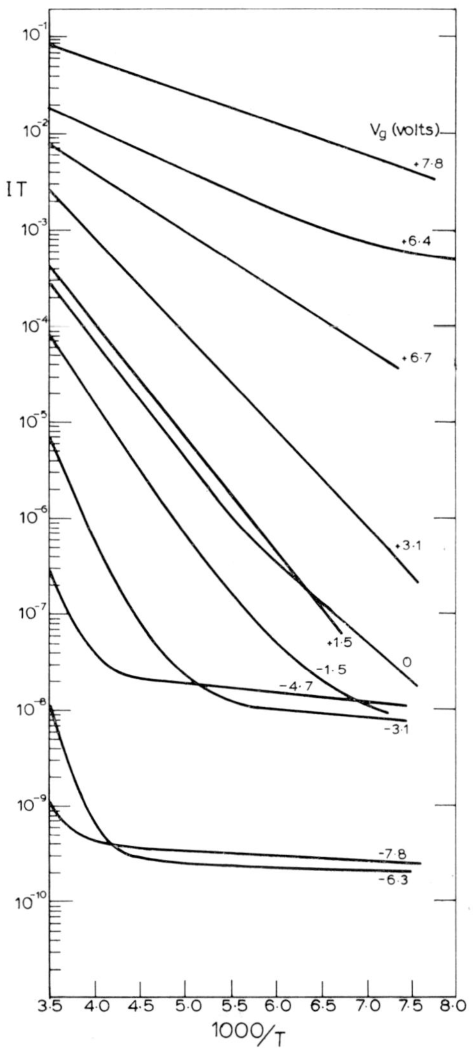

Figure 17. Ln (current and temperature) as a function of reciprocal temperature for CdSe films with gate voltage as a parameter.

$^{1000}/T$ for 1300 Å CdSe films with a mean grain size of ~1000 Å. The films were in a TFT structure and gate voltage is shown as a parameter in the figure.

The Debye length (for $V_g = 0$) at 300 K is ~250 A giving a mean space-charge penetration of ~300 A below the surface. Down to ~150 K the plot is linear and below this temperature deviates from linearity. At 150 K the space-charge penetration depth is ~600 A and therefore equal approximately to half the thickness of the film. The two-dimensional model is then applicable and the curvature of the plot can be fitted to a finite value for $\Delta\phi$. This trend is confirmed by the increasing departure from linearity as the gate voltage becomes more negative and the space-charge penetration depth increases. For positive gate voltage the penetration depth is reduced and the plots are linear over the whole temperature range. These results will be more fully discussed in a later paper.

Conclusions

There is now ample evidence that conductivity in thin, polycrystalline semiconductor films is barrier limited, characterized by a mobility which increases exponentially with temperature. Barrier heights quoted in the literature vary widely but are usually of the order of a few tenths of an electron volt at most. The absolute value of mobility is controlled mainly by the number of barriers and increases with increasing crystallite size. The barrier height may also depend on the vacuum pressure and rate during deposition in that these factors determine the amounts of gas occluded at the grain boundaries. Modulation of the barrier height may be achieved by the field effect, as in TFT structures, or by irradiation by light, as in photoconductive films. Most semiconductor films exhibit extrinsic conductivity with relatively high-carrier densities which remain fairly constant with varying temperature. The use of barrier modulation to control conductivity permits the application of polycrystalline films to a variety of devices.

References

[1] M J Stowell, *Phil Mag*, **22**, 1970, 1.
[2] S Ino, *J Phys Soc Japan*, **21**, 1966, 346.
[3] J G A Allpress and J V Sanders, *Aust J Phys*, **23**, 1970, 23.
[4] J W Matthews, *Physics of Thin Films*, p 137. (Edited by G Haas and R E Thun), Academic Press, New York (1967).
[5] M J Stowell, *Thin Solid Films*, **12**, 1972, 341.
[6] M J Stowell and T J Law, *Phil Mag*, **19**, 1969, 1257.
[7] B Lewis (private communication).
[8] K Okuyama and Y Kumagai, *J Appl Phys*, **46**, 1975, 1473.
[9] E J Weidmann and J C Anderson, *Thin Solid Films*, **7**, 1971, 265.
[10] J D Filby and S Nielson, *J Electrochem Soc*, **112**, 1965, 957.
[11] K Ishiwatari, T Oka and K Akiyama, *Japan J Appl Phys*, **6**, 1967, 1170.
[12] J F Pocza, A Barna, P B Barna, I Pozsgai and G Radnoozi, *Japan J Appl Phys*, Suppl 2, pt I, 1974, 525.
[13] S Vojdani and R Hasheman, Internal Report, Arya Mehr University, Tehran (1976).
[14] J C Anderson, *Adv Phys*, **19**, 1970, 79.
[15] K Okuyama and Y Kumagai, *Japan J Appl Phys*, **12**, 1973, 1884.
[16] E Conwell and V Weisskopf, *Phys Rev*, **77**, 1950, 388.
[17] D L Dexter and F Seitz, *Phys Rev*, **86**, 1952, 964.
[18] R E Miles, PhD Thesis, London University (1973).
[19] M Green and R E Miles, *J Phys D*, **6**, 1973, 145.
[20] J C Anderson, *Thin Solid Films*, **16**, in press.
[21] A Barna, P B Barna, Z Bodo, J F Pocza, I Pozsgai and G Radnoczi, *Thin Solid Films*, **23**, 1974, 49.
[22] K Okuyama, *Thin Solid Films*, **33**, 1976, 165.
[23] R Graeffe, PhD Thesis, Helsinki Technical University (1969).
[24] H F Van Heek, *Solid State Electron*, **11**, 1968, 456.
[25] Y Kumagai and K Okuyama, *Japan J Appl Phys*, Suppl 2, pt I, 1974, 793.

2.15 The stoichiometry of surfaces in cadmium sulphide thin films

R Hill, *Department of Physics and Physical Electronics, Newcastle upon Tyne Polytechnic, Newcastle upon Tyne, NE1 8ST, England*
I A S Edwards, *School of Chemistry, University of Newcastle upon Tyne, Newcastle upon Tyne, NE1 7RU, England*

The sulphur to cadmium ratio in cadmium sulphide thin films has been compared to that of a clean single crystal using Auger spectroscopy. Thermally evaporated CdS films were found to be deficient in sulphur in their outer layers, but were stoichiometric after etching. Electron beam evaporated CdS films were found to be stoichiometric.
It was found that baking these films at 190°C in argon or nitrogen at atmospheric pressure increased the sulphur to cadmium ratio, probably due to sulphur diffusing to the surface and being physisorbed there.
From measurements of the temperature dependence of the copper sulphide layer in CdS–Cu_2S cells it was found that chalcocite, djurleite and diginite copper sulphide phases are present even in efficient cells. A model is proposed for the structure of the copper sulphide layer in which the copper deficient phases are the result of excess sulphur on the CdS surfaces and vary in thickness according to the degree of non-stoichiometry of the CdS surfaces.

Introduction

Photovoltaic devices based on CdS–Cu_2S heterojunctions have been actively investigated for more than 20 years, and progress up to last year has been discussed at the CdS Workshop.[1] It is still found that power output efficiencies for thin film CdS–Cu_2S cells are below 8%, whilst theoretical models containing quite reasonable assumptions about the heterojunction and the carrier transport parameters show that these cells could have efficiencies of up to 15%.

CdS–Cu_2S junctions are usually formed by a chemical replacement of cadmium near the surface by copper atoms. This can be done either by a 'plating' process where the CdS film is dipped into a solution containing a copper salt or by a 'dry barrier' process where a layer of CuCl is deposited onto the surface of the CdS film by vacuum evaporation followed by an interdiffusion process at about 190°C. In either case the replacement reaction $CdS + 2CuCl = Cu_2S + CdCl_2$ takes place. It seems to be generally assumed[1] that the sulphur lattice remains undisturbed during the replacement reaction and only cadmium and copper diffuse through the anion lattice. This assumption is probably valid to a good approximation, although some diffusion of sulphur will also occur, particularly during the baking of the cell up to 200°C which occurs in both processes.

The stoichiometry of the copper sulphide layer is of vital importance to the efficiency of the solar cell, since a reduction in the copper to sulphur ratio from 2.00 to 1.99 can reduce the cell efficiency by a factor of about two.[2]

The object of this work was to investigate the stoichiometry of the surfaces of cadmium sulphide thin films prepared under different conditions and to investigate the effects of changes in this stoichiometry on the performance of CdS–Cu_2S solar cells.

The surface stoichiometry of CdS films

The surface properties of CdS have been actively investigated for some time, and a critical review of this work has been published.[3] Most of this work has been performed using single crystals of CdS, in order to ensure repeatable results although the characteristics of even single crystal surfaces depend to some extent on the details of the growth procedure.[4] The surface properties of CdS films have been studied in less detail, but properties such as oxygen adsorption,[5] surface topology[6] and surface orientation[7] have been studied, the effects of oxygen adsorption on the densities of CdS surface states has been reviewed by Many,[8] who points out that the small photodesorption cross-section of oxygen on CdS surfaces suggests that there is a thin insulating layer on the CdS surface.

The surfaces of ZnS films have been studied at Balzers High Vacuum AG using SIMS (Pulker, private communication) and it was found that the surfaces were covered by an approximately monomolecular layer of ZnS_2. A similar layer of CdS_2 could exist on the surfaces of CdS films, and in view of the critical importance of CdS stoichiometry on solar cell efficiency, it was felt that this possibility should be investigated. Unfortunately, a SIMS apparatus was not available to us, and an Auger spectrometer was used. It is difficult to obtain quantitative data using Auger techniques, but it is possible to measure changes in the sulphur to cadmium ratio at the surface for films prepared under different conditions and subjected to different post-deposition treatments. The CdS films used in this study were prepared either by thermal evaporation or by electron beam evaporation. The thermally evaporated films were deposited onto molybdenum substrates at 300°C in a 'Hot Wall' system with a wall temperature of 220°C, at a pressure of about 10^{-5} torr. The electron beam evaporated films were deposited on molybdenum substrates at 260°C at pressures of about 10^{-7} torr. The thermally evaporated films were between 5–20 μm thick and showed very sharp optical absorption edges when undoped. SEM studies showed that the films had the usual collumnar structure with crystallite sizes of a few microns. The electron mobility in films doped to give resistivities of 100–1000 Ωcm was about 100 $cm^2\ V^{-1}\ s^{-1}$ with electron diffusion lengths of a few microns. The electron beam evaporated films were about 0.5 μm thick and so were not suitable for producing heterojunctions. They also had a very sharp optical absorption edges and electron mobilities of 100–150 $cm^2 V^{-1} s^{-1}$; and were of high resistivity (10^8 Ωcm). The substrates for use in the Auger spectrometer were about 1 cm diameter and were surrounded by other molybdenum or glass substrates so that many tests could be carried out on nearly identical CdS films, including formation into CdS–Cu_2S cells.

The CdS deposition parameters were varied so that the formation of both good (efficiency $\geq 4\%$) or bad (efficiency $< 1\%$) cells could be studied. The sulphur to cadmium ratio at the surface of the films was studied using a Vacuum Generators Leed/Auger spectrometer. A slice from a single crystal of CdS grown by a closed tube vapour transport technique[9] was studied in order to determine the relative amplitudes of the sulphur and cadmium peaks given by stoichiometric CdS. The slice was cleaned by etching in 5N HCl, and whilst under vacuum, was baked to 250°C and subject to ion bombardment. The cadmium to sulphur ratios of the films could then be related to the value obtained from the single crystal. These ratios were measured for CdS films deposited by both electron beam and thermal evaporation, with and without post-deposition etch in 5N HCl and also for films which had a 7 mm diameter CdS–Cu_2S heterojunction produced in their centres. These latter films were made by evaporation of CuCl through a 7 mm diameter mask concentrically onto the CdS film and producing the heterojunction by baking to 190°C to cause the replacement reaction and subsequently washing off the $CdCl_2$. The observed sulphur to cadmium ratios are shown in Table 1. These results show that

Table 1. Cd surface stoichiometry.

CdS conditions	Ratio of Auger peak heights sulphur/cadmium
Clean single crystal slice	2.3 ± 0.1
Thermally deposited films	1.8 to 2.0
Thermally deposited films after etching	2.2 ± 0.1
Electron beam evaporated films	2.3 ± 0.1
Unetched thermally evaporated films after dry barrier formation	2.2 ± 0.1
Etched thermally evaporated films after dry barrier formation	2.6 ± 0.1

the films deposited by thermal evaporation have surfaces which are sulphur deficient whilst the electron beam evaporated films are close to stoichiometry. This is probable due to the long cooling time needed in the thermal evaporation system, where the substrates take about 5 h to cool from 300°C as opposed to the 10 min or so needed for cooling in the electron beam system. Since the films were under vacuum during the cooling time, some sulphur evaporation would occur. After about 0.2 μm of the thermally evaporated films had been etched away, the sulphur to cadmium ratio was again equal to that of the single crystal to within the experimental errors. For those CdS films which had been subjected to the dry barrier process of CdS–Cu_2S junction formation, the sulphur to cadmium ratios increased from less than 2.0 to 2.2 ± 0.1 for unetched films and from 2.2 ± 0.1 to 2.6 ± 0.1 for etched films. These results would suggest that the sulphur concentration near the film surface increases during the barrier process. The vacuum evaporation of CuCl is done with the CdS film at temperatures below 60°C, so the only part of the dry barrier process which could give rise to significant sulphur diffusion is the baking at 190°C which was carried out with the films in argon or oxygen-free nitrogen at atmospheric pressure. It was not possible to heat CdS films in the Auger spectrometer to 190°C when the chamber was filled with argon or nitrogen at atmospheric pressure without damaging the heater filament, so the baking conditions could not be reproduced exactly. Back-filling the Auger spectrometer chamber with argon at 10^{-4} torr and heating the CdS films to 190°C did appear to increase the sulphur to cadmium ratio slightly, rather than cause a decrease which was observed when heating to 190°C at 10^{-9} torr.

The increase in sulphur concentration at the surface of the CdS films can be due only to diffusion of sulphur from the interior of the films, and there must therefore be a sulphur concentration gradient within the anion lattice. The increase in the sulphur surface concentration after the dry barrier process is probably due to free sulphur condensing out in a physisorbed state on the surface. ESCA measurements performed on some dry barriered samples by Dr L Hazel of the University of Surrey have shown that the 2 *s* peak in the sulphur spectrum is split, and that light ion etching reduces this peak to the usual singlet. This is precisely the behaviour which would be expected if free sulphur were present on the CdS surface, but a good deal of additional work is necessary before this suggestion can be confirmed.

The effect of CdS surface stoichiometry on CdS–Cu_2S cells

Copper sulphide can exist in a number of well-defined phases[10] and the effects of these phases on CdS–Cu_2S cell efficiency is well known.[1,2] A very convenient way of detecting the existence of these various phases is to measure the resistivity of the copper sulphide layer as a function of temperature.

Part of the phase diagram of copper sulphide is shown in Figure 1, where it can be seen that the chalcocite phase has a transition temperature of 103°C Djurleite has a transition temperature of 93°C whilst diginite has a transition temperature of between 76 and 83°C. The relative CdS–Cu_xS cell efficiency for the different phases is[2] chalcocite 100 (say) djurleite 60 diginite 10. It might be expected therefore that good cells would show the chalcocite transition at 103°C whilst bad cells would show the diginite transition.

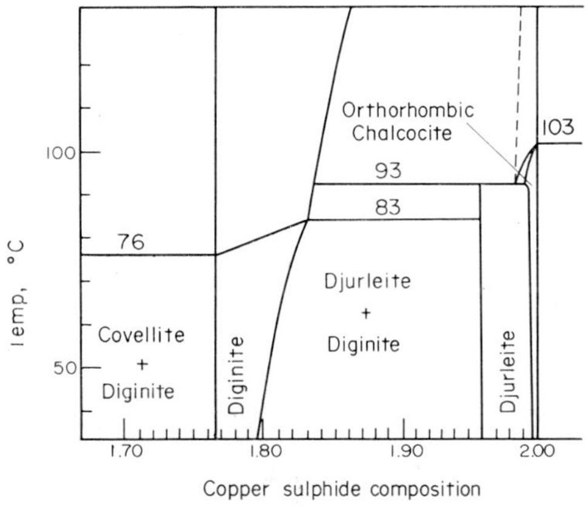

Figure 1. Part of the phase diagram of copper sulphide Cu_xS after Roseboom.[10]

The resistivity of the copper sulphide layer was measured in an argon atmosphere using Van der Pauw techniques whilst the junction was reverse biased. Typical results for good, medium and bad cells are shown in Figure 2. A good cell is one where point probe open circuit voltage and short circuit current under AM1 illumination correspond to those for cells with efficiencies greater than 4%, a bad cell is one for which point probe measurements correspond to those for cells of less than 1% efficiency.

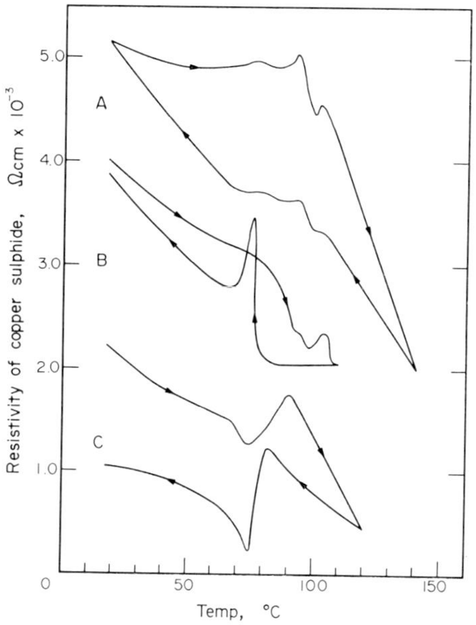

Figure 2. The temperature dependence of the resistivity of copper sulphide in CdS–Cu_2S cells for (A) a good cell (B) a medium cell and (C) a bad cell. For curve C, the ordinate is in units of 10^{-4} Ωcm.
The arrows on the curves show the direction of temperature change.

It can be seen from Figure 2 that the chalcocite transition is observed for the good and the medium cells but not for the bad cells whilst the djurleite and diginite transitions are observed for all the cells although the dignite transition is very weak in the good cells. The range of resistivities shown in Figure 2 is more typical of the diginite composition ranges than djurleite ($\rho \sim 10^{-2}$ cm) or chalcocite ($\rho \sim 1$ cm),[11] although the Van der Pauw method is strictly appropriate only for a homogeneous sheet, and the actual value of resistivity is unlikely to be very accurate. It was found that etching of the CdS films prior to dry barrier formation always reduced the cell efficiency, increased the diginite transition and reduced the chalcocite transition, as would be expected from the data in Table 1. The low cross-film resistivity and the occurrence of both diginite and djurleite transitions in good cells shows that the copper sulphide is not uniform in phase. The existence of excess sulphur on the CdS film surfaces suggested by the data in Table 1 and the ESCA measurements would lead to a model such as that shown in Figure 3. This represents the CdS crystallites after dry barrier formation, and shows an outer skin of diginite caused by the excess sulphur on the original CdS outer surfaces, with the copper sulphide phases changing through djurleite to chalcocite deeper in the copper sulphide layer. A major difference between good and bad cells would then be in the relative volumes of the copper sulphide phases.

This model explains the results so far obtained on CdS–Cu_2S cells formed by dry barrier techniques on CdS deposited at

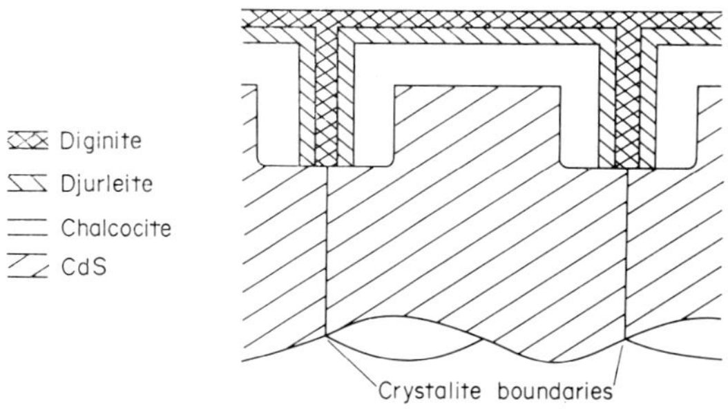

Figure 3. Proposed model of a CdS–Cu_2S junction. The drawing is not to scale, since the copper sulphide layer is about 1500 Å thick whilst the CdS crystallites are a few microns across.

300°C, i.e. with approximately zero porosity.[12] It is not necessarily appropriate for plated cells although identical results have been obtained for the resistivity measurements using cells plated by conventional Cleavite[13] techniques and by the techniques suggested by Aerschodt and Reinhartz.[14] It has not been found possible to investigate the unreacted CdS surface after plating since no mask could be found which would resist the plating solutions and also leave the CdS surface free of contamination or damage. The plating solutions are acidic and therefore etch the CdS during plating. It might then be expected that there would be some excess sulphur on the CdS surfaces after baking. The Institute of Energy Conversion have measured the variation of copper, sulphur and cadmium ratios with depth in their CdS–Cu_2S cells using Auger analysis.[15] It does appear from their data that the copper sulphide surface is slightly rich in sulphur, in agreement with the model shown in Figure 3, and that there is excess sulphur at the CdS/Cu_2S interface, possibly as a result of the continuous etching in the plating bath and the subsequent activation bake.

Conclusion

This work has shown that the surface stoichiometry of CdS films varies with different deposition methods and post-deposition treatment. These data do not permit a definite statement to be made concerning the existence of a CdS_2 monolayer on the surfaces of CdS crystallites, but they do suggest strongly that excess sulphur can occur on these surfaces and that this excess sulphur plays an important part in determining the composition of the copper sulphide layer in CdS–Cu_2S cells. The films deposited by thermal evaporation had surfaces which were sulphur deficient, although this was probably due to the prolonged heating in vacuum which the films suffered during the cooling down period. The dry barrier formation techniques resulted in an increase in the sulphur to cadmium ratio at the surface probably due to sulphur diffusion during the baking of the films to 190°C in a gas at atmospheric pressure. It seems likely that this sulphur is physisorbed on the crystallite surfaces, and it would be expected to have an activation energy for desorption of about 0.1 eV or about 3kT at the baking temperature. In vacuum therefore the rate of desorption should be quite large, but the rate should be considerably reduced by the other adsorbed gas molecules at atmospheric pressure.

By varying the deposition parameters of the thermally evaporated films, it was possible to produce good, medium and bad CdS–Cu_2S cells on unetched CdS. In these cases, no difference in the surface sulphur concentration was observed outside the limits of experimental error in the initial CdS film or the unreacted CdS after junction formation, and the splitting of the sulphur 2*s* peak was observed for both good and bad cells. Although non-stoichiometry of the CdS is only one of many reasons why a CdS–Cu_2S cell may have low efficiency, this does suggest that the outer surfaces of good and bad cells are similar.

The results of the measurements of copper sulphide resistivity as a function of temperature are in agreement with the measurements of Bougnot *et al*[14] on single crystal copper sulphide, and show the existence of all three phases of cuprous sulphide even on cells with quite high efficiency. This suggests that the model shown in Figure 3 is valid for all the cells produced by dry barrier techniques, and the resistivity measurements in these laboratories on plated cells suggest that the model may be valid for these cells also. The high efficiency cells would have very thin diginite and djurleite layers, whilst bad cells would have dignite and djurleite layers extending through most or all of the copper sulphide layer.

Acknowledgements

We would like to thank the following members of the Department of Physics and Physical Electronics, Newcastle upon Tyne Polytechnic for their invaluable assistance: Mr A N Casperd for preparing the thermally evaporated CdS films and the dry barrier cells; Mr M W McEwan for the resistivity measurements on the dry barrier cells; and Mr D Ridley for the resistivity measurements on plated cells.

References

[1] *International Workshop on CdS solar cells and other abrupt heterojunctions*, University of Delaware, USA, May 1975, NSF-RANN AER75-15858.
[2] M Makayama, *Japan J Appl Phys*, **10**, 1971, 1415.
[3] C G Scott and C E Reed, II–VI and III–V Compounds in *Surface of Phosphors and Semiconductors*. (Edited by C G Scott and C E Reed), Academic Press, London (1975).
[4] E H Weber, *Phys Stat Solidi*, **1a**, 1970, 665.
[5] S Baidyaroy and P Mark, *Surf Sci*, **30**, 1972, 53.
[6] M H Christmann, K A Jones and K H Olsen, *J Appl Phys*, **45**, 1974, 4295.
[7] C Laermans, L Michiels and A de Bock, *Thin Solid Films*, **15**, 1973, 317.
[8] A Many, *Crit Rev Solid State Sci*, **4**, 1974, 515.
[9] R Hill and R B Lauer, *Mat Res Bull*, **2**, 1967, 861.
[10] E H Roseboom, *Econ Geol*, **61**, 1966, 641.
[11] J Bougnot, F Gustavino, H Luquet and D Sodini, *Phys Stat Sol*, **8a**, 1971, K93. N Kakayama, *J Phys Soc Japan*, **25**, 1968, 290.
[12] S Ogura, *Thin Solid Films*, **30**, 1975, 3.
[13] Cleavite Corp Electron Res Div Cleveland USA. NASA Report Contract No AF33(615)—5224.
[14] A E V Aerschedt and K K Reinhartz, *Solar Cells*—Proceedings of Colloquium organized by ECOSEC 1970, 95.
[15] Institute of Energy Conversion University of Delaware, USA. NSF/RANN/SE/GI–34872/PR74–4.

2.16 Technology of large area Cu_2S–CdS solar cells

H W Schock, G Bilger, W H Bloss, G H Hewig and F Pfisterer, *University of Stuttgart, Institut für Physikalische Elektronik, Stuttgart, Germany*

Solar energy can be converted in different ways: the most attractive application is the direct conversion of solar radiation to electricity. For large scale application large area photovoltaic devices are necessary. Highly efficient solar cells can be produced on the basis of thin semiconducting films. An automized pilot line production for CdS layers, which are evaporated under high vacuum conditions, and their subsequent treatment to get an encapsulated thin film solar cell, are described. Investigation methods and a theoretical model of the heterojunction are reported.

Introduction

For terrestrial applications the Cu_2S–CdS thin film solar cell has a good chance to become a serious competitor of the silicon solar cell because the thin film structure renders it possible to produce low-cost large-area solar cells. A pilot line production for 7×7 cm^2 cells is installed to study the reproducibility of the process steps and to investigate the applicability of thin film solar cells for solar energy conversion on earth.

Evaporation of the CdS-layer

One of the main steps in the fabrication process of Cu_2S–CdS solar cells is the evaporation of CdS to form *n*-type semiconducting layers of a thickness of 20–30 μm. Ordinary glass with an evaporated silver layer serves as substrate material and back contact.

As the quality of the CdS-layer has a great influence on the performance of the final solar cell, we accomplished reproducible evaporation conditions. A schematic view of the arrangement for evaporation is shown in Figure 1. The graphite source is radiation heated by a tantalum cylinder. A thermocouple is built in coaxially to measure exactly the temperature of the crucible. The source is baffled by a quartz frit with a porosity of 100 μm. Thus it is possible to obtain spatter-free CdS-layers using several kinds of CdS-powders. For substrate heating a graphite meander is used.

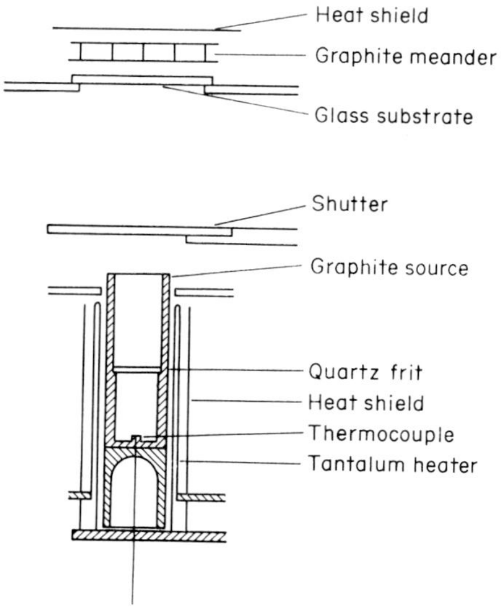

Figure 1. Schematic diagram of the evaporation arrangement.

Figure 2 shows the interior of the vacuum system. Three sources are operating simultaneously for the pilot line production of 7×7 cm^2 solar cells. The temperatures of the sources and substrates are controlled individually. A control unit determines the running off of the complete process such as the heating of the sources and of the substrates, the opening and closing of the shutter and the changing of the substrates. The thickness of the CdS-layers is determined by the source and the substrate temperatures, and the duration of the evaporation. No rate control or thickness monitoring is applied during the evaporation. It takes 20 minutes to evaporate a CdS-layer of a thickness of 30 μm at a source temperature of 950°C and a substrate temperature of 250°C. The thickness of the layers is reproducible within a limit of $\pm 10\%$. On these conditions the CdS is growing as hexagonal crystallites with a diameter of about 0.5 μm with the *c*-axis perpendicular to the substrate. It is possible to produce 12 CdS-films within one cycle of about 2 h.

Etching and barrier formation

The further fabrication steps are etching and barrier formation by chemical treatments. To increase the surface roughness and thereby the light absorption of the final cell, the CdS-layer is etched in a dilute HCl solution, which reacts preferentially along the grain boundaries.

The *p–n* junction is formed by dipping the CdS-layers in a solution of CuCl, at a temperature of 90°C.[1] During the dipping Cd is removed and replaced by two single valued copper ions. After a dipping time of about 10 s the thickness of the *p*-type Cu_2S-layer is about 2000 Å. Various parameters are influencing this process. Especially the cuprous ion concentration and the acidity of the solution are important. Optimum values seem to be 3 g/l CuCl and a pH value of 5.

Copper treatment

An additional step to improve the cells is the evaporation of a thin copper layer of about 50 Å thickness on to the cuprous

Figure 2. The evaporation plant for the pilot line production of CdS films.

sulphide.[2] During an additional heat treatment in air, copper is diffusing into the Cu_xS and is filling the copper vacancies in this layer. Furthermore a surface layer which acts as a mirror for minority carriers is believed to be formed.[3]

Encapsulation

For terrestrial solar cell operation it is necessary to seal the cell hermetically. The structure of a complete cell is shown in Figure 3. The transparent front contact is formed by a gold plated copper grid.

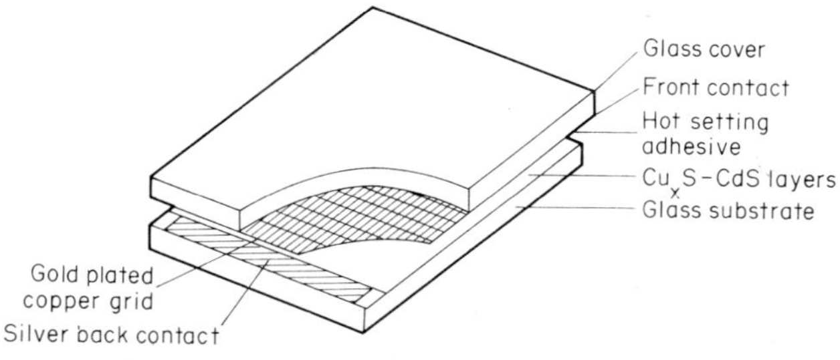

Figure 3. Structure of an encapsulated Cu_2S–CdS solar cell.

A glass plate, which is coated with a hot setting adhesive is pressed on the cell to seal it hermetically against oxygen and water vapour. Figure 4 shows the press for the encapsulation of an array of four 7 × 7 cm^2 cells.

Figure 4. The press for the encapsulation of the cells.

Homogeneity of the cells

The difficulties to get homogeneous cells are increasing with cell area. The homogeneity of the cells can be tested by scanning the cells with a light spot.[4] We have done this by means of a slow scan flying spot tube or by scanning the cells with a semiconductor laser beam. Figure 5 shows the block diagram of the flying spot. The beam intensity of a slow scan monitor is modulated by the output signal of the cell. Photographic pictures are taken from the monitor screen, because the scanning speed is too low to observe the monitor picture directly. A typical picture of a 7×7 cm^2 cell of about 3% efficiency scanned with the laser light spot is shown in Figure 6. The bright regions show high efficiency. Furthermore, the distribution of the short circuit current of one single line can be plotted with an XY recorder. It can be shown that the output current of the dark regions is only half of that of the bright regions. Thus it can be expected that homogeneous cells show much higher efficiencies.

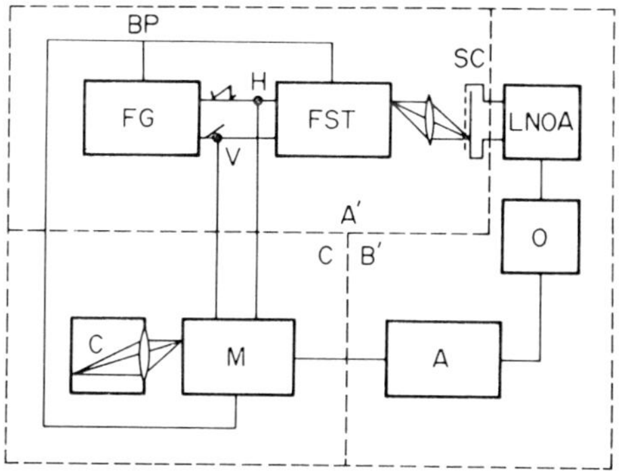

FG Function generator
BP Blanking pulses
H Horizontal deflection
V Vertical deflection
LNOA Low-noise operational amplifier
O Oscilloscope
A Amplifier
M Monitor
C Camera
SC Solar cell
FST Flying spot tube

Figure 5. The flying spot systems.

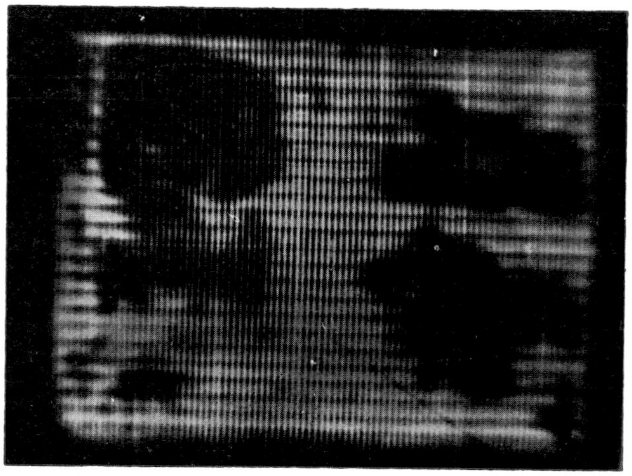

Figure 6. The distribution of the short circuit current of a 7×7 cm^2 solar cell.

Structure of the Cu_2S–CdS solar cell

For the development of the Cu_2S–CdS cells it is necessary to investigate the shape of the *p–n* junction, which is mainly determined by the crystalline structure of the CdS film.

The evaporated CdS film consists of small crystallites with the *c*-axis oriented perpendicular to the substrate. The Cu–ion dipping solution reacts chiefly along the grain boundaries. Thus a comb-like structure of the cross-section of the *p–n* junction results. This special structure influences the parameters of the cell.

For the investigation of the structure the cells were lapped at an angle of 40°.[5] After lapping the cells were etched in a dilute nitric acid solution, which attacks CdS especially along the grain boundaries, whereas Cu_2S is not attacked. A light microscope picture of a bevelled Cu_2S–CdS *p–n* junction together with a schematic view of the corresponding cross-section is shown in Figure 7. In the bright central region I the bevelled surface has not reacted during the etching, because the grain boundaries are covered with Cu_2S. In the dark region II a granular structure of the CdS appears. The angle of the lapping together with the width of the bright region I corresponds to a penetration depth of the Cu_2S of the order of about 2 μm along the grain boundaries.

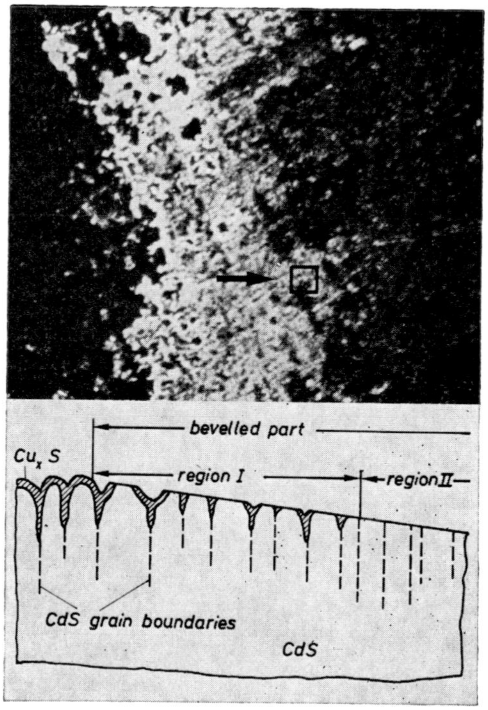

Figure 7. Light microscope picture of a bevelled Cu_2S–CdS cell.

Another way to make the structure of the cell visible is to investigate the lapped and etched cross-section of the junction by means of a scanning electron microscope.

The upper part of Figure 8 shows the ordinary secondary emission picture of the cross-section adjacent to the surface of the cell. It is possible to see the single grains, but there is no

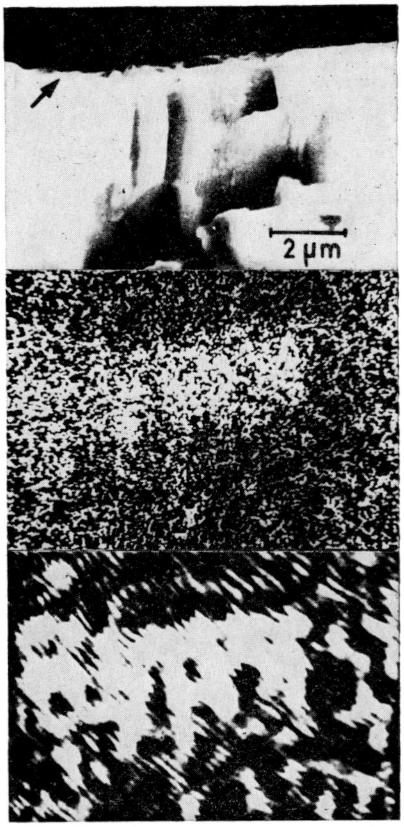

Figure 8. SEM picture of the cross-section of a Cu_2S–CdS cell.

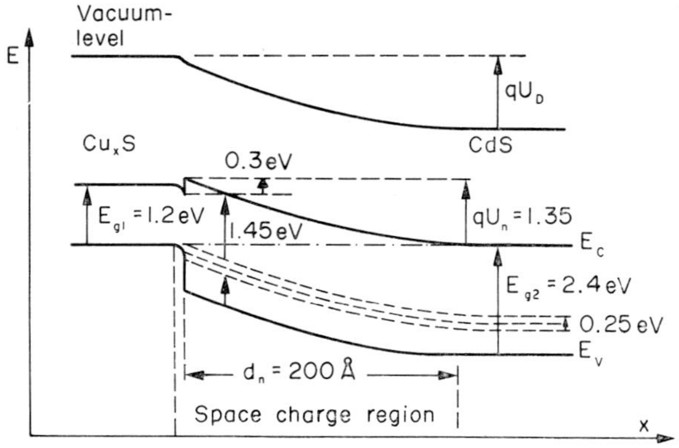

Figure 9. Band diagram of the Cu_2S–CdS heterojunction.

prominent difference between CdS and Cu_2S. In the central part of the slide the distribution of copper in the same region is shown as a higher density of the light spots. It has been obtained by X-ray analysis using the copper Kα-line. Due to the noisy signal, the copper rich regions were not clearly resolved in this picture. So a coherent image processing system was applied to reduce the noise and to increase the intensity gradient of the picture. The result of this operation is shown in the lower part of the figure. It can be seen that the structure follows the predicted shape. Again the copper extends from the surface downward to about 2.5 μm.

Thus the cell can be regarded as a combination of a planar cell and a vertical multijunction cell. The effect of the increased junction area must be taken into account for the calculation of the reverse current, the open circuit voltage, and the collection efficiency.[6,7]

Theoretical model of the Cu_2S–CdS heterojunction

Other studies concerned the properties of the heterojunction Cu_2S–CdS. Special optical effects have to be explained by a theoretical model. Several theories are existing.[8–10] We are proposing a model which includes tunnelling processes.[11] Assuming the electron affinity of CdS to be smaller than that of Cu_2S, then a potential peak in the conduction band results, which occurs at the heterojunction. Heat treatment causes Cu atoms to diffuse into CdS. These copper atoms act as acceptors. The energy levels are in the range 0.7–0.95 eV above the valence band of CdS as shown in Figure 9. The occupation of a copper acceptor with an electron from the valence band due to the absorption of photon causes an additional negative space charge in the junction resulting in a larger width of the potential peak, which leads to a smaller transmission coefficient. Photons with an energy larger than 1.45 eV are able to ionize an acceptor, which results in an additional positive space charge in the junction and, therefore, a larger transmission coefficient is observed.

The shape of the potential in the junction was calculated in the frame of the Schottky-approximation, assuming a constant space charge density in the p and n region in the junction:

$$\Delta U = -\frac{\rho}{\varepsilon\,\varepsilon_0}$$

where $\rho = q \cdot N_{D,A}$ is the constant space charge density and $N_{D,A}$ is the doping concentration of donors in CdS and acceptors in Cu_2S, respectively. The width of the space charge density in the p and n region (d_n and d_p) can be determined:

$$d_{n,p} = \left[\frac{2\,\varepsilon_{n,p}\,\varepsilon_0\,U_{n,p}}{q\,N_{D,A}}\right]^{1/2}$$

$U_{n,p}$ corresponds to the diffusion potential in the n- and p-region. Using the diffusion potential U_D and the space charge neutrality of the junction, d_n and d_p can be determined. The doping concentrations of Cu_2S and CdS are assumed to be 10^{20} 1/cm^3 and 10^{19} 1/cm^3, whereas the dielectric constants ε_p and ε_n are assumed to be 6 and 7, respectively. If the diffusion potential in n-CdS is 1.35 V, the equations above yield the following results:

$$d_n = 95 \text{ Å}$$
$$d_p = 9.5 \text{ Å}$$
$$U_p = 0.12 \text{ V}$$
$$\Delta E_c = 0.27 \text{ eV}$$

The doping levels seem to be very high and should influence the performance of the cell in an extreme manner. But it has to be considered that the doping level of the bulk material may be lower. The high doping levels assumed occur perhaps due to the difference of the lattice constants of Cu_2S and CdS and therefore in the junction region the doping levels are determined by the resulting stress in the interface.

To calculate the quantum mechanical transmission coefficient a stepwise constant potential was assumed as shown in Figure 10. The calculated parabolic shape was approximated by using constant potential steps of 0.05 eV. Therefore the curve of the potential was approximated by single points.

Thermal relaxation of the electrons in the junction is not involved in the calculation method. For electrons tunnelling from Cu_2S into CdS, i.e. minority carriers corresponding to the short circuit current, the width of the layer in which the electrons could thermalize is only 10 Å and the depth of the potential is in maximum equal to 0.15 eV. Therefore, thermal relaxation should not strongly affect the results.

The method of calculating the transmission coefficient was proofed in the case of a potential with a behaviour like secans hyperbolicus, which can be solved analytically. In this case, the differences between the exact values and the calculated values, assuming a stepwise constant potential, are smaller than 1 %. Experimental data of several optical measurements can be explained quantitatively by this model.

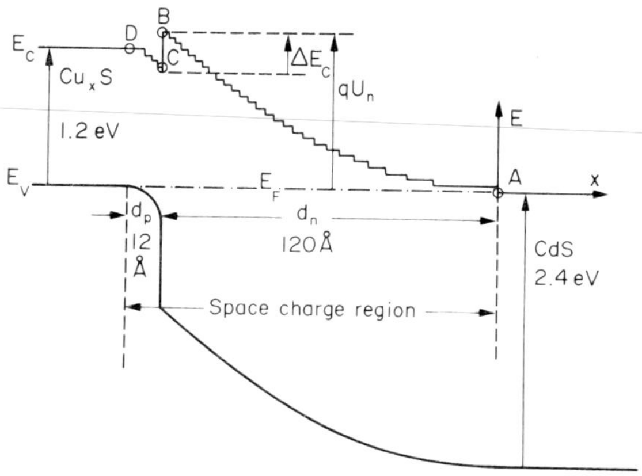

Figure 10. Approximated potential.

Test generator

The solar cells of our pilot line production have been assembled to a generator which has now been installed on the roof of our laboratory in Stuttgart. It is shown in Figure 11. The generator has a power output of about 20 W at an illumination of 100 mW/cm^2.

Figure 11. A 20 W Cu_2S–CdS photovoltaic generator.

Conclusions

The first results concerning the gain of the pilot line production and the stability of the cells are quite promising, so it seems to be possible to produce this thin film device in an industrial scale and thus to solve energy problems in a first step at remote places and after proving its reliability and further costs reductions to make a contribution to the energy supply of industrialized countries.

Acknowledgement

This work is supported by the Federal Ministry for Research and Technology of the F R G.

References

[1] L R Shiozawa, A Augustine, G A Sullivan, J M Smith and W R Cook Jr, Final Report; Contact AF 33(615)–5524, 1 June 1966–31 May 1969.
[2] K Bogus and S Mattes, *Proc IXth IEEE Photov Spec Conf*, p 91, Silver Spring (1972).
[3] F Pfisterer, G H Hewig and W H Bloss, *Proc XIth IEEE Photov Spec Conf*, p 460, Scottsdale (1975).
[4] G H Hewig and F Pfisterer, *Int Congr The Sun in the Service of Mankind*, Proc Sect, p 208, Paris (1973).
[5] M K Mukherjee, F Pfisterer, G H Hewig and H W Schock, *Proc Conf Solar Electricity*, Toulouse (1976).
[6] W P Rahilly, *Proc IXth Photov Spec Conf*, Silver Spring (1972).
[7] S M Dhariwal, L S Kothari and S C Jaim, *J Phys D*, **8**, 1975, 1321.
[8] W D Gill, R H Bube, *J Appl Phys*, **41**, 1970, 9.
[9] L R Shiozawa, *Conf Rec VIIth IEEE Photov Spec Conf*, Pasadena (1968).
[10] K W Böer and J Phillips, *Proc IXth IEEE Photov Spec Conf*, Silver Spring (1972).
[11] G H Hewig, Thesis, Univ Stuttgart (1974).

2.17 Joule heating in ZnS electroluminescent films

D May,* *Division of Materials Science, Thames Polytechnic, Wellington St, London SE18, England*

The current (i)*–voltage* (V) *characteristics of ZnS electroluminescent films have been interpreted taking into account Joule heating. Graphs of log* i *vs log* V *for dc data showed a square law behaviour up to a critical power which was sample-dependent. At higher powers, the data lay on a straight line of increased slope (typically 2.5). A comparison of the* i–V *characteristics of a sample under dc and pulsed voltage excitation showed that the pulsed characteristics followed a square law in the voltage range which gave enhanced conduction under dc conditions. By assuming a simple law of cooling to be obeyed, the activation energy relating the pulsed and dc currents at corresponding voltages was calculated. It was also possible to explain the variation in critical power from sample to sample. Estimates of the temperature at the critical power suggested a phase change associated with the copper sulphide present in the devices. Examination of the* i–V *data reported by other workers proved to be consistent with those reported here. It is argued that this approach may therefore have a more general validity.*

Introduction

A few authors have considered the influence of Joule heating in their discussions of the dc current (*i*)–voltage (V) characteristics of electroluminescent ZnS films, one of the earliest such publications being by Goldberg and Nickerson.[1] However, no quantitative assessment of its influence seems to have been reported to date. In this paper the gross *i*–V characteristics of ZnS:Mn, Cu films are considered, taking into account the influence of Joule heating. The detailed *i*–V characteristics of such films have been reported elsewhere.[2,3]

Film preparation

Two types of film were prepared: ZnS: Mn, Cu films activated by the embedding technique of Thornton[4]; and Cu_xS–ZnS: Mn, Cu heterojunctions using a method similar to that of Vlasenko and Gergell.[5] The ZnS films were 1–1.5 μm thick and the Cu_xS film $\sim$0.1 μm thick. The films were deposited using an electron-beam evaporator at a pressure of $\sim 10^{-6}$ torr ($\sim 10^{-4}$ Pa) Substrate temperatures were $\simeq$95°C for the embedding ZnS film, but $\simeq$240°C for the heterojunction ZnS film. For the Cu_xS deposition, the substrate was unheated, but heat from the crucible caused the substrate temperature to rise to $\simeq$60°C. Film test areas had typical dimensions of 1 cm $\times$ 2 mm. Reported measurements were made at room temperature unless otherwise stated.

Measurements

Direct current measurements of the *i*–V characteristics were mostly made using AVO meters (Mark 8), although for lower current ranges a Levell meter type TM9BP was sometimes used. Power was supplied using a Kingshill model 1500 stabilized supply.

Pulsed *i*–V characteristics were obtained by applying rectangular, unidirectional voltage pulses of 5 ms width, frequency 30 Hz and pulse height V_p, using a modified Advance pulse generator PG 55A, and observing the resulting current pulse on an Advance OS2100 oscilloscope. The pulses were made as short as possible to minimize heating effects, but could not be made less than 5 ms owing to the slow current decay to a steady value, which corresponded to the dc value for the equivalent dc voltage.

Results

Graphs of log *i* vs log V for the dc data gave straight lines of slope close to two up to a certain voltage after the usual initial ohmic range (Figure 1). However, above this critical voltage (V_c), the data lay on a straight line of increased slope, mostly in the range 2.4–2.7. This suggested the possibility of Joule

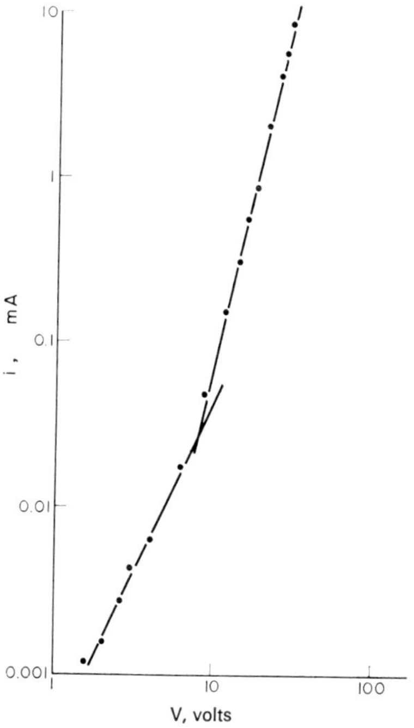

Figure 1. Graph of log *i* vs log V under dc conditions for a heterojunction sample.

* Now at Department of Science, Tottenham College of Technology, High Road, London N15, England.

heating, causing a current rise when the applied voltage exceeded a critical value V_c. Consecutive measurements of i–V characteristics of a sample under both dc and pulsed excitation showed that the pulsed characteristics followed the square law at pulse height voltages V_p greater than V_c (Figures 2 and 3).

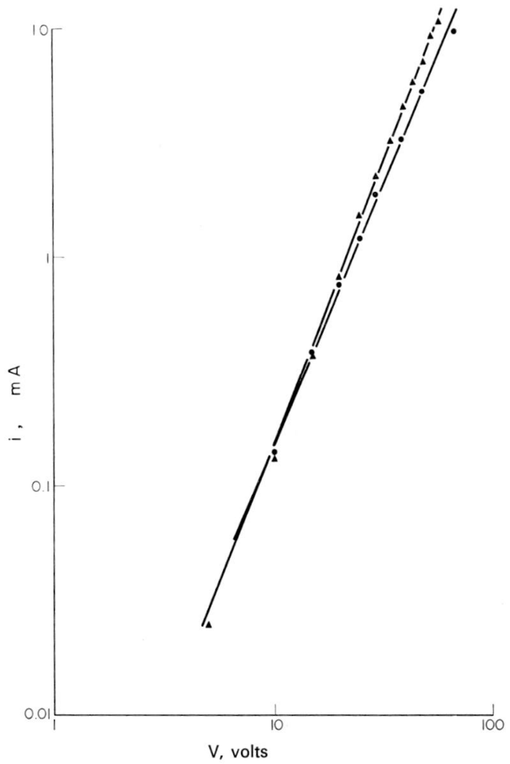

Figure 2. Comparison of dc and pulsed data. ● Pulsed, ▲ dc. The slope under pulsed conditions is slightly more than two, due to Wannier–Stark peaks influencing the data.[2,3]

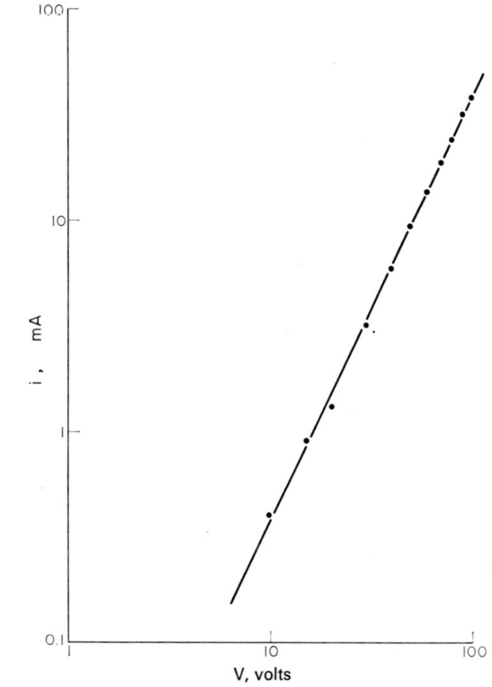

Figure 3. i–V characteristic under pulsed conditions. The sample is the same as that of Figure 2, but the data was taken earlier in the history of the sample, so that Wannier–Stark effects[2,3] are less apparent; the slope is very close to two.

(It was necessary to measure dc and pulsed characteristics consecutively since, with increased use, the device conductance fell. This is the ageing problem reported by authors such as Goldberg and Nickerson.[1])

Analysis of results

To interpret the results it was necessary to make certain assumptions, which led to an equation which could then be tested using the data.

(A) The dependence of dc current on temperature.

The data at voltages V less than V_c fit a square law. At a voltage V ($>V_c$), it is supposed that i is raised to a value i_2, corresponding to a film temperature T_2; whereas under pulsed conditions ($V_p = V$) the steady current has a value i_1, at a film temperature T_1. (In general, $T_1 \neq T_0$, where T_0 is room-temperature.)

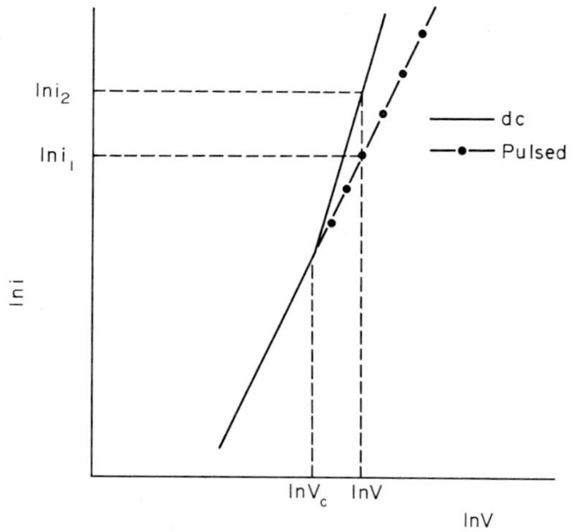

Figure 4. Schematic form of ln i − ln V characteristic. At $V < V_c$, the dc and pulsed characteristics coincide.

$$\text{Hence } i_2 = aV^2 \exp\left(\frac{-\phi}{2kT_2}\right)$$

$$\text{and } i_1 = aV^2 \exp\left(\frac{-\phi}{2kT_1}\right)$$

where a is a constant.

We suppose here that there is an exponential dependence on temperature, with activation energy ϕ.

$$\text{Then } \ln\left(\frac{i_2}{i_1}\right) = -\frac{\phi}{2k}\left[\frac{1}{T_1} - \frac{1}{T_2}\right] \tag{1}$$

(B) A law of cooling is assumed.

(i) A particularly simple choice is

$$i_2 V = M(T_2 - T_0) \tag{2}$$

where M is a constant.

This is Newton's law of cooling and supposes that heat is lost steadily to the surroundings by conduction. A similar approach has been used[6] in discussing thermal breakdown in insulators.

(ii) A law of cooling in the absence of a steady draught is the 5/4 power law, which is probably more appropriate under the experimental conditions.

$$i_2V = N(T_2 - T_0)^{5/4}$$

where N is a constant.

For simplicity we first consider Newton's law of cooling.

Substituting for T_2 from equation (2) into equation (1) we have

$$\frac{1}{\ln\left(\frac{i_2}{i_1}\right)} = \frac{2kT_1}{\phi}\left[\frac{i_2 V + MT_0}{i_2V + M(T_0 - T_1)}\right] \quad (3)$$

If the pulse width and frequency are such as to avoid any heating while the voltage is on, then $T_1 = T_0$. Then equation (3) becomes

$$\frac{1}{\ln\left(\frac{i_2}{i_1}\right)} = \frac{2kT_0}{\phi} + \frac{2kT_0^2 M}{\phi} \cdot \frac{1}{i_2V}$$

This is the equation of a straight line. It proves more convenient to plot a slightly different form of the equation, viz.:

$$\frac{i_2V}{\ln(i_2/i_1)} = \frac{2kT_0}{\phi}(i_2V) + \frac{2kT_0^2 M}{\phi} \quad (4)$$

Inserting the data into equation (4), a satisfactorily straight line results (Figure 5). The slope enables the activation energy ϕ to be calculated. $\phi \simeq 2.6\, kT_0$ (i.e. $\simeq 0.065$ eV). Hence M may be calculated.

We now estimate the temperature T_c at the critical voltage V_c. Writing $(i_2V)_c$ as the corresponding critical power, we have from equation (2)

$$(i_2V)_c = M(T_c - T_0) \quad (5)$$

Hence

$$T_c = 54°\text{C}. \quad (T_0 = 25°\text{C})$$

If heating were to occur under pulsed excitation, $T_1 > T_0$. However, provided $T_1 < T_c$, the results suggest that the current does not rise above i_1. So the worst case to consider is $T_1 = T_c$. Applying this to equation (3) and using equation (5) we find

$$\frac{1}{\ln\left(\frac{i_2}{i_1}\right)} = \frac{2kT_c}{\phi}\left[\frac{i_2V + MT_0}{i_2V - (i_2V)_c}\right]$$

$$\therefore \frac{i_2V - (i_2V)_c}{\ln\left(\frac{i_2}{i_1}\right)} = \frac{2kT_c}{\phi}[i_2V - (i_2V)_c] + \frac{2kT_c}{\phi}[(i_2V)_c + MT_0]$$

Again, this is the equation for a straight line. However, using the same data as before, the resulting graph is less satisfactory than that in Figure 5 (Figure 6). This shows that putting $T_1 = T_0$ is a far better approximation than putting $T_1 = T_c$. We thus assume that $T_1 = T_0$.

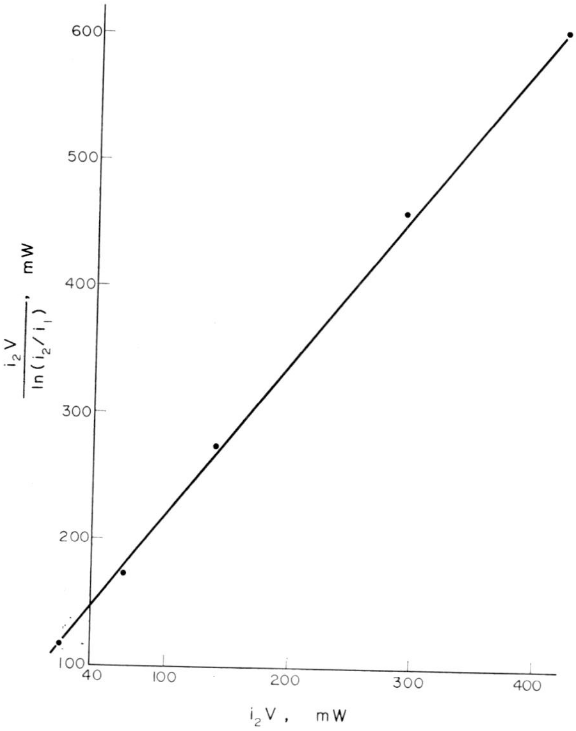

Figure 5. Graph to test Joule heating assumptions, using Newton's law of cooling and $T_1 = T_0$. Note that the derivation of the equation does not depend on any particular form of i–V relationship at a given temperature.

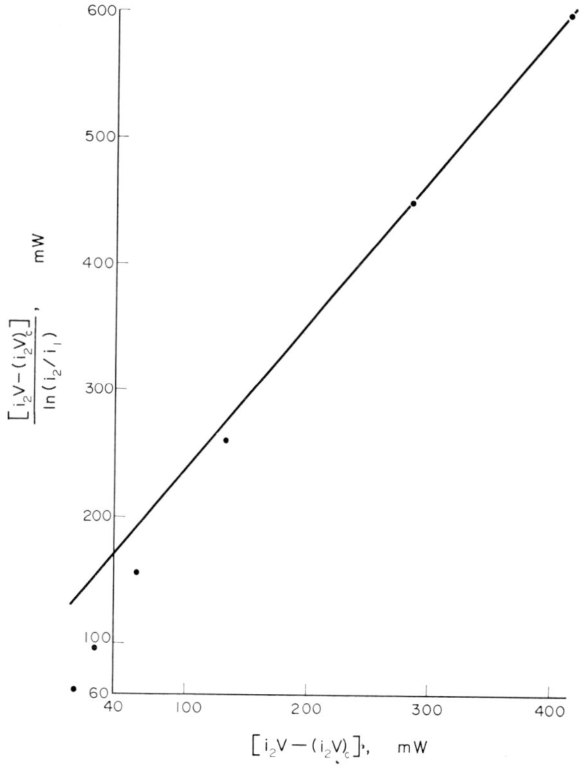

Figure 6. Graph to test Joule heating assumptions using Newton's law of cooling and $T_1 = T_c$.

We now consider the 5/4 power law of cooling. A similar approach to the above yields

$$\frac{1}{\ln\left(\frac{i_2}{i_1}\right)} = \frac{2kT_1}{\phi} \cdot \frac{(i_2 V)^{4/5} + N^{4/5}\, T_0}{(i_2 V)^{4/5} + N^{4/5}\,(T_0 - T_1)}$$

For $T_1 = T_0$, and multiplying the equation by $(i_2 V)^{4/5}$

$$\frac{(i_2 V)^{4/5}}{\ln(i_2/i_1)} = \frac{2kT_0}{\phi} \cdot (i_2 V)^{4/5} + \frac{2kT_0^{\,2}\, N^{4/5}}{\phi}$$

This also yields a satisfactory straight line (Figure 7). The value of ϕ is similar to that obtained above: $\phi \simeq 2.2kT_0 \simeq 0.055$ eV.

Using $(i_2 V)_c = N(T_c - T_o)^{5/4}$ we find $T_c \simeq 59°C$.

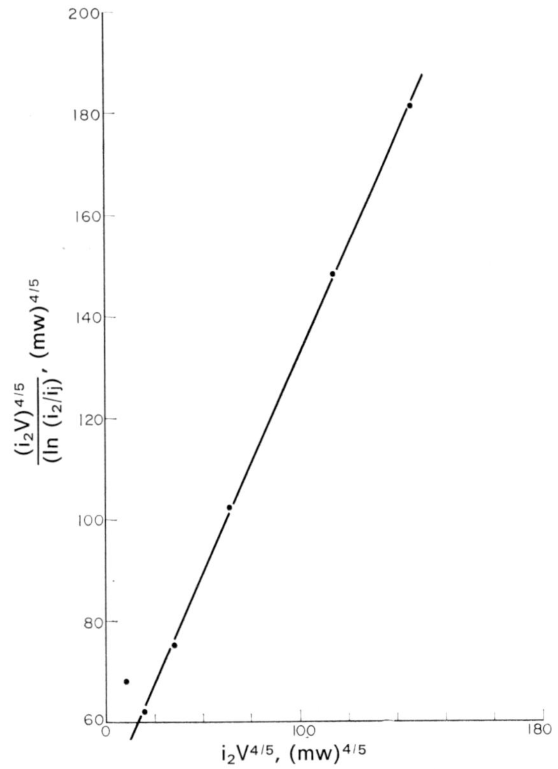

Figure 7. Graph to test Joule heating assumptions using the 5/4 power law of cooling and $T_1 = T_0$.

From the two methods we may estimate the film temperature at various voltages. At 37 V we find $T_2 = 296°C$ using Newton's law of cooling, whereas the 5/4 power law yields $T_2 = 231°C$ at 37 V. We conclude that the method is not a very sensitive one for establishing film temperatures.

Other reported data

Preliminary examination of i–V data reported by other workers suggests that Joule heating is a widespread feature of direct current electroluminescence (dcel) i–V data. From the papers of Goldberg and Nickerson[1] and of Vlasenko and Gergell,[5] i–V data have been derived and re-plotted in the log i–log V form. (Figures 8 and 9.) Graphs with two linear sections are obtained (after an ohmic section, if reported). The slope at lower voltages is close to two, but above a certain voltage exceeds two; in fact both slopes exceed 3.

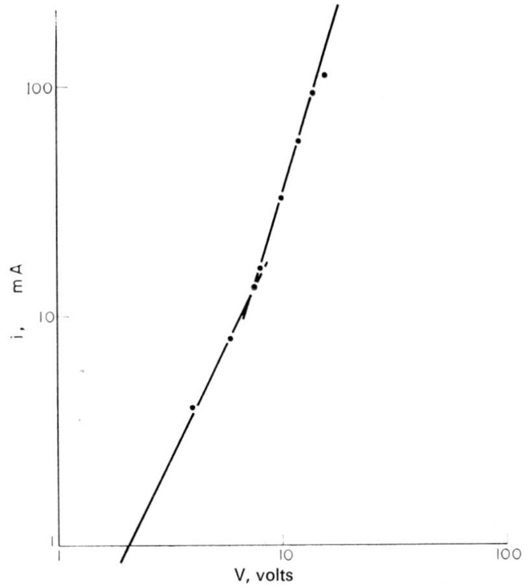

Figure 8. i–V data from Goldberg and Nickerson.[1]

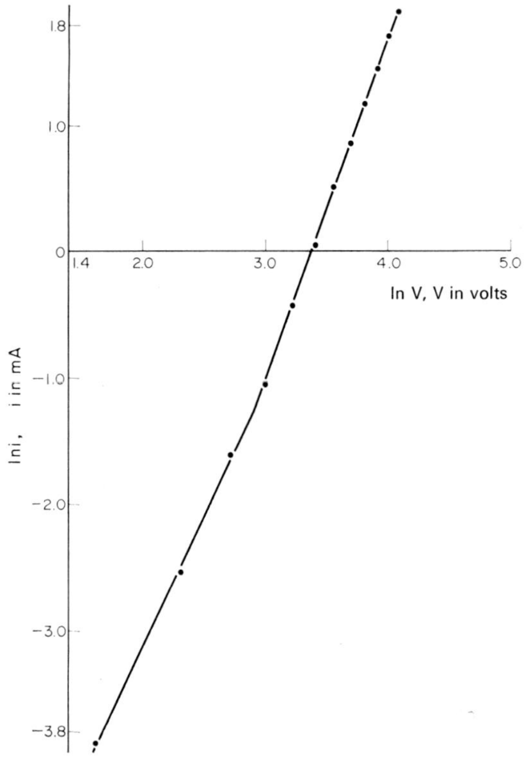

Figure 9. i–V data from Vlasenko and Gergell.[5]

Applying the data to equation (4), satisfactory lines again result (Figures 10 and 11). Values of ϕ and T_c have been calculated. For the data of Goldberg and Nickerson[1] an activation energy of 0.064 eV has been found and $T_c \simeq 60°C$, close to the values reported above. However, the data of Vlasenko and Gergell[5] yield an activation energy of about 0.16 eV and $T_c \simeq 25°C$. This may be related to the low critical power of 0.24 mW, and the less satisfactory line (Figure 11). Using the 5/4 power law we find $\phi \simeq 0.16$ eV and $T_c \simeq 36°C$, in better agreement with the other data. Goldberg and Nickerson's data then yields $\phi \simeq 0.08$ eV and $T_c \simeq 53°C$ (Figures 12 and 13).

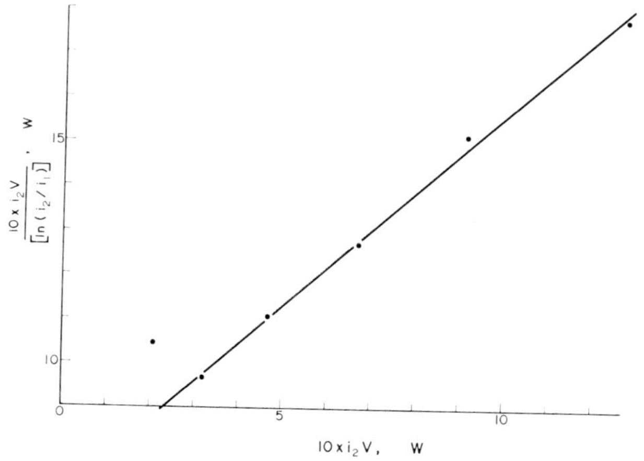

Figure 10. Graph to test Joule heating assumptions applied to *i*–V data from Goldberg and Nickerson,[1] using Newton's law of cooling and $T_1 = T_0$.

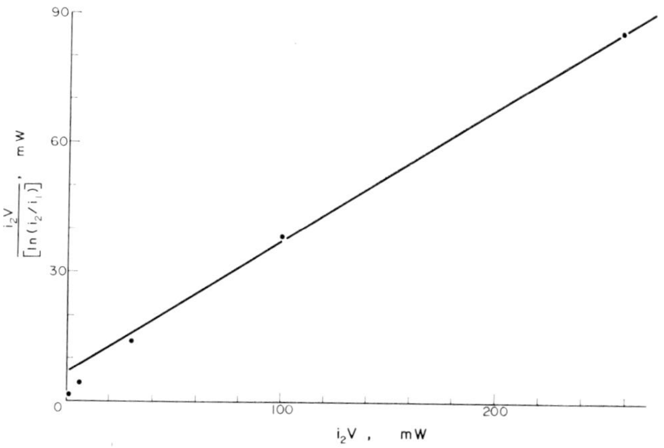

Figure 11. Graph to test Joule heating assumptions applied to *i*–V data from Vlasenko and Gergell,[5] using Newton's law of cooling and $T_1 = T_0$.

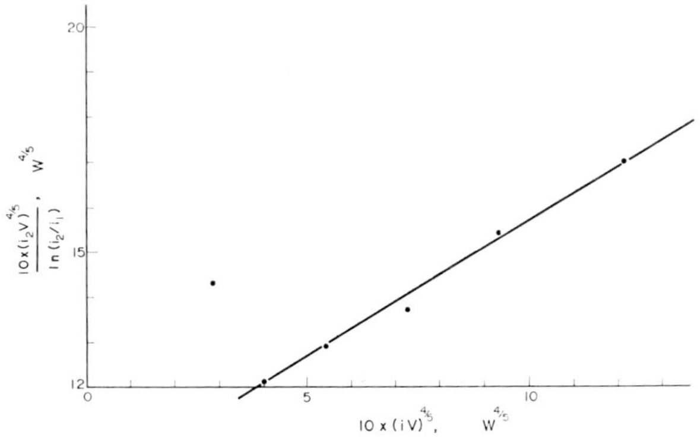

Figure 12. Graph to test Joule heating assumptions applied to *i*–V data from Goldberg and Nickerson[1], using the 5/4 power law of cooling and $T_1 = T_0$.

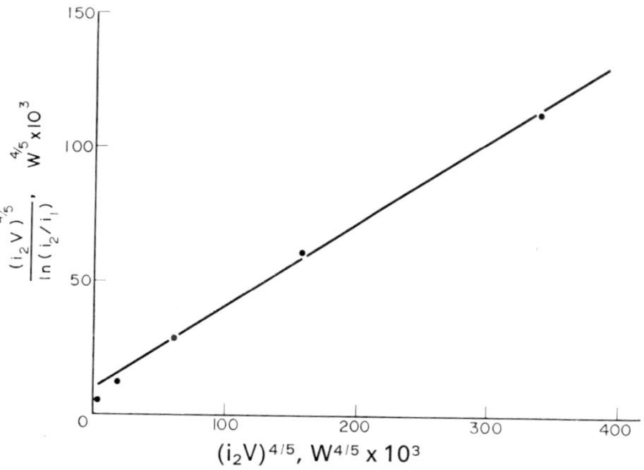

Figure 13. Graph to test Joule heating assumptions applied to *i*–V data from Vlasenko and Gergell,[5] using the 5/4 power law of cooling and $T_1 = T_0$.

Significance of T_c

To gain some insight into the significance of T_c, measurements of dc current at a fixed voltage as a function of temperature have been made. The measurements were both on powder panels (which also show Joule heating in their *i*–V characteristics) as well as on thin films. The results indicate a change in device conductance as a function of temperature at $\simeq$80°C on the cooling cycle. Okamoto and Kawai[7] have studied the electrical conduction of copper sulphides in relation to their phase transitions. They report a change in electrical conductivity at 80°C when cooling high digenite, which they associate with its phase change to digenite. There are also indications of a kink in our log $G - 1/T$ graphs (G = conductance) in the region of 50–60°C. Although no change in conductance has been reported in this range by Okamoto and Kawai,[7] Donnay *et al*[8] have reported a change in crystal structure of digenite in the temperature range 60–65°C. We suggest that T_c may be identified with such a phase change. (The values calculated above must be upper limits, since they ignore latent heat involved in the phase change.) (Results to be published.)

Discussion

The value of the critical power $(i_2V)_c$ varies considerably from sample to sample, from 7 to 100 mW. This is related to the variation in M from sample to sample. If a test area lies close to the substrate edge, its ability to cool is expected to be larger than if it is more centrally placed. Similarly, if the sample and substrate size are relatively small, $(i_2V)_c$ would be expected to be small. This proves to be so for sample 33.1 which was on a particularly small substrate and for which $(i_2V)_c \simeq 7$ mW. On balance, the results suggest that the 5/4 power law of cooling is to be preferred.

It seems likely from the evidence for Joule heating that changes in the electrical properties of the Cu_xS phase occur as the applied voltage is raised. This complicates the development of a band model of the Cu_xS–ZnS heterojunction. In the construction of such a model it becomes important to know the Cu_xS phase present under operating conditions. For example, Vlasenko and Gergell[5] prepared Cu_xS–ZnS:Mn, Cu, Cl film heterojunctions and measured their properties. They reported values of $(E_G + E_F)$ for the Cu_xS films to be 1.7 eV and 1.25 eV according to their conductivity. (E_G is the bandgap and E_F the Fermi level.) However, no mention is made of the condition of the Cu_xS films when the measurements were made. If the measurements were made on an as-evaporated film, we now see that the results may bear little relation to the Cu_xS film properties in the operating device.

Conclusions

Experimental evidence for the existence of Joule heating in dcel ZnS films has been found. A semi-theoretical approach to the data involving a law of cooling has enabled the activation energy associated with the current increase to be estimated. The method proves to be rather an insensitive one for establishing film temperatures. Preliminary work on the effect of externally heating the films indicates a change in electrical properties at $\simeq$80°C (cooling cycle) and possibly in the 50–60°C range. This suggests phase changes in the copper sulphide. Consideration of other reported *i*–V data shows that they are consistent with this approach. It is therefore concluded that the *i*–V characteristics of ZnS films showing dcel are strongly influenced by Joule heating which causes change(s) in the copper sulphide phase.

Acknowledgements

The author wishes to thank the Ministry of Defence (Procurement Executive) for a research assistantship, during which this work was begun. Thanks are also due to present and former members of the dcel research team, including E Erdil, J Mayo, P J F Smith, N J Werring and the author's supervisor, A Vecht.

References

[1] P Goldberg and J W Nickerson, *J Appl Phys*, **34**, 1963, 1601.
[2] D May and A Vecht, *J Phys C Solid State Phys*, **8**, 1975, L505.
[3] D May and A Vecht, *Thin Solid Films*, **36**, 1976, 139.
[4] W A Thornton, US Patent No 3044 902 (1962).
[5] N A Vlasenko and A N Gergell, *Phys Stat Sol*, **26**, 1968, K77.
[6] J J O'Dwyer, *The Theory of Electrical Conduction and Breakdown in Solid Dielectrics*, p 192, Clarendon Press, Oxford (1973).
[7] K Okamoto and S Kawai, *Japan J Appl Phys*, **12**, 1973, 1130.
[8] G Donnay, J D H Donnay and G Kullerud, *Am Mineral*, **43**, 1958, 228.

2.18 ac electrical properties of magnesium and aluminum fluoride thin films

Y Danto and J Salardenne, *Laboratoire de Recherche en Electrotechnique et Physique de Solide, Université de Bordeaux 1, 351 Cours de la Libération, 33405 Talence, France*

ac electrical properties of thin evaporated films of magnesium and aluminum fluorides have been investigated within a frequency range from 5 Hz to 50 kHz. The composition of the films was determined by nuclear analysis (backscattering and α, X reactions), and their structures by X-ray diffraction. All measurements were carried out in a vacuum chamber. Results are related neither to the film structure nor, in the case of films previously submitted to atmospheric conditions, to the impurity rates. For both materials, the classical law giving the conductivity versus frequency, $\sigma \sim \omega^n$, is verified with $n < 1$. These results have been compared with those derived from thermally stimulated depolarization (TSD) measurements. The energy distribution of the dipoles, deduced from TSD results does not seem to be large enough to agree with an interpretation based on the Debye model. Another explanation according to a new model proposed by Jonscher,[3] can be considered.

Introduction

This work belongs to a more general study of electrical and optical properties of ionic compound thin films such as MgF_2 and AlF_3.[1,2] We are concerned here with the dielectric properties of these materials. The composition and the structure of the films are determined by nuclear analysis (backscattering and α, X reactions) and by X-ray diffraction. With a rigorous control of preparation parameters, the main characteristics of the films (stoichiometry, impurity rates, structure) can be predetermined.

The classical law $\sigma \sim \omega^n$ giving the dependance of the ac conductivity versus frequency is generally discussed in terms of the Debye model with a relaxation time distribution.[12,21] Another approach to the dynamics of dielectric polarization, recently developed[3] can also be used. It seems interesting to compare these ac conductivity measurements with the results of the thermally stimulated depolarization (TSD) which gives the energy distribution of the dipoles, in order to examine the compatibility of both methods.

1. Experimental apparatus—films characterization

Samples used in this work are metal–insulator–metal 'sandwich' structures, obtained by thermal vaporization in vacuum (10^{-7} torr). The substrates, made of thin mica slides, are cleaved just before utilization. Temperature can be regulated between 77–600 K. Structure geometry is obtained with a system of movable masks, the surface of each sample being 4 mm^2. Thickness of MgF_2 and AlF_3 films is controlled with a piezoelectric balance, and thickness of metallic electrodes (gold or aluminum) by measuring their resistance during evaporation.

The structure of the films depends on substrate temperature and condensation rate. If the condensation rate is less than 10 Å s^{-1}, good crystallization is obtained with a substrate temperature greater than 490 K for MgF_2 and 450 K for AlF_3. In the case of MgF_2, a marked preferential orientation appears with the (111) plane parallel to the substrate.[4] However, deposition of MgF_2 at temperature less than 400 K gives amorphous films having a poor adherence to the substrate. AlF_3 films obtained with high substrate temperatures have been found very sensitive to impurities and give frequent short-circuits.[6,7] For these reasons, the electrical properties have been studied only with crystallized MgF_2 and amorphous AlF_3 films.

The composition of these films has been systematically examined with check samples. Two methods of nuclear analysis were simultaneously used: Rutherford backscattering of α particles, and α, X reactions. Each method has a preferential range of applicability, especially for sensitivity and separation power.[5,6] In a previous study, we showed the strong influence of preparation conditions upon the nature and the concentration of impurities in the films. Thus, it was possible to control these conditions, specially to use those corresponding to a minimum impurity concentration. Consequently, the main precautions are:

utilization of powders of great purity, submitted to a pre-treatment under fluorine;

utilization of crucibles and surrounding metallic protection made of molybdenum, with which both fluorides do not react.

Pre-heating of powder under vacuum, limiting the effect of preferential evaporation of the coumpounds which form with the impurity elements.

Under these conditions, the main impurities contained in the films are Na, K, Ca; the concentrations are about 100 ppm. These elements are present in the whole thickness of the films, with a decreasing concentration from the film-substrate interface to the external surface. They are likely oxidized into compounds of the $NaMgF_2$ type for MgF_2 and of the $NaAlF_4$ type for AlF_3, and probably form aggregates in the host material (MgF_2 or AlF_3),[1,2] Moreover, both methods of analysis enable measurement of the ratio fluorine/basic metal. The results for the two materials are given in Table 1, which shows a deficiency of fluorine, particularly at the interface layer-substrate. A gas analysis performed with a mass spectro-

Table 1. Stoichiometry of Mg F_2 and Al F_3 films.

Film thickness (Å)	300	500	1000	3000
Fluorine : magnesium ratio	1.93 ±0.01	1.95 ±0.01	1.98 ±0.01	1.98 ±0.01
Fluorine : aluminum ratio	2.61 ±0.03	2.84 ±0.01	2.91 ±0.01	2.92 ±0.01

meter during evaporation clearly showed the presence of hydrofluoric acid (HF).

2. ac conduction

2.1. Experimental results. Measurements have been carried out for frequencies between 5 Hz and 100 kHz with a resistance-capacitance bridge, General Radio 1616, connected to a variable frequency detector PAR 124. In fact, most of data were limited to 50 kHz, in order to avoid secondary effects due to electrode resistance, which cause anomalous increasing of dielectric losses.

The measured conductivity at ambiant temperature and for weak electric fields is about $10^{-15}\ \Omega^{-1}\ m^{-1}$ for MgF_2 and $10^{-14}\ \Omega^{-1}\ m^{-1}$ for AlF_3.

For an Au–MgF_2–Au structure, Figure 1 gives the variations

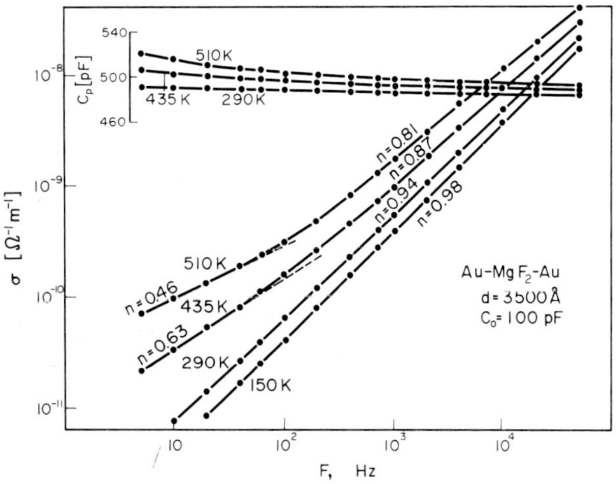

Figure 1. Au–MgF_2–Au structure—variations of conductivity (σ), and capacity (Cp) against frequency for different temperatures.

of conductivity and capacity versus frequency for different temperatures. A good fit is obtained with the general law:

$$\sigma(\omega) = A(T)\omega^n \tag{1}$$

with the exponent n varying from values close to unity at low temperatures to lower values close to 0.8 for highest temperatures. Moreover, in the low frequency and high temperature ranges a further decrease of n is observed.

The curves obtained for Al–AlF_3–Al structures (Figure 2) verify the same general law $\sigma(\omega) \sim \omega^n$, but in this case, values of n are very close to unity, whatever the temperature and the frequency.[8]

The corresponding variations of the capacity are characterized by a weak decrease with increasing frequency; this decreasing is more apparent when the value of n is weak.

The variations of the tangent of the loss angle (tan δ) are represented on Figures 2 and 3.

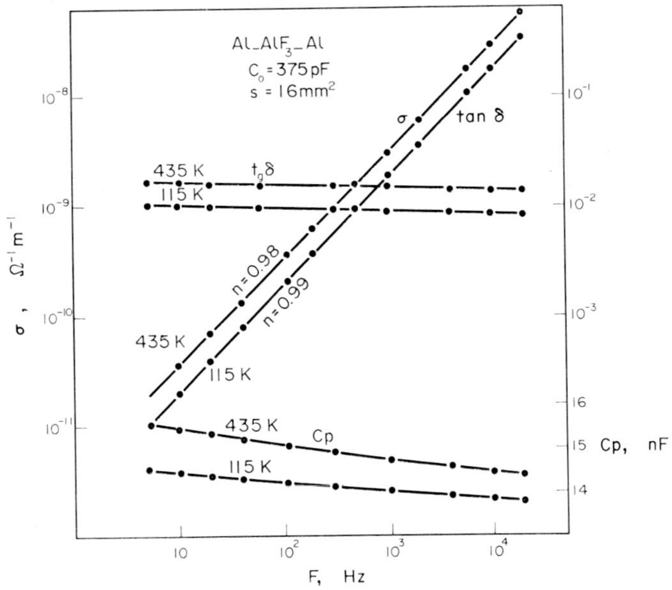

Figure 2. Al–AlF_3–Al structure—variations of conductivity (σ), capacity (Cp) and loss angle tangent (tan δ) against frequency for different temperatures.

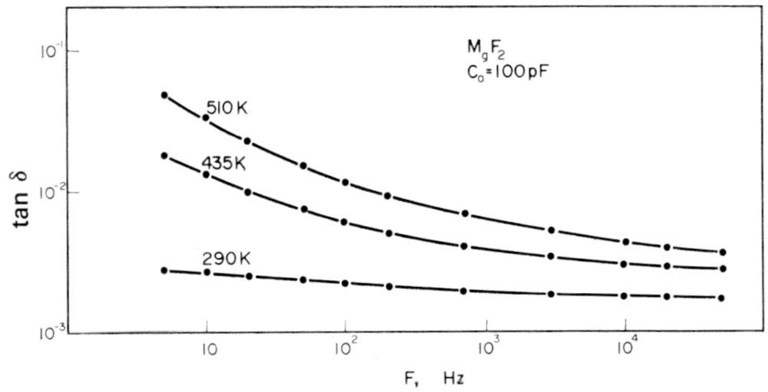

Figure 3. Au–MgF_2–Au structure—curves showing the variations of tan δ versus frequency for different temperatures.

2.2. Kramers–Krönig relations. Taking into account the preceding results concerning the conductivity, the dielectric losses, given by the imaginary part of the permittivity $\varepsilon(\omega)$ or of the complex susceptibility $\chi(\omega)$, are given by the following frequency law:

$$\chi''(\omega) = \varepsilon''(\omega) = B(T)\,\omega^{n-1}\ (n < 1) \tag{2}$$

If the system is linear, the imaginary and real parts of $\chi(\omega)$ are connected by Kramers–Krönig relations. If $\chi''(\omega)$ agrees the law given above, these relations imply:

$$\chi'(\omega) = \varepsilon'(\omega) - 1 = \chi''(\omega)\tan\frac{n\pi}{2} \tag{3}$$

Figure 5a and b, relative to MgF_2 and AlF_3 respectively, show in the considered frequency range a good fit with the relation (3), giving a constant ratio between $\chi''(\omega)$ and $\chi'(\omega)$. Experimental values of $\chi'(\omega)$ have been deduced from capacity measurements, where the part corresponding to 'high frequency' polarization mechanism (giving negligible losses) was excluded.

2.3. Discussion on the basis of the Debye model with a distribution of the relaxation times. First of all, taking into account results obtained in the dc conduction study, which show a

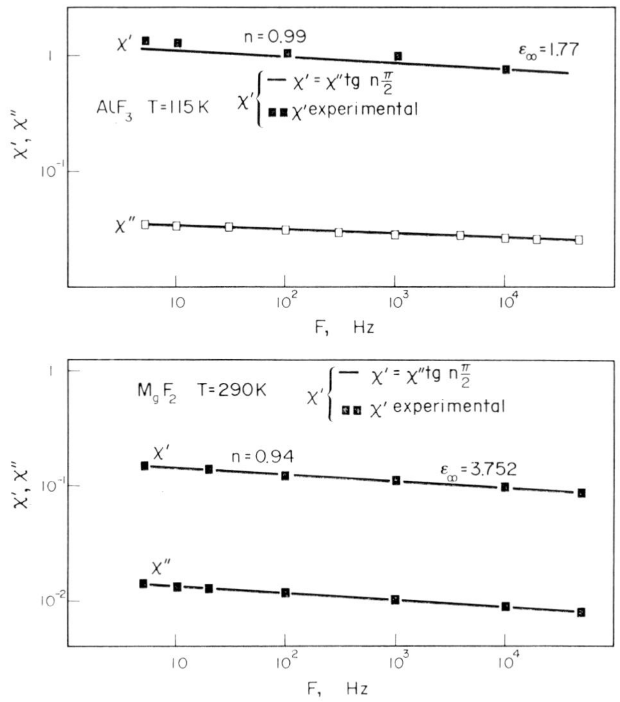

Figure 4. Relation between the real and the imaginary part of dielectric susceptibility (χ' and χ'') deduced from the Kramers–Krönig transformations with the observed law (2) giving the experimental values of χ''.

'hopping' conduction in localized states at low temperatures, the possibility of dielectric losses resulting from the same mechanism has been examined. Pollak and Geballe[12] treated this problem by using an expression of relaxation time derived from the Miller and Abrahams[13] model for hopping conduction, with a random spatial distribution of sites. In the same way, Nott[14] found a conductivity given by the expression:

$$\sigma(\omega) = \log_N{}^m\left(\frac{\Omega_0}{\omega}\right) \tag{4}$$

where Ω_0 is the vibration frequency of phonons, and m an exponent greater than unity. In the simplest case of hopping in isolated pairs of centres, $m = 4$, and $\sigma(\omega)$ can be written in the approximated form:

$$\sigma(\omega) = A\omega^n \text{ with } n \simeq 0.8$$

This value of n is clearly less than those obtained for MgF_2 at low temperatures (Figure 1) and for AlF_3 (Figure 2).

Values of m less than 4 correspond to the multiple hopping hypothesis. Expression (1) is still obtained, but n is closer to unity. However, experimental values for the two materials ($0.97 < n < 0.99$ at $T = 150$ K) lead to an evaluated value of Ω_0 having no physical sense ($\Omega_0 \simeq 310^{25}$ s^{-1} for $m = 1$); therefore, this process of losses does not seem to be preponderant in this kind of material.

It becomes necessary to consider another type of transition and particularly the classical double potential well model. In this case, one considers an electronic or ionic charge hopping over the potential barrier between two energy states separated by a distance r. The relaxation time is generally a function of r, of the potential barrier height E_B and of the energy difference ΔE between the two states. If ν_{12} and ν_{21} are the transition frequencies in each direction, τ is given by:

$$\tau = (\upsilon_{12} + \upsilon_{21})^{-1} \tag{5}$$

Considering a distribution of τ values is equivalent to supposing a distribution of r, ΔE and E_B. The real part of the conductivity can be written, according to Pollak and Pike.[11]:

$$\sigma(\omega) = N.\omega$$

$$\int_\tau\int_{\Delta E}\int_{E_B} \alpha(r,T)\frac{\omega\tau}{1+\omega^2\tau^2}p(r,\Delta E, E_B)\, r^2 \mathrm{d}r\, \mathrm{d}\Delta E\, \mathrm{d}E_B \tag{6}$$

where N is the concentration of the centres and $\alpha(r, T)$ is the elementary polarizability when an electric field is applied. $p(r, E, E_B)$ represents the probability density to find a pair of centres with given parameters r, ΔE and E_B.

The relaxation time can generally be written in the simple form:

$$\tau = \tau_0 \exp\frac{E_B}{kT}$$

where τ_0 is a constant equal to the inverse of the vibration transition frequency between the two sites. In this case, one can remark from relation (6) that if the function $p(E_B)$ is weakly dependent of E_B and r nearly a constant, the general law $\sigma = A\omega^n$ is obtained with a value of n close to unity.

In these conditions, to justify the results concerning AlF_3 ionic vibrations can be supposed on very short distances, around a constant mean value.[11] This hypothesis is taken from a model proposed by Anderson Halperin and Varma concerning anomalous thermal properties of glasses. Pollak and Pike have given a solution of equation (6), the integration of which turning mainly on E_B and ΔE. They obtained:

$$\sigma(\omega) = K\, N\omega T \tan\mathrm{h}\, \frac{E}{kT}\left(\frac{1}{\overline{\Delta E}\cdot\overline{E_B}}\right) \tag{7}$$

where $\overline{\Delta E}$ and $\overline{E_B}$ are the mean values of the amplitudes of the distribution for ΔE and E_B. An estimation of these two values can be deduced from the curves shown Figure 2. Taking an hopping distance of about 2 to 3 Å, we obtain: $N \simeq 10^{+19}$ cm^{-3}, which seems an acceptable value.

In the case of MgF_2, the gradual decreasing of the exponent n with increasing temperature in the middle frequency domain, can be interpreted by considering a model proposed by Pike[17]: this author assumed a coulombic dependence between E_B and the distance r, with limited hops between nearest neighbours. In this model, the quantity $1-n$ is assumed to increase linearly with temperature. However, only a qualitative fit has been obtained with our experimental results.

The origin of dielectric losses in this temperature and frequency domain can arise from: electronic hops between two interacting coulombic wells induced by fluorine vacancies or by impurities (like sodium) present in the films; ionic participation.

It can be noticed that for AlF_3 amorphous films, losses are somewhat larger than for crystallized films of MgF_2.

3. Thermally stimulated depolarization

The M–I–M structures are submitted to a constant electric field under high temperature condition ($T > 500$ K), then they are quickly cooled down to liquid nitrogen temperature. The current obtained with the short-circuited sample is then recorded during reheating. Two main types of polarization can be considered: dipolar polarization; space charge polarization.

In both cases, the expression of the thermally stimulated current is given by the general formula:

$$I = I_0 \exp\left[\left(-\frac{E}{kT}\right) - \frac{1}{B}\int_{T_0}^{T} C(T) \exp\left(-\frac{E}{kT}\right) dt\right] \qquad (8)$$

E is the activation energy of the process and B is the heating speed given by:

$$T = Bt + T_0$$

For a dipolar polarization, where it is possible to define a relaxation time, we have:

$$C(T) = \frac{1}{\tau_0}$$

If a space charge process is considered, the term $C(T)$ depends on the polarization field and on the trap density.

The experimental curves of the current versus temperature for MgF_2 and AlF_3 are represented on Figure 5. For both materials,

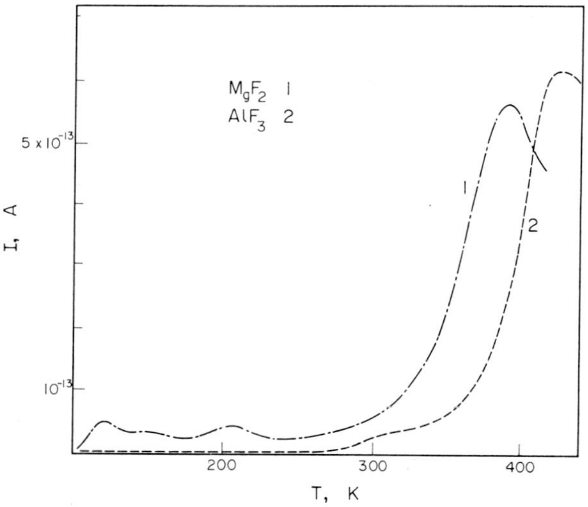

Figure 5. Thermally stimulated currents curves obtained with a MgF_2 layer (curve 1) and a AlF_3 layer (curve 2).

the position of the current maxima is not depending on the amplitude of the applied field. This seems to be an indication that the dipolar mechanism is prevailing.[19] The determination of the energies was carried out with the classical Garlick and Gibson method, based on the study of the slope of the low temperature side of the current peak, and allowing the quenching technique. Moreover, if the second exponential term in expression (8) is neglected, it turns out:

$$\log_N I = -\frac{E}{kT} + \log_N I_0$$

and the dipole density can be evaluated from term I_0. Figure 6 gives the energy repartition obtained in the considered temperature range.

4. Comparison between the two kinds of results—discussion

Comparison of these TSD results with those concerning ac measurements leads to two main remarks:

(1) From Figure 5, the energy levels obtained at high temperature are distributed like a quasi-continuum between 0.7 and 1.3 eV.

On the other hand, energy values E_B corresponding to $\omega\tau = 1$ are given by:

$$E_B = kT \log_N \frac{1}{\omega\tau_0}$$

In the case of MgF_2 (Figure 1), with τ_0 of the order of magnitude 10^{-12} s, one obtains, between 10 and 100 Hz and between 435 and 510 K:

$$0.75 \text{ eV} < E_B < 1 \text{ eV}$$

Thus, the energy distribution in this frequency and temperature domain is in good agreement with the distribution energy levels obtained with the TSD technique. Furthermore, the mean density of states in this region is appreciably larger than the density calculated for the lowest energy levels (Figure 6).

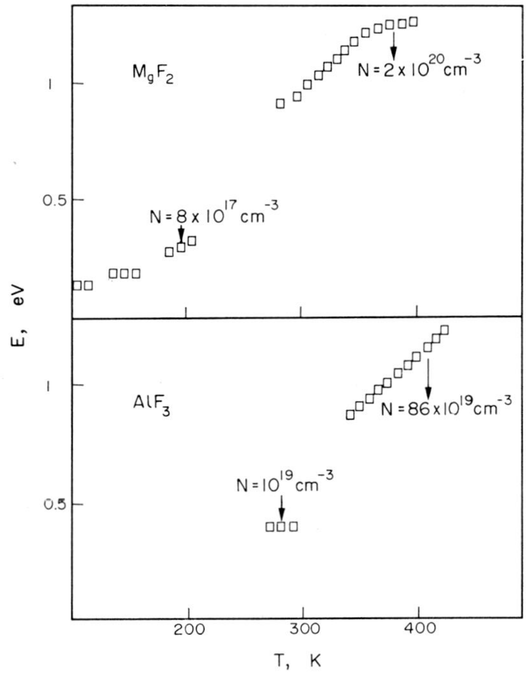

Figure 6. Energy spectrum of dipoles versus temperature for MgF_2 and AlF_3.

This can be connected to an increase of the function $p(E_B)$ defined above. So, the marked decrease of exponent n of the law (1) in the high temperatures and low frequencies range would be explained by a more important density of 'high energy' dipoles. The fact that an ionic dc conduction with an activation energy of 1 eV is observed in the same region of temperature can give an indication of the existence of ionic dipoles in that energy range.

(2) The second remark is more restrictive. As it appears on Figure 6, there are no levels between 0.4 and 0.7 eV. If both methods of measurements are equivalent as far as the nature of dipoles giving the polarization is concerned, this absence of levels involves a difficulty for the existence of a general distribution of relaxation times in the whole range of frequencies and temperatures. However, in the Debye relaxation process, only a large distribution of the relaxation times can justify the experimental law (1).

In these conditions, the non-Debye losses model recently proposed by Jonscher[3] can be considered. In this model, as in the former, a localized charge can hop from one site to another, but this is accompanied by a displacement of the surrounding atoms, which involves a partial screening of the charge. This process is analogous to the formation and transit of a polaron, and leads to a lowering of the carrier energy. Jonscher pointed out that a simple linear relation exists between the energy lost per cycle and the energy stoked corresponding to the polarization of the system. This leads to the expression:

$$\frac{\chi''}{\chi'} = \frac{1-s}{s} = \cot\frac{n\pi}{2}$$

where factor s is less than unity and represents the non-screened part of the charge.

This hypothesis is plausible for MgF_2 layers: the Debye temperature of this material is about 600 K and the evolution of the n value observed on the curves shown in Figure 1 could be related to an increase of screening with temperature.

5. Study of samples with high impurity concentration

The analysis described in paragraph 1 allowed us to control the preparation parameters acting on the nature and the concentration of impurities. By increasing the pre-heating time, and keeping the other factors constant, it was possible to get an increase of the Na, K, Ca concentrations. They become in the range of 1–5% for a pre-heating time of 1 min or less. At the same time, this increase of impurity concentrations involves a more important lack of fluorine, with a concentration of the fluorine vacancies of about 10^{21} cm^{-3}.

For these samples, it is necessary to make a distinction between the measurements carried out directly under the preparation vacuum and those made when the films have been submitted to atmospheric conditions.

The results are the same for both studied materials. In the first case, the ac conductivity increases slightly, and agrees with the law $\sigma = A(T)\omega^n$, with $n \simeq 0.8$. This decrease of n could result from an increase of the electronic participation to the dielectric losses. In the second case, the ac conductivity behaves identically as in the purest films. This effect can be explained by the neutralization of fluorine vacancies by OH^- ions. This hypothesis is correlated to the infrared spectrometry measurements (1).

Conclusion

The dependence of the ac conductivity versus frequency for magnesium and aluminum fluoride thin films is well represented by the classical law:

$$\sigma = A(T)\omega^n$$

In AlF_3 the exponent n is close to 1 for all temperatures and frequencies used. In MgF_2 the same result is only obtained at low temperature ($T \simeq 150$ K with $n = 0.98$) but n decreases as temperature increases. For the highest temperatures, a further decrease of n can be observed at low frequencies. An interpretation based on the Debye model can be proposed, assuming a distribution of the relaxation times. Pollak and Geballe's[12] model concerning hopping of electrons between random localized states cannot explain, in both cases, the observed results. For AlF_3, one can use Pollak and Pike's[12] model, in which the hopping of carriers takes place over potential barriers between two wells, with a constant hopping length. For MgF_2, Pike's[17] model with randomly distributed sites, accounts qualitatively for the results.

The energy distribution of the dipoles has been obtained by the study of the thermally stimulated depolarization currents. For the low frequency range and for high temperatures, the agreement between the ac and TSD results is good. But the energy distribution derived from TSD measurements does not seem to be large enough to agree with an interpretation of the frequency dependence of ac conductivity, based on the Debye model. Another explanation, based on Jonscher's[3] model, can be considered.

The impurity rates affect the ac conductivity only in the case of layers studied directly in the preparation vacuum, and σ is then slightly increased; furthermore, the $\sigma \sim \omega^n$ law is still obtained, but with $n \simeq 0.8$. An increasing of the electronic participation to the dielectric losses, is probably responsible for these effects.

References

[1] Y Danto, Thesis, University of Bordeaux I (1976).
[2] A Barrière, Thesis, University of Bordeaux I (1976).
[3] A K Jonscher, *Nature*, **253**, 1975, 717.
[4] A Barrière, Y Danto and J Salardenne, *Thin Solid Films*, **18**, 1973, 201.
[5] B Saboya, Thesis, University of Bordeaux I (1973).
[6] B Saboya, J F Chemin, J Roturier, A Barrière, Y Danto and J Salardenne, *J Appl Phys*, **8**, 1975, 1008.
[7] J Salardenne, Y Danto and A Barrière, *Ann Chim*, **10**, 1975, 201.
[8] A Barrière, Y Danto, J Salardenne, *Thin Solid Films*, **26**, 1975, 273.
[9] A Erdelyi, *Tables of Integral Transforms*, Vol II, McGraw-Hill, New York (1954).
[10] H Fröhlich, *Theory of Dielectrics*, Oxford University Press, Oxford (1958).
[11] M Pollak and G E Pike, *Phys Rev Lett*, **28**, 1972, 1449.
[12] M Pollak and T H Geballe, *Phys Rev*, **122**, 1961, 1742.
[13] A Miller and E Abrahams, *Phys Rev*, **120**, 1960, 745.
[14] N F Nott and E A Davis, *Electronic Processes in Non-Crystalline Materials*, Clarendon Press, Oxford (1971).
[15] M Pollak, *Phil Mag*, **23**, 1971, 519.
[16] P W Anderson, B I Halperin, C M Varma, *Phil Mag*, **25**, 1972, 1.
[17] G E Pike, *Phys Rev B*, **6**, 1972, 1572.
[18] G F J Garlick, A F Gibson, *Proc R Soc*, **60**, 1948, 574.
[19] G Seve and M Lassabatere, *Thin Solid Films*, **15**, 1973, 285.
[20] A Barrière, Y Danto, J Salardenne, A Schallon, *Thin Solid Films*, **12**, 1972, 539.
[21] P Debye, *Polar Molecules*, Dover, New York (1945).

2.19 Conduction phenomena in $Nb–Nb_2O_5–Au$ thin film structures fabricated by various techniques

E Sciri and R Chabicovsky, *Institut für Industrielle Elektronik, Technische Universität Wien, Austria*

Conduction mechanisms have been studied in niobium–niobium oxide–gold thin film sandwich structures with the oxide prepared by anodic oxidation and reactive sputtering, respectively. Different I–V characteristics have been obtained with diodes having gold electrodes made by both evaporation and sputtering. Significant rectification was found only, when evaporation techniques were used. For this type of diode the $Nb–Nb_2O_5$ interface is generally assumed to act as ohmic contact, whereas the $Nb_2O_5–Au$ interface behaves like a Schottky contact. Conduction mechanisms relevant to the various portions of the I–V characteristics are discussed. Current drifts are also treated, which are attributed to the migration of ionized traps. A theoretical model explaining the drift phenomena is presented.

1. Introduction

It is well known that some MOM (metal–metaloxide–metal) thin film structures show interesting properties like rectification, negative resistance, optoelectronic effects and switching behaviour[1–7]. However, various phenomena are not well understood at present.

Niobium oxide, which has been investigated by several authors[1,2,10,12] is a convenient material for MOM sandwich structures. The purpose of our experimental and theoretical work was to obtain insight into the fundamental conduction processes and drift phenomena in $Nb–Nb_2O_5–Au$ thin film devices manufactured by various techniques.

2. Sample preparation

The glass substrates, 50 × 50 × 0.85 mm, were cleaned ultrasonically by a multi-step procedure. After a 10 min glow discharge treatment in the vacuum chamber a 3000 Å niobium film was sputtered onto the cooled substrates. An argon pressure of 2.10^{-3} mbar and a dc target voltage of 1 kV was used. The oxide films were prepared by both anodic oxidation and reactive sputtering. The anodic oxidation was carried out in a 2.5 % borax solution with anodization voltages of 5–50 V yielding oxide films with a thickness between 150 and 1600 Å[8,9]. Reactively sputtered oxide films were fabricated in an argon–oxygen atmosphere with 10, 20 and 30 % oxygen content. Films in the thickness range 300–900 Å were produced. The top layer was evaporated or sputtered onto the oxide through a mechanical mask. These gold electrodes had a diameter of 2.8 mm and a thickness of about 600 Å. All measurements were carried out in air and at room temperature.

3. Experimental results and discussion

It was found that sputtering of the gold electrodes leads to small or negligible rectification, whereas evaporation yields pronounced rectification. This behaviour was detected not only in structures with anodic oxide (Figure 1), but also in samples with reactively sputtered oxide (Figure 2). The measured I–V characteristics of the two types of structures are qualitatively similar. This indicates that the same conduction process occurs in these samples. The much lower conductivity in reactively sputtered oxide devices is caused by the essentially lower content of impurities acting as donors. The following considerations are valid for both anodic oxide and reactively sputtered oxide devices. Some basic electrical properties of the investigated MOM structures are summarized in Table 1.

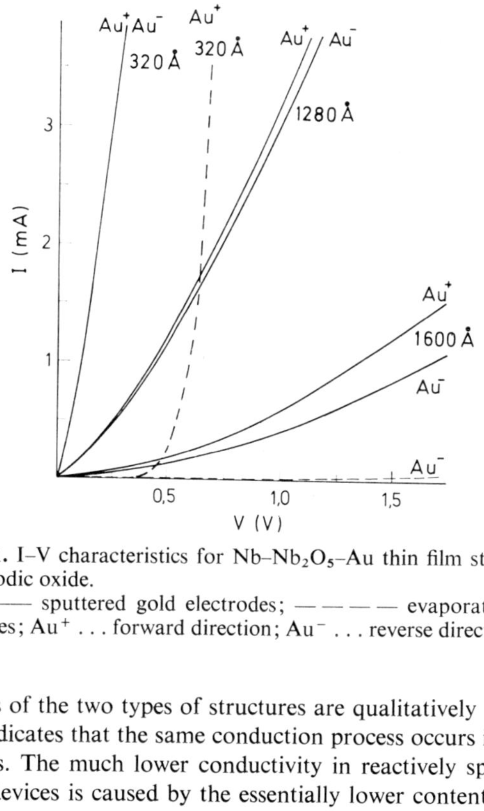

Figure 1. I–V characteristics for $Nb–Nb_2O_5–Au$ thin film structures with anodic oxide.
——— sputtered gold electrodes; — — — — evaporated gold electrodes; Au^+ . . . forward direction; Au^- . . . reverse direction.

3.1. Samples with evaporated gold electrodes. The band diagram of $Nb–Nb_2O_5–Au$ sandwich structures with anodic oxide and evaporated top electrodes was proposed by Hickmott[1] and improved by Schwartz[2] (Figure 3). This diagram gives detailed information about the energy levels and fundamental properties of the interfaces. However, the process of current flow in these structures is not yet completely explained.

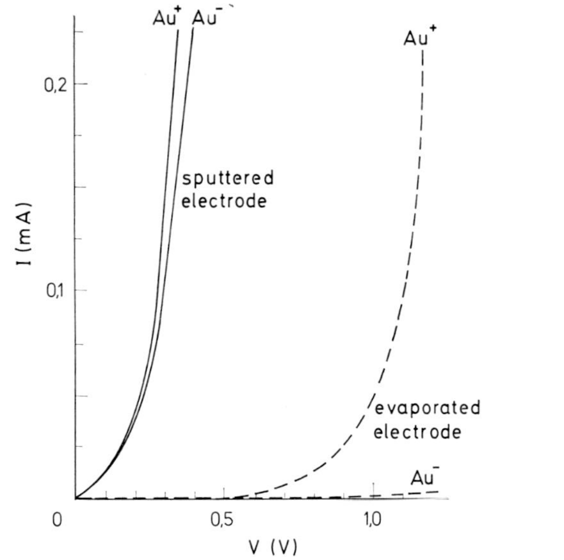

Figure 2. I–V characteristics for $Nb–Nb_2O_5–Au$ thin film structures with reactively sputtered oxide (300 Å).

Table 1. Rectification, drift behaviour and breakdown voltage of $Nb–Nb_2O_5–Au$ structures manufactured by various techniques.

Top electrode fabricated by	Evaporation			Sputtering
Niobium oxide fabricated by	Anodic oxidation	Reactive sputtering	Anodic oxidation	Reactive sputtering
Rectification	Significant	Significant	Small or negligible	Negligible
Current drift	Medium	Strong	Weak	Medium
Breakdown voltage (for $d = 1600$ Å)	>10 V	<2.5 V	>10 V	<2.5 V

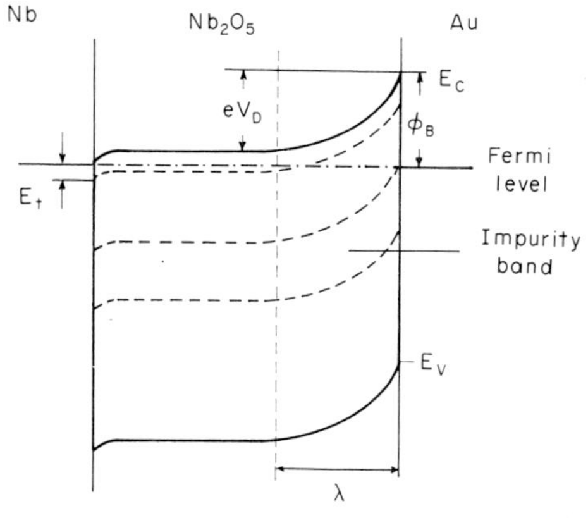

Figure 3. Band diagram for $Nb–Nb_2O_5–Au$ thin film structures. The band gap of niobium oxide is about 3.3 eV and the impurity band is located between 1–1.6 eV below the conduction band edge.[1] $E_t = 0.16$ eV was obtained for the trap level.

The task was to study the conduction mechanisms in the full voltage range for both the forward and reverse direction. In the low voltage range the I–V curves of the samples are similar to those of usual semiconductor Schottky barriers and the current is due to Schottky emission (Figure 4). In the higher voltage range beyond 0.8 V the samples exhibit space charge limitation in the forward direction. However, the Pool–Frenkel mechanism was found to be the dominating process beyond 0.3 V in the reverse direction (Figures 4 and 5).

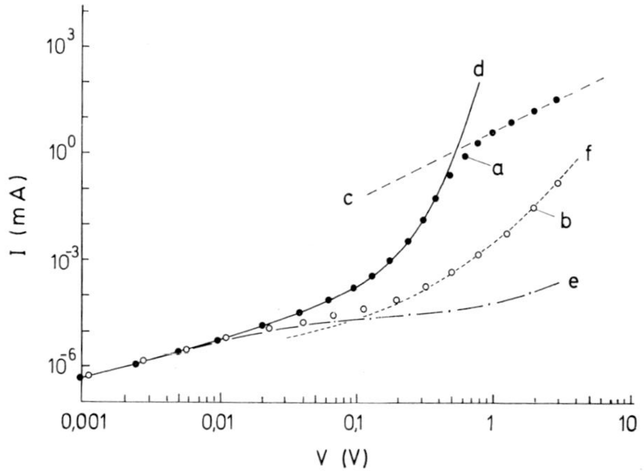

Figure 4. Measured I–V characteristics of a $Nb–Nb_2O_5–Au$ structure with anodic oxide (1600 Å) and evaporated gold electrodes for forward and reverse bias, compared with different conduction mechanisms. ● ● ● ● (curve a): measured forward current; o o o o (curve b): measured reverse current; — — — — (curve c): space charge law (equation 1.3); ———— (curve d): Schottky emission in forward direction (equation 1.1); —·—·—·— (curve e): Schottky emission in reverse direction (equation 1.1); - - - - - (curve f): Pool–Frenkel effect (equation 1.8).

The theoretical curves are based on the following data: $A = 7$ mm², $N_D = 1.9 \times 10^{19}$ cm^{-3}, $N_t = 10^{21}$ cm^{-3}, $E_t = 0.16$ eV, $\phi_B = 0.96$ eV, $V_D = 0.9$ V, $\mu_n = 1$ cm²/Vs,[16] $\varepsilon_r = 44$,[12] $\varepsilon_r^* = 5.76$.[2]

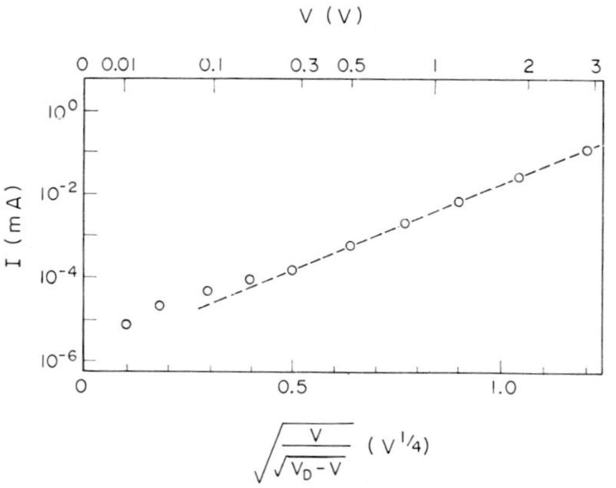

Figure 5. Measured I–V curves of a $Nb–Nb_2O_5–Au$ structure with anodic oxide (1600 Å) and evaporated gold electrodes, for reverse direction. The plot reveals the agreement with the Pool–Frenkel effect for voltages above 0.3 V ($V_D = 0.9$ V).

A better understanding of the different conduction processes in the MOM structures is achieved by dividing the I–V characteristics in the following sections.

Forward bias (gold electrode positive).

(a) Electron emission over the forward biased barrier at voltages up to 0.8 V. The current flow over a Schottky barrier can be expressed by[3,15]

$$I = \frac{4A\pi emk^2T^2}{h^3}\exp\left(-\frac{\phi_B - \Delta\phi}{kT}\right)\left(\exp\frac{eV}{nkT} - 1\right) \qquad (1.1)$$

with

$$\Delta\phi = \left[\frac{e^7 N_D(V_D - V)}{2(8\pi)^2(\varepsilon_o\varepsilon_r{}^*)^3}\right]^{1/4} \tag{1.2}$$

where I is the current, A the diode area, e the elementary charge, m the electron mass, k the Boltzmann constant, h the Planck constant, T the temperature, ϕ_B the barrier height, N_D the donor density, ε_o the vacuum dielectric constant, $\varepsilon_r{}^*$ the high frequency dielectric constant, V the applied voltage, V_D the diffusion voltage and $\Delta\phi$ the barrier lowering by image forces.

In the forward voltage range $\Delta\phi$ is >0 for $0 < V < \phi_B/e$ (barrier exists) and $\Delta\phi = 0$ for $V \geqslant \phi_B/e$ (barrier disappears). Factor n takes into consideration the deviation from ideal diode behaviour caused by recombination in the depletion region ($n = 1 \ldots 2$).

Figure 4 shows very good agreement between the theoretical and experimental I–V curve in the forward bias section below 0.5 V. The evaluation of equation (1.1) was carried out with the following data: $\phi_B = 0.96$ eV, $\varepsilon_r{}^* = 5.76$, $n = 1.98$, $N_D = 1.9 \times 10^{19}$ cm^{-3}, $T = 300$ K, $A = 7$ mm^2. N_D was obtained from the reverse bias I–V curve and factor n from a log $(I/I_o + 1)$ versus eV/kT plot. The value of n very near to $n = 2$ indicates that most of holes recombine with electrons in the depletion region. The existence of holes in niobium oxide and hole injection into the impurity band was previously demonstrated[1].

In the voltage range 0.5–0.8 V the current flow dominated by Schottky emission changes to a space charge limited current.

(b) Space charge limited current at forward bias beyond 0.8 V. When the voltage exceeds 0.8 V, the Schottky barrier disappears and the electric field can be assumed to be uniform in the oxide. Now both interfaces act as ohmic contacts. The $Nb–Nb_2O_5$ interface provides electron injection into the conduction band and the $Nb_2O_5–Au$ interface enables hole injection into the impurity band ('double injection'). However, the holes recombine within a small region in the vicinity of the gold electrode because of the smaller hole mobility. Therefore, in the greater part of the oxide the current is carried only by electrons. It was found that the measured I–V characteristics correspond very well to the space charge law $I \propto V^2$, indicating a single carrier conduction (Figure 4).

With the existence of traps in the oxide, the space charge limited current is given by[13]:

$$I = \frac{9A\mu_n\varepsilon_o\varepsilon_r N_c}{8N_t}\exp\left(-\frac{E_t}{kT}\right)\cdot\frac{V^2}{L^3} \tag{1.3}$$

where N_c is the state density in the conduction band, N_t the trap density, E_t the trap level counted from the conduction band edge, L the oxide thickness, μ_n the electron mobility, ε_0 the vacuum dielectric constant and ε_r the relative dielectric constant.

By using this equation, $E_t = 0.16$ eV was found from log I versus $1/kT$ plots. Then, with $\mu_n = 1$ cm^2/Vs[16], $\varepsilon_r = 44$[12] and $L = 1600$ Å, equation (1.3) yields

$$N_t = 10^{21}\ \text{cm}^{-3}$$

Reverse bias (gold electrode negative).

(a) Schottky emission at low voltages up to 0.1 V. Figure 4 shows good agreement between the measured reverse current (curve b) and the current calculated from equation (1.1) by using the same values of ϕ_B, N_D, $\varepsilon_r{}^*$, n and T as mentioned before. However, at voltages beyond 0.1 V the measured current deviates strongly from equation (1.1). In the voltage range 0.1–0.3 V the process of current flow gradually changes from Schottky emission to the Pool–Frenkel mechanism.

(b) Pool–Frenkel emission at reverse bias above 0.3 V. In the voltage region beyond 0.3 V the thickness λ of the depletion region (barrier) grows distinctly with increasing reverse bias. With donors and traps uniformly distributed in the oxide, λ is given by[15]:

$$\lambda = \left[\frac{2\varepsilon_o\varepsilon_r(V_D - V)}{eN_D}\right]^{1/2} \tag{1.4}$$

where V_D is the diffusion voltage (Figure 3).

A significant electric field exists only in the depletion region. Supposing a uniform field therein, this results in

$$F = \frac{V}{\lambda} = \left[\frac{eN_D}{2\varepsilon_o\varepsilon_r}\right]^{1/2}\cdot\frac{V}{(V_D - V)^{1/2}} \tag{1.5}$$

This field is high enough to activate Pool–Frenkel emission within the depletion region. For amorphous materials the current corresponding to the Pool–Frenkel effect is given by[14]

$$I = I_o \exp\left(\frac{\beta}{kT}\sqrt{F}\right) \tag{1.6}$$

with

$$\beta = \left[\frac{e^3}{\pi\varepsilon_o\varepsilon_r{}^*}\right]^{1/2} \tag{1.7}$$

Combining equations 1.5 and 1.6 leads to

$$I = I_o \exp\left(\alpha\cdot\frac{V^{1/2}}{[V_D - V]^{1/4}}\right) \tag{1.8}$$

where

$$\alpha = \frac{1}{kT}\left[\frac{e^3}{\pi\varepsilon_o\varepsilon_r{}^*}\right]^{1/2}\left[\frac{eN_D}{2\varepsilon_o\varepsilon_r}\right]^{1/4} \tag{1.9}$$

A log I versus $V^{1/2}/[V_D - V]^{1/4}$ plot of measured curves (Figure 5) shows good agreement with theory for voltages beyond 0.3 V, if V_D is assumed to be 0.9 V ($<\phi_B$). The slope of the straight line in Figure 5 yields $\alpha = 9.67$. With $\varepsilon_r = 44$ and $T = 300$ K, equation 1.9 results in

$$N_D = 1.9 \times 10^{19}\ \text{cm}^{-3}.$$

With this value the barrier thickness λ and the electric field F at the onset of the Pool–Frenkel mechanism at 0.3 V have been calculated. Equation (1.4) yields λ ($V = 0.3$ V) = 176 Å corresponding to the data estimated by Hickmott[1]. From equation (1.5) the field strength $F(V = 0.3\ \text{V}) = 1.7 \times 10^5$ V/cm is obtained.

The dominating role of the Pool–Frenkel mechanism for reverse bias beyond 0.3 V indicates, that the current is carried by electrons, which are thermionically emitted from the impurity band into the conduction band. The empty states remaining in the impurity band contribute to the current as hole conduction. Another explanation is, that electrons get into the conduction band by thermally assisted tunneling from the gold electrode and pass through the rest of the barrier by Pool–Frenkel emission.

3.2. Samples with sputtered gold electrodes. These structures show no significant rectification (Figures 1 and 2) indicating that only a very low or negligible barrier exists. It is believed that sputtered gold atoms penetrating into the oxide surface during the sputtering process cause gold doping in the vicinity of the surface. Implantation is not expected for such a low energy range (1 kV) used in the sputtering process. However, Fuschillo *et al*[10] have shown, that penetration of sputtered niobium atoms into niobium oxide is possible by a combined diffusion–implantation mechanism. The current in samples with sputtered gold

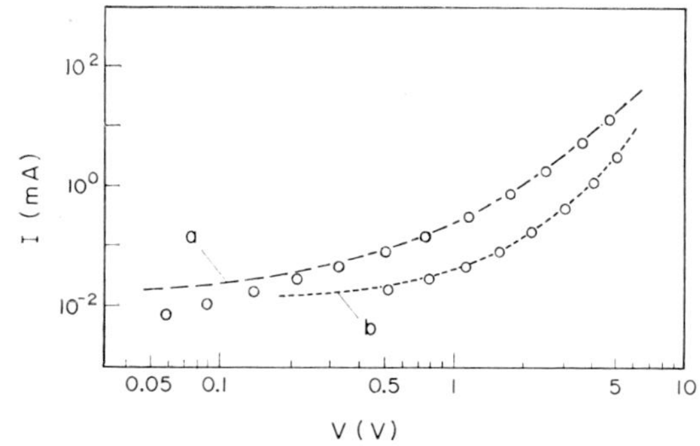

Figure 6. Typical measured I–V characteristics of Nb–Nb_2O_5–Au structures with anodic oxide (1600 Å) and sputtered gold electrodes, corresponding in one case to the Pool–Frenkel effect and in the other case to the Pool mechanism.
○ ○ ○ ○ experimental results; ———— (curve a): Pool–Frenkel effect (equation 1.6); ---- (curve b): Pool mechanism (equation 1.10). The theoretical curves are based on the data $\varepsilon_r^* = 5.76^2$, $K = 0.51\ V^{-1}$.

electrodes is limited by bulk processes in the oxide. Many of the measured I–V characteristics fit to the Pool–Frenkel effect (equation 1.6) both in the forward and reverse direction. In some cases the measured curves follow the relation for the Pool type emission mechanism[14] (Figure 6):

$$I = I_o \exp(KV) \tag{1.10}$$

3.3. Drift phenomena in structures with anodic oxide and evaporated gold electrodes. Many of the recorded I–V characteristics show a significant current drift (Figure 7). In order to get more

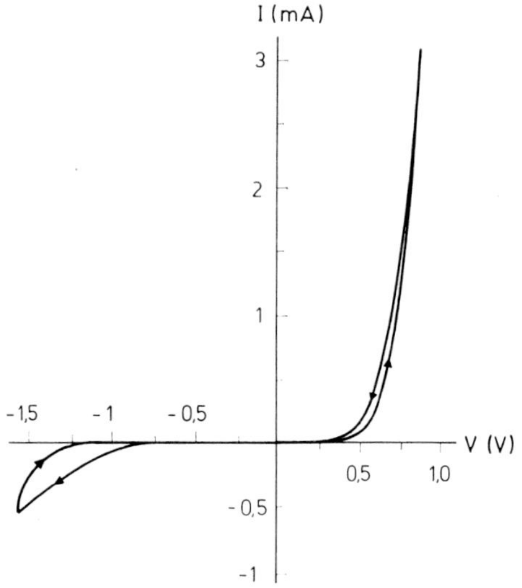

Figure 7. Drift effect in Nb–Nb_2O_5–Au diodes with anodic oxide (1600 Å) and evaporated gold electrodes (cycle time: 4 s).

information on this effect a step voltage was applied to the diodes and the current response was measured (Figure 8). Generally, drift effects could be caused by self heating of the samples. However, in the tested structures a current induced temperature rise was found to be negligible. The authors' explanation for the observed drift is that moving ionized traps influence the electronic conduction process in the oxide. They assume that the migrating traps are gold atoms or other impurities.

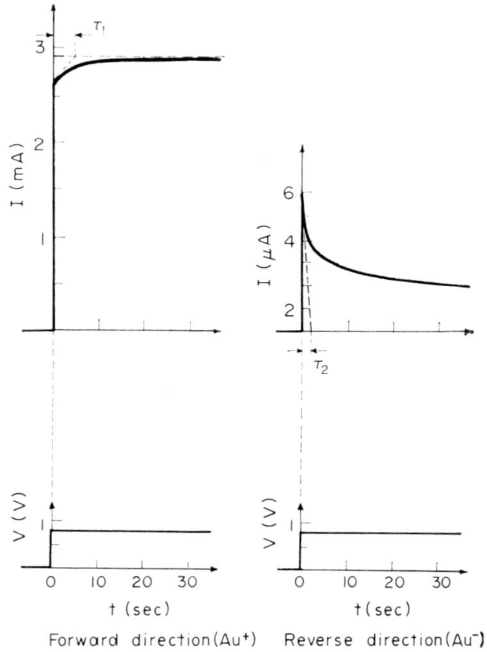

Figure 8. Current versus time curves of a MOM structure with anodic niobium oxide (1600 Å) and evaporated gold electrodes for forward and reverse bias. The voltage was switched on after long turn off periods ($t^* \to \infty$).

In order to obtain a quantitative picture of the observed drift phenomenon, the current response for a forward voltage beyond 0.8 V was calculated. A theory for the reverse bias can be derived in a similar way.

At forward voltages beyond 0.8 V no barrier exists and the electric field is taken to be uniform in the oxide layer. Moreover, the calculation was simplified by considering a trap density averaged over the oxide thickness. When an electric field is applied to the sample in forward direction (gold electrode positive), negative charged traps move slowly towards the positive electrode. Traps, which have reached the electrode, can no more influence the electronic conduction in the bulk of the oxide. Movable and fixed traps are supposed to exist in the oxide. The total trap density is

$$N_t = N_f + N_m(t) \tag{2.1}$$

where N_f is the average density of fixed traps and $N_m(t)$ is the average density of movable ionized traps. When a constant electric field is applied in forward direction, $N_m(t)$ should decrease. A polarization effect is supposed and therefore an exponential decay of the average trap density in the oxide layer is defined:

$$N_m(t) = N_{mo} \exp\left(-\frac{t}{\tau}\right) \tag{2.2}$$

where τ has the meaning of a transit time of movable traps passing through the oxide layer with the thickness L.

$$\tau = \frac{L}{\mu_i F} \tag{2.3}$$

where μ_i is the ion mobility.

When the voltage is turned off, the ionized traps sited at the oxide–gold interface return into the oxide, thus increasing the trap density $N_m(t)$ with the time constant τ^* (Figure 9). Figure 9 shows the trap density as a function of time, when the applied

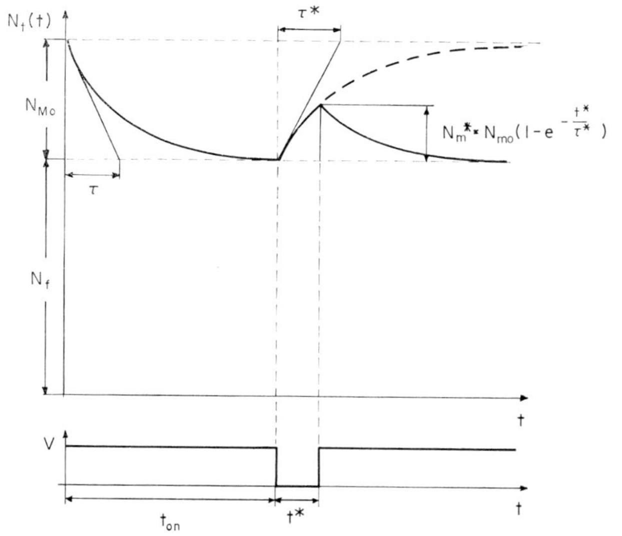

Figure 9. Trap density versus time, when a forward bias voltage is switched on and off ($t_{on} \gg 3\tau$).

voltage is turned on for a long period ($t_{on} \gg 3\tau$), turned off and turned on again. When the voltage is switched on after the turn off period t^*, the density $N_m(t)$ is given by

$$N_m(t,t^*) = N_{mo}\left[1 - \exp\left(-\frac{t^*}{\tau^*}\right)\right]\exp\left(-\frac{t}{\tau}\right) \tag{2.4}$$

The total trap density is

$$N_t(t,t^*) = N_{mo}\left\{a + \exp\left(-\frac{t}{\tau}\right)\left[1 - \exp\left(-\frac{t^*}{\tau^*}\right)\right]\right\} \tag{2.5}$$

with

$$a = \frac{N_f}{N_{mo}} \tag{2.6}$$

The space charge limited current depends on the trap density N_t. Considering that ion migration—compared with the quickness of electron emission processes or thermal velocity of electrons—is a very slow process, equation (1.3) for the steady state space charge limited current is taken and N_t is replaced by $N_t(t,t^*)$ from equation (2.5.) This results in

$$I(V,t,t^*) = \text{const}\frac{V^2}{N_{mo}\left\{a + \exp\left(-\frac{t}{\tau}\right)\left[1 - \exp\left(-\frac{t^*}{\tau^*}\right)\right]\right\}} \tag{2.7}$$

When a voltage step is applied, the current normalized to $I(t = \infty)$ is given by

$$\frac{I(t)}{I(t=\infty)} = \frac{a}{a + \exp\left(-\frac{t}{\tau}\right)\left[1 - \exp\left(-\frac{t^*}{\tau^*}\right)\right]} \tag{2.8}$$

The derivative of this function with respect to time at $t = 0$ is defined as a new time constant τ_1 (Figure 8):

$$\tau_1 = \left[1 + \frac{1 - \exp\left(-\frac{t^*}{\tau^*}\right)}{a}\right] . \tau \tag{2.9}$$

From the experimental drift curves (Figure 8) the following data were found:

$$\frac{I(t=0)}{I(t=\infty)} = 0.9 \quad \text{and} \quad \tau_1 = 4\text{ s } (t^* = \infty).$$

Finally equations (2.8) and (2.9) yield

$$a = \frac{N_f}{N_{mo}} = 10.2 \quad \text{and} \quad \tau = 3.64\text{ s}.$$

With the data used in the experiments ($d = 1600$ Å, $V = 0.8$ V) the mobility of the traps is obtained by

$$\mu_i = \frac{L}{\tau F} = 0.9 \times 10^{-10}\text{ cm}^2/\text{Vs}.$$

This value corresponds very well with $\mu_i = 10^{-10}$ cm^2/Vs previously reported for metal oxides[11].

Moreover, good agreement between measured and calculated time constants τ_1 was found for several turn off periods t^*.

Conclusion

The electrical behaviour of Nb–Nb_2O_5–Au structures with anodic niobium oxide is similar to those with reactively sputtered oxide. However, electric properties depend strongly on the fabrication technique of the top electrodes. Electrical conduction in structures with evaporated electrodes is governed by different mechanisms in different voltage ranges. The I–V characteristics of the investigated samples show that Schottky emission is dominating at forward bias below 0.8 V and at reverse bias below 0.1 V. In the forward voltage range above 0.8 V the current is limited by space charge and in the reverse range beyond 0.3 V the current flow is due to the Pool–Frenkel mechanism. Structures with sputtered gold electrodes exhibit no Schottky barrier. The conduction process for both polarities is attributed to the Pool–Frenkel effect or to the Pool type emission mechanism.

The observed drift effects have been explained by trap migration in the oxide. The measured current response was compared with that calculated from the theory and good agreement was found.

Acknowledgement

This work is supported by the Fonds zur Förderung der wissenschaftlichen Forschung, Austria.

References

[1] T W Hickmott, *Thin Solid Films*, **3,** 1969, 85.
[2] K J Schwartz, Y I Chiou and H Thompson Jr, *Thin Solid Films*, **6,** 1970, 81.

[3] L I Maissel and R Glang, *Handbook of Thin Film Technology*, McGraw-Hill, New York (1970).
[4] W S Chan and C K Loh, *Thin Solid Films*, **6**, 1970, 91.
[5] Y Katsuta, A E Hill, A M Pahle and J H Calderwood, *Thin Solid Films*, **18**, 1973, 53.
[6] B Jankovski, *Proc Int Conf Thin Films*, Budapest, 1975.
[7] G Dittmer, *Thin Solid Films*, **9**, 1972, 141.
[8] F Kover and M J Musselin, *Thin Solid Films*, **2**, 1968, 211.
[9] L Young, *Anodic Oxide Films*, Academic Press, New York (1961).
[10] N Fuschillo, B Lalevic and N K Annamalai, *J Vac Sci Technol*, **12**, 1975, 52.
[11] A T Joffé, *Physik der Halbleiter*, Akademie Verlag, Berlin (1958).
[12] B Lalevic, N Fuschillo and N Annamalai, *Thin Solid Films*, **23**, 1974, 249.
[13] M A Lampert and P Mark, *Current Injection in Solids*, Academic Press, New York (1970).
[14] M H Hill, *Phil Mag*, **23**, 1971, 59.
[15] E Spenke, *Elektronische Halbleiter*, Springer Verlag, Berlin (1965).
[16] Y Sasaki, *J Phys Chem Solids*, **13**, 1960, 177.

2.20 Size effects in thin films of cadmium arsenide

L Żdanowicz, G Pocztowski and W Żdanowicz, *Institute of Solid State Physics Polish Academy of Sciences, Zabrze, Poland*

Electrical resistivity of cadmium arsenide films of different thicknesses in the range 0.02–10 μm were measured in situ. *The results are compared with the simple Fuchs–Sondheimer model and also with Mayadas and Shatzkes model for polycrystalline films. Good fit of experimental data to the M–S model was obtained assuming that the electron mean free path* L = *696 Å for the examined film, that the reflection parameter* k = *0.1 and* p = *0. The electron microscopic images of examined films are discussed in connection with resistivity data. Also the thickness dependence of room temperature effective mobility is compared with the theoretical model.*

1. Introduction

The electrical properties of thin films usually differ markedly from those of bulk material, because of essential differences in the microstructure of thin films and bulk material[1]. Surface states as well as dimensions and shapes of crystallites introduce new scattering mechanisms and consequently make the thin-film parameters depend in some way on thickness.

For several years it has been a common practice to analyse electrical resistivity data of polycrystalline films with the help of the Fuchs–Sondheimer theory (F–S theory) for size effects[2,3]. These authors assumed that in thin polycrystalline metal films the following electron scattering mechanisms were operative: (i) an isotropic background scattering due to phonons and point defects (as in the bulk); (ii) a scattering because of the external film surfaces. For film thicknesses comparable with the mean free path (mfp) of carriers L, the limitation of carrier mobility by diffusive surface scattering becomes important. The F–S model can be summarized by the following equations:

$$\frac{\rho_f}{\rho_0}=\frac{4}{3}\frac{L}{d}\frac{(1-p)}{(1+p)}\left(\ln\frac{L}{d}\right)^{-1} \quad \text{for} \quad d<L \tag{1a}$$

and

$$\frac{\rho_f}{\rho_0}=1+\frac{3}{8}(1-p)\frac{L}{d}, \quad \text{for} \quad d\geqq L \tag{1b}$$

where ρ_f is the resistivity of the film; ρ_0 is the resistivity of the bulk material; d thickness of film, L the mean free path and p is the specularity parameter which represents the probability for an electron to be specularly reflected at the film boundary ($0\leq p\leq 1$). However, the experimental data for resistivity and carrier mobility gives good qualitative but rather poor quantitative fitting to the F–S model.

It has been suggested in recent years that the electrical resistivity of thin polycrystalline films with granular structures is affected not only by isotropic electron and surface scattering mechanisms but also by grain boundary scattering. In order to take this into account Mayadas and Shatzkes[4] calculated the electrical resistivity of polycrystalline and monocrystalline thin metallic films (M–S model) by adding to the two electron scattering mechanisms of the F–S model a third one associated with the presence of crystallite boundaries (M–S model). They assumed a column-like structure for polycrystalline films and assumed that only intergranular planes perpendicular to the applied electric field cause additional electron scattering and that the average interplanar distance is equivalent to the average crystallite size D.

The general M–S equations can be given by:

$$\frac{\rho_f}{\rho_g}=\left[1-\frac{A}{f(\alpha)}\right]^{-1} \tag{2}$$

where

$$A=\frac{6}{\pi k}(1-p)\int_0^{\pi/2}\mathrm{d}\varphi\int_1^{\infty}\frac{\cos^2\varphi}{H^2(t,\varphi)}\left(\frac{1}{t^3}-\frac{1}{t^5}\right)\times\frac{1-\exp\{-ktH(t,\varphi)\}}{1-p\exp\{-ktH(t,\varphi)\}}\,\mathrm{d}t \tag{2a}$$

with

$$H(t,\varphi)=1+\frac{\alpha}{\left(1-\dfrac{1}{t^2}\right)^{1/2}\cos\varphi}$$

ρ_f—the resistivity of film; ρ_g—the resistivity of polycrystalline film infinitely thick; $k=d/L$, d—film thickness, L = mfp of the electrons, p—specularity parameter,

$$\alpha=\frac{L}{d}\frac{r}{1-r} \tag{2b}$$

$$f(\alpha)=1-\tfrac{3}{2}\alpha+3\alpha^2-3\alpha^3\ln\left(1+\frac{1}{\alpha}\right) \tag{2c}$$

and r is the electron reflection coefficient characterizing the electron reflecting power of the potential barrier between crystallites; r can take values from 0 to 1 and make it possible to fit the experimental data to the theory.

When $\alpha=0$, equation (2) reduces to the F–S equation.

If we assume that the crystallite diameter is equal to film thickness ($D=d$), then the ratio between ρ_f and the resistivity of a single crystal film infinitely thick ρ_0 is:

$$\rho_f/\rho_0=[f(\alpha)-A]^{-1} \tag{3}$$

in which A, α and $f(\alpha)$ have the same meaning as in equation (2).

When $k>10$ or $r\to 0$ equation (3) reduces to F–S model. As equations (2) and (3) cannot be evaluated analytically, Mola and Heras[5] have tabulated some linearized expressions in a form convenient for comparison with experiment.

It is of special interest both from a purely scientific and a practical point of view to compare the theoretical models presented above with experimental data for film resistivity because such a comparison can help to elucidate the real film structure. The purpose of the work described here is to explore the relation of film thickness to the microstructure and electrical properties of vacuum deposited cadmium arsenide thin films[6–8].

2. Experimental details

Thin Cd_3As_2 films have been obtained by direct thermal deposition of the compound in a vacuum of 2–8 $\times$ 10^{-6} torr[6] on to mica cleaved in air, optically polished glass or Si substrates. The temperatures of the substrates during deposition were $T_s \simeq 430 + 480$ K; the deposition rates, $\nu = 0.5 + 200$ Å/s. The film thickness range under investigation was $d = 0.02 + 10$ μm. In order to derive meaningful conclusions from the experimental data, a number of experiments with film resistance measurements were carried out in very clean conditions in vacuum chamber *in situ* during the process of film deposition. The measured values of resistivity were recorded with an X–Y recorder. The film thickness in such cases were evaluated by measuring the final thickness of the film with an interferometer and also from the deposition time, assuming a constant deposition rate. This relation was confirmed by the authors in several independent experiments for the thickness range under investigation. The carrier mobility of the films was measured at room temperature after removing the film from the vacuum. The structure and microstructure of the films were examined using

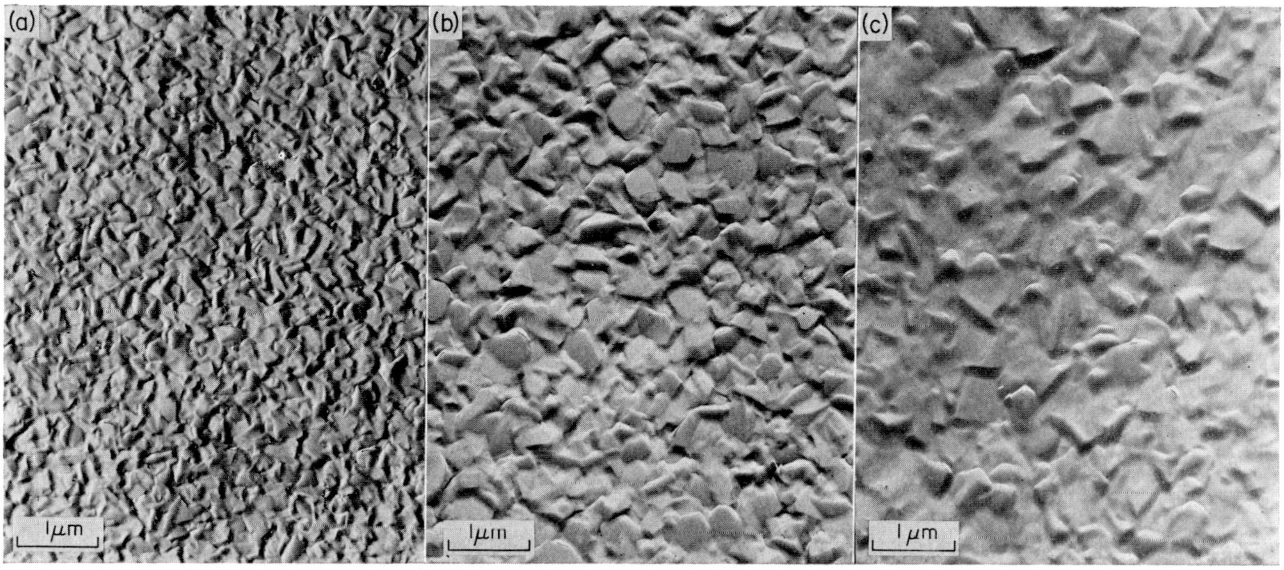

Figure 1. Electron microscopic images (replicas) of polycrystalline Cd_3As_2 films on mica; temperature of substrate $T_s = 443$ K, deposition rate 20 Å/s; (a) $d = 0.72$ μm; (b) $d = 1.2$ μm, (c) $d = 2.4$ μm.

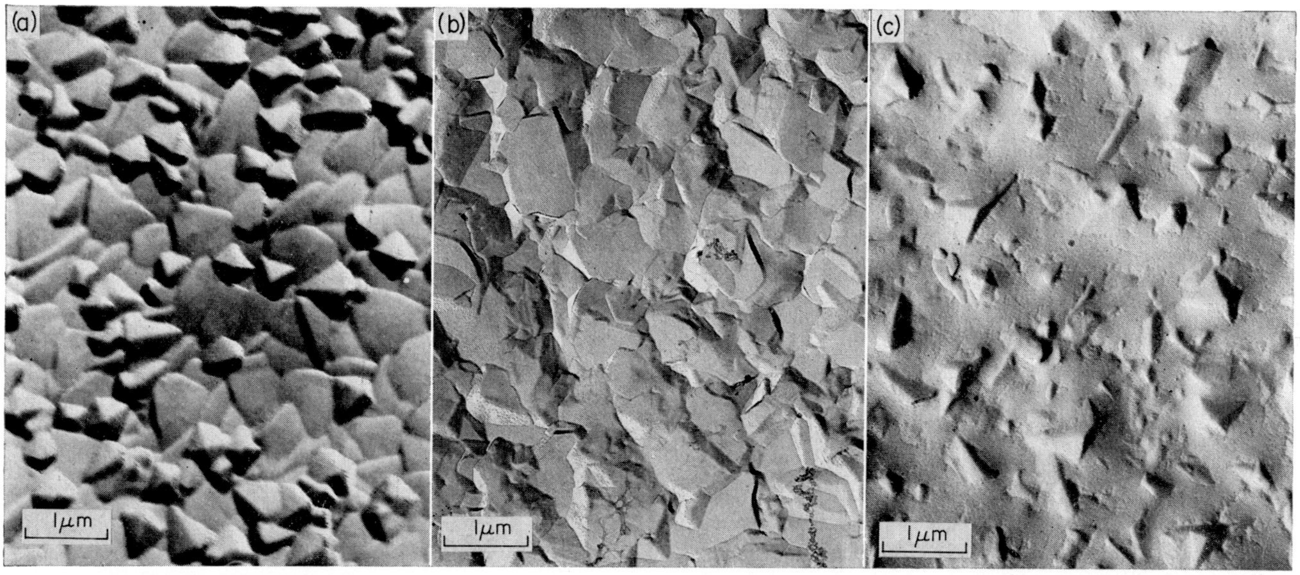

Figure 2. Electron microscopic images (replicas) or polycrystalline Cd_3As_2 films on mica; $T_s = 460$ K; deposition rate 60 Å/S, (a) $d = 3.87$ μm; (b) $d = 5.60$ μm; (c) $d = 8.6$ μm.

Figure 3. Microstructures of polycrystalline Cd_3As_2 films (replicas); (a) on Si substrate, $T_s = 490$ K, deposition rate 20 Å/s; $d = 2.65$ μm. (b) Film presented in Figure 2(a) after annealing; $T_{an} = 750$ K.

Jeol JEM–120 electron microscope, with the help of electron diffraction and of platinum shadowed carbon replicas (Figures 1–3). Sometimes the deposited films were annealed at 750–770 K in air after covering the film with a SiO_x protective layer (called 'protected recrystallization') in order to complete the crystallization process inside the film[8] (Fig. 3(b)). In this paper the results of resistivity measurements are given for as-deposited films, only.

3. Structure of films

The films obtained under the conditions described above were polycrystalline or textured[7]. The electron-microscopic images of the microstructure of these films are presented in Figures 1–3 at different stages of film growth. As was pointed out in our earlier work[9], during the initial stages of growth separate crystalline islands were formed; they grew in size and then coalesced by grain boundary migration. This process occurs very rapidly during the first stages of growth and then it disappears[9]. Subsequent growth of grains occurs as a result of a reduction of surface roughness by surface diffusion of the atoms to 'channels' between the crystallites (Figure 1(a), (b), (c)). The replicas of film cross-section confirm that the crystallites grow as 'columns' from the substrate to the film surface. At further growth, the density of the 'column aggregates' diminishes mainly as a result of the columnar crystallites aggregating into larger single-blocks (Figure 2(a), (b), (c)). At higher temperatures and on a suitable monocrystalline substrate (Si) the coalescence results in the production of large crystallites having shapes close to the equilibrium form (Figure 3(a)).

Figure 4 shows the dependence of the average crystallite diameters D as function of as-deposited film thickness d; the value of D was experimentally evaluated from the electron-microscopic images. For larger film thicknesses ($d \geqq 6$ μm), D tends to saturate, while for small grain sizes ($d \leq 1.2$ μm, see inset in Figure 4), D appears to be a linear function of film thickness.

The crystallite sizes after 'protected recrystallization' of the film at 750 K markedly increase (~4 μm in Figure 3(b)) and

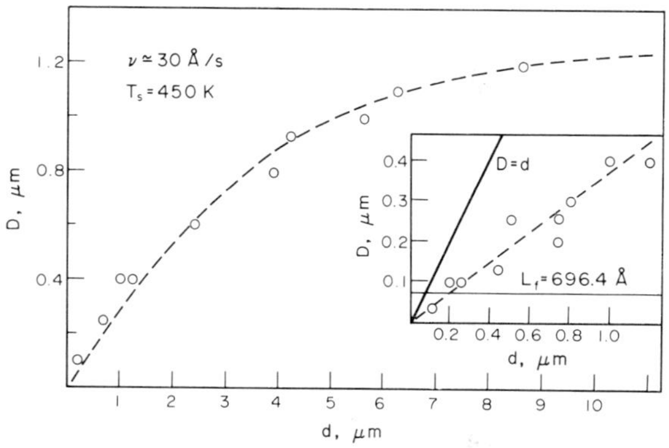

Figure 4. Average crystallite diameter D as function of film thickness for several Cd_3As_2 films ($T_s \simeq 450$ K).

suggest that during the high-temperature recrystallization process, the film undergoes a liquid-like state (melting point for bulk Cd_3As_2 $t_m = 721$°C).

4. Electrical properties of the films

Representative results of *in situ* measurements of resistance R for Cd_3As_2 film obtained during the growth of the films are presented in Figure 5. The dependence $R(d)$ plotted on a log–log scale clearly reveals two linear regions, the first characterized by a slope with $\tan \alpha_1 \simeq -2$ and the second by $\tan \alpha_2 = -1$.

This is very clear experimental support for the two dependences $R = f(d)$ predicted in the F–S model and shown in equations (1a) and (1b). According to these relations the resistance of a film of a length a and width b should vary with the film thickness d as follows[10]:

$$R(d) = \rho_0 \frac{4}{3} \frac{aL}{bd^2} \left(\frac{1-p}{1+p}\right) \left(\ln \frac{L}{d}\right)^{-1}, \quad \text{for} \quad d < L \tag{4a}$$

and

$$R(d) = \rho_0 \frac{a}{bd} \left(1 + \frac{3}{8}(1-p)\frac{L}{d}\right), \quad \text{for} \quad d \geq L \tag{4b}$$

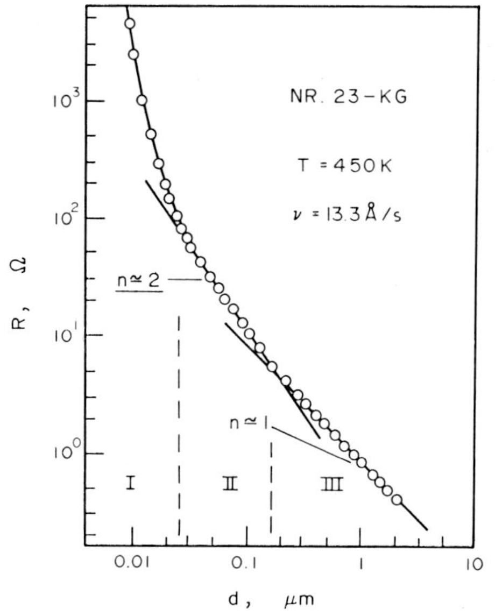

Figure 5. Thickness dependence of resistance R for film $N_0 = 23$-KG, measured *in situ* in the process of film deposition. $T_s = 450$ K.

Assuming for Cd_3As_2 $m^* = 0.05\ m_o$ and $\mu_b = 10^4\ cm^2/V\ s$ we obtain:

$$L_b = (3\mu/4e)(2\pi m^* k_B T)^{1/2} = 1640\ \text{Å} \qquad (5)$$

at 300 K.

Thus, according to the F–S model the film thickness during film growth is proportional to d^{-n} with $n = 2$ for $d < L$, at the early stages of film growth and $n = 1$ for $d \geq L$ with increasing thickness of film.

The transition from the former to the latter dependence occur at a film thickness in the range 1000–1500 Å. The same dependence with the same value for the transition thicknesses were obtained in[10] for antimony films.

The same dependence is also presented in Figure 6 as an experimental relation of the ρ_f/ρ_o ratio versus the $k = d/L$ at $T = 450$ K. For comparison, the theoretical curves calculated according to the F–S and M–S models are also given.

In the calculations, the mean free path L was evaluated from

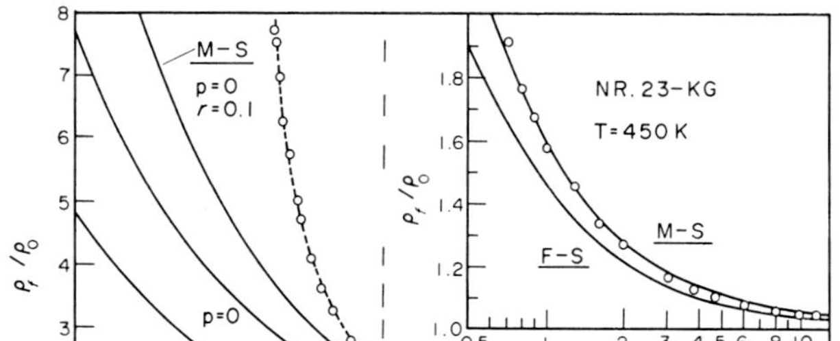

Figure 6. The dependence of ρ_f/ρ_0 on film thickness for film presented in Figure 5.
F–S curves were calculated according to Fuchs–Sondheimer theory (equation 1), M–S curve—according to Mayadas–Shatzkes theory assuming $d = D$ (equation 3) .[5]

relation (5) assuming that the electron mobility of the examined film $\mu_f = 3470\ cm^2/V$ s at 450 K and that ρ_o is the resistivity value for the final thickness $d = 10\ \mu m$ (at the end of deposition). The calculated mfp value was $L_f = 696$ Å. The M–S model calculations were performed according to equation 3 using the data tabulated by Mola and Heras[5] for the reflectivity parameter $r = 0.1$ and $p = 0$ assuming $D = d$.

It is also possible to calculate the room temperature effective carrier mobility μ_f assuming that the current carriers in both film and bulk are electrons and that their densities are identical[6]; then $\sigma_f/\sigma_b = \mu_f/\mu_b$[11].

The results are presented in Figure 7.

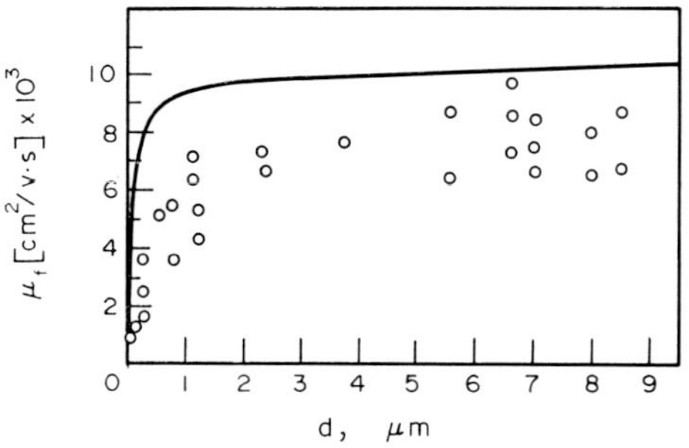

Figure 7. Thickness dependence of effective electron mobility for Cd_3As_2 films (experimental points). Continuous curve was calculated according to F–S model assuming $n = 2 \times 10^{18}\ cm^{-3}$ and $\mu_b = 10^4$ cm^2/V s.

5. Discussion

It is evident from the data in Figures 6 and 7 that the film resistivity ρ_f and the effective carrier mobility μ_f depend more strongly on film thickness than predicted by the F–S theory. Observation of the actual microstructure of the films makes it possible to seek correlation between the structure and the electrical properties of the films.

For better interpretation of results it is useful to divide the $R(d)$ curve in Figure 5 into 3 regions[10,13], which correspond to the same thicknesses in the ρ_f/ρ_o curve in Figure 6.

It is clear, that the transitions between the regions are connected with changes in the mechanism of film growth. The sharp change of resistivity in the first region where $d < 300$ Å corresponds to a discrete, not completely continuous, structure of the film during the initial stages of growth. The average film thickness is less than mfp of electrons ($k = d/L < 0.5$). In the second region ($0.5 \leqq d/L \leqq 1.5$) the coalescence process and filling of 'channels' is occurring and the film becomes continuous. The fit of the experimental data to the F–S theory is better in this region.

The best fit experiment with both theories is observed in the third region $|d/L \gg 1|$. Some deviations of experimental points from the F–S curve in Figure 6 can be attributed to additional scattering of electrons at the intercrystalline barriers. The fit of experimental points to the M–S curve in Figure 6 in the 2nd and 3rd regions is excellent. It is therefore evident, that the assumptions of the M–S model are applicable to the polycrystalline structures described above. The values $k = 0.1$ and $p = 0$ (completely diffusive scattering of electrons at the outer surfaces) results in satisfactory agreement with the theory in both regions.

In the first region, where the film is not continuous and the 'physical origin' of grain boundaries is different, the value of the k parameter characterizing the electron scattering mechanism at the grain boundaries must also be different because the mean free path of electrons in this region changes as a result of very small crystallite sizes.

A similar disagreement with the M–S theory in this region was also obtained by Mola, Borrajo and Heras[12] for cobalt films. It seems, however, that neither the F–S nor M–S theories are applicable for the first region. However, it is interesting that both theories prepared for the case of metal films with simple structure and spherical Fermi surfaces can be applied to this complex semiconductor compound. It must be pointed out that the Fermi surface for Cd_3As_2 is also nearly spherical (one ellipsoid with axes ratio 1.18[15]) and the conduction band is approximately parabolic.

For the very thin Cd_3As_2—films of thicknesses comparable with the de Broglie wavelength ($d \sim 1000$ Å) and of particularly high quality (rate of deposition $\nu \sim 10$ Å/s, extremely) high mobility of electrons $\mu_H \simeq 1000$ cm^2/V s for $d \simeq 0.1$ μm the quantum size effect superimposes the classical size effect described above. This shows up as the oscillations of resistivity and carrier mobility values depending on the film thickness. The oscillation period experimentally observed was $\Delta d \simeq 100$ Å. The details of this effect in Cd_3As_2 films were described elsewhere[14].

Acknowledgements

Our thanks are due to Mr M Niedźwiedź for his help in electron microscopic observations and to Mr J Korona for his valued assistance in the experiment. The financial support was from Warsaw Institute of Physics Polish Academy of Sciences.

References

[1] A Many, Y Goldstein and N B Grover, *Semiconductor Surfaces*, North-Holland, Amsterdam (1971).
[2] K Fuchs, *Proc Camb Phil Soc*, **34,** 1938, 100.
[3] E H Sondheimer, *Adv Phys*, **1,** 1952, 1.
[4] A F Mayadas and M Shatzkes, *Phys Rev*, **B1,** 1970, 1382.
[5] E E Mola and J M Heras, *Electrocomp Sci Technol*, **1,** 1974, 77.
[6] L Żdanowicz, *Acta Physica Pol*, **31,** 1967, 1021.
[7] L Żdanowicz and S Miotkowska, *Thin Solid Films*, **29,** 1975, 177.
[8] W Żdanowicz and L Żdanowicz, *Exper Tech Phys*, **18,** 1970, 185.
[9] L Żdanowicz, S Miotkowska and M Niedźwiedź, *Thin Solid Films*, **34,** 1976, 41.
[10] A H Abou El Ela and S Mahmoud, *Appl Phys*, **7,** 1975, 45.
[11] H H Wieder, *J Vac Sci Technol*, **9,** 1972, 1193.
[12] E E Mola, J Borrajo and J M Heras, *Surf Sci*, **34,** 1973, 561.
[13] W N Wigdorovich, G A Ukhlinow and D W Yakashvili, *Soviet Solid State Phys*, **16,** 1972, 2744.
[14] L Żdanowicz, W Żdanowicz and G Pocztowski, *Thin Solid Films*, **28,** 1975, 345.
[15] I Rosenman, *J Phys Soc Japan Suppl*, **21,** 1966, 370.

2 Electrical properties of thin films—special topics

2.21 Epitaxial germanium–silicon structures obtained in high pure vacuum

L N Aleksandrov and R N Lovyagin, *Institute of Semiconductor Physics, Academy of Sciences of the USSR, Novosibirsk, USSR*

The main aspects of the formation of epitaxial germanium–silicon structures by condensation in UHV by the methods of molecular beams and ion sputtering are considered. The influence of the conditions of preparation of the substrate surface and of the condensation on a film structure and film-structure interfaces are studied. The role of carbide particles and pseudomorphism in the formation of the film structure and surface morphology is shown. The conditions for obtaining an atomically clean silicon surface without carbide particles are determined. Data on electrophysical properties of Ge films and junctions, in particular p–n *junctions which are formed on the silicon surface, are given. The concentration of carriers (electrons or holes in the film ranged from* 10^{16} *to* 10^{18} cm^{-3}*, the mobility from 600 to 1000 (293 K), and from 2000 to 6000* cm^2/Vs *(77 K). The structures had a maximum of spectral photosensitivity in the region of 0.9–1* μm*.*

1. Introduction

Junctions of epitaxial pairs of semiconductors with different widths of a forbidden band or with deep impurity centres in the *p*- or *n*-regions of homojunctions are of great interest for producing a base for developments in micro- and optoelectronics.

The processes of obtaining and investigating properties of many heteroepitaxial systems have been studied in many works, some of which discussed by Milnes and Feucht[1] as well as in other surveys[2,3].

The choice of vacuum methods of epitaxy is influenced by many factors: the possibility to prepare atomically clean crystal–substrate surfaces and their preservation in the UHV up to the moment of vapour supply, the small delay in the transition from one condensable material to another, the possibility of mass-spectrometric control of vapour composition, the reduction of epitaxy temperature, which permits provision of sharper transitions in doping impurities. Investigations, carried out recently[3–11], allowed one to obtain some ideas of the influence of residual gaseous components in the vacuum on the substrate surface condition (cleanliness, orientation, morphology) and of the temperature on the mechanism of growth of homoepitaxial layers in vacuum, their structure and properties. It was shown that the layer-by-layer growth of silicon in the case of homoepitaxy is possible immediately after supplying the vapour to an atomically clean substrate surface of silicon in the case of ion sputtering[3,7,8]. It is also known[9–11] that, even in the case of homoepitaxy, carbide contamination on the substrate surface before condensation results in the formation of three-dimensional growth islands, which coalesce at later stages of condensation and after that the layer-by-layer growth is observed.

In the present paper we consider the regularities of formation of heteroepitaxial semiconductor structures in vacuum, using the Ge–Si system with a large (about 4%) mismatch of lattice parameters, as well as the conditions of formation of structures with prescribed electrophysical and photoelectric properties, permitting the production of photodiodes with a wide range of spectral sensitivity.

2. Experimental

Wafers of *n*-silicon ($\rho = 0.3$, 7.5 or 15 ohm cm) of $20 \times 5 \times 0.4$ mm^3 size, oriented along (111) $\pm 3°$ after chemicomechanical polishing, were carefully degreased and placed between tantalum contacts in a vacuum chamber, which were annealed in high pure vacuum at 1300–1400°C. Then low temperature (200°C) heating of the glass vacuum chamber (for 3 h) followed at a pressure of 10^{-6} torr, which was achieved with the aid of a diffusion pump filled with oil VM-5. Oil vapours were cut off with the aid of three successive traps with liquid nitrogen. The experimental installation was shown in[12] a flange joint with aluminium seal, surrounded by a groove, in which the pressure of 10^{-3} torr is produced, provides a reliable and handy joint. The residual atmosphere in the vacuum chamber was analysed with an Omegatron (IDPO-1) mass spectrometer which permitted measurement of the partial pressure of gas components with mass numbers of 2 to 100 at the level of 5×10^{-11} torr. The mass spectrum of residual gases in the vacuum chamber 3–4 h, after switching on the system is shown in Figure 1. With silicon being sputtered on the chamber walls, oxygen is reliably gettered up to the level of 5×10^{-11} torr. In this residual atmosphere the silicon substrate was heated at 1300°C for 10–20 min. After the

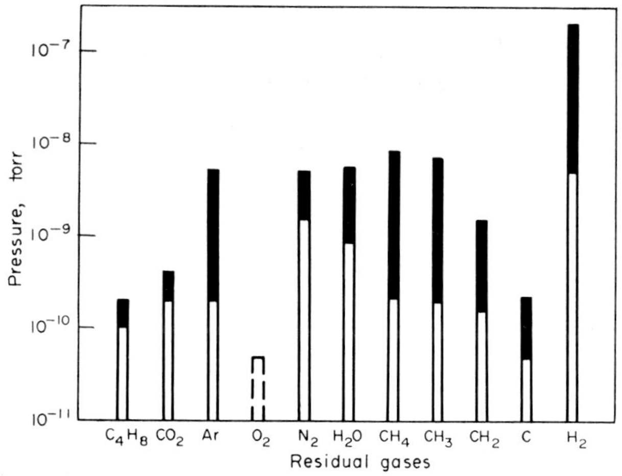

Figure 1. Partial pressures of the residual gases in the vacuum chamber before epitaxy with switch on (black columns) and switch off heaters.

temperature decreased to that required for epitaxial growth, a molecular flow of germanium was supplied to the silicon substrate surface. In the case of ion sputtering with an independent discharge in especially pure argon, the voltage was supplied to the target, which was cleaned beforehand, before admitting argon into the chamber. At the potential difference used between the target and the substrate, the electric field strength was about 1 kV/cm. After applying the potential, argon was let into the chamber up to 5×10^{-4} torr for 1–2 s. Other characteristics of experiments on ion sputtering of germanium on silicon were described previously[14]. The substrate temperature during epitaxy (T_s) ranged from 480 to 700°C, film thickness (d_f) from less than 10 Å to 3 μm, and the rate of film deposition (R_d) was about 7 Å/s for ion sputtering and about 50 Å/s for thermal evaporation of germanium. The layout of the apparatus used is shown in Figure 2. Generally the time necessary for an experiment amounted to 6–7 h.

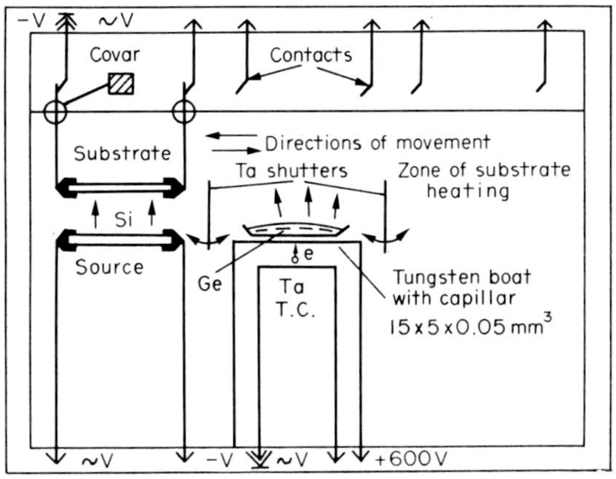

Figure 2. Schematic diagram of the apparatus for condensation of germanium on silicon. Ta tc thermocathode (foil $15 \times 4 \times 0.012$ mm^3).

3. Results and Discussion

Figure 3 depicts the microrelief of a silicon surface, oriented near {111} with some deviations from this plane within ±3° along directions, which do not coincide to any extent with ⟨112⟩. Such a microrelief results from sublimation of surface silicon layers at about 1300°C for 10–20 min. It is easy to notice that the silicon surface contains a considerable density of trihedral angles at steps (the mean step height is 30 Å, their directions are close to ⟨110⟩). These angles are the most probable sinks for adatoms into a chemisorbed state, since the distance between them is less than a length of surface diffusion of adatoms[8,9]. However, such a homogeneous surface is not always obtained. Previously, when heating in vacuum, whose residual atmosphere contained traces of oxygen, we and other experimenters[6,7,11] observed triangular etching pits, whose dimensions depended on the partial pressure of oxygen. Carbon, migrating on the surface, gathered readily at the bottom of the pits and near them. After reliable gettering of oxygen to a level less than 5×10^{-11} torr, the etching pits were not observed after heating. One more factor which prevents obtaining a surface with the regular structure connected with the crystallographic peculiarities and crystal orientation, are carbon and carbon-containing components on the substrate surface, after processing the substrate in air. If the substrate temperature is slowly increased over 800°C, carbon atoms together with silicon adatoms are free to migrate along singular terraces of the surface, assembling in high-energy places, microregions with a great number of quasi-unsaturated bonds. These are the places, where silicon carbide pins are formed, which are so-called pinning points of etching steps. According to our observations, the density of such carbide particles depends considerably on the quality of the surface polishing. As in the experiments with substrates where the quality of polishing of both sides was different, the surface with the worst treatment contained a number of pins, which was many orders of magnitude higher than the density of particles on the surface with good treatment. The density of pinning points, which could be realized at heteroepitaxy by us, ranged from 10^2 to 10^5 cm^{-2}.

In the region of localization of a carbide particle, one should expect great stresses, which can result in formation of dislocations. But in any case the structures of the thermal etching steps on both sides of the carbide particle differed considerably. One of the regions, neighbouring this carbide particle, contains a great number of kinks on steps. As will be shown below, the origin of germanium islands, limited by rapidly growing faces, occurred in this region.

Surface morphology and structure of continuous germanium–silicon films. Films with $d_f \geqslant 400$ Å obtained at a relatively low substrate temperature ($T_s = 480$°C) had a surface, on which one could observe large regions with a terrace-shaped relief, arising after coalescence of 'islands'. The relatively smooth surface remained till $T_s = 650$°C. Figure 4 shows the relief of a rather thick germanium film ($d_f = 3$ μm) on silicon obtained at $T_s = 590$°C. The step height was about 200 Å, the distance between them about 0.5 μm. Growing films in the temperature range of more than 650°C, one could observe an enlargement of the sandwich surface relief with formation of neighbouring large polyhedrons of about 1 μm size. A similar change of the surface morphology occurred after heating of the substrate-film sandwich, obtained at $T = 480$°C, up to 700°C for several minutes[12].

It is interesting to study what happens to silicon on the substrate–film interface with an increase of temperature. The silicon surface morphology after selective etching of the germanium film in a mixture of hydrofluoric acid and hydrogen peroxide, in which silicon was not etched, was studied for this purpose. After

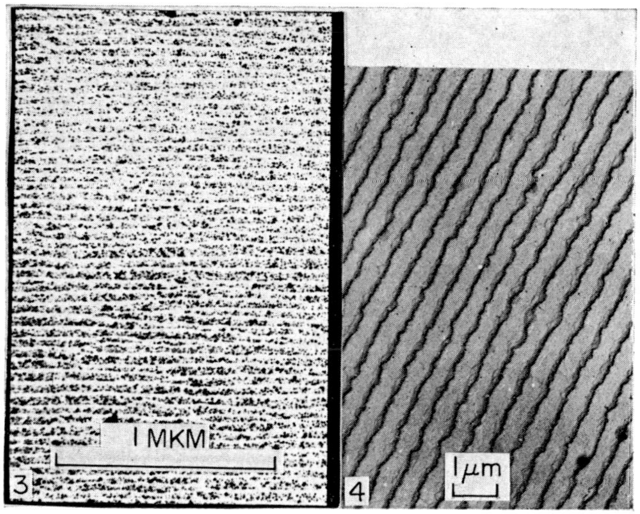

Figure 3. Relief of the silicon surface after heating in vacuum at 1300°C, 10 min, replica with small-angle shading by gold. Orientation along (111) ± 3°.
Figure 4. The surface relief of thick germanium film (3 μm), T_s = 590°C.

removal of the film grown at T_s = 480°C one could observe a step-by-step relief similar to that which the pre-epitaxial surface showed before on the silicon surface (Figure 5a). With an increase of temperature of growth or heating of already grown low-temperature films up to 650°C and above, the silicon surface under the film broke somewhat, and grooves of about 30 Å depth in the form of a grid with hexagonal cells (of about 1 μm size) arose (Figure 5b, c). It is known that at temperature less than 600°C plastic flow of silicon is negligible, so usually it deformed elastically until damage occurred[15]. The emergence of the groove grid is probably facilitated by the plastic flow of silicon in the range of temperature above 600°C and connected with stresses at the interface under consideration.

The dislocation density of the films of more than 0.2 μm thickness (T_s = 650°C) reached 10^{10} cm^{-2}. In this case, there were practically no microdoubles and packing defects in the film. The measured moiré period amounted to 50–70 Å and was independent of the thickness of the silicon substrate under the film.

Peculiarities of building-in of adatoms with carbide particles into the crystal lattice on the surface. The analysis of the surface microrelief with different contents of silicon carbide particles has shown that competitive regions of capture of adatoms into a chemisorption state arise on the growth surface (Figure 6). These regions could be seen when examining the growth surfaces at an initial stage[12–14]. But in this case the following detail was noticed: germaium islands, rapidly growing on the active regions near carbide particles, consumed considerably more quantity of material through microkinks and peninsulas on the steps. More rapid building-in of adatoms into large germanium islands can most probably explain a considerably smaller barrier to building-in of germanium atoms to a 'native' lattice as compared to the barrier of adatom addition to stressed sections of the pseudomorphic germanium film. In this connection one can affirm that in the case of the initial stages of heteroepitaxy energy constants, characterizing the adatom behaviour on the growth surface, such as surface diffusion energy, desorption energy, must be a function of the film thickness until the film lattice parameter reaches the value corresponding to the bulk crystal. The energetic barrier of the adatom building-in into the lattice must decrease with increase in film thickness and, according to an experiment by Adamsky[4] under pure conditions, this barrier is close to zero in the case of germanium homoepitaxy on the surface (100). Of interest is the following peculiarity of growth of germanium islands near carbide particles: islands, somewhat winding around a carbide particle, grow mainly in the direction of initial stages on the silicon substrate surface. This nature of growth is most probably explained by considerable activity of relevant microfaces of the island surfaces rather than by adatom diffusion anisotropy along and across steps. But this is the subject of a special investigation.

Electrophysical properties of films and junctions. Some electrophysical and photoelectrical properties of germanium films and junctions, obtained on silicon by thermal evaporation of germanium in the system used for ion sputtering, are tabulated in Table 1. These results should not be considered as optimum, since more suitable epitaxy conditions apparently can be realized. As a result of high-temperature treatment of *n*-type silicon in vacuum before epitaxy, a *p*-layer formed to a depth of

Figure 5. Morphology of the silicon substrate surface after selective etching of germanium film obtained at T_s 480 (a), 650 (b) and 750°C.

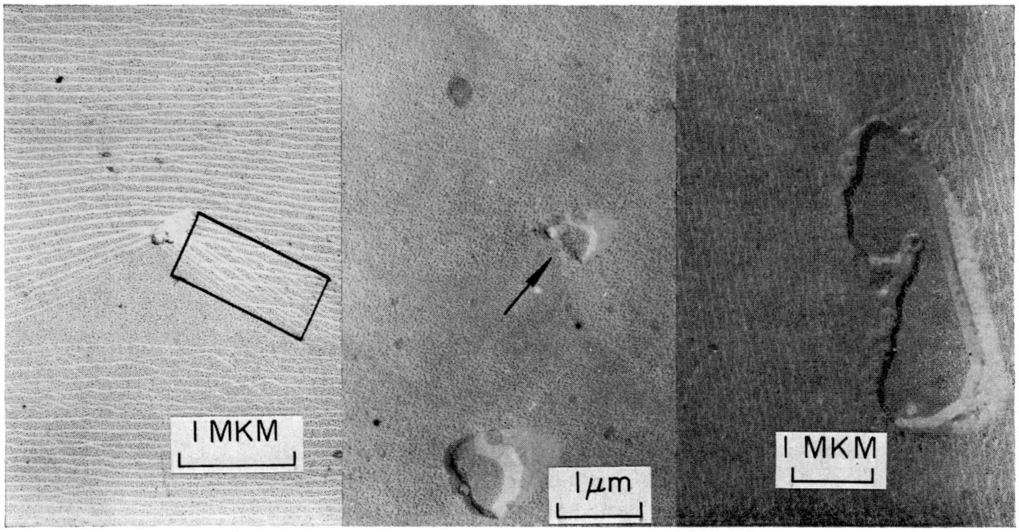

Figure 6. Competing regions of growth on contaminated silicon surface. The nature of growth of islands from carbide particles. The active region of building-in of germanium adatoms is put in rectangle (a). The arrow points to the formation of island near a carbide particle (b), growth island (c).

Table 1. Some properties of germanium films deposited on silicon by the vacuum evaporation, thicknesses of the films are from 1.5 to 3 μm

Substrate resistivity, ohm cm, orientation	Hole mobility, $cm^2 V^{-1} s^{-1}$	Hole concentration cm^{-3}	Built-in potential/ photovoltage, V	Reverse current density at −3 V, A/cm^2
n Si, 15, (111)	700 (296 K)	6×10^{16} (296 K)	0.55/0.4	10^{-6}
n Si, 15, (111)	1000 (295 K)	5×10^{16} (295 K)	0.45	10^{-6}
n Si, 7.5, (100)	600 (295 K)	2×10^{16} (295 K)	0.55/0.35	5×10^{-7}
	2000 (77 K)	6×10^{15} (77 K)		
n Si, 0.35, (111)			0.75/0.3	3×10^{-7}

$T_s = 550°C$; deposition rate (R_D) equal 0.3 μm/min.

0.5–1.0 μm in the surface region at the times and heating temperatures used. The formation of a *p*-layer during thermal treatment of silicon in UHV was previously observed by Allen *et al.*[16] However, there is no reliable information on the nature of formation of this layer, and the properties of *p–n* junctions obtained by this method are not considered. The probable mechanism of this layer formation seems to us to be connected with evaporation of a donor impurity from the surface region of the substrate. The formation of vacancies on the substrate surface and their propagation deep into the crystal accelertae impurity diffusion and promote depletion of donor impurities in the surface-space regions. Vacancies, remaining after the donor impurity evaporation, give acceptor levels that lead to the *p*-layer formation. The mass-spectrometric analysis of the *p*-layer for the concentration of elements of the third group has demonstrated their absence up to 10^{17} cm^{-3}, though the concentration of holes in these layers exceeded 10^{17} cm^{-3}. Phosphor was investigated as a donor impurity in silicon crystals. The partial pressure of the phosphor vapour at the heating temperature (1300°C) was 10^4 times greater than the silicon vapour pressure, it provided the conditions for evaporation of phosphor from the surface-space layer. The concentration C_s on the surface is less than the initial concentration C_s of phosphor in the bulk volume; it caused a diffusion flow to the surface. In this case the profile of distribution of phosphor in the surface space region is determined by the difference in the effective diffusion coefficients D_1 in volume (diffusion activation energy is 3.69 eV) and D_2 in the vacancy-enriched region (this activation energy is for vacancy formation as little as that in volume and can be less than 1 eV[17]). Under our conditions $D_1/D_2 \approx 10^8$, i.e. the concentration C_s penetrates practically the whole surface space region of width x_0, and further $C(x, t)$ is described as $C_v \operatorname{erf}[(x - x_0)/2\sqrt{D_1 t}]$, according to the solution of the problem of diffusion in a semi-infinitive body with an interface, since $C_s \ll C_v$. Study of voltage-current and voltage-capacity characteristics of mesadiodes on the basis of these *p–n* junctions, using a contact to silicon from epitaxial germanium or molybdenum, has showed that a straight branch of voltage–current characteristics corresponds to a theoretical model of Sah *et al*;[18] densities of reverse current at a voltage of 3 V range from 10^{-6} to 10^{-7} A cm^{-2}, and the contact potential difference corresponds to that calculated for volume impurity concentrations in *p*- and *n*-regions of the junction, determined from Hall measurements. The current density at large reverse voltage was mainly determined by the junction surface state. A newly prepared mesadiode had a reverse current which was an order of magnitude higher than that resulting from ageing at room temperature for several days. The elimination of surface leakage is possible by the deposition of low-temperature oxides on the mesa structures. Of special interest are the photoelectrical properties of these diodes. The open-circuit photoelectromotive force of the structures under investigation ranged from 0.45 to 0.52 V. Maximum spectral sensitivity was shifted to a long-wave region and corresponded to 0.9–1.0 μm at room temperature (Figure 7). The integral sensitivity of the obtained

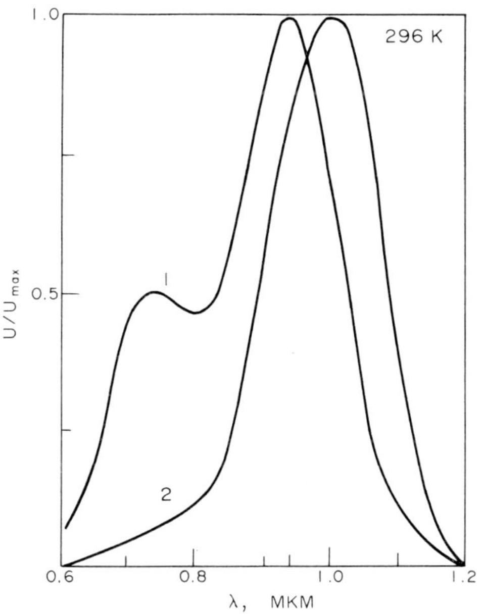

Figure 7. Spectral characteristic of photodiode Mo–Si–*p*–Si–*n* at the *n*-silicon with resistivity 0.3 (1) and 7.5 ohm cm (2).

photodiodes amounts to 9 mA/L, and the dark current at a bias of -1 V ranges from 10^{-8} to 10^{-9} A (for devices with an area of about 10 mm^2). The excellent characteristics of the obtained photodiodes and the maximum sensitivity shift towards the long-wave region point to the possibility of a simple way of producing *p–n* junctions at shallow depths (down to 1 μm) and with atomically clean *p*-type surfaces.

Of interest is the extension of the region of spectral sensitivity of such homojunctions at the expense of generation of current carriers in the narrow-gap material epitaxially grown on one of *p–n* junctions regions, or at the expense of creation of deep impurity levels in one of the junction regions. The qualitative homojunction in a wide gap material permits rather small dark currents; it provides high sensitivity of photoreceivers on the basis of such structures.

4. Conclusions

The results obtained from investigations of structural, electrophysical, and photoelectrical properties of epitaxial systems on the basis of silicon permit us to conclude that the development and improvement of vacuum methods of obtaining multi-layer structures in the form of devices, using epitaxial processes together with the development of methods of ion doping, vacuum soldering of contacts, junction protection by low-temperature oxide films, obtained by reactive sputtering and other methods, can provide production of high-sensitive photoreceivers and other active instruments for optoelectron devices.

The most essential problem, which is important at present, is doping of submicron layers in multilayer structures with programmed space distribution of a doping impurity. There is every reason to hope that this problem will be solved.

Acknowledgements

We thank Prof I G Yudelevich for mass-spectrometric analysis of silicon layers and Dr M P Sinyukov and P A Simonov for optical measurements of epitaxial structures and discussions of the results obtained.

References

[1] A G Milnes and D Feucht, *Heterojunctions and Metal-Semiconductor Junctions*, Academic Press, New York (1972).
[2] L N Aleksandrov, *Mikroelektronika*, Part 2, p 120, (1972).
[3] L N Aleksandrov and R N Lovyagin, *Obzory po elektron teknike*, **2,** 1974, 193.
[4] R F Adamsky, *J Appl Phys*, **40,** 1969, 4301.
[5] I H Khan, *J Appl Phys*, **44,** 1973, 14.
[6] B A Joyce, J H Neave and B E Watts, *Surf Sci*, **15,** 1969, 1.
[7] O P Pchelyakov, R N Lovyagin, E A Krivorotov, A L Toropov, L N Aleksandrov and S I Stenin, *Phys Stat Solidi*, (a) **17,** 1973, 33.
[8] L N Aleksandrov and R N Lovyagin, *Thin Solid Films*, **20,** 1974, 1.
[9] H C Abbink, R M Broudy and G P McCarthy, *J Appl Phys*, **39,** 1968, 4673.
[10] A G Cullis and G R Booker, *J Cryst Growth*, **9,** 1971, 132.
[11] D I D Thomas, *Phys Stat Sol*, **13,** 1966, 35.
[12] L N Aleksandrov and R N Lovyagin, *Phys Stat Solidi*, (a) **37,** 1976, 1.
[13] L N Aleksandrov, R N Lovyagin, O P Pchelyakov and S I Stenin, *J Cryst. Growth*, **24/25,** 1974, 298.
[14] O P Pchelyakov, L N Aleksandrov, R N Lovyagin and S I Stenin, *Kristallographia*, **21,** 1976, 811.
[15] W R Runyan, *Technology Semiconductor Silicon*, McGraw-Hill, New York (1955).
[16] P G Allen, T M Buck and J T Law, *J Appl Phys*, **31,** 1960, 979.
[17] *Atomic Diffusion in Semiconductors*. Edited by D Shaw, Plenum Press, London (1973).
[18] C T Sah, R N Noyse and W Shockley, *Proc IEEE*, **49,** 1957, 1228.

2.22 Electrical and piezoresistive properties of $Pb_{1-x}Sn_xTe$ thin films

A Rogalski, J Rybiński and M J Malachowski: *Institute of Quantum Electronic, 01–489 Warsaw 49, Poland*

Some of the problems of producing $Pb_{1-x}Sn_xTe$ films by hot-wall technique are discussed in relation to the design of the deposition conditions. Electrical and piezoresistive properties of the p- and n-type films were measured. Temperature dependence of the resistivity, Hall constant, Hall mobility and gauge factor of the films were measured and discussed. Dependence of the conductivity on applied strain to the p-type and n-type films were measured. A probable explanation of the mechanism of the physical phenomena in strained $Pb_{1-x}Sn_xTe$ films is described.

1. Introduction

Recently, the hot-wall technique has rapidly become a recognized approach to prepare thin films of $Pb_{1-x}Sn_xTe$ for infrared device applications[1–4]. The technique is known to be capable of producing high quality films with desired composition. Advantages of the technique are low growth temperatures and relatively short growth periods. In this paper we report on application of the hot-wall technique for thin-film piezoresistive device fabrication. It is well known, that, the gauge factor of the semiconductor materials decreases as temperature increases. We have observed that the gauge factor of *p*-type $Pb_{1-x}Sn_xTe$ films is nearly temperature independent. The results obtained are illustrated and discussed.

2. Film preparation

The films were prepared by thermal evaporation of PbTe + SnTe in a 2×10^{-6} torr vaccum in the hot-wall condition, condensing the material in the form of a film composition ratio corresponding to $x = 0.18$ on to silicon and mica substrates. The best electrical properties have been found if the substrates were heated at temperatures in the range of 260–280°C. The films were deposited at a rate of 3 μm h^{-1}. The films, both *p*- and *n*-type had a Hall mobility approximately 10^4 cm^2/Vs at 77 K. The films of the best properties have been obtained after annealing or diffusing cadmium. Annealing was performed at 450°C for 10–30 h in an evacuated sealed quartz ampoule containing the film and the alloy with excess of the metal. Diffusion of cadmium was obtained by annealing the film in the presence of metal-rich bulk material Cd–In (1% of Cd) for 4 h in an evacuated sealed quartz ampoule[5]. Temperature of the film region of ampoule was 450°C and Cd–In alloy region was 180°C.

Ohmic contacts were made by evaporating a layer of Au and In on to the *p*- and *n*-type layers respectively. The evaporation of contacts were made after annealing to reduce diffusion of the contact metals into the surfaces of the film. Electrical leads of silver wire were joined with a silver paste to the evaporated metal films.

3. Results and discussion

Typical properties of $Pb_{1-x}Sn_xTe$ films on Si or mica substrates at 77 K are listed in Table 1. Carrier concentrations and mobilities range between 5.1×10^{16} and 1.6×10^{18} cm^{-3} and 6×10^3 to 3.6×10^4 cm^2/Vs respectively.

Figures 1–3 show the dependence on temperature of the resistance, Hall mobility and Hall constant of the samples. The films behave in the similar manner to those described in references 1 and 3. Concentration of the carriers in *n*-type films, as can be seen from Figure 2, increases with temperature rise and the mobility decreases with temperature rise (Figure 3). The concentration of the carriers in *p*-type films behaves in the opposite manner.

Table 1. Properties of $Pb_{1-x}Sn_xTe$ films at 77 K

Sample number	Composition x	Substrate	Substrate temperature °C	Thickness μm	Type	Concentration cm^{-3}	Hall mobility cm^2/V s
1	0.18	Si	270	3.5	*p*	1.6×10^{18}	6000
2	0.17	mica	265	2.2	*n*	2.5×10^{17}	14300
3*	0.18	mica	270	1.1	*n*	2.6×10^{17}	20000
4†	0.18	mica	260	3.2	*n*	8.7×10^{16}	36000
5‡	0.18	mica	265	2.0	*n*	5.1×10^{16}	16200
6	0.18	mica	275	2.7	*p*	6.1×10^{17}	6300

* After isothermic annealing of the film at 450°C for 15 h.
†,‡ After diffusion of cadmium in the film. Cd–In was heated at temperatures 180 and 170°C, respectively.

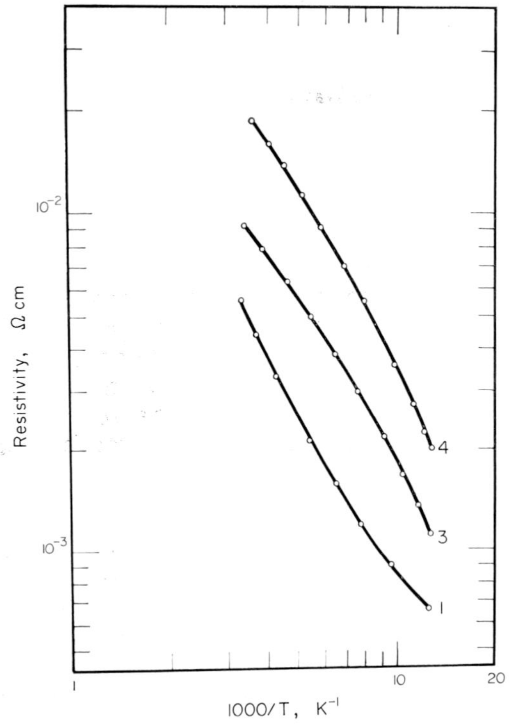

Figure 1. Temperature dependence of resistivity of the films described in Table 1.

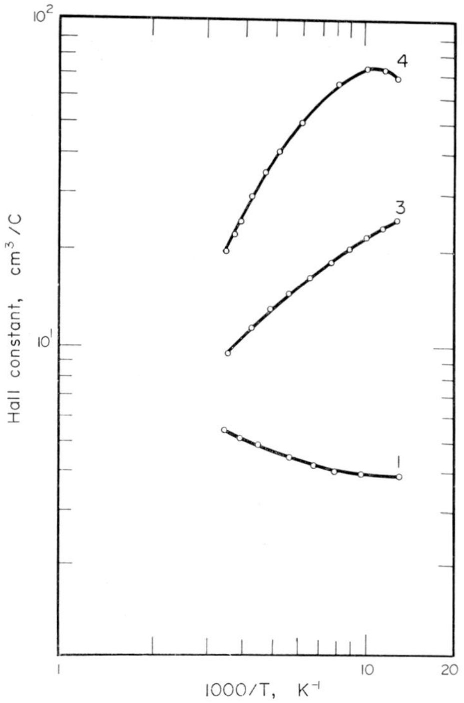

Figure 3. Temperature dependence of Hall constant of the films described in Table 1.

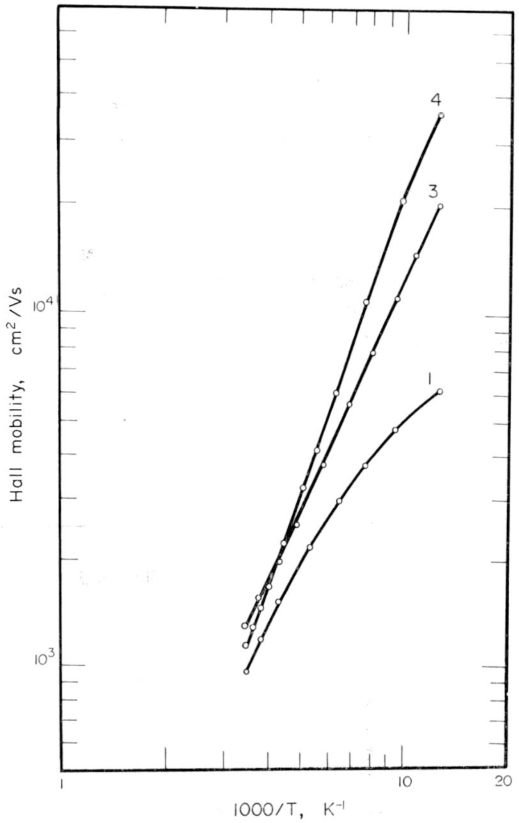

Figure 2. Temperature dependence of Hall mobility of the films described in Table 1.

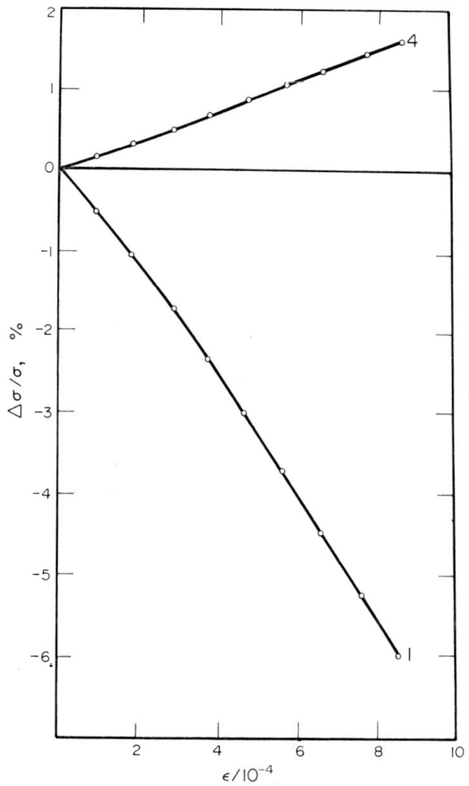

Figure 4. Conductivity–tensile strain dependence of the two films *p*- and *n*-type described in Table 1.

The resistance–strain data were acquired from tests performed in air in the 200–350 K temperature range in the manner described in the paper[5]. The desired strain in the test films was obtained by the controlled elastic bending of the strip substrate held firmly at one end and deflected at the other end shaped in the manner to introduce isotropic strains.

In Figure 4 are shown conductivity–strain dependence of *n*- and *p*-type films. The applied tensile strain causes the conductivity of the *p*-type films to decrease whereas for a compressive strain the conductivity increase. The *n*-type films behaviour is opposite to that of *p*-type.

Changes in the conductivity of the films caused by an applied strain are affected by changes of mobility of the carriers, because the concentration of the carriers calculated from the Hall constant measurements does not depend on the applied strain.

Figure 5 illustrates decrease of the gauge factor of the *n*-type films with increasing temperature. The gauge factor of *p*-type films is essentially temperature independent in the range of 200 to 350 K, and reaches a value of 58 at room temperature. Gauge factor of *n*-type films is about 20 times at the same temperature.

Discussion of the differences of the piezoresistance behaviour of the $Pb_{1-x}Sn_xTe$ *n*- and *p*-type films can be based on the results obtained on the PbTe[7,8]. If temperature dependence of the piezoresistive coefficients of $Pb_{1-x}Sn_xTe$ are similar to that of PbTe, this probably results the opposite strain dependence of the conductivity of *n*- and *p*-type films, as in PbTe[7]. *n*-type films have lower gauge factor than that of the *p*-type films because piezoresistive coefficients of the *n*-type films are lower.

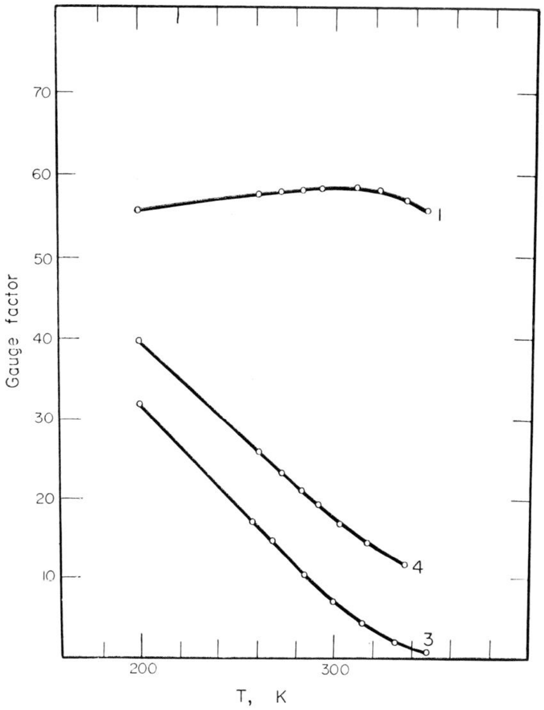

Figure 5. Temperature dependence of gauge factor of the three films described in Table 1 ($\varepsilon = 5 \times 10^{-4}$).

4. Conclusions

We have demonstrated that hot-substrate technique can produce high-quality $Pb_{1-x}Sn_xTe$ films. Their electrical properties have indicated that properties of the films are comparable to those of best bulk crystals. This films as was find elsewhere have suitable properties for infrared device application. In this paper we have demonstrated possible application of the films as piezoresistive device. Because of the small thickness of the films the mechanical load on the substrate will be completely negligible. The *p*-type film having comparatively high value of gauge factor, which is nearly temperature independent can find an interesting applications.

References

[1] J R Dixon and R F Bis, *Phys Rev*, **176**, 1968, 942.
[2] R F Bis, J R Dixon and J R Lownej, *J Vac Sci Technol*, **9**, 1972 226.
[3] I Kasai, J Hornung and J Baars, *J Electron Mater*, **4**, 1975, 299.
[4] I Kasai, D W Bassett and J Hornung, *J Appl Phys*, **47**, 1976, 3167.
[5] M Granveand and Y Petroff, *Phys Stat Solidi* (a), **3**, 1970, 629.
[6] H Holloway, E M Logothetis and E Wilkes, *J Appl Phys*, **41**, 1970. 3543.
[7] Yu W Ilisavskij, *Fiz Tverd Tela*, **4**, 1962, 918.
[8] R Ito and K S Shogenji, *J Phys Soc Japan*, **18**, 1963, 1343.

2.23 The deposition by evaporation of Cu–Al alloy films

F d'Heurle, C Alliota, J Angilello, V Brusic, J Dempsey and D Irmischer, *IBM Thomas J Watson Research Center, Yorktown Heights, NY 10598, USA.*

Thin films of Cu-Al alloys from pure Cu to pure Al were deposited by vacuum evaporation from single alloyed sources heated by electron bombardment. Information about the composition of the films obtained by electron microprobe was completed by lattice parameter measurements and structure information obtained by X-ray diffraction, as well as electrical resistivity measurements. The anticipated compositions of the films were calculated from the vapor pressures of the pure elements and thermodynamic data for liquid Cu–Al alloys at 1373 K. The agreement between the calculated values and the experimentally observed compositions is good only for alloys with low Al concentrations: less than 20 at.% Al in the source material. For the remainder of the Cu–Al system, with more than 20 at.% Al in the source, the films were found to contain less Al than was calculated. This discrepancy might be due to too high values for the reported vapor pressures of Al.

1. Introduction

The deposition of alloy films often requires the use of specialized techniques, which may necessitate particular equipment and controls, or set limitations to the quality of the films which are obtained. The flash evaporation of small alloy granules, for example, is not compatible with good source outgassing practices. The evaporation of a source until exhaustion makes it difficult to monitor film thickness adequately. Sputtering in either the dc or rf modes obliges one to fabricate special cathodes and may be prohibited in cases where the substrates cannot be exposed to electrically charged particles. Obviously, whenever practical, the deposition of alloy films from a single source by usual evaporation means has the advantage of greater simplicity.

The conditions of alloy evaporation are presented in the classic texts on thin film deposition.[1–3] Problems encountered in evaporating specific alloys continue to be the object of relatively recent publications: on Ag–Cu[4,5] and on Ni–Fe.[6] One article treats effects resulting from diffusion limitations in the liquid sources on the evaporation of Cu–Be and Pb–Sn alloys.[7] The motivation for the present study of the Cu–Al system stems from different origins. Al alloys containing a small concentration of Cu have become useful in electronic applications because of their enhanced resistance to electromigration failures.[8,9] (A very recent paper[10] treats specifically the problems associated with the deposition of such alloys). Cu alloys containing Al are well known for their oxidation resistant properties. Some of these can be used for decorative purposes because of their rich yellow color.[11] Finally, the existence of thermodynamic data[12] on liquid Cu–Al alloys provides the possibility of anticipating the composition of the vapor over Cu–Al sources and comparing that composition with the experimentally determined alloy concentration in the resulting films.

2. Experimental procedures: film deposition

The vacuum system used is a conventional oil diffusion pump system. The backstreaming of oil vapor is limited by an optically dense baffle cooled to about −50°C. Moreover, a liquid N_2 trap and a Ti sublimation pump has been added to one side of the vacuum chamber. The source is an Airco-Temescal device where heat is provided by an electron beam which is deflected through a 180° bend. The water cooled Cu hearth includes a cavity in the form of an inverted truncated cone, about 2.5 cm in diameter and 1.5 cm deep, for the source material. The alloy to be evaporated was placed directly in contact with the Cu hearth without any auxiliary crucible. The source to substrate distance is approximately 30 cm. The substrates were most usually oxidized Si wafers. These were backed with Ga and mounted on a temperature controlled substrate holder with Mo pedestals to avoid reaction with the Ga. All (but three depositions which were carried on at room temperature) depositions were made on to substrates maintained at 200°C. The pressure which was lowered to the range $1–2 \times 10^{-5}$ Pa prior to deposition increased by a factor of 10, to the low 10^{-4} Pa range during evaporation. The depositions were carried at a rate of 20×10^{-10} m s^{-1}, after outgassing for about 2 min under conditions corresponding to a rate of 40×10^{-10} m s^{-1}. The intended film thickness of 5000×10^{-10} m was achieved within limits of 20%. The alloy ingots were cast in the proper shape in an arc furnace after several meltings to insure homogenization. With one exception, all the films to be considered below were obtained from the first evaporation of any given ingot. In one case five films were deposited successively from one ingot in order to ascertain the variations of film composition under conditions of repeated evaporations. During each evaporation the electron beam about 1 cm^2 in cross-section was swept simultaneously in the x and y directions over the surface of the source material. The surfaces could be observed to be totally liquid. It is difficult to ascertain whether the sources as a whole were liquified. However, in all cases the evaporating surfaces were well above the melting temperatures, and materials such as Cu or Al cannot sustain much temperature gradient. The main barrier to heat flow is obviously to be found at the non-wetting interface between the source material and the Cu hearth. Thus it can be assumed that most, if not all, of the source material was liquefied during evaporation.

The composition of the films was determined by electron microprobe analysis. Some of the results obtained were confirmed through X-ray diffraction which indicates the crystalline structure of the films and their lattice parameter, thereby providing ancillary information about their composition. In order to compare lattice parameters with values found in the literature[13] it is necessary to correct the film values for deviations due to stresses. These were measured where appropriate by usual substrate bending measurements.[14,16] Finally the resistivity of the films was obtained with a four point probe.

3. Experimental results

(1) Electron microprobe analysis. The film compositions were determined with an ARL-EMX-SN electron microprobe. The relative concentrations were obtained by monitoring the intensity of both the Al K_α and the Cu L_α lines. The use of two acclerating voltages, 5 kV with an average penetration of about 1000×10^{-10} m and 8 kV with an average penetration of about 2000×10^{-10} m (both within the film thickness of about 5000×10^{-10} m) gave a measure of the homogeneity of the films. When compared, the results obtained with the two accelerating potentials were generally within ± 1 at.% of each other without any discernible systematic variation.

The Al concentration of the films obtained upon the first evaporation from a series of alloy ingots is plotted in Figure 1 as a function of the composition of the ingots. As will be shown later, the compositions obtained by the electron microprobe are generally consistent within ± 1 at.% Al with indications derived from the other analytical means. They will be used henceforth as representing the true film composition. The results shown in Figure 1 are indicative of relatively well behaved evaporations, where the compositions of the films vary smoothly with the composition of the source material. However, the curve relating film to source composition differs widely from linearity, indicating large deviations from Raoult's law.

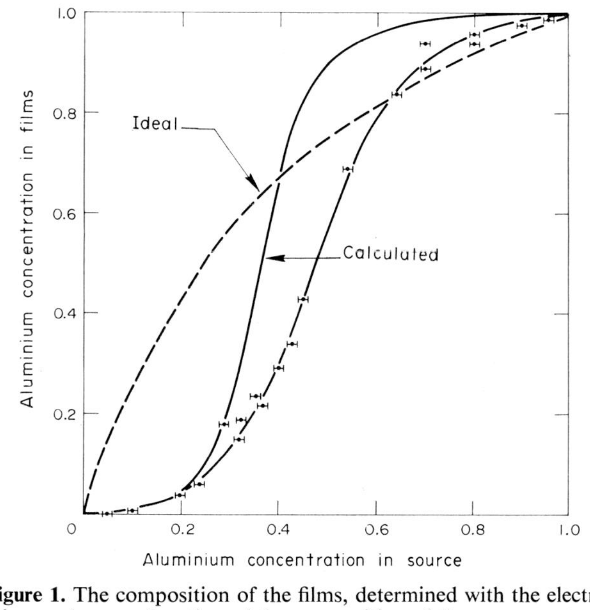

Figure 1. The composition of the films, determined with the electron microprobe, as a function of the composition of the source material. The calculated and ideal compositions are also shown as two curves, respectively, continuous and dashed.

The first film obtained from an ingot containing initially 35.5 at.% Al had a concentration of 25 at.% Al. The films obtained from the same ingot in four successive evaporations had respective Al concentrations of 25, 26, 28 and 29 at.%. The initial weight of the ingot 37 g, was reduced by 2 g during each film deposition.

(2) Electrical resistivity. Below 300°C the equilibrium diagram of the Cu–Al system[17] displays two terminal solid solutions (with up to approximately 20 at.% Al on the Cu side, and a narrow range of solubility on the Al side) as well as many intermediate phases: γ_2 at about 33 at.% Al, δ at 37 at.% Al, η_2 at 43 at.% Al, ζ_2 at 48 at.% Al, and the well known θ Al_2Cu, which forms precipitates in Al alloys containing Cu. One might therefore anticipate discontinuities in the relationship between the resistivity of Cu–Al films and their composition. However, in the cases at hand the resistivities plotted in Figure 2 as a function of film composition more or less follow the parabolic curve anticipated for alloy systems with complete solid solution. The values (crosses on Figure 2) at 29.5 at.% and 34 at.% Al

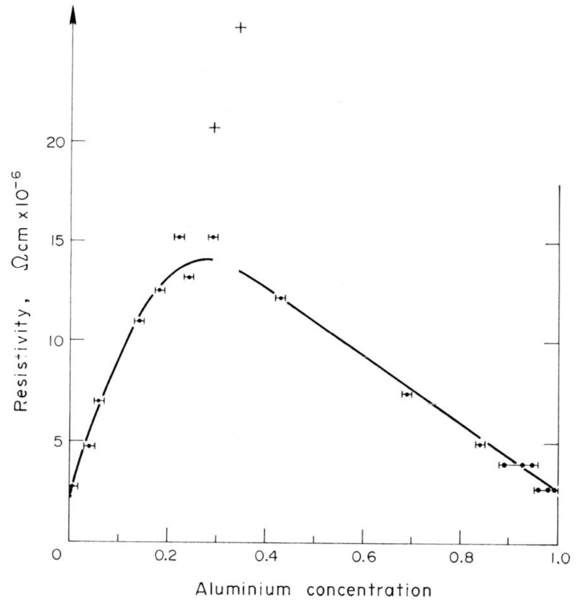

Figure 2. The resistivity of the films as a function of their composition.

are markedly higher. These correspond to the γ_2 phase. A resistivity of 7.5 $\mu\Omega$ cm for a film containing 69 at.% Al is quite close to the reported value for Al_2Cu:6.5 $\mu\Omega$ cm.[16] At one extreme, films containing more than 95 at.% Al had resistivities of the order of 3 $\mu\Omega$ cm, close to the value for pure Al, 2.6 $\mu\Omega$ cm. The resistivities of films rich in Cu are also plotted on Figure 3 (crosses) as a function of the Al concentration. The film values are close to what is anticipated for bulk (as opposed to films) samples of equal composition[19] which are represented by discrete dots and a continuous curve. Below approximately 5 at.% Al the resistivity values follow a straight line corresponding to the residual resistivity for Al in Cu, 0.95 $\mu\Omega$ cm per at.%.[20]

(3) X-ray diffraction. All the films obtained were analyzed on a horizontal X-ray diffractometer equipped with a Cu tube and a pyrolytic graphite bent crystal monochromator adjusted for

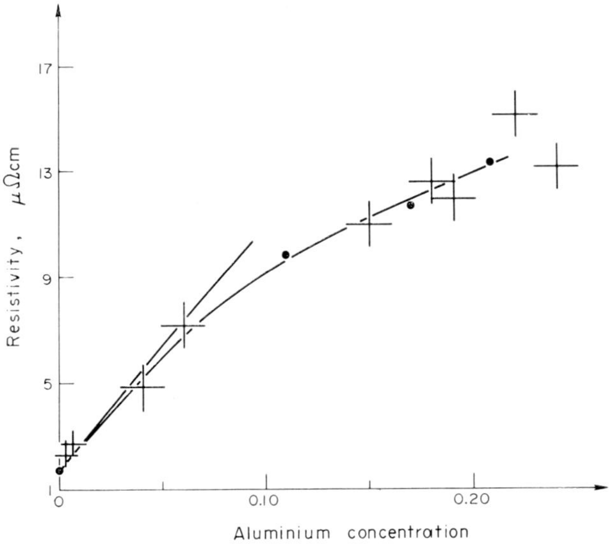

Figure 3. The resistivity of the Cu-rich films as a function of their composition.

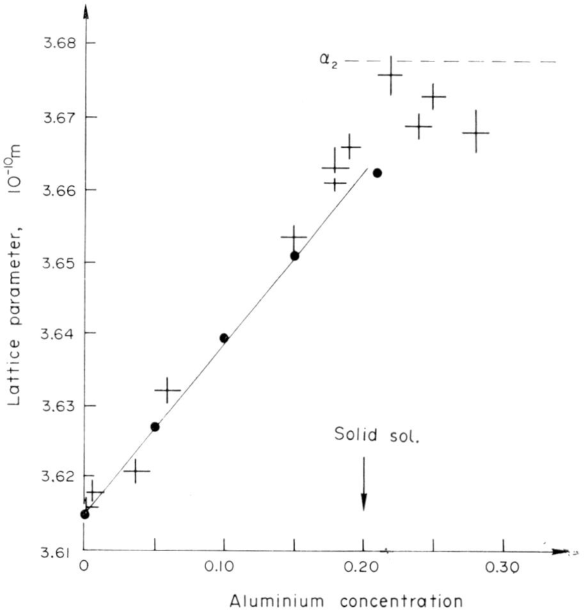

Figure 4. The lattice parameter of the Cu-rich films as a function of their composition.

the Cu K_α lines and located between the sample and the detector. Although the films were generally deposited at 200°C, the diffraction patterns displayed broad peaks, indicative of relatively small grain size and internal microstresses. Peaks at higher angles than the (222) diffraction for the Cu solid solutions at 94° (2 θ) were usually too weak to be observable. The Cu K_{α_1}–Cu K_{α_2} doublet was not resolved. Occasionally diffraction patterns were obtained at a grazing angle on films: these showed somewhat greater sensitivity in revealing weak diffraction lines, but because of the geometry had poor accuracy for spacing measurements. The results of the diffraction studies will be presented below classified according to different ranges of Al concentration.

(a) Cu-rich alloys up to 29 at. % Al. All of the films in which the Cu solid solution could be detected in the diffraction patterns displayed a marked (111) preferred orientation. The lattice parameters which were often derived entirely from the (222) diffraction line were corrected for stress (see Appendix). The corrected values are plotted on Figure 4 as a function of the Al concentration. The values reported for bulk samples[13] are also plotted on Figure 4 as plain dots joined by a straight line. Notice that the parameter reported for a sample containing 21 at.% Al lies somewhat below the line and fixes the solubility limit at 20 at.% Al. The film parameters correspond within the limits of experimental errors to the parameters for bulk samples. The high values found for films containing more than 20 at.% Al could indicate that in the films, as deposited, the solubility limit could extend beyond 20 at%. Al. Indeed in the range between 20 and 29 at. % Al the observed lattice parameters lie about half way between the value for the saturated Cu α solid solution and the value for another fcc phase α_2 whose existence at low temperatures and at about 23 at. % Al was the object of a special study.[21] Evidence for the isolated presence of this α_2 phase could not be detected in the films. A surprising aspect of the diffraction patterns for films containing more than 18 at.% Al is the appearance at about 40° (2 θ) of a diffraction line which could not be unambiguously identified. The extra line is clearly visible in the patterns shown in Figure 5b and c for films containing 25 and 29 at.% Al, but traces of it are also to be found in the pattern for a film containing 18 at.% Al (Figure 5a). It corresponds to a spacing of 2.24×10^{-10} m, so that it might be the (220) diffraction of a cubic phase with the β–Mn structure, a lattice parameter of 6.25×10^{-10} m and a composition approximating Cu_4Al.[22] The extra line disappears upon annealing at 500°C followed by rapid cooling. Diffraction lines for the γ_2 phase which are not visible in a film containing 24 at.% Al after deposition become distinct upon annealing at 325°C. The unequivocal identification of phases in the Cu–Al films being studied is made more complicated by the relatively high state of disorder and the presence of preferred orientations which hide some diffraction lines.

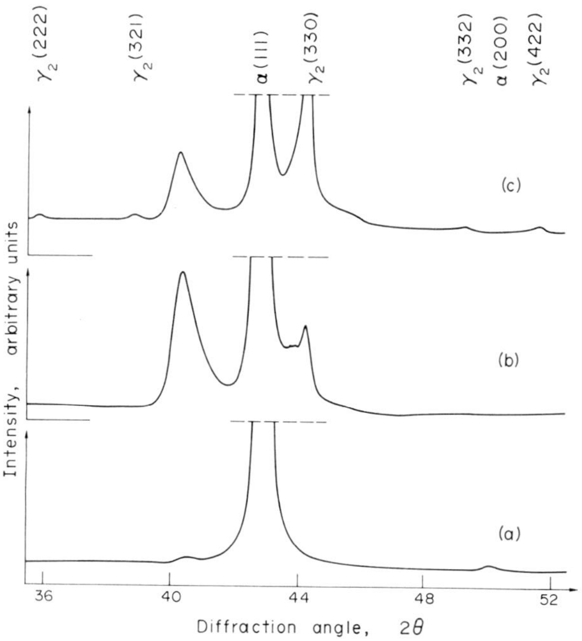

Figure 5. The diffraction patterns for three films containing respectively 18 at.% Al (a), 24 at.% Al (b), and 29 at.% Al (c) from about 35 to 50° (2θ).

(b) Films containing 29.5 and 35 at.% Al. The diffraction features, lattice spacings and relative intensities, for two films containing 29.5 and 34 at.% Al are listed in Table 1 together with the reported[23] features for the γ_2 phase (with a cubic structure and a lattice parameter of 8.703 $\times$ 10^{-10} m). Twelve diffraction peaks are identifiable as belonging to γ_2. Most probably the peak corresponding to a spacing of 2.11 $\times$ 10^{-10}m for the film with 29.5 at.% Al belongs to the α solid solution.

Table 1. Cubic γ_2 phase.

29.5 at.% Al d exp. 10^{-10} m	I/I₀	34 at.% Al d exp. 10^{-10} m	I/I₀	γ2 d cal.* 10^{-10} m	I/I₀	hkl
		3.917	16	3.890	4	210
3.570	100	3.576	100	3.553	8	211
2.955	21	2.955	40	2.901	8	300
2.533	21	2.533	16	2.511	2	222
2.324	15	2.324	32	2.326	3	321
2.113†	6			2.111	1	322
2.070	30	2.066	32	2.051	100	330
1.865	42	1.865	32	1.855	4	332
1.783	18	1.784	24	1.776	4	422
1.683	9			1.675	1	511
1.451	30	1.452	24	1.450	8	600
1.258	27	1.256	16	1.256	3	444
1.188	24	1.188	27	1.184	12	633

* Ref 23.
† This is also approximately equal to the (111) spacing of the Cu—20 at.% Al solid solution.

This would be in keeping with the lower resistivity shown on Figure 2 (as compared to the film with 34 at.% Al), and with the reported lower limit of Al concentration for this phase: approximately 30 at.%.[15] By means of a least square technique and the usual Nelson–Riley function the measured lattice spacings yield a lattice parameter value of 8.663 $\times$ 10^{-10} m (uncorrected for stress) which is in good agreement with the published value (above).

The calculated peak intensities in Table 1 were derived for Cu K_α radiation and for a powder sample.[23] By matching the experimental intensities with the intensities expected from a randomly oriented sample one sees that the films exhibit some degree of preferred orientation. The planes which are most favored to lie parallel to the substrate are the (211) planes. The (111) planes at an angle of only 10.5° from (211) also give greater intensities than anticipated.

(c) Mid range—about 50 at.% Al. The next film in order of increasing Al content was analyzed as having a concentration of 43 at.% Al. It was anticipated therefore that its structure would be characteristic of the ζ_2 phase which corresponds to this composition.[17] It is reported to be monoclinic with unit cell[13] dimensions: $a = 7.074 \times 10^{-10}$ m, $b = 4.078 \times 10^{-10}$ m $c = 10.02 \times 10^{-10}$ m and $\beta = 90.633°$. The lattice spacings for that structure were calculated and found to be incompatible with the spacings of the film.

The diffraction lines for the 43 at.% Al film were indexed on the spacings calculated for the η_2 phase, which is reported[17] to be stable at a concentration of 48 at.% Al. It has an orthorhombic structure[13] with unit cell dimensions of: $a = 6.894 \times 10^{-10}$ m, $b = 4.088 \times 10^{-10}$ m and $c = 9.89 \times 10^{-10}$ m. On the basis of the indices thus obtained, the unit cell characteristic of the film material was derived by a least square computer technique according to the experimentally determined spacings. The new cell dimensions, which are: $a = 6.891 \times 10^{-10}$ m, $b = 4.091 \times 10^{-10}$ m and $c = 9.920 \times 10^{-10}$ m, show close agreement with the previously reported values. On Table 2 one may compare the experimentally observed spacings with those which are calculated according to the new cell dimensions.

Table 2. Orthorhombic η_2 phase.

2θ	d exp. 10^{-10} m	I/I₀	hkl	d calc* 10^{-10} m
15.57	5.691	100	101	5.660
23.42	3.798	32	011	3.782
25.19	3.535	12	110	3.518
25.74	3.461	28	200	3.445
31.46	2.844	12	202	2.830
35.09	2.557	7	211	2.547
40.19	2.244	40	301	2.238
43.82	2.066	7	213	2.061
66.04	1.415	9	404	1.415
76.94	1.239	4	415	1.240
82.93	1.164	8	424	1.164
83.63	1.156	5	514	1.156

* *d* calculated from $a = 6.891$, $b = 4.091$, $c = 9.920$.

Further work outside the scope of the present study would be required to determine whether the microprobe analysis in the present case is in error by about 4 at.% Al, or whether in the present conditions of film nucleation and growth the η_2 phase would be more stable than ζ_2.

(d) Al-rich films: from 66 to 100 at.% Al. A film which was analyzed as containing 69 at.% Al displayed nine diffraction lines, all of which could be indexed according to the structure of the θ phase, Al_2Cu.[24] Indeed, a film with 69 at.% Al should be almost entirely pure θ. This is a tetragonal phase with unit cell dimensions: $a = 6.063 \times 10^{-10}$ m and $c = 4.872 \times 10^{-10}$ m. Comparison of diffraction results on films with the standard spacings can be made from Table 3. There one shall also find the spacings obtained for four films with compositions between 84 at.% and 95 at.% Al. All of the spacings listed were found for at least one of these films, all four of which displayed the (111), (200) and (220) diffraction peaks for Al. With respect to preferred orientation, it was noticed that the (200) diffraction peak for the θ phase, which should not be weak since it has a relative intensity of 35, was not detected in any of the film patterns.

Three films with analyzed Al concentrations of 94, 98, and 99 at.% respectively had been deposited on substrates at room temperature. Prior to being investigated by X-ray diffraction they were annealed for 4 h in a 15% H_2–85% N_2 mixture at 200°C, and rapidly cooled. The diffraction patterns for these films displayed strong (111) Al peaks, and several peaks indicative of a second phase which could not be indexed either according to the θ structure, or according to the θ' structure,[25] tetragonal with $a = 4.04 \times 10^{-10}$ m and $b = 5.8 \times 10^{-10}$ m. Since the very low resistivities of these films indicated that they were mostly Al, in agreement with the analytical results, the question of their exact structure was not pursued any further.

Table 3. θ phase and aluminium.

69 at.% Al		84–95 at.% Al	Al_2Cu*			
d exp 10^{-10} m	I/I_0	*d* exp 10^{-10} m	*d* 10^{-10} m	I/I_0	hkl	Al hkl
4.283	100	4.293	3.304	100	110	
2.368	1	2.366	2.374	70	211	
		2.338				111
2.140	20	2.139	2.146	35	220	
2.117	30	2.115	2.121	90	112	
		2.022				200
		1.916	1.919	70	310	
1.898	10	1.898	1.901	60	202	
1.607	1	1.607	1.611	13	222	
		1.432				220
1.390	2		1.395	9	213	
		1.234	1.234	20	332	
		1.218	1.219	8	004	
1.170	1		1.17	9	114	
1.042	2		1.040	2	530	
		1.026	1.028	17	314	
		0.928	.925	3	523	
		0.904	.906	6	424	

* Ref 24.

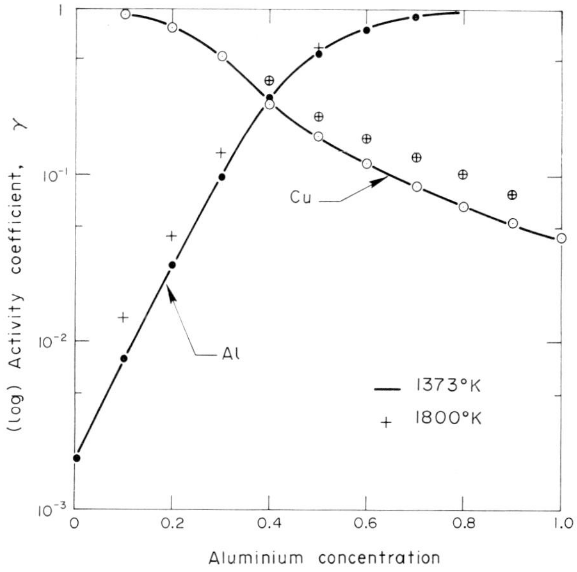

Figure 6. The activity coefficients (γ) for both Cu and Al plotted on a logarithmic scale as a function of composition at 1373 K,[12] and at 1800 K.

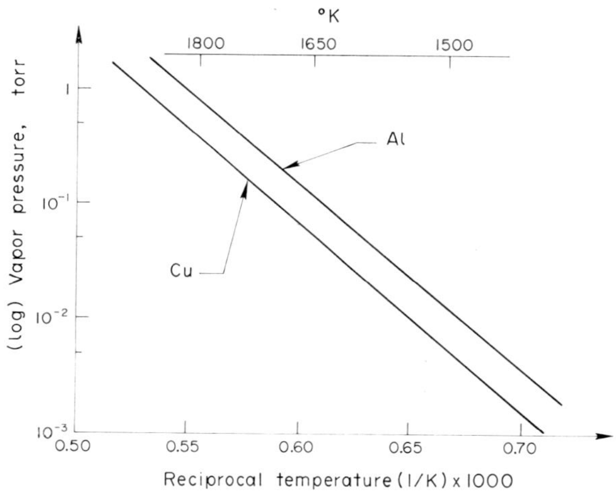

Figure 7. Arrhenius plot of the vapor pressures of pure Cu and Al in the range of temperature from 1400 to 1900 K.[26]

4. Calculated film composition

For a pure material the rate of evaporation is given by Langmuir's equation:

$$r = 4.4 \times 10^{-4}\, p\, (TM)^{-0.5} \tag{1}$$

where r is the rate of evaporation in mole cm^{-2} s^{-1}, p is the vapor pressure in Pascal, T and M are the absolute temperature and the atomic weight respectively. For an alloy the equation must be modified slightly to:

$$r_i = 4.4 \times 10^{-4}\, a_i\, p_i\, (T M_i)^{-0.5} \tag{2}$$

where r_i is the rate of evaporation for species i and a_i is the activity of species i in the alloy at the evaporation temperature. The activity is the product of the concentration times the activity coefficient ($a_i = c_i \gamma_i$). The activity coefficient expresses the departure of the alloy system from ideality, since in an ideal solution, where there is little chemical affinity between the elements, the activity tends to be equal to the concentration and the activity coefficient tends toward one. For the Cu–Al system the deviations from ideality are quite wide. On Figure 6 the logarithms of the activity coefficients at 1373 K were taken from the literature[12] and plotted as a function of the Al concentration. From equation (2) above it can be seen that the composition of the vapor C_{Al}/C_{Cu} as a function of the composition of the liquid source c_{Al}/c_{Cu} will be given by:

$$C_{Al}/C_{Cu} = (c_{Al}/c_{Cu})\,(\gamma_{Al}/\gamma_{Cu})\,(p_{Al}/p_{Cu})\,(M_{Cu}/M_{Al})^{0.5} \tag{3}$$

In the range of temperatures which are of interest here, the ratio of the vapor pressures of Al and Cu[26] is about 2, as seen on Figure 7. Thus the product $(p_{Al} p_{Cu})\,(M_{Cu}/M_{Al})^{0.5}$ is equal to 3.08. In order to complete the calculations of vapor composition according to equation (3), one still needs the activity coefficients at the proper temperatures.

Short of a direct measure of the temperature of the source during evaporation, which given the evaporation conditions would require an uneasy procedure, one may evaluate that temperature from a knowledge of the rates of deposition and the geometry of the system. For a planar source of unit dimension (1 cm^2) and for a substrate lying parallel to the source at a distance h from the source the rate of film thickness growth t is given by:[27]

$$t = r\,(M/\rho)\,(\pi h^2)^{-1} \tag{4}$$

(where ρ is the density of the material deposited). In the source used the area of evaporation, estimated to be equal to the area of the electron beam impact, is close to 1 cm^2. By means of relations (1) and (4), and the vapor pressure charts of Figure 7, the temperature of the source for the evaporation of pure Cu at a rate of 20×10^{-10} m s^{-1} is evaluated as approximately 1800 K. For Al the temperature would be lower, about 1500 K.

A complete set of thermodynamic data for Cu–Al exists only for one temperature 1373 K.[12] Exact values at higher temperatures cannot be derived from these data alone. However, if one assumes (hypothesis A) that the entropy of the solution (and its fractions partitioned to both constituents) is constant, one may evaluate the activity coefficients from the relations below:

$$\overline{\Delta F}_i^{xs} = \overline{\Delta H}_i = T\overline{\Delta S}_i^{xs} \quad (5)$$

and

$$RT \ln \gamma_i = \overline{\Delta F}_i^{xs} \quad (6)$$

where, $\overline{\Delta F}_i^{xs}$ is the relative partial molar excess free energy, $\overline{\Delta H}_i$ is the relative partial molar enthalpy, and $\overline{\Delta S}_i^{xs}$ is the relative partial molar excess entropy for species i. Alternatively, one may assume[4] (hypothesis B) that the free energy of the solution remains constant with temperature and calculate the activity coefficients directly from equation (6) alone. Strictly neither hypothesis is entirely valid. However, the first approach (hypothesis A) which does not neglect the entropy contribution to the free energy is formally more correct. It was used for all subsequent derivations.

The activity coefficients for both Cu and Al were calculated for 1800 K according to relations (5) and (6) from the published values[12] of γ, $\overline{\Delta H}$ and $\overline{\Delta S}^{xs}$ at 1373 K. These 1800 K values are also plotted on Figure 6 together with the values of 1373 K. In order to calculate the concentration of the vapor from relation (3) it was assumed that the temperature of evaporation varied linearly with the Al concentration in the source material. The values of the activities a_i for both Al and Cu, the value of $a_{Al} \times 3.08$, and the final Al and Cu concentrations C_i in the vapor are listed in Table 4. The calculated vapor (or film)

Table 4. Calculated vapor composition.

Al c liquid	Al *a*	Al 3.08 *a*	Cu *a*	Al C vapor	Cu C vapor
0.1	0.0015	0.0046	0.843	0.005	0.995
0.2	0.0094	0.0289	0.611	0.045	0.955
0.3	0.0393	0.1210	0.367	0.248	0.752
0.4	0.1356	0.4176	0.184	0.694	0.306
0.5	0.279	0.8593	0.099	0.897	0.103
0.6	0.447	1.377	0.056	0.961	0.039
0.7	0.610	1.877	0.029	0.985	0.015
0.8	0.758	2.334	0.014	0.994	0.006
0.9	0.889	2.738	0.005	0.998	0.002

compositions, and the ideal compositions (assuming that the γs are both unity) have been superimposed on Figure 1 together with the experimental results. In order to show more clearly how the Cu–Al system behaves in the limits of low and high Al concentrations, the same set of film composition values have also been plotted on a logarithmic scale on Figure 8. Admittedly the assumption made above that the evaporation temperature varies linearly with the source composition is not strictly correct. From the knowledge of the activities and the vapor pressures one may determine more or less exactly the evaporation temperature for each composition. When this is done for example for the alloy with 50% Al the evaporation temperature is found to be close to 1725 K. One may then make a new estimate of the calculated composition of the film. However, since the activities vary slowly with temperature, the new presumably more exact composition ($C_{Al} = 0.893$) is not significantly different from that shown in Table 4 ($C_{Al} = 0.897$).

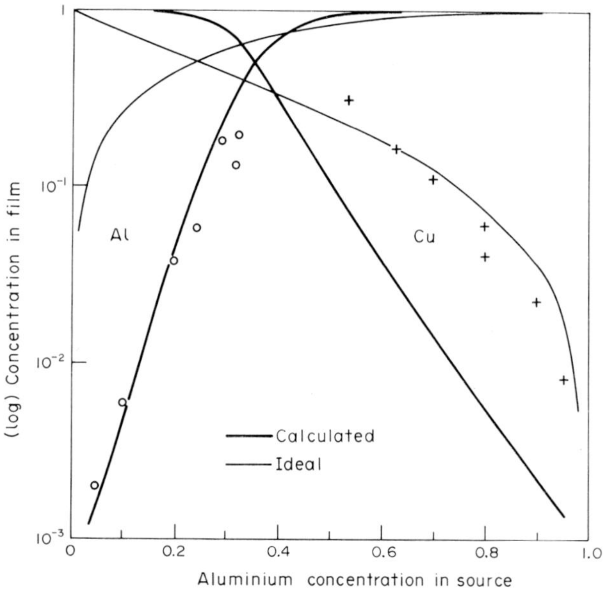

Figure 8. The experimental and calculated compositions of the films plotted on a logarithmic scale as a function of the Al concentration in the source material.

5. Discussion

Except in the domain of low Al concentrations where the fit is quite good, the experimental data vary systematically from the calculated compositions. Above 20 at.% Al the physical system behaves as if the source material was richer in Cu than it actually is. A possible explanation would be that the limitations of diffusion in the source material prevents equilibration of the evaporating surface.[7] This could only be valid for alloy sources containing more than about 43 at.% Al where the films were found to contain more Al than the sources, below that concentration the liquid diffusion explanation fails. There may exist some errors in the vapor pressure data for the pure materials: a much better fit between theoretical results and experiment would obtain if the vapor pressure of pure Al was assumed to be equal to that of Cu. However, a literature search yielded higher values for the vapor pressure of Al[28] than those which were used,[26] but not lower ones. It may be noted that Al has a tendency to form suboxides either AlO or Al_2O which have high vapor pressures. Given the high reactivity of Al with O_2 it is possible that some degree of surface oxidation may interfere with the experimental derivation of thermodynamic data and (or) with the evaporation results: the measured total pressure of about 1–2 $\times 10^{-4}$ Pa during evaporation was determined in the evaporation chamber itself (not at the pump). Two assumptions have been implicitly made throughout: (a) that the composition of the vapor is the same as that of the film, i.e. the sticking coefficient is unity, and (b) that the evaporation coefficient α is almost unity (which is why it is not entered in Langmuir's equation 1). The evaporation coefficient relates the rate of evaporation to the equilibrium vapor pressure; in some special cases, especially for evaporation from a solid[29] it may be smaller than one, however there exists no readily available reason why it should not be one in the system at hand. Electrical effects may play a role but have not been evaluated. One effect which is certainly in the right direction is the exchange of momentum between the incoming electrons and the Al and Cu atoms. The Al atoms being lighter will have more recoil than the Cu atoms; however, this effect is not likely to have sufficient magnitude to explain the observed discrepancy between theory and experiment. With an electron beam current of about 0.25 A it is found that the flux of incoming electrons is of the same order of magnitude at the flux of evaporated atoms. However, even with an acceleration voltage of 10,000 V the momentum of the electrons is about 10^{-12} times smaller than the thermal momentum of the atoms, much too small to be a matter of consequence. The relative ionization efficiency of Cu and Al, and the motion of charged ions may also have to be considered.

Acknowledgements

We are greatly indebted to our colleagues, F Cardone, S Ellmann, J Karasinsky and R MacInnes for their help in carrying some of the experimental work. The discussion of thermodynamic points with L M Foster and D Jepsen was very enlightening and most appreciated.

Appendix

Stresses and strains in thin films. When observed at room temperature, thin films are usually in a state of stress. The stresses are of two different origins: intrinsic stresses due to the process of film nucleation and growth, and extrinsic stresses due to the difference in thermal expansion between the films and the substrates. Independent stress measurements were carried on films with low Al content films (up to 20 at.%) by the wafer bending and Newton ring technique.[14–16] The films were deposited on specially thinned and optically polished glass wafers heated to 200°C. The measurements indicated that the intrinsic stresses were negligible and that the extrinsic tensile stresses amounted to 2.6×10^8 Pa. When heated to 500°C, the same films could sustain tensile stresses twice as high in the range $5\text{–}6 \times 10^8$ Pa.

In the case of the films being considered here (Figure 3) the substrates were oxidized Si wafers and the deposition temperature was also 200°C. The intrinsic stresses should be close to zero, and the extrinsic stresses should be due to the difference in the expansion coefficients of the metal, 18×10^{-6}/°C for Cu–Al[30] and the Si substrates, 3.2×10^{-6}/°C.[31] The strains in the plane of the films ε_xa and ε_y are equal, and given by

$$\varepsilon_x = \Delta a \, \Delta T$$

where Δa is the difference in expansion coefficients (14.8×10^{-6}/°C) and ΔT is the magnitude of the temperature excursion, 175°C (200–25) in the case at hand. The planar stresses, $\sigma_x = \sigma_y$, are given by

$$\sigma_x = \frac{E}{1-\nu}\varepsilon_x$$

where ν and E are the Poisson ratio and the Young's modulus for Cu–Al, respectively. Exact values for these two quantities could not be obtained, however, from the values given for Cu and Cu–Zn,[32] one can estimate that for Cu–Al ν should be close to 0.35 and E to 11×10^{10} Pa. Thus σ_x would be about 4.4×10^8 Pa, a value smaller than the maximum sustained stress of $5\text{–}6 \times 10^8$ Pa.

When measuring lattice spacings from a diffractometer, the mirror like geometry insures that all measurements, regardless of the diffraction angle, correspond to a direction normal to the plane of the film, i.e. the spacings are those of planes laying parallel to the plane of the film. In the presence of a film stress the experimentally determined spacings will need to be corrected for a strain ε_z with a magnitude given by

$$\varepsilon_z = -\frac{2\nu}{1-\nu}\varepsilon_x$$

where the negative sign indicates that the lattice parameters need to be increased for tensile stresses and decreased for compressive stresses. For the films on Si substrates deposited at a substrate temperature of 200°C ε_z should be equal to 2.8×10^{-3} so that the measured lattice parameters of about 3.66×10^{-10} m should be augmented by 0.010^{-10} m.

References

[1] L Holland, *Vacuum Deposition of Thin Films*, p 182, John Wiley, New York (1961).
[2] L Maissel and R Glang, *Handbook of Thin Film Technology*, pp 1–73, McGraw-Hill, New York (1970).
[3] R Berry, P Hall and M Harris, *Thin Film Technology*, p 149, Van Nostrand–Reinhold, New York (1968).
[4] T Santala and C Adams, *J Vac Sci Technol*, **7**, 1970, S 22.
[5] D G Ast and D J Krenitsky, *J Vac Sci Technol*, **13**, 1976, 969.
[6] J McCloskey, *J Vac Sci Technol*, **8**, 1971, 626.
[7] C Ziling and V Pcholkin, *Thin Solid Films*, **32**, 1976, 39.
[8] I Ames, F d'Heurle and R Horstmann, *IBM J Res Devel*, **14**, 1970, 461.
[9] F d'Heurle and R Rosenberg, *Physics of Thin Films*, **8**, 1973, 257, Academic Press, New York.
[10] L C Hecht, *J Vac Sci Technol*, **14**, 1977, 648.
[11] ref 1, p. 190.
[12] R Hultgren, P Desai, D Hawkins, M Gleiser and K Kelley, *Selected Values of the Thermodynamic Properties of Binary Alloys*, p 154, American Society for Metals, Metals Park, Ohio (1973).
[13] W B Pearson, *A Handbook of Lattice Spacings and Structure of Metals and Alloys*, p 327, Pergamon Press, Oxford (1958).
[14] J D Finegan and R W Hofman, *J Appl Phys*, **30**, 1957, 597.
[15] R Glang, R Holmwood and R Rosenfeld, *Rev Sci Instrum*, **36**, 1965, 7.
[16] A Blachman, *Met Trans*, **2**, 1971, 699.
[17] *Metals Handbook*, **8**, 1973, 259, American Society for Metals, Metals Park, Ohio.
[18] J Gniewek and C Wasik, *J Appl Phys*, **42**, 1971, 2151.
[19] C Smithells, *Metals Reference Book*, **3**, 1967, 692, Plenum Press, New York.
[20] F Blatt, *Physics of Ekectronic Conduction in Solids*, p 199, McGraw-Hill, New York (1968).
[21] R Jewett, The 'X' Phase of the Copper Aluminium System, PhD Thesis, University of Wisconsin (1971).
[22] J Leach, *J Inst Met*, **92**, 1963, 64.
[23] NBS Mono, 25, sect **11**, 1974, 79.
[24] Joint Committee on Powder Diffraction Standards, Powder Diffraction File 25-12 (1975).
[25] J Silcock and T Heal, *Acta Cryst*, **9**, 1956, 680.
[26] R Honig, *RCA Review*, **23**, 1962, 567.
[27] ref 1, p 145.
[28] A N Nesmeyanov, *Vapor Pressures of the Elements*, p 452, Academic Press, New York (1963).
[29] G Rosenblatt, *Treatise on Solid State Chemistry*, **6A**, 1976, 165, Plenum Press, New York.
[30] *Metals Handbook*, p 930, American Society for Metals, Metals Park, Ohio (1968).
[31] L Maissel, *J Appl Phys*, **31**, 1960, 211.
[32] ref 19, p 708.

2.24 Structural and electrical properties of discontinuous gold films on glass

T G Andersson and S H Norrman, *Department of Physics, Chalmers University of Technology, Fack, S-402 20 Göteborg, Sweden*

In the present work thin gold films are investigated in the thickness interval 10–160 Å. The films are deposited in UHV on to glass substrates at room temperature and with an electric field in the substrate plane. In the thickness interval 10–45 Å the film is discontinuous and the size distribution of the islands as determined from electron micrographs is described with a log-normal distribution function. The onset of the in situ *electrical conduction at the average thickness 45 Å makes it possible to measure the film resistance as a function of the film thickness during the condensation. At the thickness 53 Å stable metallic continuous paths are formed and the film achieves metallic properties. The fraction of the surface covered with material and the island density are measured 20 h after the deposition. Both these entities change rapidly at the stage when the film grows to be metallic continuous.*

1. Introduction

In this paper the growth of gold films on glass substrates are investigated by use of electron microscopy and by resistance measurements. Before the presentation of the results both growth and electrical properties of films are briefly discussed in general terms.

During the condensation of a metal on to a glass surface several growth stages are observed. The nucleation occurs simultaneously with an initial diffusion of metal into the interior of the substrate.[1] Later the film growth proceeds by island growth since the metal generally is more bound to itself than to the substrate.[2] The coalescence growth which follows next is a result of liquid-like coalescence after binary collisions of islands on the surface.[3,4] When a critical surface fraction is covered with metal there is a large scale coalescence (LSC) and a metallic continuous film is formed. Here generally the film is semicontinuous (the substrate surface is not completely covered).

At the discontinuous stage the electrical conduction is commonly described by a thermal activation of electrons over an electrostatic barrier and a subsequent tunelling through the inter-island potential barrier.[5] The conduction in evaporated metallic continuous films is similar to the conduction in a bulk metal. However, there is a reduction of the conduction due to excess impurities and grain boundaries and also from an increased scattering of the conduction electrons at the parallel film surfaces.

In the present paper gold films are investigated in the thickness interval 10–160 Å. The plan of the paper is as follows. First the experimental details are presented. Next follows the size distribution of islands in the thickness interval preceding the onset of conduction. From this event the resistance as a function of thickness is measured and the resistivity is compared with values for materials with different conduction mechanisms. Finally the surface coverage and island density are analysed as a function of the film thickness.

2. Experimental details

The experimental technique is very important for the formation of a thin film as it determines its growth and structure. In order to achieve reproducible results the films in the present work were formed as follows. The films were vacuum deposited from an Al–oxide crucible heated by a W-wire. Gold of purity 99.99% was deposited in uhv at a pressure less than 4×10^{-9} torr during the deposition and with a background pressure less than 1×10^{-9} torr. The residual gases in the vacuum system as measured by a Micromass 1 mass spectrometer were found to be H_2O (mass number 18), CO and N_2 (28) and Ar (40).

The deposition rates were in the interval 0.2–0.5 Å/s. No difference in the film structure has been found between films deposited within this range. In order to control the number of charged particles in the evaporation beam the source was always biased with +100 V for suppressing the electron current. Thus no negatively charged particles were emitted from the source.

A temperature range 77–500 K on the substrate holder is obtained by cooling with liquid nitrogen and by resistive heating. A standard thermocouple (iron–constantan) fixed to the substrate holder measured the temperature. A calibration of the temperature of the substrate was performed by placing a thermocouple on the substrate surface. All substrates were cleaned by the following procedure: (a) initially surface grease was removed by short ultrasonic cleaning in freon or isopropyl alcohol, (b) a vapour decrease in isopropyl alcohol before deposition of the electrodes, (c) a new vapour decrease in isopropyl alcohol followed by mounting in the uhv-system and baking at 275°C for at least 12 h when baking the uhv-system, and (d) a baking at 300°C for 30 min in a pressure of 1×10^{-9} torr or less. The last procedure was performed immediately before the film deposition.

The electrical measurements were performed according to Figure 1. Two gold electrodes were evaporated onto the glass substrate in a separate vacuum chamber before mounting in the uhv-system. Measurements were made by applying a fixed

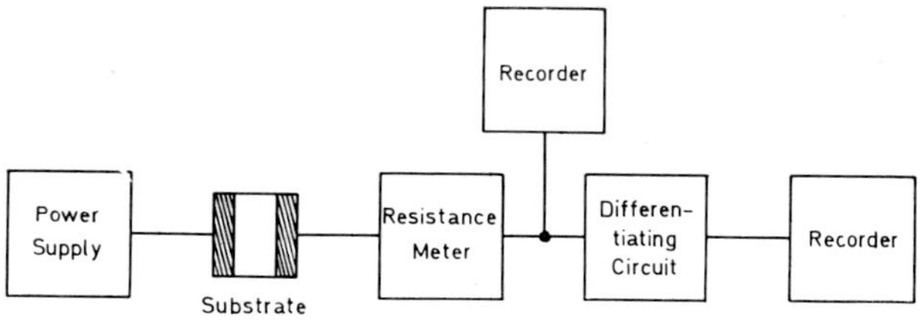

Figure 1. Block diagram over the equipment for electrical measurements. The differentiating circuit makes it possible to measure the slope of the resistance curve and determine the LSC. The size of the substrate is 0.5 × 0.5 in. and the film is 0.5 × 0.5 mm.

voltage, $-U_0 \pm 0.1\%$, to one electrode and connecting the other to the input of an electrometer amplifier in a resistance meter.[6] The resistance meter is capable of registering currents from 0.1 pA to 1 mA (i.e. resistances in the range $10^{14}-10^4\ \Omega$ when $U_0 = 10$ V) continuously. Resistances as low as 10 Ω could be measured by changing the applied film voltage to −10 mV. By maintaining a fixed voltage over the film, during the *in situ* measurements, and allowing the current to increase during film growth, the Joule heating was minimized. The value of U_0 was 10 V in the measurements.

A discontinuous film requires a backing support in order to maintain the structure when it is investigated in an electron microscope. A stabilizing carbon layer was evaporated on top of the gold film in a separate vacuum system. The substrate was dissolved in diluted HF acid and the subsequent free film was photographed in a Phillips EM 300 electron microscope operating at 100 keV. Figure 2 shows a sequence of photographs, on six films with different thicknesses, taken 20 h after the deposition.

3. Coalescence growth

The sizes of the islands shown in Figure 2 were analysed regarding them as having elliptical images with major and minor axes a and b, respectively. In order to simplify the treatment of the sizes an 'equivalent projected diameter' was defined, the geometric mean $x = (ab)^{1/2}$ which is the diameter in a circular image. This makes it possible to use a theory for describing size distributions. The theory was originally developed for describing gas evaporated ultrafine particles.[3] Here the particles were assumed to grow by liquid-like coalescence after binary collisions. In the case of a thin film this is named coalescence growth.

The island growth is considered to proceed as a series of discrete binary collisions between islands moving on the substrate area.[4] Thus the final discontinuous film has grown by a finite number j of island collisions. At each collision and coalescence the change of island volume is assumed to be a random fraction of the volume V_j after coalescence.

$$V_j - V_{j-1} = y_j V_j \tag{1}$$

where y_j, $j = 1, 2, 3, \ldots,$ is a set of mutually independent random variates, also independent of V_j. After j collisions the initial island volume V_0 has changed to V_j.

$$V_j = V_0 \prod_{k=1}^{j} \frac{1}{1 - y_k} \tag{2}$$

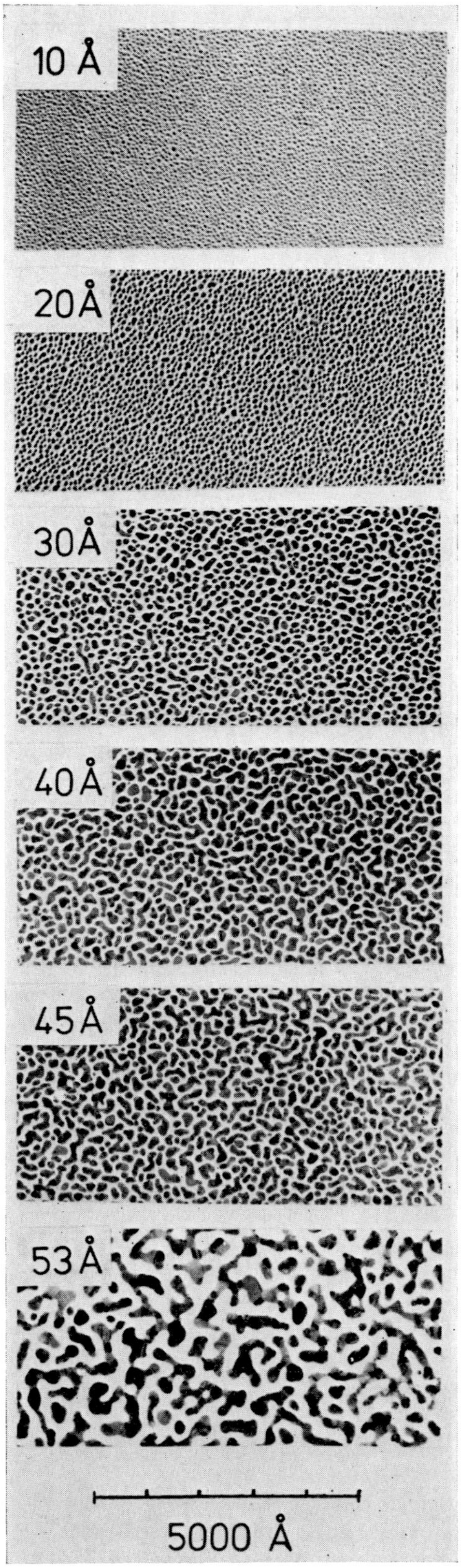

Figure 2. Electron micrographs of discontinuous gold films at six different values of the average film thickness. The figures represent regions in the middle of the film. At 45 Å the logarithmic film resistance was 12.7 at the deposition stop and 6.4 at 53 Å.

The logarithm of equation (2) gives a sum of independent random variables each with the same probability distribution

$$\ln(V_j/V_0) = \sum_{k=1}^{j} \ln\left(\frac{1}{1-y_k}\right) \quad (3)$$

Now the Central Limit Theorem of mathematical statistics predict that $\ln(V_j/V_0)$ goes asymptotically to a normal distribution. It follows then that the island volumes obey a log-normal distribution. The log-normal distribution function (LNDF) for the volume is

$$f_{LN}(V) = \frac{1}{(2\pi)^{1/2}\ln\sigma_V}\exp\left\{-\frac{1}{2}\left(\frac{\ln(V/\bar{V})}{\ln\sigma_V}\right)^2\right\} \quad (4)$$

where $\bar{V}$ is the median volume and σ_V its standard deviation.

In discontinuous films, particle sizes, not volumes, are measured on electron micrographs. Spherical islands and oblate spheroids (with symmetry axis perpendicular to the surface) have an image which is circular. However, ellipsoidal islands, prolate and oblate spheroids (with symmetry-axes parallel with the surface) have an image consisting of an ellipse.

For islands whose volumes are written in the form (e.g. islands described by the geometric mean $x = (ab)^{1/2}$):

$$V = \eta x^s \quad (5)$$

where η and s are constants, the log-normal distribution function is expressed by

$$f_{LN}(x) = \frac{1}{(2\pi)^{1/2}\ln\sigma_x}\exp\left\{-\frac{1}{2}\left(\frac{\ln(x/\bar{x})}{\ln\sigma_x}\right)^2\right\} \quad (6)$$

where $\bar{x}$ is the median diameter and σ_x is the standard deviation in the log-normal distribution function.

In Figure 3 it is shown that the quantity is well described by

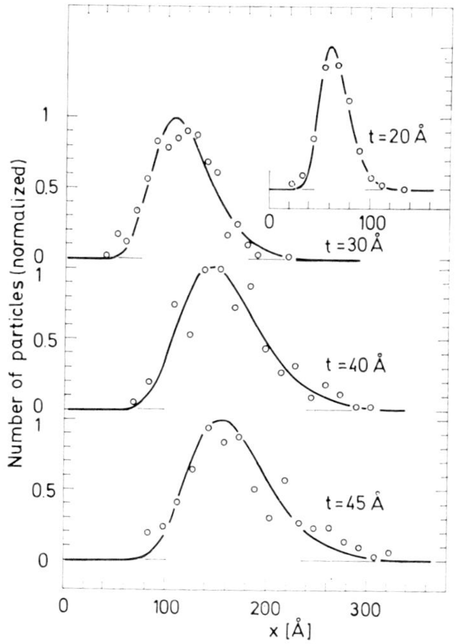

Figure 3. Size distributions for discontinuous gold films of four different thicknesses. Each of the curves is based on measurement on 300 islands. The circles represent the middle of size histograms and the curves are the fitted log-normal distribution functions. The values of $\bar{x}$ and σ_x are: 65, 1.25; 116, 1.30; 155, 1.30 and 165, 1.28 respectively.

the LNDF. Here the size distributions are shown for four different thicknesses. An electric field parallel to the surface plane has earlier been shown to increase the coalescence. Several combinations of metals and substrates have elsewhere been investigated.[7–10] The results were that an electric field induces the formation of an electrically continuous structure earlier than when there is no field applied.

All the present experiments were carried out with an electric field of 20 V/cm applied in the substrate plane and with a suppressed electron current in the beam. The results for room temperature on the substrate were that the electrical conduction in the film starts at an average thickness of 45 Å, while the metallic continuous film is formed of an average thickness of 53 Å. The electrical conduction during the deposition will now be discussed and related to structural change of the film.

4. Electrical conduction

4.1. The film resistance during the deposition. The electrical conduction during the deposition of a gold film on to a glass surface is shown in Figure 4. Here the logarithmic resistivity, ρ, is

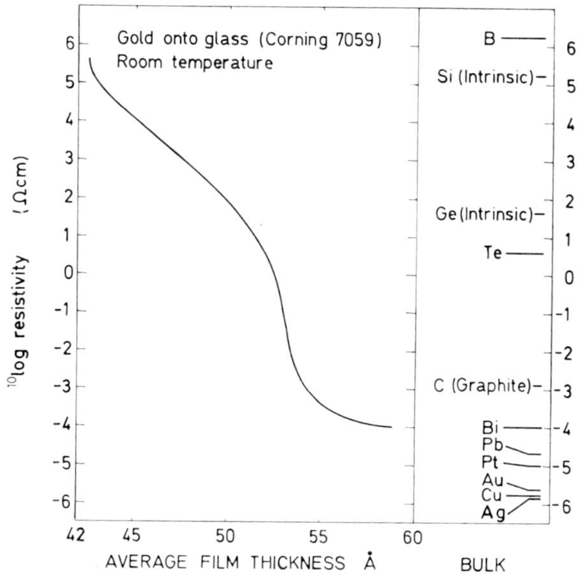

Figure 4. The variation of the film resistivity during the deposition of a gold film on to glass (Corning 7059) at room temperature. Inserted is the resistivity of some materials with activated or metallic conduction.

shown as a function of the average film thickness. Inserted is also the values of the resistivity for different materials. From the resistivity values, these materials can approximately be separated into two regions: semiconductors with $\log\rho > 0$ and metals with $\log\rho < -3$.

According to the slope of the resistivity curve this can be divided into three parts. The upper part of the curve covers resistivity values of the same order of magnitude as for the semiconducting materials. When the measurement begins the film is discontinuous (cf. Figure 2). The conduction mechanism in a discontinuous film is similar to that of a semiconductor since an electron must be activated a finite energy in order to be transferred to a neighbouring island. In a discontinuous film with activation energy δE and island separation d (in Å) the conduction θ_c can be written:[11]

$$\sigma_c = \sigma_0 \exp\left(-\frac{\delta E}{kT}\right) \exp\left(-1.03\,(m^*\bar{\phi})^{1/2} d\right) \quad (7)$$

where σ_0 is a constant, k is the Bolzmann constant, T is the absolute temperature, m^* is the electron mass of a tunnelling electron relative the free electron mass and $\bar{\phi}$ is the mean inter-island potential barrier height (in eV). The conduction depends very strongly on the separation d between two islands. Since the experimental curve is obtained during the deposition the interpretation is that the film is here discontinuous, and that the resistivity change depends on the decreased separation between the islands. In the actual thickness interval the first exponential factor in equation (7) can be assumed to be approximately constant. If the deposition is interrupted the post-deposition resistance increases to a saturation value after 20 h ageing. Here the resistance as a function of the applied electric field is first ohmic and then decreases approximately linearly with the square root of the applied field.[12] Furthermore the temperature coefficient of resistance is negative. All these properties are typical for discontinuous films.

In the resistivity range which separates semiconducting and metallic materials the rate of resistance change in the experimental curve is larger. Elsewhere it has been shown[13] (also for Al films on ceramic substrates[14] and for cermet films[15]) that this stage in the curve is associated with the formation of stable metallic continuous paths. The intermediate stage is best described by the sheet resistance value, $R_\square$, which is measured during the deposition.

If the deposition is interrupted when $R_\square < 1$ MΩ/□ the film resistance decreases as a function of time. Stable films have metallic electrical properties such as ohmic conduction and positive temperature coefficient of resistance.

In a bulk metal the intrinsic resistivity is considered to be a sum of temperature-dependent scattering of electrons at phonons and a non-temperature-dependent scattering at impurities. In comparison with a bulk there are additive contributions to the resistivity in a thin film[16] from: a larger amount of grain boundaries, different chemical composition, and different mechanical stresses. When the electron mean free path is comparable with the film thickness there is also an additive term to the total resistivity from scattering of electrons at the outer surfaces. This scattering depends on the nature of the surface such as the surface roughness and adsorbates on the surface.

When $\log \rho < -3$ in Figure 3, the film is metallic continuous. Here the conduction increases with increasing film thickness. However, the resistance value for a film is much larger than the bulk value. This depends on the filamentary character of the film, the elevated film temperature during deposition and the previously mentioned structural differences between the film and the bulk.

According to the electrical measurements, the resistivity range, $0 < \log \rho < -3$, determine the LSC. This result is supported by structural investigation.

4.2. The change in film structure at LSC. The area fraction covered with metal as a function of film thickness is shown in Figure 5. The films which have a non-metal conduction were found to cover between 47 and 55 % of the substrate. Films with a metallic conduction have a surface coverage between 71 and 78 % in the thickness range 53–85 Å. At critical thickness 53 Å there is a step in the surface coverage. All the films are aged during 20 h after the deposition stop. A change in the structure can occur during this time. An interpretation of the difference in the surface coverage is that the film temperature attained from the evaporation source causes an agglomeration of the discontinuous film while the semicontinuous films increase their surface coverage. The mean value of the surface coverage, on the six films nearest the transition at 53 Å, is 60 %. This value can be compared to theoretical calculations on percolation using randomly distributed overlapping spheres.[17] These calculations predict continuous metallic paths to be formed when 68 % of the surface is covered. Figure 5 also shows that the films are semicontinuous within a thickness interval extending more than 100 Å.

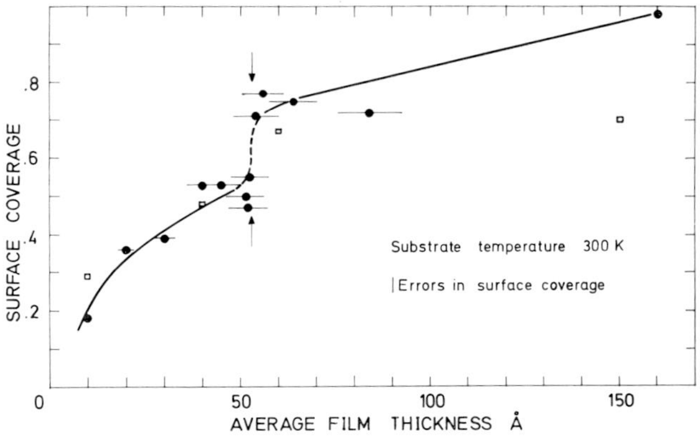

Figure 5. The surface fraction covered with metal as a function of the average film thickness. The critical thickness is marked with arrows.

If the surface coverage changes very suddenly this must be observed also in the island density. In Figure 6 the island density is plotted as a function of the film thickness. The island density decreases smoothly until the film thickness 53 Å where there is a reduction of approximately 90 %.

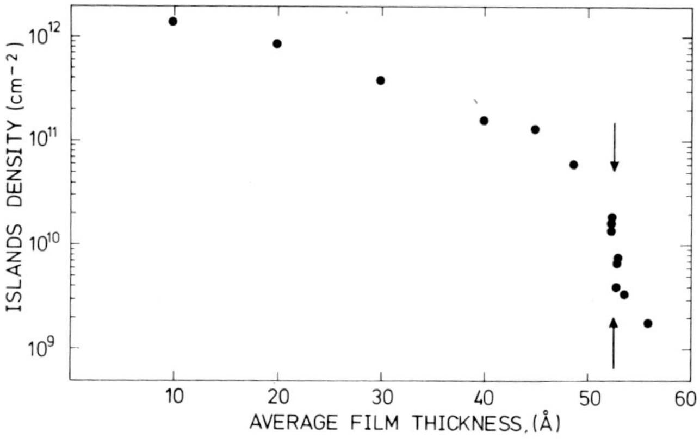

Figure 6. The island density as a function of average film thickness. There is a large reduction of the number density of islands at the average film thickness 53 Å.

Summary

Gold films deposited on glass were discontinuous in the interval 10–45 Å. The size distribution of islands was well described by a log-normal distribution function in the thickness range 20–45 Å. At 45 Å the film starts to conduct current when an electric field is applied in the substrate surface from the beginning of the deposition. Analyses of the resistance curve as a function of thickness, the fractional surface area covered with material and the island density show that the film grows to be metallic continuous at the average thickness 53 Å. However, the surface is not completely covered until the film thickness is 160 Å.

Acknowledgements

We gratefully acknowledge Professor G Brogren for support during the course of the work. Dr C G Granqvist is thanked for many valuable discussions and for reading the manuscript.

References

[1] J F Hamilton, P C Logel and R C Baetzold, *Thin Solid Films*, **9**, 1976, 233.
[2] J A Venables, *Thin Solid Films*, **32**, 1976, 135.
[3] C G Granqvist and R A Buhrman, *J Appl Phys*, **47**, 1976, 2200.
[4] T Andersson and C G Granqvist, *J Appl Phys*, in press.
[5] T J Coutts, *Electrical Conduction in Thin Metal Films*, Elsevier, Amsterdam (1974).
[6] T Andersson, L Yström and B Persson, *J Phys E: Sci Instrum*, **8**, 1975, 398.
[7] K L Chopra, *J Appl Phys*, **37**, 1966, 2249.
[8] D I Kennedy, R E Hayes and R W Alsford, *J Appl Phys*, **38**, 1967, 1986.
[9] R M Hill, *Thin Solid Films*, **1**, 1967, 241.
[10] E Ahilea and A A Hirsch, *J Appl Phys*, **42**, 1971, 5601.
[11] T Andersson, *J Phys D*, **9**, 1976, 973.
[12] T Andersson, *J Appl Phys*, **47**, 1976, 1752.
[13] T Andersson, PhD thesis, (unpublished).
[14] J Fridrich and J Kohout, *Thin Solid Films*, **7**, 1971, R49.
[15] B Abeles, P Sheng, M D Coutts and Y Arie, *Adv Phys*, **24**, 1975, 407.
[16] P Wissman, *Surface Physics*, *Springer Tracts in Modern Physics*, No 77, Springer-Verlag, Berlin (1975).
[17] G E Pike and C H Seager, *Phys Rev*, **B10**, 1974, 1421.

2.25 Thickness dependent effects of thermal ageing in thin conductive films

A J Tosser,* C R Tellier* and J Launey, *Université de Haute-Normandie, Laboratoire d'Electronique et d'Automatique, B.P. 4006–76066 Le Havre Cedex, France*

Al and Zn films, 110–900 Å thick, are dc or rf sputtered; SnO_2 films in the 1500–8000 Å thickness-range are rf sputtered. Thickness dependence variations in electrical resistance are consistent with a two-layer model: the first layer, less than 150 Å thick, is the thinner continuous layer that may be obtained; the second layer has a surface smoothness increasing with thickness up to a value near 1300 Å, above this thickness the surface ordering progressively decreases. Mayadas–Shatzkes conduction (with a constant grain diameter) occurs within metal films; thus the electronic reflection factor on grain boundaries and the electronic specular reflection factor on interfaces are determined. Thermal ageing slightly modifies the grains but improves the interface specular factor and induces thickness dependent ageing effects. Variations in the tcr of metallic films are consistent with the two-layer model and the approximate M–S equations.

1. Introduction

Thermal ageing is a convenient procedure to obtain a good reproducibility of electrical characteristics in thin films and has been extensively studied in the case of evaporated films[1,2] but only separate indications have been given about dc or rf sputtered thin films[3–10]; we describe in this paper a general aspect of the thickness dependent ageing effects in thin sputtered conductive alms.

2. General phenomena

Thin Al, Zn and SnO_2 films were rf or dc sputtered on to vycor slides. The vacuum apparatus has been previously described[7,8,11]; a programmed and regulated furnace was used to anneal the films *in situ* at a stabilized temperature (up to 414 K for metallic films; up to 675 K for SnO_2 films) for various times.

In the case of rf sputtered metallic films (Al[9] and Zn; 110–900 Å thick) a typical thickness dependence is observed in the variation of the ratio R_s/R_0, where R_0 is the initial value of the electrical resistance of the layer and R_s the stabilized value obtained after thermal ageing (Figure 1); however, R_0^{-1} and R_s^{-1} are linear functions of the thickness d for the low values of the sheet resistance (i.e. less than 8 Ω sq^{-1} in Al films); furthermore, R_s becomes much more important than R_0 in the vicinity of a critical thickness d_1, and even may be infinite.

These phenomena agree with the two-layer model we have previously proposed for dc sputtered Zn films[8] and rf sputtered Al-films[9]; in this model, the metallic film is not continuous up to the first critical layer of 114 Å in Al[9] and 140 Å in Zn; above this first critical thickness the smoothness of the film surface is progressively improved; a perfect smoothness is obtained for the second critical thickness d_2 of about 1200 Å thick for Al films.[9] Similar results are obtained with dc sputtered zinc films[8] ($d_1 \approx 140$ Å, $d_2 \approx 1020$ Å) but the defect introduced resisitivity is far more important than in rf sputtered films.

These general features are related to the nucleation-growth model proposed by Laville Saint-Martin[12–14] for dc sputtered films. The authors have shown that the nucleation-growth of sputtered films is determined by the conditions under which metatastable 'entities' may be produced in the cold plasma and an Auger de-excitation process may occur on the substrate with available electrons; in its first steps, this process does not depend on the nature of the entities but only on the nucleation centres, i.e. on the nature of the substrate and its surface state.

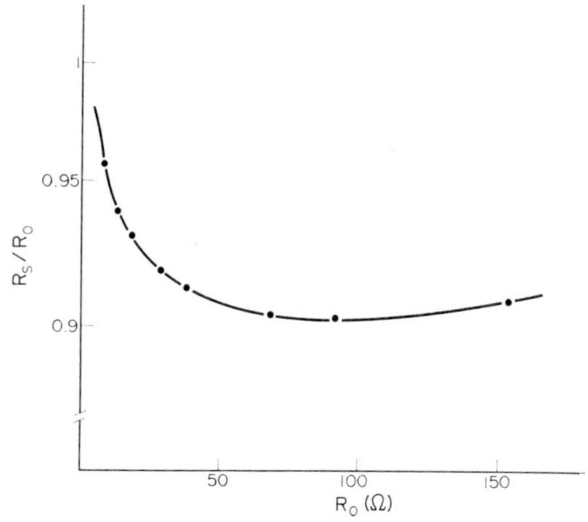

Figure 1. Resistances of thin rf sputtered Al films (48 mm long, 9 mm wide) before ageing (R_0) and after ageing (R_s).

In good agreement with the nature of this nucleation-growth of these films, we have observed that the sputtering process may be temporarily stopped and that no alteration of the electrical properties is thus induced; in the same way, the stabilized value of the resistance R_s does not depend on the substrate temperature in the 20–90°C range and the same value R_s is obtained after thermal ageing whenever spontaneous ageing has occurred or not.

3. Ageing effects

3.1. Ageing of Al and Zn sputtered films. We consider the two-layer model (Figure 2) and attempt to give an interpretation of the observed typical resistivity recoveries previously reported

* Present address: Laboratoire d'Electronique, Université de Nancy-1, C.O. 140–54037 Nancy Cedex, France.

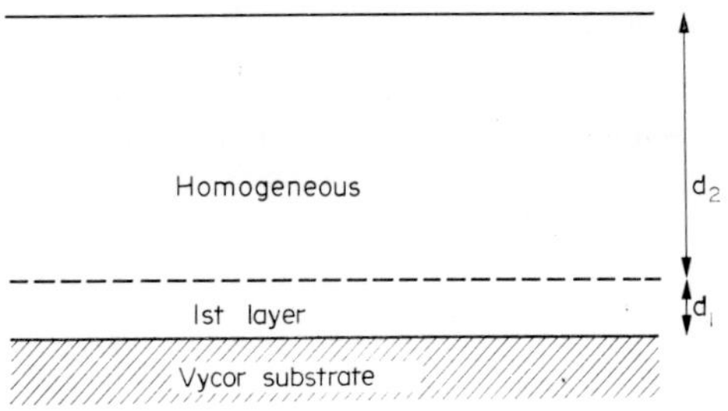

Figure 2. The thickness two-layer model of sputtered metallic films.

for Al.[10] When the film is thin, the resistivity recovery may be principally attributed to the first layer (resistance r_1); in this assumption, experimental data give a constant value of the resistance ratio r_{1s}/r_{10} of the first layer after and before ageing (the value is 0.89 for Al films[9] and a slow variation in r_{10} with R_0 is observed.

Hence we suppose that the first layer has a given thickness (equal to the first critical thickness) and that its resistance does not depend on the thickness of the second layer; the resistivity recovery of the first layer induced by ageing is then attributed to background modifications; that is consistent with the high value of the resistivity (several times the bulk value) and with the value of the activation energy of isothermal recovery whose value is near 0.8 eV for higher temperature.

We have shown[9] that the electrical conductivity in the second layer was related to a Mayadas–Shatzkes process[15] with simultaneous background, grain boundaries and surface scatterings; since the layer is quite homogeneous, with regard to the electrical properties, and since the calculated resistivity ρ_g of an infinitely thick polycrystalline film was far from the bulk value, we advanced that the mean grain diameter was constant[9] and equal to the first critical thickness.

The linearized expressions of the resistivity proposed by Mola and Heras[16] fit with the experimental data (Figure 3) and allow the calculation of the resistivity ρ_g. From Mayadas–Shatzkes[15] exact values of ρ_0/ρ_g (where ρ_0 is the bulk resistivity) we may deduce the exact values of the grain boundary reflection coefficient R before and after ageing (R_0 and R_s respectively), they differ slightly.

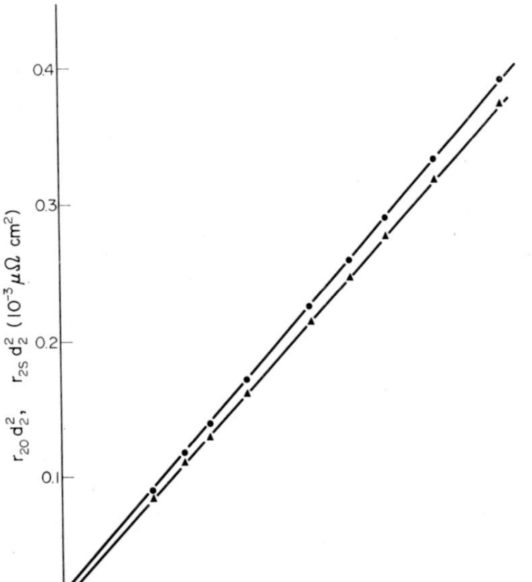

Figure 3. Variations with the second layer thickness d_2 of the product of the film resistance by thickness square power. ●: before ageing, ▲: after ageing.

In the case of Al films, the calculated values of R respectively are:

$R_0 = 0.449$ before ageing

$R_s = 0.442$ after ageing

It thus appears that ageing has a negligible effect on grains and the observed resistivity recovery of the second layer is attributed to a surface phenomenon; that also corresponds to the activation energy of the isothermal recovery whose value is less than 1 eV (0.32 to 0.42 eV for Al films 150 to 600 Å thick).[10]

If we assume that the linearization parameters tabulated by Mola and Heras[17] are linear with $(1-p)$, where p is the usual Fuchs–Sondheimer[18] electronic specular reflection factor, we calculate approximate values of p before and after ageing (p_0 and p_s respectively) which differ markedly. This assumption is valid since the approximate expressions of the electrical resistivity of polycrystalline films (Appendix 1) obtained in this way[19] constitute an adequate approximation of the tabulated exact values.[17]

The values of the specular electronic reflection factor p in Al films are:

$p_0 = 0.175$

and

$p_s = 0.505$

These facts indicate that a surface re-ordering is induced by ageing. Since the interfaces are different, p must be considered as a mean reflection factor. If we assume that the layer interface is slightly modified by ageing and that the electronic reflection on the surface interface is diffuse before ageing, the values of the specular reflection factor p_1 and p_2 (on the lower and on the top interface respectively) are according to Juretschke.[20]

Before ageing

$0.38 < p_{10} < 0.424$

$p_{20} = 0$

After ageing

$0.38 < p_{1s} < 0.424$

$p_{2s} \approx 0.67$

It thus appears that the resistivity recovery of the second layer is mainly due to the top surface re-ordering.

This agrees with the observed variations in $\alpha' = r_{2s}/r_{20}$ with r_{20} (Figure 4). Ageing has a low efficiency, i.e. $\alpha' \approx 1$, for thicker films which cannot be reordered because of their surface smoothness and for thinner films which are heavily connected to the grain-boundaries of the first layer, from which nucleation-growth goes on.[12]

We have compared the resistance ratio of the second layer r_{2s}/r_{20} for a given metal thickness obtained either after one sputtering process or after repeated sputterings with intermediate ageings. We observe that the ageing is less efficient when the layer (thicker than the first critical thickness) has been deposited on an aged layer instead of the substrate and we assume that it may be attributed to a better surface ordering; when the superimposed layers are thinner than the first critical layer, the effect of ageing is more important than on the critical layer grown on the substrate, but this comparison does not seem significant since the physical conditions of nucleation are very different; it only shows that an only critical layer cannot be so easily reordered than a superimposed layer of the same thickness grown from a metallic substrate, in good agreement with the nucleation-growth process.

3.2. Ageing of dc sputtered Zn films.[8] The two-layer model gives a convenient description of the electrical phenomena but the high density of defects modifies the quantitative aspect. In the first layer the resistivity is modified by ageing whereas the resistivity of the second layer is slightly modified, since the only ageing effect is the enhancement of the surface smoothness.

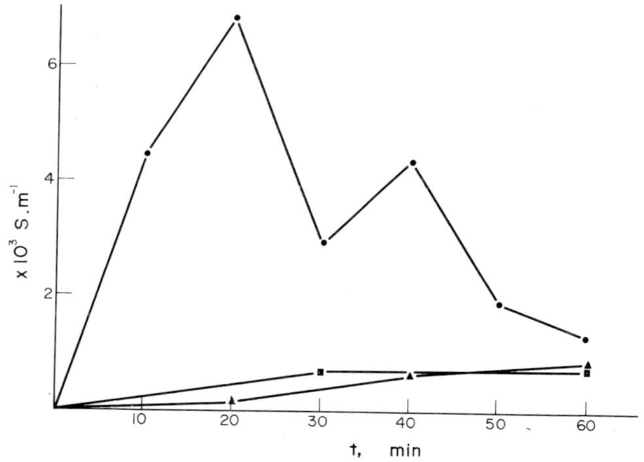

Figure 5. Variations in the SnO_2 conductivity with sputtering time and fractioned sputtering, for several elementary sputtering times.

3.3. Ageing of SnO_2 films. According to the theoretical considerations[14] a similar phenomenon of nucleation-growth of SnO_2 films may occur, but the surface electron supply is less important than for metallic films; hence the Auger de-excitation process cannot be operated with the same efficiency and these conductive layers may be less ordered which allows a higher and variable density of defects: we have effectively observed[7] that the conductivity of aged films markedly varies with the sputtering time (Figure 5). The minimum sheet resistance corresponds to 900 Å; nevertheless, a constant sputtering rate (approximately 92 Å/min) was obtained for conductive films of more than 300 Å thick; since the tcr of annealed layers is approximately $-5.10^{-3}/°C$ in the whole range of thickness (600–5000 Å), it suggests that the defect introduced resistivity is temperature independent, as observed in Al films.[10]

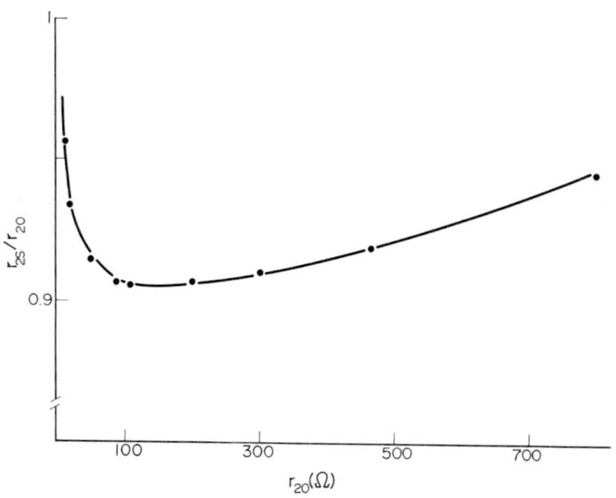

Figure 4. Variations in the ratio of the second layer Al resistances, after and before ageing, with the second layer resistance before ageing.

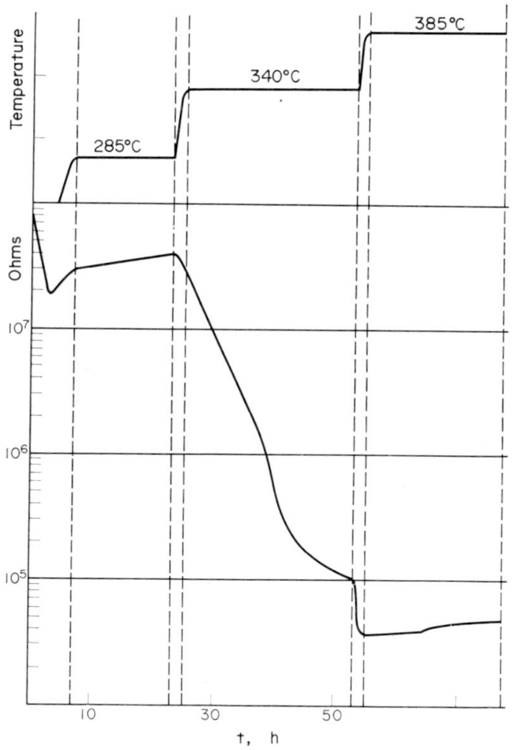

Figure 6. Variations in the SnO_2 resistivity with ageing temperature T.

An important irreversible decrease in resistivity occurs at an ageing temperature of about 330°C (Figure 6) which is slightly thickness dependent. We observe that repeated sputtering with intermediate ageings gives lower values of resistance than a simple sputtering of same total duration; but for thicker films (near 5000 Å), the value of the total resistance is the same whether repeated sputterings and intermediate ageings have been used or not (Figure 4).

The conductivity improvement increases when the value of the elementary sputtering time is in the vicinity of 10 min; that corresponds to a minimum sheet resistance ($\approx 1\ k\ \Omega$sq) after ageing. Furthermore, the important decrease in the total resistance induced by ageing is less marked when several layers are superimposed and the experimental data are consistent with the hypothesis that only the last superimposed layer is modified by ageing. The spontaneous evolution in moist air has a lower magnitude when the number of elementary sputtering process increases; it even becomes negligible for 6 or 7 elementary sputterings. All these data are consistent with the nucleation-growth of SnO_2 films if we assume, in the same way as above, that ordering progressively decreases when the thickness increases above the second critical thickness, and that the thickness dependent effect of ageing is mainly a partial re-ordering.

4. Discussion

The two-layer model may also be used to calculate the tcr of metal films. For the first layer tcr, β_1 is extrapolated from the plot of total tcr β versus total thickness d; for the top layer tcr β_2 is then calculated from β and β_1. The variations in β_2 with the thickness d_2 of the second layer agree with the theoretical expressions that may be deduced (Appendix 2) from the Mayadas–Shatzkes model and give the grain tcr β_g, i.e. the tcr of an infinitely thick polycrystalline film, and an approximate value of p_s.

In the case of Al films we calculate:

$$p_s = 0.425$$

and

$$\beta_g = 11.04 \,.\, 10^{-4}\ \mathrm{K}^{-1}$$

That gives

$$\beta_0 = 4.13 \,.\, 10^{-3}\ \mathrm{K}^{-1},$$

in good agreement with the usual value of 4.29 10^{-3} K^{-1}.

Hence it seems that the two-layer model may be implemented to describe the electrical phenomena occurring in thin conductive films.

Conclusions

Thin sputtered metallic Al or Zn films grow according to the two-layer model: the first layer of a given thickness, equal to the first critical thickness, is modified by thermal ageing; the second layer, less than the second critical layer thick, has a surface smoothness increasing with thickness and improved by ageing in a thickness dependent manner. These films are polycrystalline. Thin sputtered SnO_2 films also grow according to the two-layer model and a disordered structure appears on films thicker than the second critical layer.

Appendix 1

The resistivity of a polycrystalline film is

$$\rho_f = \rho_g[1 - A/F(\alpha)]^{-1}$$

Where ρ_g is the resistivity of an infinitely thick polycrystalline film.

The $F(\alpha)$ function is defined as the ratio ρ_g/ρ_0 of the bulk resistivity ρ_0 to the grain boundary resistivity ρ_g. It was shown[15] that:

$$F(\alpha) = 1 - \tfrac{3}{2}\alpha + 3\alpha^2 - 3\alpha^3 \ln\left(1 + \frac{1}{\alpha}\right) \tag{2}$$

The crystallite size d_g is related to the bulk mean free path l_0 and to the grain-boundary reflection coefficient R by equation (3)

$$\alpha = \frac{l_0}{d_g}\frac{R}{1 - R} \tag{3}$$

The A term which takes into account the effect of the film thickness (Fuchs–Sondheimer effect) is given by:

$$A = \frac{6}{\pi k}(1 - p)\int_0^{\pi/2} \mathrm{d}\phi \int_1^{\infty} \frac{\cos^2\phi}{H^2(t, \phi)} \left(\frac{1_3}{t} - \frac{1_5}{t}\right) \frac{1 - \exp[-kt\,H(t, \phi)]}{1 - p\exp[-kt\,H(t, \phi)]}\,\mathrm{d}t \tag{4}$$

Where $k = \mathrm{d}/l_0$ is the reduced thickness and

$$H(t, \phi) = 1 + \frac{\alpha}{\cos\phi(1 - 1/t^2)^{1/2}}$$

In case of a thin polycrystalline metallic film, equation (1) can be rewritten in the approximate linearized form[19]

$$k\rho_f/\rho_g \approx k + (1 - p)h(\alpha) \tag{5}$$

Appendix 2

For an infinitely thick polycrystalline film (i.e. without Fuchs–Sondheimer effect) the temperature coefficient β_g of grain boundary resistivity is related to the bulk tcr β_0 by the expression

$$\beta_g \approx \beta_0\left(1 + \frac{G(\alpha)}{F(\alpha)}\right) \tag{6}$$

where $G(\alpha) = \alpha\dfrac{\partial F(\alpha)}{\partial\alpha}$

Mayadas and Shatzkes[15] have shown that the total film resistivity ρ_f of a polycrystalline film for which the grain diameter is constant and independent of thickness can be approximated by a Fuchs–Sondheimer equation if the bulk mean free path l_0 is replaced by an 'intrinsic mean free path' l_g defined as

$$l_g \approx (\rho_0/\rho_g)l_0$$

Hence, the tcr of polycrystalline metallic film resistance may be

$$\mathrm{d}/\beta_f \approx \frac{1}{\beta_g}[\mathrm{d} + \tfrac{3}{8}\,l_g(1 - p)]$$

References

1 R Suri, A P Thakoor and K L Chopra, *J Appl Phys*, **46**, 1975, 2574.
2 G J Van Gurp, *J Appl Phys*, **46**, 1975, 1922.
3 M Adamov, B Perovic and T Nenadovic, *Thin Solid Films*, **24**, 1974, 89.
4 T T Sheng, R B Marcus, F Alexander and W A Reed, **14**, 1972, 289.
5 A Tosser, G Fleury and H Murray, *Thin Solid Films*, **15**, 1973, 259.
6 H Murray and A Tosser, *Thin Solid Films*, **24**, 1974, 165.
7 J Launey and A Tosser, *Electrocomp Sci Technol*, **2**, 1976, 223.
8 C Tellier and A Tosser, *Thin Solid Films*, 1976, to be published.
9 C Tellier and A Tosser, *Thin Solid Films*, to be published.
10 C Tellier and A Tosser, *Electrocomp Sci Technol*, to be published.
11 H Murray and A Tosser, *Thin Solid Films*, **17**, 1973, 75.
12 B Laville Saint-Martin and G Perny, *Thin Solid Films*, **4**, 1969, 319.
13 B Laville Saint-Martin, *Thin Solid Films*, **5**, 1970, 169.
14 B. Laville Saint-Martin, *Thin Solid Films*, **5**, 1970, 359.
15 A F Mayadas and M Shatzkes, *Phys Rev B*, **1**, 1970, 1382.
16 E E Mola and J M Heras, *Thin Solid Films*, **18**, 1973, 137.
17 E E Mola, J M Heras, *Electrocomp Sci Technol*, **1**, 1974, 77.
18 T J Coutts, *Thin Solid Films*, **2**, 1968, 337.
19 C R Tellier and A J Tosser, *Thin Solid Films*, **33L**, 1976, 19.
20 H J Juretschke, *J Appl Phys*, **37**, 1966, 435.

2.26 Nonlinearity measurements of thin films

S Hellström and H Wesemeyer, *Materials Laboratory, L M Ericsson Telephone Co; Tyresö, Sweden*

Measurement of the generated third harmonic component of voltage has proved to be a most sensitive method for the detection of nonlinearities and disturbances in the conduction path of thin film resistors. Inhomogenities and contact failures of thin film capacitors are easily revealed by the same method.

Tantalum thin film resistors and capacitors, today important for the realization of many hybrid circuits, are currently checked at the L M Ericsson Materials Laboratory by measurement in an equipment for determination of the third harmonic of voltage. Method and results with different material composition as well as relation to other thin film properties will be presented.

1. Introduction

By the advent of thin film technique a great deal of new, reliable electronic circuitry have been realized. But this technique requests control of processes in manufacture and failure detection. Besides conventional methods the revealing of nonlinearity by the detection of the generated 3rd harmonic voltage component has proved to be very valuable. Some applications will be presented here.

2. Resistors

Measurements. The detection of nonlinearities in resistors and their contacts by feeding the component under investigation with a pure sinusoidal current and measuring the fundamental voltage component as well as its harmonic components has been used for a long time (1–5). In the present application both the revealing of bad contacts and check of material properties will be treated.

Basic relations. The pure sinusoidal current in a nonlinear resistor will give rise to harmonics in the voltage. In general it has been found that for small currents the 3rd harmonic is predominant. This can be demonstrated by the expansion of voltage in powers of current for nonlinear I–V-relations:

$$V = \alpha_1 I + \alpha_3 I^3 + \alpha_5 I^5 + \ldots$$

Even powers are omitted to get the expression consistent with current reversal.

Assume

$$I = I_0 \sin \omega t$$

(load current, fundamental frequency $\nu = \omega/2\pi$).

Thus

$$V = \alpha_1 I_0 \sin \omega t + \alpha_3 (I_0 \sin \omega t)^3 + , . .$$

Neglecting at small currents the 5th and higher powers:

$$V = \alpha_1 I_0 \sin \omega t + \alpha_3 I_0{}^3 (\sin \omega t)^3$$

Trigonometrically:

$$(\sin \omega t)^3 = 1/4\ (3 \sin \omega t - \sin 3\ \omega t)$$

and

$$V = \underbrace{\alpha_1 I_0}_{V_1} \sin \omega t + \underbrace{3/4\alpha_3 \,.\, I_0{}^3}_{\ll \alpha_1 I_0} \sin \omega t - \underbrace{\tfrac{1}{4}\alpha_3 I_0{}^3}_{V_3} \sin 3\ \omega t$$

If

$$|\alpha_3| \ll |\alpha_1| : V_3 \approx \frac{\alpha_3}{4} \cdot \frac{V_1{}^3}{\alpha_1{}^3}$$

Typical values in cases under consideration: for $V_1 \sim 1V$ is $V_3 \sim \mu V$, so the above relationship should be applicable.

Because of the significance of the 3rd harmonic voltage value, as shown above, there was developed first a prototype device at L M Ericsson Labs., then a commercial available instrument for the test, on this base, of resistors, inductors and capacitors.[6] Distortion in the μV-range are readily detected, which implies a control method of very high sensitivity.

Since 1969 such an instrument is in use at the Materials Lab. of L M Ericsson for testing and evaluation of tantalum thin film components.[7] A block scheme of the instrument is given in Figure 1. To systemize the measurements we use the following

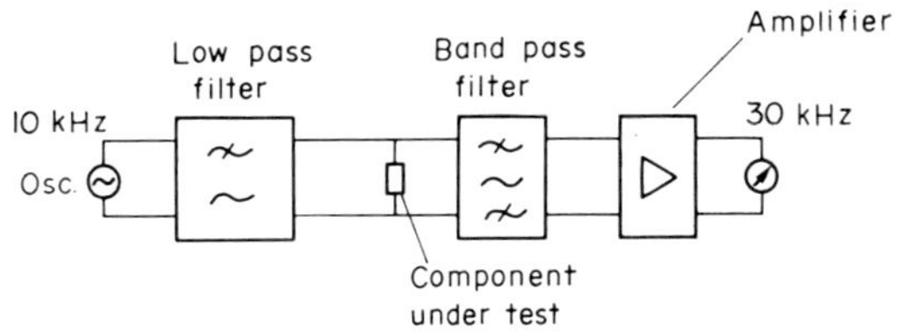

Figure 1. Block-scheme of test-equipment for voltage 3rd harmonic determination.

definitions of harmonic value in contrast to 'the Third Harmonic Index' = 20 log $V_3/V_1{}^3$, defined by others[4, 5]:

$$V_3 = 10^6 \cdot \frac{V_1{}^n}{V_0{}^{n-1}}, \tag{1}$$

which gives the same dimensions on both sides. V_3 is given in μV when V_1 has volt as unit. The parameters n and V_0 can be found from a series of measurements. A normalized value, $\bar{V}_3$, for intercomparison of different resistors is defined as the 3rd harmonic of 1 kΩ of the resistor material at a load of 1 mA:

$$\overline{V}_3 = 10^6 \cdot \left(\frac{R}{V_0}\right)^{n-1}, \tag{2}$$

where R = resistance in kΩ. The relation (1) is most easily analysed by interpretation of the logarithmic transform:

$$\log V_3 = n \cdot \log V_1 - (n-1) \cdot \log V_0 + 6 \tag{3}$$

The linear relationship is checked by a plot in a log–log-diagram. The parameters n and V_0 can be obtained graphically or analytically by linear regression. As mentioned, n is for weak currents around the value 3. Plots according to relation (3) are given in Figure 2(a) for two test-resistors and for a capacitor in Figure 2(b) to demonstrate the validity in the case of a dielectric. Figure 2(c) shows the resistance pattern.

Basically, there are two origins of nonlinearity in thin film resistors. Insufficient adhesion of contact material to resistor-layer or an insulating barrier between them can cause serious distortion revealed by an usually very strong 3rd harmonic component. Secondly the resistor material itself may give rise

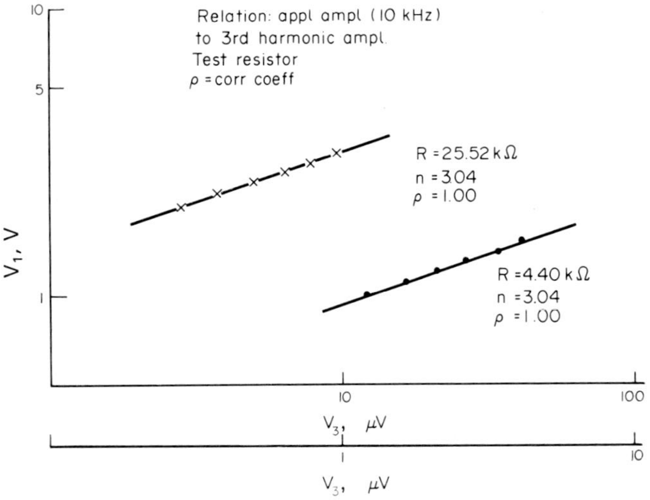

Figure 2(a). Relation of 3rd harmonic to fundamental voltage amplitude for meander (upper V_3-scale) and linear test-resistors. Resistor pattern according to Figure 2(c). (b) Relation of 3rd harmonic to fundamental voltage amplitude for a tantalum pentoxide thin film capacitor. (c) Test-resistor pattern.

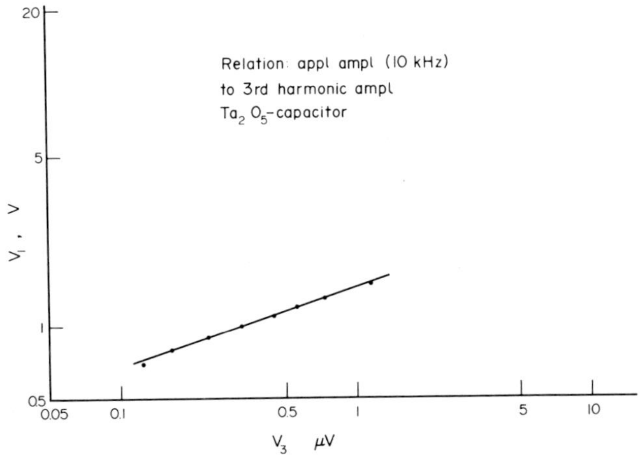

Figure 2(b).

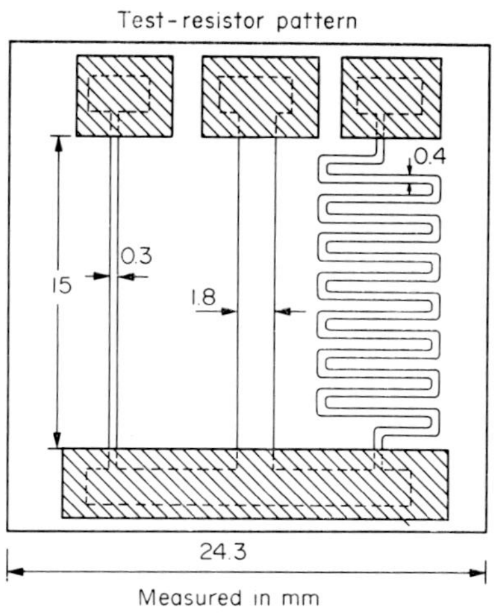

Figure 2(c).

to nonlinearities due to admixtures of special components as nitrogen and/or oxygen in tantalum thin films, to achieve predicted properties as surface resistivity and temperature dependence. Contribution from this latter is mostly rather weak but quite detectable. Next contact measurements are considered.

3. Contact characteristics of tantalum thin film resistors and capacitors

Contacts on tantalum resistors and top-electrodes of tantalum–pentoxide capacitors are formed by NiCr- and Au-layers. They are routinely checked by measurements of the voltage 3rd harmonic. This procedure is based on a suite of measurements performed with a special contact-material pattern. This was performed partly to establish the best evaporation–deposition condition of top-electrodes for tantalum–pentoxide thin film capacitors and contacts partly to gather basic data for routine check of thin film contacts in NiCr–Au.

Figure 3 shows the first pattern used. To reveal any contact distortion they were loaded, normally, with 10 mA. Measuring across all resistor parts as well as between widely spaced contacts made it possible to separate distortion from contacts and that from resistor material. See Tables 1 and 2. Inspection of these gives a difference of at least 10 times higher 3rd harmonic

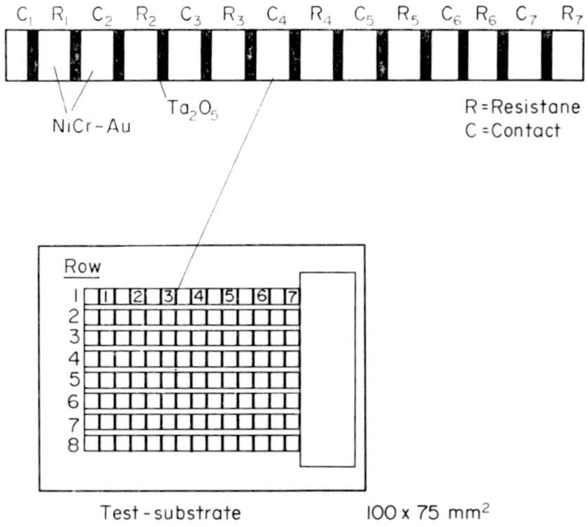

Figure 3. Test-pattern of NiCr–Au contacts and top-electrodes for thin film tantalum dielectrics.

Table 1. 3rd harmonic values of contacts and resistances of substrate with test-pattern. Load current: 10 mA.

No	Resistance (incl. contacts) (Ω)	3rd harmonic (μV)	Exponent n
1	8.52	0.115	2.95
2	8.24	0.173	2.93
3	8.54	0.121	2.92
4	8.08	0.094	2.63
5	8.22	0.091	3.03
6	8.53	0.051	2.79
7	7.98	0.038	2.63
			Mean-value 2.84 ± 0.06

Table 2. 3rd harmonic values for contact electronic transitions (separated from resistances).

No (contact)	3rd harmonic (μV) for resistance row no 2	4	6	8
2	0	0.0004	0.0011	0.0028
3	0.001	0.0002	0.00015	0.0043
4	0	0	0.0021	0.0052
5	0	0	0.0028	0.0050
6	0.0015	0.0005	0.0029	0.0006
7	0.00045	0.0028	0.0040	0.00095

value for resistances than for contacts, indicating very good contacts. The contacts are investigated by other methods, the direct pull method and heat treatment for instance, to confirm their quality. These contacts had been deposited at 150°C. Other temperatures were also investigated to determine dependence of contact function upon deposition temperature. The interfacial resistance was found to be about 10 mΩ.

On a substrate contact layers were deposited on tantalum as before but without preheatment. Measurement of the third harmonic resulted in five times higher values than in the preceding case.

A new pattern chosen to make able measurements of several contact bridges in series was applied to investigate the influence of initial and final substrate temperature on the layers at the evaporation–deposition process. The layers were deposited by a sputtering process to meet requirements for resistors or capacitors. The measurements of the contacts were repeated after one heat treatment at 150°C for 2 h and a second treatment at 250°C for 4 h. Even with low initial substrate temperatures acceptable contacts could be achieved, provided the contact layer was sufficiently thick (≥ 1μm), as checked by the 3rd harmonic method. Table 3 and Figure 4(a) and 4(b) give information about some characteristic results. From the results as a whole it could be derived that an initial substrate temperature of 100°C and final temperatures in the range 200–250°C gave the most reliable contact layers on tantalum thin films.

Besides conclusions given, effects from the orientation of the substrate during processing were observed as well as the healing of bad contacts by heat treatment (a drastic lowering in 3rd harmonic values). No influence from the tantalum layer composition could be detected. Now a contact pattern with 30

Table 3. Resistance-values, exponent, dimension factor V_0, and correlation coefficient, ρ, from 3rd harmonic measurements of electronic contact transitions. Index 1 = after deposition, 2 = after heating 2 h at 150°C and 3 = after heating 4 h at 250°C.

Sample and measure no		Resistance (Ω)	Exponent n	Dimension-factor V_0 (V)	Correlation coefficient ρ
321:1	R*	980	2.63	1820	0.99722
321:2		1004	2.84	3810	0.99936
321:3		1013	2.88	4650	0.99936
322:1A	R	777	2.88	288	0.99876
322:2A		776	2.83	364	0.99908
322:3A		787	2.86	276	0.99955
322:1B		777	1.83	144×10^3	0.99994
322:2B		776	1.91	73×10^3	0.99947
322:3B		787	1.08	12×10^3	0.99959
325:1	R	860	2.99	1440	0.99966
325:2		862	2.92	1570	0.99978
325:3		863	2.93	1680	0.99985
327:1	R	708	2.79	3200	0.99894
327:2		710	2.74	3620	0.99909
327:3		709	2.75	4010	0.99899
368:1	C	60	3.00	233	0.99995
368:2		61	3.02	213	0.99998
368:3		56	2.92	688	0.99966

* R = resistor layer, C = capacitor layer.

contact bridges in series on an 1 × 1 in. substrate giving 60 electronic transitions is used routinely for evaluation of the NiCr–Au-contact layers. See Figure 5.

The testing of whole circuits can be realized with 3rd harmonic measurements, the instrument mentioned also admitting partly automatic operation.

Further, evaluation of other types of contact layers should easily be achieved by this method.

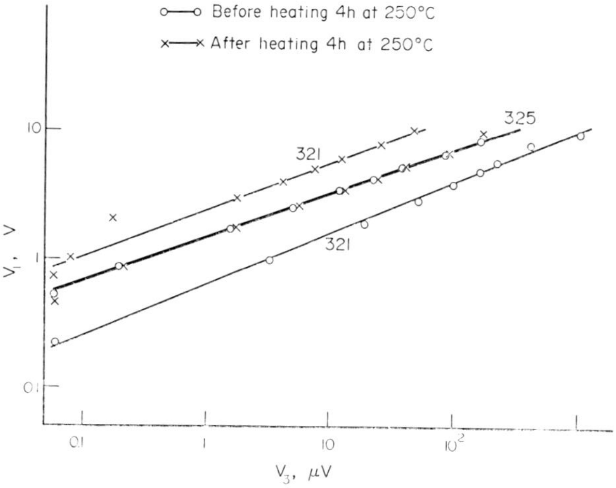

Figure 4(a). 3rd harmonic voltage measurements of a series of NiCr–Au contacts on tantalum, before and after heat treatment 4 h at 250°C. Initial substrate temperature 110°C and final 200°C. Contact layer thickness 0.27 μm. Samples 321 and 325. Compare with Table 3. (b) 3rd harmonic voltage measurements of a series of NiCr–Au contacts on tantalum, before and after heat treatment 4 h at 250°C. Initial substrate temperature 85°C and final 165°C. Contact layer thickness 1.24 μm. Samples 322, 327 and 368. Compare with Table 3.

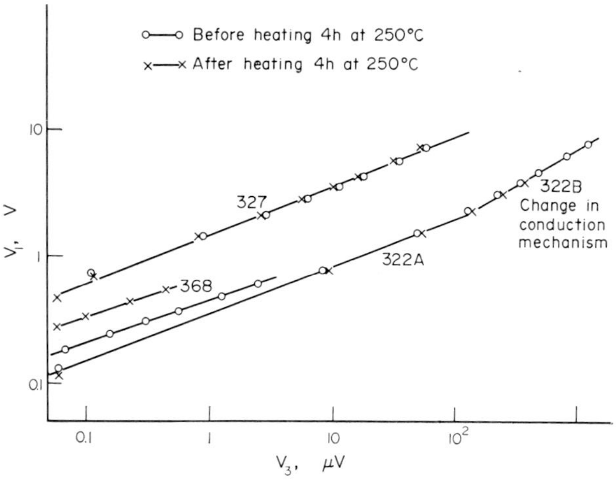

Figure 4(b).

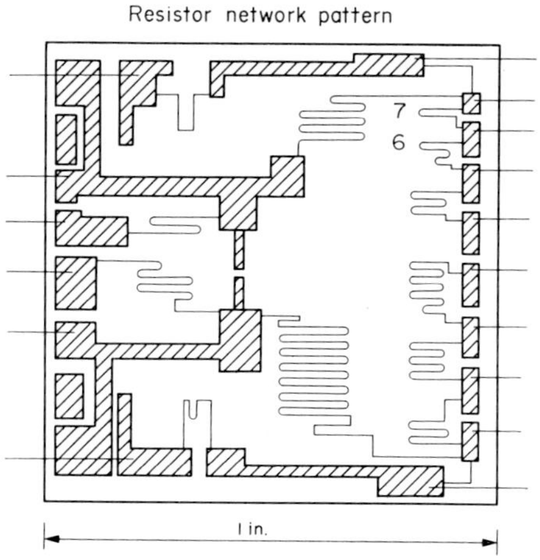

Figure 6. Resistive thin film network pattern for investigation of temperature coefficients.

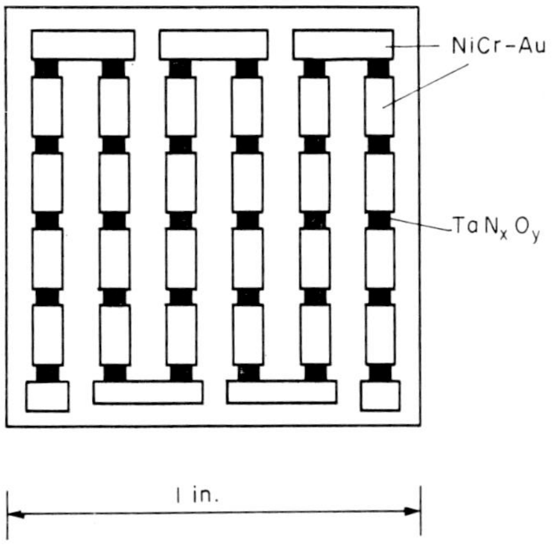

Figure 5. Pattern of contact-tantalum bridges giving 60 electronic transitions. Used for routine check of thin film contacts.

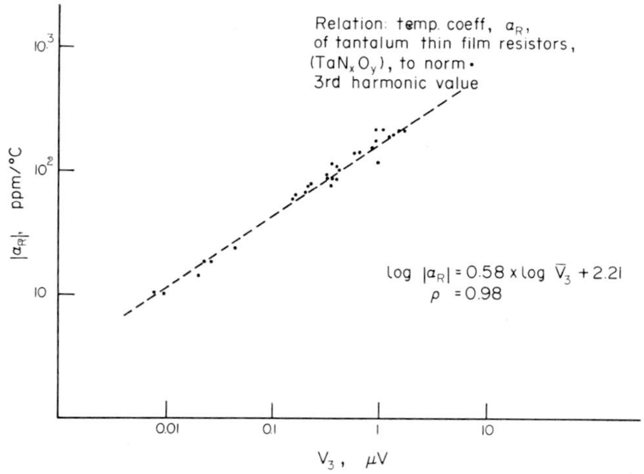

Figure 7. Relation between absolute value of temperature coefficient and 3rd harmonic voltage value for resistor 7 in Figure 6. Temperature coefficients are negative.

4. Nonlinearities from tantalum layer properties

Not only malfunction of thin film networks but also material properties of the thin film components are able to investigate by measuring the voltage 3rd harmonic.

Already in our early investigations,[7,8] we assumed some relation should exist between the detected 3rd harmonic signal and other nonlinearity dependent parameters of the material. The temperature coefficient of resistors and capacitors made from tantalum thin layers being important in their application became object to comparison with 3rd harmonic measurement. In such a case other contributions than from nonlinearity material properties must be excluded. Figure 6 shows a resistive network configuration used on this purpose. A series of temperature coefficient determinations and 3rd harmonic measurements were fulfilled. Figure 7 demonstrates the relation between these two parameters for resistor number 7 in the network, while Figure 8 refers to resistor number 6. These two resistors showed best reproducibility and were also chosen for standard temperature coefficient values. A very good correlation is proved between logarithmic values of the two parameters. This is probably due to equal origin in nonlinearities from grain barriers in the material and corresponds to the basic 3rd harmonic generation comparable to the more used signal to noise figure.

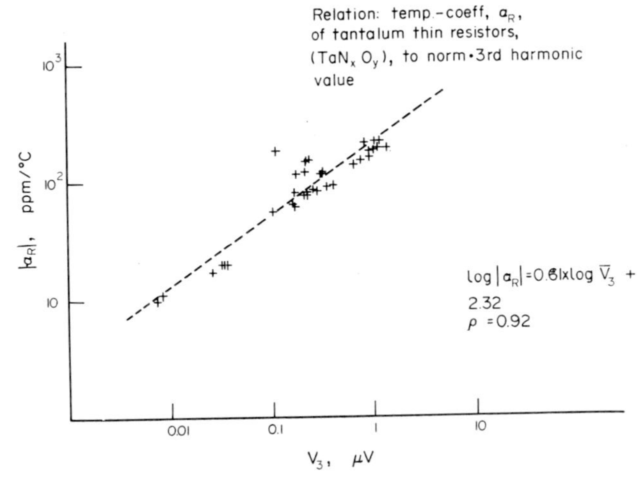

Figure 8. Relation between absolute value of temperature coefficient and 3rd harmonic voltage value for resistor 6 in Figure 7. Temperature coefficients are negative.

This result made it possible to facilitate the check of resistor temperature coefficients. In the case of a large output of networks the temperature coefficient determination is time-consuming, then only a few are measured and the rest compared through 3rd harmonic values. The advantage is evident. The relationship of temperature coefficient and 3rd harmonic value was also studied by others.[9]

Figure 9 gives a similar result with tantalum capacitors as that obtained for resistors. The experimental results confirm the fact, that the temperature dependence of the capacitance is

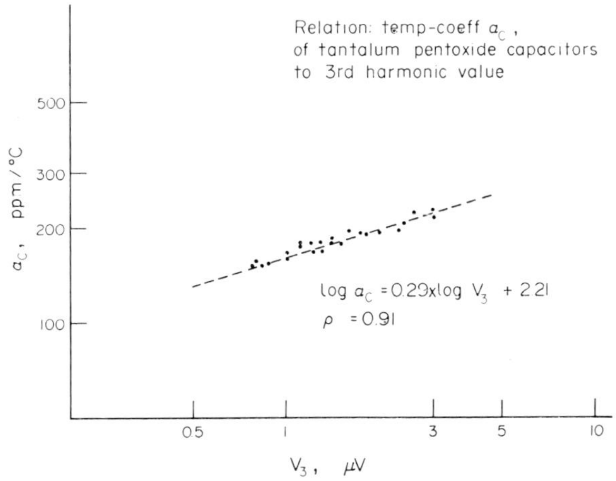

Figure 9. Relation between temperature coefficient and 3rd harmonic voltage value for thin film tantalum pentoxide capacitors.

mainly due to dielectric properties. Dielectric characteristics might be studied in this way. This is also very important as accordingly temperature coefficients of capacitance can, to certain limits, be checked by 3rd harmonic measurements. Both resistive and capacitive networks are currently checked and studied by 3rd harmonic voltage measurements. Investigation on relationship between 3rd harmonic voltage value and dissipation factor is in progress.

5. Results

The results given indicate a very sensitive method for checking and investigation of different thin film component properties. Care must, however, be taken to separate distortion of different origin.

This nonlinearity method is readily applicable both in production[5,10] and on laboratory scale. It has also attained international recognition as a test-method and is as such standardized by IEC in terms of the 3rd harmonic index.[11]

As an example of application in component research the method was tried for evaluation and prediction of life-length properties.[12] In this case, however, the first promising results have been contradicted by later experiments.[13]

6. Acknowledgements

We are indebted to Messrs S Boija, U Karlsén and H-G Westerlund of the L M Ericsson Materials Lab. for their assistance in measurements and thin film preparation. For permission to publish this work we are grateful to the Head of the L M Ericsson Materials Laboratory, E Edman.

References

[1] C E Mulders, *Tijdschri Ned Radiogenoot*, **22**, 1957, 337.
[2] V Peterson, *FKO-meddelande*, no 29, p 110, Royal Swedish Academy of Engineering Sciences, Stockholm (1958).
[3] G H Millard, *Proc Instn elect Engrs*, **106B**, 1959, 31.
[4] P L Kirby, *Elect Engng*, **37**, 1965, 722.
[5] V Peterson and P-O Harris, *Electronics*, **39**, 1966, 93.
[6] Radiometer, Copenhagen, *Instruction Manual for Component Linearity Test Equipment, Type CLT 1a.*
[7] S Hellström, L M Ericsson Telephone Co, Materials Lab, Internal Rep No 728240/Mgf 281 (1970) (in Swedish).
[8] S Hellström, L M Ericsson Telephone Co, Materials Lab, Internal Rep No Mgf 415 (1972) (in Swedish).
[9] H P Lorenz and H W Pölzberger, *NTZ*, **27**, 1974, H5, 190.
[10] E Martinsson, *Electl Components*, 1967, 371.
[11] Int Electrot Comm, IEC-report, Publication 440, First edition (1973).
[12] J C Andersson and V Rysanek, *Radio Electron Engng*, **39**, 1970, 321.
[13] J C Andersson, Private communication (1976).

2.27 1/f noise voltage in a thin film structure

S Husa, *Electronics Research Laboratory, The University of Trondheim, The Norwegian Institute of Technology, Trondheim, Norway*

1/f noise in current carrying thin film devices is caused by fluctuation of the electrical conductivity of the film material. Describing its effect on the voltage distribution in a device requires a two-dimensional analysis. This may be done by means of a theory, presented in this article, where the fluctuation is treated as current dipoles.
The geometrical dependence of the noise intensity across electrodes on Hall generator structures, made from InSb thin films, is measured. This is in fair agreement with predictions based on the theory.

Introduction

Noise measurements on thin films are usually carried out in a manner that gives easily interpretable results. An example of such an experiment is the measurement of the fluctuating component of the current running through a film when a constant voltage is applied.[1]

This method involves no field analysis. It has the disadvantage of being sensitive to contact noise. Another example of simplifying arrangements is to measure noise voltage across separate voltage electrodes with a displacement which is large compared to the width of the sample, which carries a constant current. This case requires a one-dimensional analysis.[2–4] The predictions will not be valid in the case of closely spaced electrodes or electrodes aligned at each side of the thin film specimen like in a Hall generator geometry. There is therefore a need for a two-dimensional theory describing the modulation of the voltage distribution at a thin film due to conductivity fluctuations in the film material. Such a theory will be sketched in this article and its predictions will be compared with experiment. It may be of value as the Hall generator is becoming an important electronic sensor which may be used in applications where its noise properties represent a limiting factor.

Noise in thin films

Description of noise in an electronic device can be divided into two parts: the derivation of the spectrum of the microscopic fluctuation process and the determination of the observable macroscopic noise when the microscopic effect is sufficiently well described. This article will mainly be concerned with the latter. Examples of this general procedure may be found in Van der Ziels book on Fluctuation Phenomena in Semiconductors.[5]

A one-dimensional theory relating the microscopic fluctuation to the observable noise voltage across a pair of electrodes simply states that the noise voltage is caused by the fluctuating resistance of the fraction of the specimen which lies between the contacts.

The result of this treatment is an expression

$$\overline{v_n^2} = c\frac{I_0 V}{n^2} \tag{1}$$

where $\overline{v_n^2}$ is the intensity of the noise voltage, c is a constant, I_0 is the applied constant current, V is the dc voltage on which the noise is measured, n is the carrier density.

One observes that the theory predicts a noise which vanishes when no dc voltage is present across the electrodes, but as previously mentioned, this type of treatment is only justified if the separation between the electrodes is large compared to the sample width.

The two-dimensional theory to follow applies to noiseca used by local conductivity fluctuations in the material. The volume of coherence in this connection, V_c, is small compared to the sample dimension. A constant current is running through an element of length l, width w, and thickness t. As the theory is two dimensional $t \ll l, w$. The current sets up a stationary current field, $i_o(r)$, in the sample.

Local conductivity fluctuations will perturb this field. The perturbation is representated by a field $i_n(r)$. Since the current is constant I_o, the field $i_n(r)$ must be dipole-like. At some distance the source of such a field may be represented by a current dipole:

$$P_{i,n} = i_n(r') \cdot V_c \tag{2}$$

where r' is the position of the fluctuation.

The field is then found by solving the field problem, taking into account the boundary conditions represented by the sides of the specimen.

The fluctuation field $i_n(r)$ will produce an electric field

$$E_n(r) = i_n(r)/\sigma \tag{3}$$

in the sample. σ is the conductivity of the material. $E_n(r)$ will also be a dipole-like field, as if originating from an electrical dipole of strength

$$P_{E,n} = i_n(r')V_c/\sigma \tag{4}$$

The connection between $\Delta\sigma$, the fluctuation, and the current dipole $P_{i,n}$, will in the general case depend upon a number of factors which may be represented as a constant C

$$P_{i,n} = C\Delta\sigma i_0(r')V_c \tag{5}$$

or equivalently

$$P_{E,n} = C\Delta\sigma i_0(r')V_c/\sigma \tag{6}$$

$$P_{E,n} = C\Delta\sigma E_0(r')V_c \tag{7}$$

This equation shows that it is generally the electric field or the applied dc voltage and not the applied current which is significant to the noise voltage, which is often without justification referred to as 'current noise'.

The constant C cannot be discussed further, as it is strongly related to the microscopic process which will remain unspecified in this article.

The noise intensity across a pair of contacts i and j located at r_i and r_j is simply expressed by

$$v_{n,ij}^2 = (C\Delta\sigma E_0(r')V_c \cdot (v_{r_i}(r') - v_{r_j}(r'))^2 \tag{8}$$

where $v_{r_i}(r')$ is the potential at r_i from a unit dipole at position r' when the field problem is solved taking boundaries into account. The total noise intensity due to fluctuations in the sample is expressed by an integral

$$\overline{v_{n,ij}^2} = C^2\overline{\Delta\sigma^2}E_0{}^2V_c{}^2 \int (V_{r_i}(r) - v_{r_j}(r))^2 \frac{\mathrm{d\,Vol}}{V_c} \tag{9}$$

where the field E_0 is assumed constant for reasons of simplicity. This expression will serve as a basis when the theory is compared against the experiment to be described in the next section.

The experiment

InSb thin film Hall generator specimens were processed by conventional photolithographic technique from films in the thickness range of 1–5 μm. The films were made on glass substrates by microzone synthesis as first described by Wieder.[6] The dimensions of the Hall generator are shown in Figure 1 together with a photo of the specimen mounted on a glass support. The generator had two current and four voltage electrodes marked 1 to 4 in Figure 1. It was thus possible to measure the noise across six different electrode pairs at each specimen. The Hall generator was fed by a constant current from a dry cell in series with a variable resistance which was always much larger than the sample resistance. The noise voltage was preamplified by a low noise amplifier, Brookdeal 454, analysed by a one octave band pass filter, tunable in the range 30 Hz–30 kHz, Bruel & Kiaer type 2112, and measured by a RMS detector with a variable time constant of 1–100 s. Rigorous shielding was necessary to remove ac and rf interference. Amplifier noise was subtracted from the measured values by the equation

$$\overline{v_n{}^2} = \overline{v^2(I_0)} - \overline{v^2(0)} \tag{10}$$

where $v(I_0)$ and $v(0)$ are the noise intensities measured with current I_0 and 0 through the sample.

The noise voltage had a spectrum which was close to *1/f* dependent within the frequency range covered by the analyser. A typical spectrum is shown in Figure 2. The curve is a plot of noise intensity per octave bandwidth against the frequency, and a *1/f* spectrum hence plots as a line $\overline{v^2}$ = const. An *1/f* noise spectrum in InSb films has also been previously reported.[2,4] The RMS noise voltage was found to be proportional to the applied dc voltage. This is in agreement with equation (9) as $\Delta\sigma$ is considered to be independent of E_0. It tells that the noise is caused by a conductivity fluctuation and not by effects associated with shot noise, as this gives rise to a noise intensity proportional to the applied dc voltage.

Measurements of noise voltage across the different electrode pairs were done in a one octave band with a centre frequency of

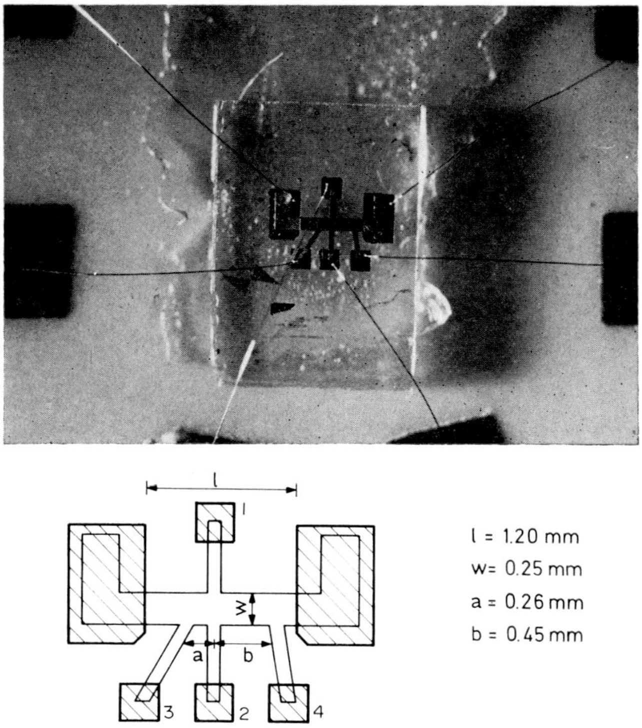

Figure 1. Geometry of the Hall effect structure.

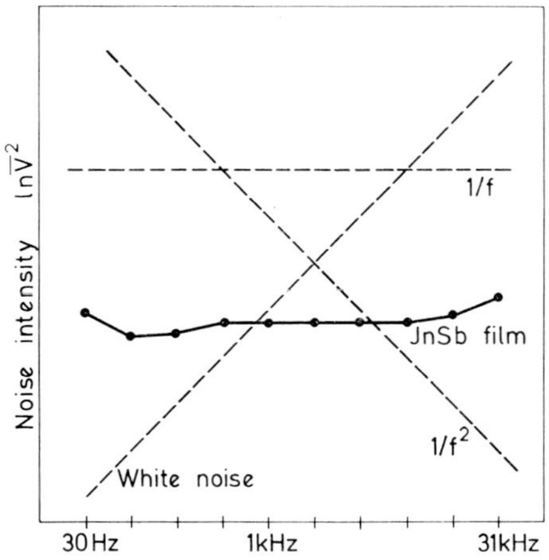

Figure 2. Measured noise spectrum.

500 Hz. Altogether 10 different elements were examined. Table 1 shows the average values of the relative intensity across the various electrode pairs together with the standard deviation of this quantity.

Discussion

The data of Table 1 may now be compared with the predictions of the theory. To do this it will be necessary to develop equation (9) further. One is here interested in the geometrical term

$$v^2 \propto \int (v_i(r) - v_j(r))^2 \mathrm{d\,Vol} \tag{11}$$

Since the film is an essentially two-dimensional structure, the voltage will be found by the line dipole expression

$$v_i = \frac{P \cdot (r - r_i)}{(r - r_i)^2} \tag{12}$$

Table 1. Measured values of noise intensity across different electrode pairs.

Electrode pair	1,2	1,3	1,4	2,3	2,4	3,4
Relative noise intensity						
Average value	0.046	0.126	0.173	1.041	0.157	0.356
Standard deviation	0.03	0.05	0.05	0.05	0.05	0.06

where P is the strength of the line dipole

$$P = \frac{1}{t} P_n \tag{13}$$

P_n is defined in (4).

Squaring (11) gives

$$v_n^2 \propto \int (v_i^2(r) + v_j^2(r))\,\mathrm{dA} - 2\int v_i(r)v_j(r)\mathrm{dA} \tag{14}$$

As $r \rightarrow r_i$ or $r \rightarrow r_j$ the first term will diverge.

This is due to three factors: use of a dipole expression at short distance, the assumption of a constant field E_0 close to the electrode and an ideal point contact.

For calculation purposes the integral might be terminated at a distance d from the electrode and the problem would be eliminated. Assigning a value to d would lead into discussion of the way electrode geometry affects the noise. Fortunately, this is unnecessary in this case since the first quadratic term will get its main contributions from the regions close to the electrodes. It may hence be regarded as a constant which is independent of the electrode position. The second term gets its contribution from the entire volume. It will not diverge as $r - r_i \rightarrow 0$, and is easily calculated numerically. When this has been done, the constant value of the first term may be chosen to make the best fit with the observed noise intensities.

The field problem is solved by imaging the line dipole with respect to the sides of the element. The number of images will be infinite, and Figure 3 indicates their arrangement. The numerical treatment considered the dipole and the 23 most significant images. The different electrode combinations resulted in values shown on an arbitrary scale in Table 2. Assigning a proper value to the constant first term and calculating the relative intensities gives results which may be compared with the measured values. It is seen that the predicted intensities do all fall within the limits of the experimental error.

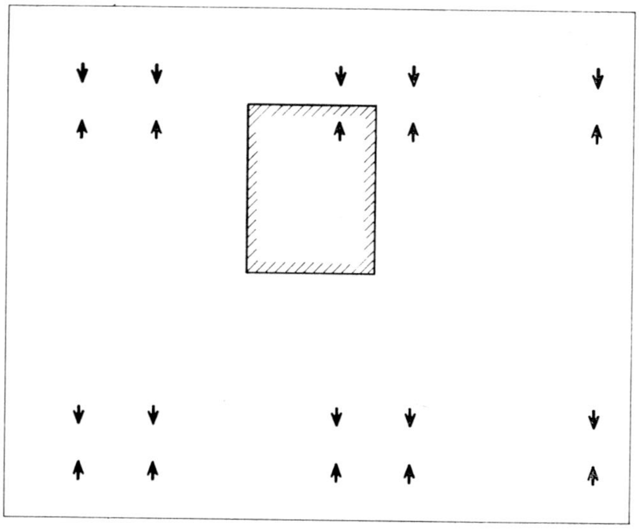

Figure 3. Pattern of the most significant imaged dipoles.

Conclusion

The fluctuation of the electrical potential distribution in a Hall effect structure is analysed by a two-dimensional model representing the fluctuation by the creation of fluctuating current dipoles. The predictions of this analysis are in fair agreement with experimental results.

References

[1] C Borough and N A Jordan, *Solid-State Electron*, **10**, 1967, 503.
[2] M Epstein, *J Appl Phys*, **36**, 1965, 2590.
[3] J J Brophy, *Phys Rev*, **106**, 1957, 675.
[4] J Shigeta, N Kotera and T Oi, *J Appl Phys*, **47**, 1976, 621.
[5] A van der Ziel, *Fluctuation Phenomena in Semiconductors*, Butterworths, London (1959).
[6] N M Davies, H H Wieder, *Proc 8th Electron and Laser Beam Symp*, p 385, University of Michigan (1966).

Table 2. Calculated values of noise intensity across different electrode pairs.

Electrode pair	1,2	1,3	1,4	2,3	2,4	3,4
Electrode dependent term	−0.38	+0.15	+0.48	+0.22	+0.49	+1.0
Constant term (Arbitrary units)	+0.5	+0.5	+0.5	+0.5	+0.5	+0.5
Relative values of total noise intensity	0.03	0.12	0.195	0.15	0.195	0.31

2.28 Current noise in W_2O_{5-x} films due to surface adsorption of water

A Masoero, P Mazzetti and A Stepanescu, *Instituto Elettrotecnico Nazionale Galileo Ferraris, Gruppo Nazionale Struttura della Materia del CNR–U.R. 24 10125–Torino, Italy*

Films of W_2O_{5-x}, obtained by vacuum evaporation of W_2O_5 on a heated insulating substrate, are electrically conducting, transparent to visible light, and have a conductivity which can be reversibly changed by single heating in an oxygen atmosphere or in vacuum. An interesting feature of these films represented by the rather low current noise, is strongly enhanced when exposed to vapours of polar molecules, like H_2O, in a suitable range of temperatures, the excess noise being generated by the dynamical adsorption–desorption processes of the vapour.
The power spectrum of the excess noise can be measured and used to get information about this process.

1. Introduction

The difficulties are well known in obtaining information on the dynamical equilibrium adsorption–desorption processes of a gas which is in contact with the surface of a metal or an insulator under suitable conditions of temperature and pressure: important dynamical parameters, like the average sticking time of a molecule on the surface, cannot be easily measured in general, and particularly under equilibrium conditions.[1]

In the present paper, a completely new method is proposed for studying adsorption–desorption processes of polar molecules at the surface of semiconducting films, making use of current-noise spectrum measurements. The method is in principle very simple: each molecule, when adsorbed, traps one carrier in the semiconducting film, and the fluctuation in the number of adsorbed molecules gives rise to a corresponding fluctuation in the conductivity of the films. The excess noise due to molecular adsorption can actually be detected only if the semiconducting film has a low intrinsic current noise and is sufficiently thin, so that surface noise effects are predominant over bulk ones. Furthermore, it must be checked that the excess noise is not generated by a change in the surface density of generation–recombination centres due to same sort of chemical reaction with the gas.[2,3]

Preliminary results obtained using W_2O_{5-x} semiconducting films, show that these conditions can be met for a wide range of polar vapours under different conditions of temperature and pressure.

This material has interesting optical and electrical properties and behaves like a semiconductor having a very low current noise, whose doping can be reversibly changed by single heat treatments. At present experiments were limited to films of this compound only, therefore no conclusions can be drawn as to the possibility that similar effects may be shown by other semiconducting or metallic films.

2. Experimental

Tungsten oxide films of the type W_2O_{5-x} can be obtained by vacuum evaporation of W_2O_5 on to a heated (300–350°C) insulating substrate, for instance fused silica or sapphire, or by heating in vaccum films of W_2O_5 evaporated onto a rt substrate.

When evaporated on a heated substrate, the films show an extremely good adherence to the substrate, in addition to excellent mechanical properties.

Interesting features of these films are their transparency to light and their conductivity, which can be reversibly changed between practically zero up to a maximum of about $10^4\ \Omega^{-1} \times m^{-1}$, by simply heating the film in an oxygen atmosphere or in vacuum.

It appears that the film loses oxygen when heated in vacuum and behaves as a semiconductor, whose doping, due to an excess of W atoms, can be reversibly changed by oxygen diffusion within the film.

As compared with usual semiconductors, current noise associated with electrical conduction in these films is very small.

In Figure 1 the total current noise $\langle \Delta G^2/G^2 \rangle$ in vacuum in a frequency band between 3 and 1000 Hz is shown versus the film resistance, for films having different thickness and resistivity.

The excess noise due to surface adsorption–desorption processes of water vapour is clearly shown in Figure 2, which represents typical power spectra of current noise in W_2O_{5-x} films having a thickness of about 50 Å, both in vacuum and in water vapour atmosphere at 5×10^{-4} torr. In this experiment the vapour was generated by spontaneous evaporation of distilled water introduced in the vacuum system by means of an input leak valve.

Excess noise spectra can be obtained by subtraction of the power spectra taken in the presence of water vapour and in vacuum.

Figure 3 shows a typical excess noise spectrum taken at different temperatures 29°C and −70°C at a vapour pressure of 10^{-2} and 5×10^{-4} torr respectively. It should be noted that all these effects are reversible, i.e. when the water vapour is evacuated current noise reduces again to the original value.

Finally in Figure 4 total noise intensity versus temperature is reported for water vapour at a pressure slightly below the critical value at each temperature.

Results similar to those reported above have also been

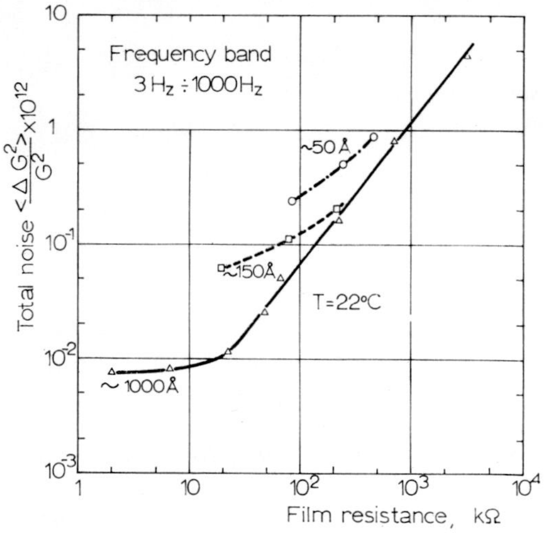

Figure 1. Current noise of W_2O_{5-x} films having different thickness and conductivity. Conductivity is changed by submitting the films to different heat treatment.

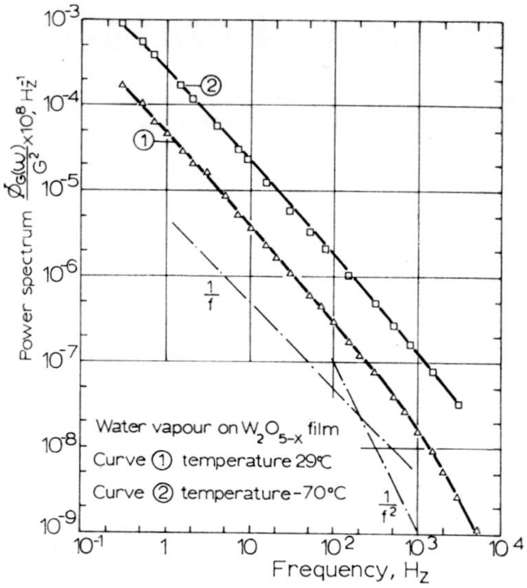

Figure 3. Excess noise spectra in a W_2O_{5-x} film at 29 and -70°C exposed to a water pressure of 10^{-2} and 5×10^{-4} torr respectively.

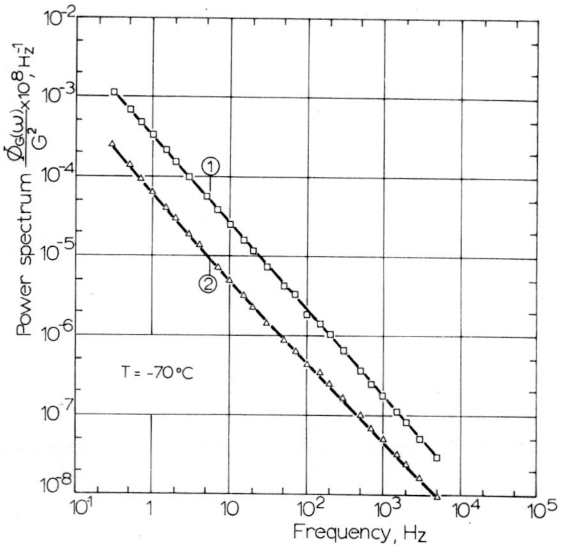

Figure 2. Power spectra of current noise in W_2O_{5-x} films exposed to a water vapour (curve 1) and in vacuum (curve 2). Vapour pressure was about 5×10^{-4} torr and temperature -70°C.

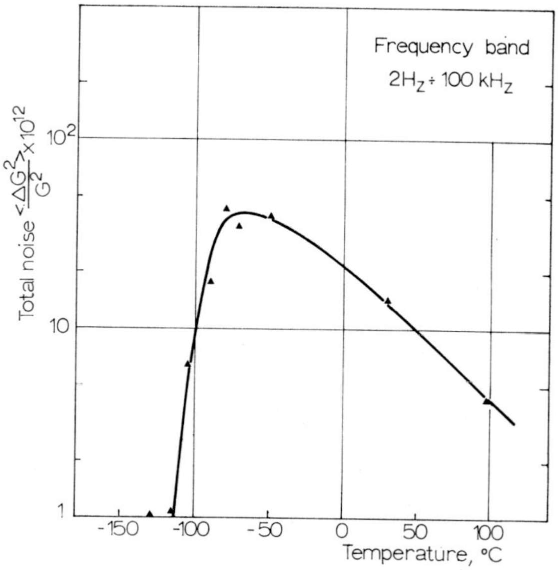

Figure 4. Excess noise intensity vs temperature in a W_2O_{5-x} film exposed to water vapour pressure slightly below the critical value at each temperature.

obtained with other polar vapours: C_2H_5OH and CH_3COCH_3. Gases like O_2, N_2, etc. dried on liquid nitrogen do not give rise to any excess noise in the films.

3. Discussion

The results reported in Figure 4 show that the excess noise is actually generated by adsorption–desorption processes in dynamical equilibrium.

The rapid drop of the noise at temperatures below -100°C and the slow drop above -40°C can be easily explained, respectively, in terms of a freezing of desorption processes and of a reduction of adsorption processes. The temperature corresponding to the maximum in this figure should correspond to the dynamical conditions of about one half monolayer adsorbed. Water condensation on the film in these experiments was always avoided by the fact that vapour pressure was kept below the critical value at each temperature.

Excess noise spectra as shown in Figure 3, are not easy to interpret. If the average sticking time for each molecule were the same and if adsorption–desorption events for each molecule were statistically independent, one should expect a shot noise spectrum of the type:

$$\phi(\omega) = \frac{\text{Const.}}{\omega^2\tau_0^2 + 1}$$

whose cut-off frequency would yield immediately the average sticking time τ_0 of the molecule.

Actual spectra may reflect the fact that the surface of the film is far from being smooth and clean and sites with different sticking time at each temperature could be present. Another possibility is that adsorbed molecules interact and the sticking time of each molecule depends on whether near sites are or are not occupied by other molecules. In any case, before making any attempt to work out a model, more experimental results are needed, a work which is now in progress in our laboratory.

References

[1] J R Anderson, *Chemisorption and Reactions on Metallic Films*, Vol. 1, Academic Press, London (1971).

[2] A L McWhorter, *l/f* noise and related surface effects in germanium, *MIT Lincoln Laboratory Report*, no 80 (May 1955).

[3] J T Law, *Proc IRE*, **42**, 1954, 1367.

3 Mechanical properties of thin films

3.1 Mechanical properties of PVD films

R F Bunshah, *Engineering Materials Department, University of California, Los Angeles, CA 90024, U.S.A.*

The mechanical properties of thin films are influenced by their structures, stability, internal stresses and thickness which makes the understanding of the mechanical behavior a complex problem. Thick films of metals and alloys seem to follow the same, well-established trends in structure and properties as bulk materials produced by more conventional wrought techniques. The state of understanding of the mechanical behavior of thick films of refractory compounds is still in its infancy and much work remains to be done for this very important class of materials.

Introduction

This paper reviews the current status of the mechanical properties of films produced by the Physical Vapor Deposition (PVD) processes of evaporation, ion plating and sputtering. The thickness of these films ranges from 100 Å to millimeters and they are used in a variety of decorative and utilitarian applications not only as coatings but as self supported shapes, i.e., sheet, foil, tubing. Although work in this area dates back many decades, the bulk of the papers in this field have been published only in the last 20 years.

PVD technology is very versatile. It enables one to deposit films of virtually every type of inorganic materials—metals, alloys, compounds and mixtures thereof—as well as some of organic materials. The deposition rates can be varied from 10 to 250,000 Å min^{-1} and even higher. The high deposition rates have come about in the last 15 years with the advent of electron beam heated evaporation sources.

Much of the early work (up to 1963) was carried out at low deposition rates and dealt mostly with thin films.* The work has been admirably summarized in various well-known treatises.[1–6] Review articles on mechanical properties of thin films have been written,[7–12] The work on thick films and bulk deposits has matured later and reviews on it have been given by Bunshah[13–16] and by Paton, Movchan and Demchishin[17] who summarized the work done at the Paton Welding Institute up to 1973. In addition the Soviet literature has numerous references to the very extensive work on thin and thick films by Palatnik and co-workers of the Kharkov Polytechnic Institute although this author cannot cite a review paper from this source.

The mechanical behavior of thin films is quite different from that of the bulk and is very much governed by the microstructure, impurity and imperfection content which in turn governs the residual stress in the film. In fact some of the incentive in the past for the extensive studies on thin films are the very high strengths as compared to cold worked bulk material and the implied promise for the development of high strength materials. This high strength was attributed in most cases to the high levels of imperfections and of residual stress; numerous mechanisms are cited in the literature for the residual stress. These deductions are often speculative because the mechanical properties of thin films are generally determined on films condensed on substrates held at low temperatures whereas the structure is studied on films condensed at high temperatures.[9] As the film gets thicker, the residual stress level decreases and the mechanical properties are very similar to those of thick films or bulk condensates. It should also be pointed out that most of the mechanical properties have been studied on films deposited by evaporation in high vacuum for the simple reason that ion plating or sputtering bring into play other process variables, such as the presence of a significant ion density in the vapor phase, electrically biased substrates, etc., which affect the structure, impurity content and residual stress levels markedly and make the correlation between structure and mechanical properties more difficult.

As a precursor to the discussion on mechanical properties, it behooves us to understand the microstructure of deposited films.

Microstructure of PVD condensates

PVD condensates deposit as single crystal films on certain crystal planes of single crystal substrates, i.e. by epitaxial growth[18]; or in the more general case the deposits are polycrystalline. In the case of films deposited by evaporation techniques, the main variables are (1) the nature of the substrate (2) the temperature of the substrate during deposition, (3) the rate of deposition and (4) the deposit thickness. Contrary to what might be intuitively expected, the deposit does not start out as a continuous film one monolayer thick and grow. Instead, three-dimensional nuclei are formed on favored sites on the substrates, e.g. cleavage, steps on a single crystal substrate, these nuclei grow laterally and in thickness (the so-called growth stage) ultimately impinging on each other to form a continuous film. Figure 1 shows the growth of gold film on rock-salt. The average thickness at which a continuous film forms depends on the deposition temperature and the deposition rate (both of which influence the surface mobility of the adatom) and varies from 10 Å for Ni condensed at 15 K to 1000 Å for Au condensed at 600 K. Once a continuous film has formed, the subsequent evolution to the final structure of the thin film is poorly understood at present. It undoubtedly depends on the factors mentioned above which in turn influence the primary variables of nucleation rate, growth rate, and

* The thickness limits for thin and thick films are somewhat arbitrary. Thin films represent thickness up to 10,000 Å and thick films from 10,000 Å and higher. Today, another viewpoint is that a film can be considered thin or thick depending on whether it exhibits surface-like properties or bulk-like properties.

surface mobility of the adatom. The problem has been tackled by Van der Drift[19] and is also the subject of a currently unpublished paper by Thornton.[20]

The microstructure and morphology of thick single phase films have been extensively studied for a wide variety of metals, alloys and refractory compounds. The structural model was first proposed by Movchan and Demchishin,[21] Figure 2 and was subsequently modified by Thornton[22] as shown in Figure 3.

Movchan and Demchishin's diagram was arrived at from their studies on deposits of pure metals and did not include the transition zone of Thornton's model, Zone *T*, which is not prominent in pure metal or single phase alloy deposits but becomes quite pronounced in deposits of refractory compounds or complex alloys produced by evaporation and in all types of deposits produced in the presence of a partial pressure of inert or reactive gas, as in sputtering or ion plating processes.

The evolution of the structural morphology is as follows, as the writer sees it today:

At low temperatures, the surface mobility of the adatoms is reduced, and the structure grows as tapered crystallites from a limited number of nuclei. It is not a full density structure but contains longitudinal porosity of the order of a few hundred ångströms width between the tapered crystallites. It also contains a high dislocation density and has a high level of residual stress. Such a structure has also been called 'Botryoidal' and corresponds to zone 1 in Figures 2 and 3.

As the substrate temperature increases, the surface mobility increases and the structural morphology first transforms to that of Zone *T*, i.e. tightly packed fibrous grains with weak grain boundaries and then to a full density columnar morphology corresponding to Zone 2 (Figure 3).

The size of the columnar grains increases as the condensation temperature increases. Finally, at still higher temperatures, the structure shows an equiaxed grain morphology, Zone 3. For pure metals and single phase alloys, T_1 is the transition temperature between Zone 1 and Zone 2 and T_2 is the transition temperature between Zone 2 and Zone 3. According to Movchan and Demchishin's original model,[22] T_1 is 0.3 T_m for metals, and 0.22–0.26 T_m for oxides; whereas T_2 is 0.45–0.4 T_m for both as shown in Figure 4. (T_m is the melting point in K.)

Thornton's modification shows that the transition temperatures may vary significantly from those stated above and in general shift to higher temperatures as the gas pressure in the synthesis process increases.

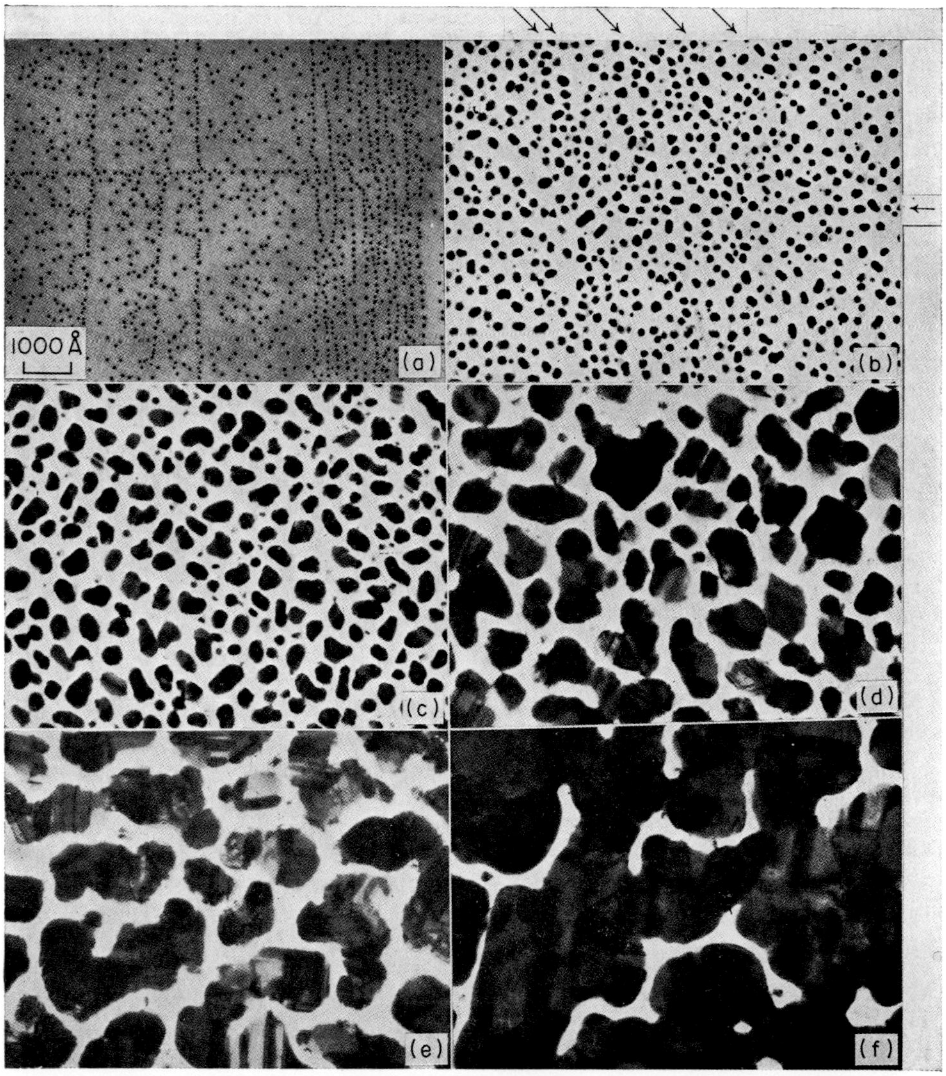

Figure 1. Sequence of micrographs illustrating the effect of increasing deposit thickness of gold on rocksalt. ×8000.[8]

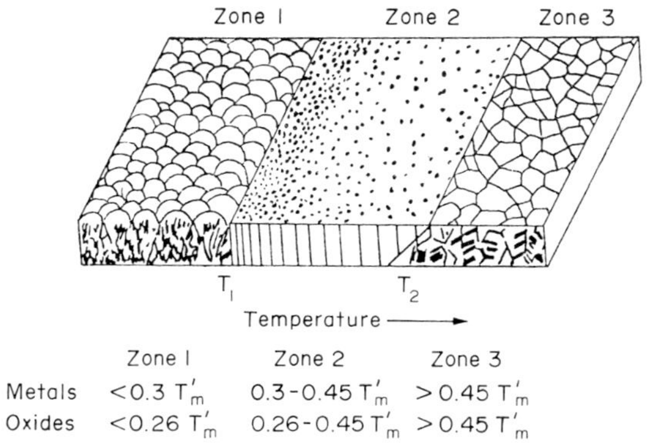

Figure 2. Structural zones in condensates.[21]

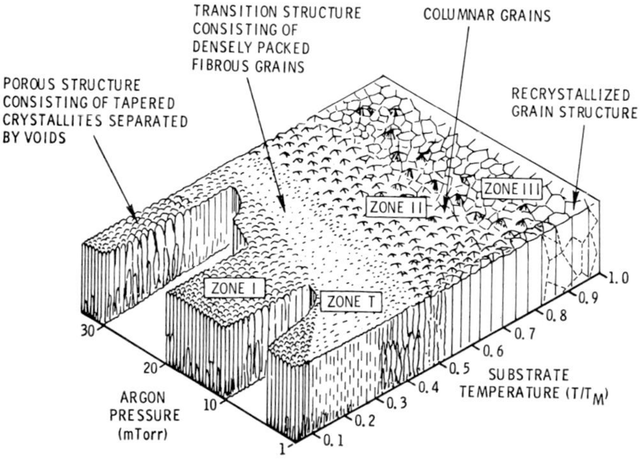

Figure 3. Structural zones in condensates.[22]

It should be emphasized that:

1. The transition from one zone to the next is not abrupt but smooth. Hence the transition temperatures should not be considered as absolute but as guidelines.
2. All zones are not found in all deposits. For example, Zone *T* is not prominent in pure metals but becomes more pronounced in complex alloys, compounds or in deposits produced at higher gas pressures. Zone 3 is not seen very often in materials with high melting points.

Figure 4 shows scanning electron micrographs of Zones 1, *T*, and 2 in TiC deposits. Other examples may be seen in the illustrations in ref [13–16].

Most thick deposits exhibit a strong preferred orientation (fiber texture) at low deposition temperatures and tend towards a more random orientation with increasing deposition temperature. Figure 5 shows the evolution of a large grained columnar morphology in a Be deposit from a much larger number of fine grains which were originally nucleated on the substrate. As growth proceeds, only those grains with a preferred growth direction survive, presumably due to considerations of the minimization of surface energy.

Mechanical property determination

A number of testing techniques have been used to determine the strength properties of thin films. They include the high speed rotor test[23] the bulge test[24–28] microtensile testing machines of the 'soft'[29–32] and the 'hard' categories[33–36] and even fixtures which can be operated in the electron microscope.[37, 38] Hoffman[10] has reviewed the test techniques and the reader could do no better than to read Hoffman's article or the original references. The basic handling problem encountered with the preparation and mechanical property testing of thin film specimens is much less severe with thick films for which many of the standard test specimens, machines and techniques can be readily used. Therefore, the spectrum of mechanical properties measured on thick films is much broader than with thin films.

Mechanical properties of thin films

The tensile properties of thin films have been reviewed.[9, 10, 36, 39] As Hoffman[10] concludes, the data reported are not very consistent even on the same material. The reader is advised to consult the references for details.

In general, the observed strength of vapor-deposited metal films consists of three parts:

$$\sigma_{OBS} = \sigma_{Bulk} + \sigma_{Imperfections} + \sigma_{Thickness}$$

where σ_{Bulk} is the inherent strength level of bulk polycrystalline material in the annealed state, $\sigma_{imperfections}$ is the contribution due to point and line defects resulting from the deposition process in excess of those normally found in the bulk annealed state and $\sigma_{thickness}$ is the contribution arising from the smallest dimension of the film and its limiting effect on grain size such that dislocation multiplication and migration are impeded.[31] Table 1 gives the strength properties of thin films of some metals and compares them to bulk values.[10] In many cases the strengths are about 200 times those of annealed bulk samples and 3 to 10 times those of hard drawn samples. The tensile strength values are given numerically as well as by fractions of the shear modulus. The ductility of the high strength films is very limited which is similar to the behavior of high strength fibers or whiskers. A principal point of contention is whether the ultimate tensile strength is a function of the film thickness or not. The discrepancy also appears to be dependent on the test method used, i.e. between the bulge test and tensile test. In many cases it appears that the strength decreases as the film thickness increases from approximately the 200–300 Å range to about 2000–4000 Å range. At the greater thickness the strength is about the same as that of heavily worked bulk material. There are several papers relating the strength properties of thin films to the 'crystallite size' and 'block structure' as influenced by the deposition temperature, stress, recovery and recrystallization process.[40–49] One manifestation of this is the phenomenon of creep or plasticity in room temperature tensile tests as exhibited by an irreversible initial loading curve but almost reversible unloading and reloading curves as long as the previous stress level is not exceeded. An example of this is shown in Figure 6 from Neugebauer[30] as the change in slope of the stress–strain curve. The possibility of this change in slope being related to an elastically soft measurement or to creep in the cement of the grips altogether be discarded.

Long term creep rates have been measured and for gold they vary from 10^{-7} to 10^{-4} min^{-1} depending on load, dimensions and the amount of prestrain.[30] The estimates of the relative elastic and plastic extension at fracture vary from completely elastic to an almost even mixture of elastic and plastic deformation.

Fracture in ductile gold single crystal films[36] results from a localized plastic deformation with resultant thinning of the film

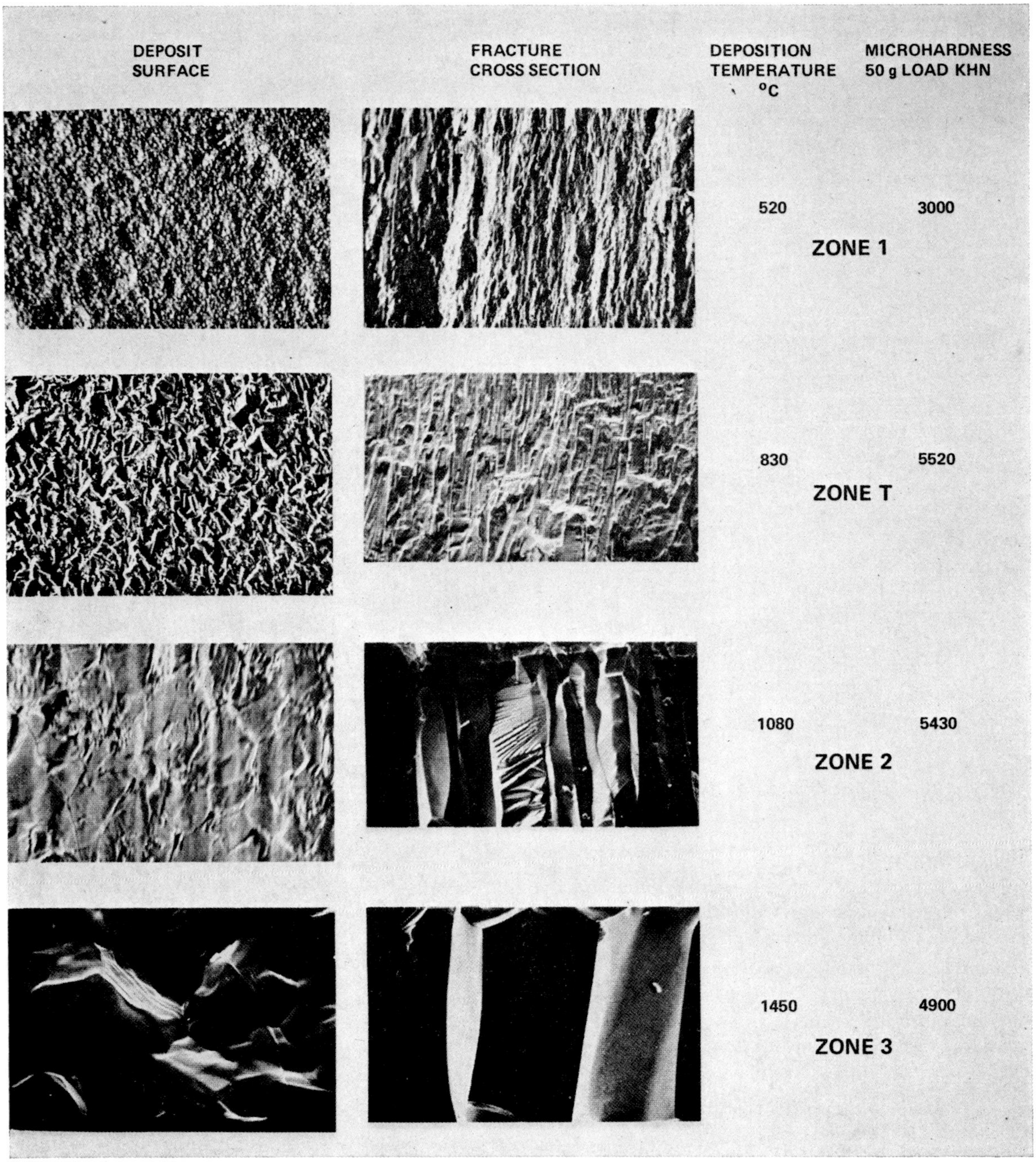

Figure 4. Structure and microhardness of TiC deposits at various substrate temperatures.[79]

and a rise in stress level. Eventually the smaller cracks formed in this manner join to cause fracture. The dislocations necessary for the deformation are not the grown-in dislocations but those which nucleate and multiply in discontinuous regions. Most observations show no necking prior to fracture. The maximum stress appears to correspond to that needed to propagate cracks from flaws existing in the specimen. In polycrystalline nickel, the fracture is the 'clean-cleavage' type.[31]

Mechanical properties of thick condensates and bulk deposits

Table 2 lists the mechanical properties of thick deposits of metals, alloys, refractory compounds and laminated structures. In many cases, the mechanical test data are quite extensive showing yield strength, ultimate tensile strength, hardness and ductility as a function of grain size, deposition temperature and test temperature. One of the features of the data is that the properties of thick deposits of metals and alloys are very similar to those of wrought materials which are produced by the conventional processes of melting, casting, mechanical working and heat treatment.

We will consider each type of material separately since the behavior of metals and alloys is vastly different from that of refractory compounds.

The early work in this area was that of Bunshah,[53,54] Bunshah and Juntz[55] and Smith[56] who deposited thick films of Be, Ti and Cu respectively and measured mechanical properties. In 1965, Palatnik and co-workers published a paper

Table 1. Strength properties of thin films.[10]

Material	Structure	Maximum tensile strength kg/mm²	Shear modulus*	Strain at fracture, %	Thickness-dependent	Reference
Au	Bulk hard-drawn	28†	G/114	‡		
	(111) crystal film	81	G/36	1.2	No	36
	(100) crystal film	27	G/110	0.5	Yes	25
	(110) (111) polycrystalline film	49	G/59	1	No	30
	Polycrystalline film	55	G/54	0.7	Yes	24
	(100) crystal film	26	G/115	3.5	No	32
	Polycrystalline film	32	G/92	2.3	No	32
Ag	Bulk hard-drawn	37†	G/75		—	—
	Polycrystalline film	59	G/47	0.7	Yes	24
	Polycrystalline film	42	G/68	0.3–0.4	—	50
Cu	Bulk hard-drawn	49†	G/98	—	—	—
	Polycrystalline film	93	G/51	1.8	Yes	51
	Polycrystalline film	88	G/54	—	No	37
	Rolled foil	19	G/256	10–15	No	52
Ni	Bulk cold-rolled	125†	G/67	—	—	—
	Polycrystalline film	210	G/40	1.8	Yes	31
Al	Bulk cold-rolled	16†	G/171	—		—
	Polycrystalline film	42	G/66	0.5–0.8		50

* Shear moduli from *AIP Handbook*, 1957.
† Bulk tensile strength from *Handbook of Chemistry and Physics*, 42nd ed., 1961.
‡ Fracture strain not quoted for bulk material.

Figure 5. Photomicrograph of a Be deposit showing the evolution of large columnar grains.

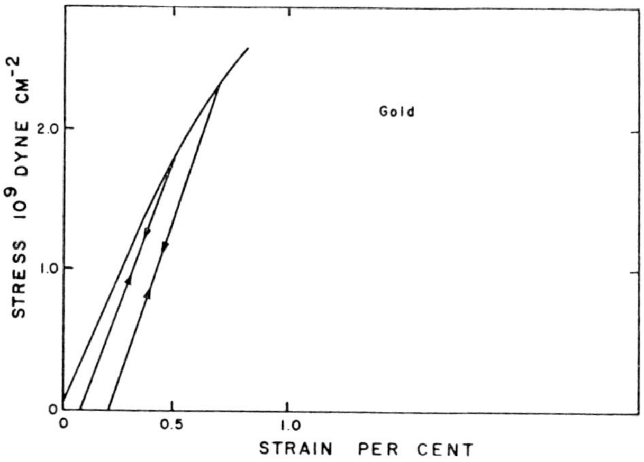

Figure 6. Typical stress–strain curve for thin film.[30]

on mechanical properties of Al condensates.[57] It is impossible to review in detail all the papers. The pertinent data are shown in Table 2 and the discussion below will concentrate on the highlights.

Metals and alloys. Movchan and Demchishin studied the tensile properties and microhardness of Ni, Ti and W condensates produced at various deposition temperatures. No tensile tests were performed on specimens deposited in Zone 1 (Figure 2). Tests on specimens deposited in Zone 2 showed high strength and low ductility at low deposition temperature. The strength decreased and the ductility increased with deposition temperature. The strength and ductility values of specimens deposited in Zone 3 showed approximately the same values as for recrystallized specimens produced from wrought material. The microhardness variation with deposition temperature for Ni, Ti and W is shown in Figure 7. The tapered crystallite morphology in Zone 1 showed a high hardness much greater than that of annealed metal. The hardness decreased rapidly with increasing deposition temperature to a fairly constant value for Zone 3 morphology which corresponds to the hardness of recrystallized metals.

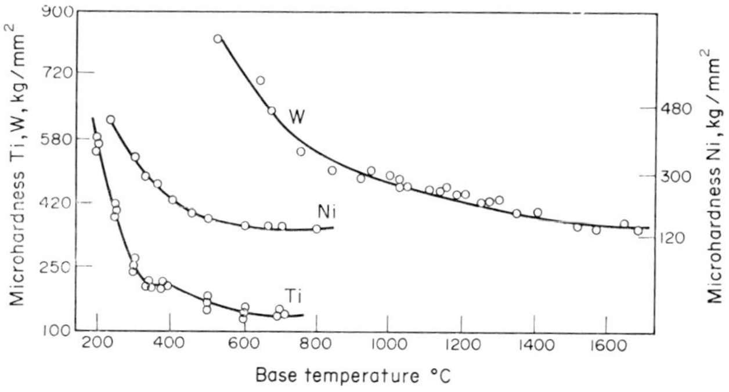

Figure 7. Variation of microhardness with deposition temperature for metals.[21]

Table 2. Mechanical properties of thick films or bulk condensates.

Material	0.2% Yield stress (kg/mm²)	Ultimate tensile strength	% Elong (or % RA)	Microhardness (kg/mm²)	Deposition temp (°C)	Grain size (μm)	Thickness (μm)	Test temp (°C)	Reference
Al				200	200		40	25	57
				160	300				
				80	400				
				8 Knoop at 0.5 g load	250–350		1–2	25	67
Cu	15					1	4–10	25	63
	28					0.7			
	40					0.1			
	45					0.07			
		18–22	35		400–800		1000	25	17
				240	120		>15	25	68
				170	180				
		50	2.38				3–20	25	69
		16	0.27						
		(annealed 500°C)							
		60				0.1	0.5	25	70
		40				0.6	3–5		
Ag	10					1	5–20	25	63
	35					0.1			
Fe		66–72		250			20	25	71
	45	46	4	120	400		1000	25	72
	20	35	20	90	550				
	15	25	18	75	700				
Be		140	0.1	240 (Knopp)	350–425		75–100	25	72
		281	0.3	215	480–550				
		316	0.45	168	625–685				
		295	1.0	170	760–790				
	62 (bend test)		>4 (bend test)			100	1000	25	54
	47 (bend test)		>5.2 (bend test)						
Ni		115	0.3			0.2	0.2–2	25	73
		66–74		250	400–420		20	25	74
				450	260				
				600	220				
	129	338	39(20)	96.6	554	26	275	25	59
	135	336	32(20)	109	425	19	375		
	155	366	35(21)	104	329	12.7	375		
	247	403	21(21)	125	260	3.7	375		
	140 (melted and wrought)	407	44(30)	91		18	357		
		138	0.6				2.6	25	27
					250	620	1000	25	21
					400	420			
					600	380			
		28	20		400		1000	25	17
		38	12		800				
Ti	189	295	25(66)		800	32	250	25	58
	225	302	20(38)		600	8	250	25	
	436	471	3(25)		450	1	250	25	
	189 (melted and wrought)	302	40(61)			50	250	25	
				560	200		1000	25	21
				200	400				
				130	500				
	280–420		20				100–250	25	75
	140–280		35					200	
Mo	14	19	(5.3	191	990	530	300		76
	24	28	(0.8)	169	704	85			
	38	44	(1.0)	186	804	19			
Nb	19		7.3	112	745	10	300		76
V	37		22	115	540	0.7	300		76
W				720	600		1000	25	21
				460	1000		1000	25	
				300	1000		1000	25	

Table 2.—*continued*

Material	0.2% Yield stress (kg/mm²)	Ultimate tensile strength	% Elong (or % RA)	Microhardness (kg/mm²)	Deposition temp (°C)	Grain size (μm)	Thickness (μm)	Test temp (°C)	Reference
Ni-20Cr	337	540	26(27)	80	950	13	375	25	60
	577	780	21(26)	112	780	2.6	375	25	
	752	871	18(19)	131	680	1.8	375	25	
	1195	1223	6(8)	180	427	0.5	375	25	
	393 (melted and wrought)	569	37(31)	78		10.4	325	25	
	25	49	25(21)		950	13	375	1000	
	40.9	60	53(41)		680	1.8	375	1000	
Ti–6Al–4V	970–1195	1068–1195	1–12				79	25	65
80Ni–20Cu		35–45	8–25		400–800		375	1000	17
50Ni–50Cu		40–55	5–20		400–800		375	1000	
20Ni–80Cu		27–42	10–30		400–800		375	1000	
80Ni–16Fe		180		650	200			25	74
(permalloy)		150		250	400			25	
Al_2O_3				380	400		300	25	21
				80	600		300	25	
				200	1000		300	25	
				1000	1400		300	25	
				500–800	1100		300	25	81
ZrO_2				450	400		300	25	21
				350	600		300	25	
				400	1000		300	25	
				1000	1400		300		
Y_2O_3		248		384	120	3.2	100	25	78
		226		256	210	5	100	25	
		215		331	304	8	100	25	
		208		320	416	14	100	25	
		205		360	538	21	100	25	
		198		502	721	58	100	25	
Ti_2O				960–1960	750			25	82
TiO				700–2015	1100			25	
SiO_2				300			10	25	83
Ta_2C				1907–2440	500			25	84
TaC				1800–2400	500			25	
SiC (rf sputter)				3500–4000			2.6	25	85
TiC (ARE process)				2710	520		50	25	79
				2955	730		50	25	
				2955	830		50	25	
				4110	1080		50	25	
				4160	1120		50	25	
				2400			5	25	80
ZrC				2500–2800	550		50–75		92
(Hf-3Zr)C				2700–3030	550		50–75		
VC				2350	550		50–75		95
NbC				2300	550		50–75		
				LAMINATES					
Cu–Ni laminate		35–55	13				2	25	86
Cu–Fe laminate	35–70	110		400–800	240–360		15–25	25	87
Cu–Ni laminate		80	1.8		300		0.1	25	88
Cu–Fe laminate		125	0.88		150		0.04	25	
Ni–Fe laminate		150	0.45		150		0.04	25	
Ni–SiO		150		300			20	25	89
		105		300			20	200	
		95		300			20	400	
		80		200			20	600	
Ti–TiC		251	I.8		500		250		90

Bunshah and co-workers studied the effect of deposition temperature on the grain size, tensile properties and hardness of Ti,[55,58] Ni,[59] Nb, V, Mo[76] and Ni–20 Cr[60] alloys for deposits made in Zones 2 and 3. They found that increasing deposition temperature produced larger grain size, low strength, higher ductility, and low hardness. Even at the lowest deposition temperature in Zone 2, the ductility was good (>20% RA for 1 μm grain diameter Ti at a yield strength of 56,000 psi). Moreover, they found that both the yield strength and hardness varied as the inverse square root of grain diameter, i.e. followed the Hall–Petch relationship,[61,62] which is

$$\sigma_{ys} = \sigma_0 + kd^{-1/2}$$

where σ_{ys} is the yield strength; d is the grain diameter and; σ_0, k are constants. Figure 8 shows an example of this relationship for Ni–20 Cr alloy.

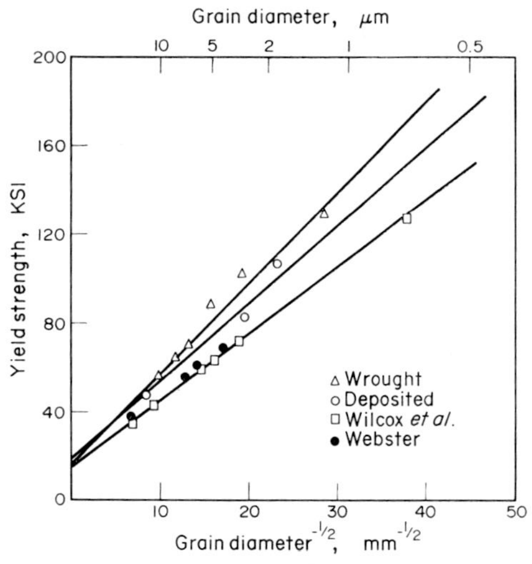

Figure 8. Variation of yield strength with grain-size for Ni–20 Cr.[60]

For all these metals and alloys the yield strength, ductility and hardness values correspond to those of the same materials produced by casting, mechanical working and recrystallization. The variation of yield-strength and hardness with grain-size, i.e. Hall–Petch type relationships were also very similar between the deposited and wrought materials, small variations being ascribable to differences in grain morphology and preferred orientations. The Ni–20 Cr alloy showed good strength at 1000°C and also obeyed the Hall–Petch relationship.

The Hall–Petch relationship is also obeyed by thick films of Cu and Ag to grain-sizes as small as 0.05 μm as shown by Nenioto, Jumbou and Suto.[63] Thus, these thick deposits behave as true engineering materials.

Chambers and Bower[64] studied the mechanical properties of 18-8 stainless steel, gold and magnesium and showed that their tensile properties were very similar to their wrought counterparts.

Smith, Kennedy and Boericke[65] studied the $(\alpha + \beta)$ type Ti-6Al-4V alloy. They showed that the tensile properties are very similar to the wrought material except for a much smaller value in per cent elongation due to premature onset of plastic instability in a tensile test at pinholes in the deposited samples. The bend ductility was however superior to the wrought material.

Paton, Movchan and Demchishin[17] studied the mechanical properties of dispersion-strengthened alloys. For Ni–NbC alloys, the yield strength increases with volume fraction of the dispersed phase and decreases with an increase of deposition temperature which in turn increases the particle size and inter-particle spacing. They also studied the steady-state creep rate of Ni–ZrO_2 dispersion strengthened alloys which is very low at a very high temperature.

The same authors[17] also showed that it is possible to produce thick deposits of all the alloys across the Cu–Ni system and that the mechanical properties vary systematically with composition as would be expected.

Movchan, Demchishin and Kooluck[66] studied Fe, Fe–NbC, and Fe–Ni–NbC alloys. Their results indicate that NbC increases the strength of Fe at 25 and 700°C. Alloying the Fe matrix with Ni produces additional strengthening at both temperatures thus showing the possibility of combining solid solution strengthening and dispersion strengthening.

Refractory compounds

Deposits of refractory compounds, oxides, nitrides and carbides are very important for wear resistant applications in industry. Their structure and properties are strongly dependent on the deposition process. Their behavior is very different from metals and alloys. It is also very hard to measure the mechanical properties of ceramics by tensile tests similar to those used for metals and alloys because of their brittle nature. A very good test to measure the fracture stress of such brittle coatings is the Hertzian fracture test which measures the fracture stress and the surface energy at the fracture surface.[77] Colen and Bunshah[78] used to measure the fracture behavior of Y_2O_3 deposits of various grain sizes.

Figure 9 shows the variation in microhardness with deposition temperature for Al_2O_3 and ZrO_2 from the work of

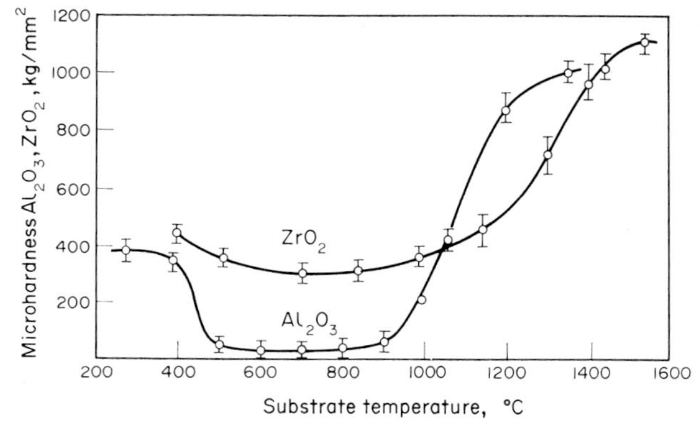

Figure 9. Variation of microhardness with deposition temperature for Al_2O_3 and ZrO_2.[21]

Movchan and Demchishin,[21] showing that the behavior of these oxide deposits is quite different in one respect from that of metals (Figure 8). The hardness falls when the structure changes from tapered crystallites (Zone 1) to columnar grains (Zone 2) as with metals. However, unlike metals the hardness increases markedly as the deposition temperature rises from 0.3 T_m to 0.5 T_m. The authors attribute this to a more 'perfect' material produce at the higher deposition temperatures due to 'volume processes of sintering'. A similar hardness curve was obtained for Y_2O_3 deposits.[78]

Figure 10 from the work of Raghuram and Bunshah[79] also shows a very marked increase in microhardness of TiC deposits on going from 0.15 T_m (500°C) to 0.3 T_m (1000°C). The hardness increases for the oxides and TiC with increasing deposition temperature. Both these sets of results may be explained by the following concept. Since the strength of ceramics is very

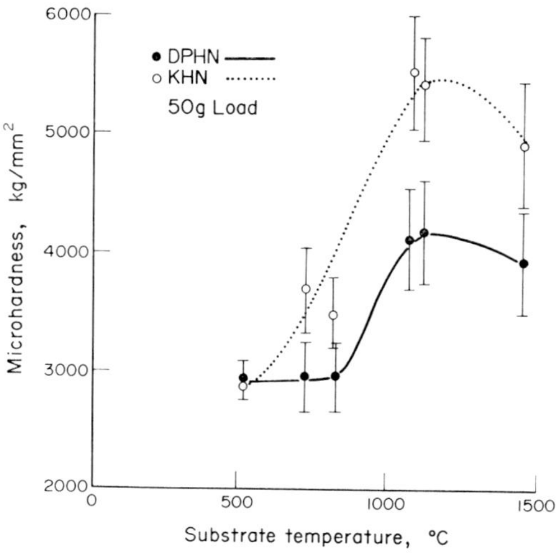

Figure 10. Variation of microhardness with deposition temperature for TiC.[79]

adversely effected by growth defects and at the higher deposition temperatures, the occurrence of these defects is markedly reduced, the hardness (or strength) increased very significantly. However, it should be noted that the absolute value of the hardness of the oxides is much lower than that of the carbides. Thus the possibility of a different explanation for the 'similar' behavior of these materials, i.e. the hardness increase with deposition temperature needs to be investigated.

The hardness data on sputtered TiC and TiN coatings are quite similar to those produced by evaporation techniques.[80]

Laminates

Laminate composites are very attractive because of the uniform properties in the plane of the sheet. In comparison to mechanical methods of producing laminate composites, e.g. bonding of sheets, physical vapor deposition techniques are very suited to the production of such composites particularly if each lamellae is to be very thin (~0.1 μm) in order to improve the strength and toughness of the composite. It also gives us the opportunity to have a high strength ceramic as one of the phases. Work to date has been very limited. In one investigation of the Cu–Ni laminates, no improvement in strength was found.[86] It is not an ideal system to investigate. In the Fe–Cu[87,88] and Ni–SiO[89] composites, the strength increased significantly. The Ni–SiO system was notable for the fact that the composite retained its high strength up to 700°C, which is reminiscent of the dispersion strengthened SAP alloys. This appears to be a most fruitful field for future research.

Damping capacity

As a final note, it may be mentioned that fine-grained films of Al and Cu have been found to have a much higher damping capacity owing to the relaxation processes associated with large angle grain boundaries. Below 180 K, the damping is equal for single crystal and polycrystalline films.[91]

Summary and conclusions

Physical Vapor Deposition (PVD) techniques are a powerful way to produce thin films, thick films and bulk deposits of a variety of materials. The mechanical properties of thin films are influenced by their structure, stability, internal stresses and thickness which makes the understanding of the mechanical behavior a complex problem. Thick films of metals and alloys seem to follow the same, well established trends in structure and properties as bulk materials produced by more conventional wrought techniques. The state of understanding of the mechanical behavior of thick films of refractory compounds is still in its infancy and much work remains to be done for this very important class of materials.

Acknowledgements

The support of the National Science Foundation under Grant DMR 75-13347 is gratefully acknowledged.

References

[1] L Holland, *Vacuum Deposition of Thin Films*, Chapman & Hall, London (1956).
[2] L Maissel and R H Clang (editors), *Handbook of Thin Film Technology*, McGraw-Hill, New York (1970).
[3] C A Neugebauer, J B Newkirk and D A Vermilyea (editors), *Structure and Properties of Thin Films*, Wiley, New York (1959).
[4] R W Hoffman and H G F Wilsdorf (editors), *Thin Films*, American Society for Metals (1964).
[5] R Niedemeyer and H Mayer (editors), *Basic Problems in Thin Film Physics*, Vandenhoeck and Ruppercht, Gottengen (1966).
[6] *Physics of Thin Films*, Academic Press, New York (1963) (onwards—in multiple volumes).
[7] J A Allen, *Rev Pure Appl Chem*, **4**, 1954, 133.
[8] G A Bassett and D W Pashley, *J Inst Metals*, **87**, 1958, 449.
[9] R W Hoffman, *Thin Films*, p 99, American Society for Metals (1964).
[10] R W Hoffman, *Physics of Thin Films*, **3**, 1966, 246, Academic Press, New York.
[11] W Buckel, *J Vac Sci Technol*, **6**, 1969, 606.
[12] K Kinosita, *Thin Solid Films*, **12**, 1972, 17.
[13] R F Bunshah, *Proc 4th Int Conf on Vacuum Metallurgy*, p 17, Iron and Steel Inst Japan (1973).
[14] R F Bunshah, *J Vac Sci Technol*, **11**, 1974, 633.
[15] R F Bunshah, *J Vac Sci Technol*, **11**, 1974, 814.
[16] R F Bunshah, *New Trends in Materials Processing*, p 200, American Society for Metals (1976).
[17] B E Paton, B A Movchan and A V Demchishin, *Proc 4th Int Conf on Vacuum Metallurgy*, p 251, Iron and Steel Inst Japan (1973).
[18] D W Pashley, *Adv Phys*, **5**, 1956, 173.
[19] A Van der Drift, *Phillips Res Rep*, **22**, 1967, 267.
[20] J A Thornton, *Ann Rev Mater Sci*, 1977, to be published.
[21] B A Movchan and A V Demshishin, *Fizika Metall*, **28**, 1969, 83.
[22] J A Thornton, *J Vac Sci Technol*, **11**, 1974, 666.
[23] J W Beams, J B Breazeale and W L Bart, *Phys Rev*, **100**, 1955, 1657.
[24] J W Beams, *Structure and Properties of Thin Films*, p 183, Wiley, New York (1959).
[25] A Catlin and W P Walker, *J Appl Phys*, **31**, 1960, 2135.
[26] S Jovanovic and C S Smith, *J Appl Phys*, **32**, 1961, 121.
[27] P I Krukover and V A Buravikhin, *Fizika Metall*, **22**, 1966, 144.
[28] D G Brandon and Z Bauer, *Israel J Technol*, **8**, 1970, 247.
[29] D M Marsh, *J Sci Instrum*, **38**, 1961, 229.
[30] C A Neugebauer, *J Appl Phys*, **31**, 1960, 1096.
[31] C D'Antonio, J Hirschorn and L Tarshis, *Trans AIME*, **227**, 1964, 1346.

[32] J M Blakely, *J Appl Phys*, **35**, 1964, 1756.
[33] D Kuhlmann-Wilsdorf and K S Raghavan, *Rev Sci Instrum*, **33**, 1962, 930.
[34] A Lawley and S Schuster, *Rev Sci Instrum*, **33**, 1962, 1178.
[35] E Orowan, *Z Phys*, **82**, 1933, 235.
[36] J W Menter and D W Pashley, in *Structures and Properties of Thin Films*, p 111, Wiley, New York (1959).
[37] H G F Wilsdorf, *Rev Sci Instrum*, **29**, 1958, 323.
[38] D W Pashley, *Proc R Soc Lond*, **A225**, 1960, 218.
[39] C A Neugebauer, in *Physics of Thin Films*, Vol 2. (Edited by G Hass and E Thun), Academic Press, New York (1964).
[40] L S Palatnik, A I Ill'inski, G V Federov and V S D'yachenko, *Izvestra VUZ. Fizika* (*Soviet Physics Journal*), **1**, 1966, 122.
[41] R L Grunes, C D'Antonio and F K Kies, *J Appl Phys*, **36**, 1965, 2735.
[42] M Ya Fuks, V V Belozerov and Yu F Boyko, *Fizika Metall*, **33**, 1972, 571.
[43] K Kinosita, K Maki, K Nakamizo and K Takenchi, *Japan J Appl Phys*, **6**, 1967, 42.
[44] M Ya Fuks, V V Belozero and Yu F Boyko, *Fizika Metall*, **33**, 1972, 571.
[45] F A Doljack and R W Hoffman, *Thin Solid Films*, **12**, 1972, 71.
[46] M Ya Fuks, L S Palatnik, V V Belozerov, Yu V Zolotnitsky and S T Roshchenko, *Fizika Metall*, **36**, 1973, 316.
[47] B Y Pines and N S Tan, *Fizika Metall*, **19**, 1965, 899.
[48] I T Aleksanyan, *Fizika Metall*, **25**, 1968, 947.
[49] R W Hoffman, *Thin Solid Films*, **34**, 1976, 185.
[50] L S Palatnik, M Ya Fuks, B T Boiko and A T Pugacheu, *Soviet Physics Dokl English Translation*, **8**, 1964, 713.
[51] A Oding and I T Aleksanyass, *Soviet Phys Dokl*, **8**, 1964, 818.
[52] A Lawley and S Schuster, *Trans AIME*, **230**, 1964, 27.
[53] R F Bunshah, *Proc ASM Golden Gate Conf*, p 31, 1962.
[54] R F Bunshah, *Metals Eng Q*, **6**, 1964, 8.
[55] R F Bunshah and R S Juntz, *Trans Vac Met Conf*, p 200, American Vacuum Society (1965).
[56] H R Smith Jr, *Proc ASM Golden Gate Conf*, p 569, 1962.
[57] L S Palatnik, G V Federov, A I Prokhavulov and A I Federenko, *Fizika Metall*, **20**, 1965, 574.
[58] R F Bunshah and R S Juntz, *Met Trans*, **4**, 1973, 21.
[59] N Kane and R F Bunshah, *Proc 4th Int Conf Vacuum Metallurgy*, p 242, Iron and Steel Inst Japan (1973).
[60] N Agarwal, N Kane and R F Bunshah, *J Vac Sci Technol*, **22**, 1975, 619.
[61] E O Hall, *Proc Phys Soc Lond*, **B64**, 1951, 747.
[62] N J Petch, *J Iron Steel Inst*, **174**, 1951, 25.
[63] M Nenioto, R Jumbou and H Suto, *Trans Japan Inst Metals*, **12**, 1971, 113.
[64] D L Chambers and W K Bower, *J Vac Sci Technol*, **7**, 1970, S62.
[65] H R Smith, K Kennedy and F J Boericke, *J Vac Sci Technol*, **7**, 1970, S48.
[66] B A Movchan, A V Demichishin and L V Kooluck, *J Vac Sci Technol*, **11**, 1974, 869.
[67] H Yamamoto and G Kamoshita, *Trans Japan Inst Metals*, **12**, 1971, 12.
[68] L S Palatnik, M Ya Fuks, A I Il'yinskiy and O G Alaverdova, *Fizika Metall*, **22**, 1966, 744.
[69] C A O Henning, F W Boswell and J M Corbett, *Acta Met*, **23**, 1975, 177.
[70] K K Ziling, L D Pokrovskiy and V Yu Pohelkin, *Fizika Metall*, **29**, 1970, 1112.
[71] L S Palatnik, A I I'lyinskiy and A G Ravlik, *Fizika Metall*, **19**, 1965, 310.
[72] G Mah and C W Nordin, *Proc 16th Annual Conf Soc of Vacuum Coaters*, p 103, Chicago (1973).
[73] C A O Henning, F W Boswell and J M Corbett, *Acta Met*, **23**, 1975, 187.
[74] L S Palatnik, A I Il'yinskiy, A G Ravhk, A A Nechitayls and G Ye Lyakh, *Fizika Metall*, **27**, 1969, 1114.
[75] C D Turk and H L Marcus, *Trans Met Soc AIME*, **232**, 1968, 2251.
[76] M A Sherman, R F Bunshah and H A Beale, **12**, 1975, 697.
[77] F C Frank and B R Lawn, *Proc R Soc Lond*, **299A**, 1967, 291.
[78] M Colen and R F Bunshah, *J Vac Sci Technol*, **13**, 1976, 536.
[79] A C Raghuram and R F Bunshah, *J Vac Sci Technol*, **9**, 1972, 1389.
[80] G Mah, C W Norden and J F Fuller, *J Vac Sci Technol*, **11**, 1974, 371.
[81] R F Bunshah and R J Schramm, *Thin Solid Films*, **41**, 1977, in press.
[82] W Grossklaus and R F Bunshah, *J Vac Sci Technol*, **12**, 1975, 593.
[83] S Furuuchi, H Sakata and K Aiwaka, *Japan J Appl Phys*, **13**, 1974, 1905.
[84] W Grossklaus and R F Bunshah, *J Vac Sci Technol*, **12**, 1975, 811.
[85] K Wasa, T Nagai and S Hayakowa, *Thin Solid Films*, **31**, 1976, 235.
[86] C A O Henning, F W Boswell and J M Corbett, *Acta Met*, **23**, 1975, 193.
[87] L S Palatnik, A I I'lynskiy and N P Sapelkin, *Soviet Phys Solid St*, **8**, 1967, 2016.
[88] I I Solonovich and V I Startsev, *Problemy Prochn*, **1**, 1973, 28.
[89] L S Palatnik, A I I'lyinskiy, N M Biletchanko and R I Sinel-nikova, *Fizika Metall*, **32**, 1971, 199.
[90] R F Bunshah, Y D Gupta and A C Raghuram, unpublished data.
[91] V S Postnikov, I V Zolothukhin and V K Belongov, *Fizika Metall*, **26**, 1968, 957.
[92] A C Raghuram, R Nimmagadda, R F Bunshah and C N J Wagner, *Thin Solid Films*, **20**, 1974, 187.
[93] R F Bunshah and A C Raghuram, *J Vac Sci Technol*, **9**, 1972, 1385.

3.2 The electromechanical behaviour of a full component (dielectric and Cu/Ni constantan alloy) for thin film strain gauge deposited upon steel-substrate

J Gouault, M Hubin, G Richon and B Eudeline, *Laboratoire de Thermoélectricité, Couches Minces, ISHN, Université de Rouen, et INSCIR Mont-Saint-Aignan, 76130—France*

We have elaborated and studied a full component (dielectric and Cu/Ni constantan alloy) in the form of thin film for use as a strain gauge on the strainable member of a pressure transducer. The dielectric layer is laid upon a steel substrate by successive evaporation of alumina and silicon monoxide. The alloy–film is deposited from separate sources by coevaporation, the nickel flux being monitored by the copper flux.
The dielectric shows a fitting insulation resistance till 300°C and the alloy exhibits a resistance drift with temperature of less than 10^{-4}/°C. We observe a good mechanical endurance during a static strain-state of 3000 micro-deformations and during the course of a sinusoïdal strain variation for more than 10^6 cycles. The value of the gauge-factor is about 2.2.

Introduction

Usually, strain gauges are bonded upon the strainable member of a mechanical transducer by cement or glue. They constitute a dielectric medium transmitting the deformations of the sensing element to the gauges, their resistance variations being dependent on the longitudinal and transversal deformations that can affect them.

But the use of cement causes a number of disadvantages like creep, relaxation hysteresis and a limited range of operational temperatures.

To avoid the use of cement, a procedure has been developed of evaporating in vacuum an alloy-film gauge of the constantan type upon a steel-substrate with a ceramic layer evaporated in between.[4–9]

Procedure for obtaining the full component of a strain gauge

1. Surface-treatment of the substrate. A face of the substrate in the form of a triangular bar is polished first with Emery paper no. 180 to 600, and finally by felt coated with diamond-paste, ranging in grain size from 6 to 1 μm, till a relief depth of less than 1/10 μm is reached.[10]

2. Preparation of the insulating film. After being carefully cleaned the steel-substrate is put into a vacuum container where a pressure of residual gas lower than 10^{-7} torr is obtained by an oil-diffusion pump. Its temperature is increased to 200°C (Figure 1a).

Then, oxygen is introduced by pipes in the vicinity of the substrate raising the vacuum pressure to 10^{-5} torr. By electron bombardment of an alumina specimen, a 1 μm thick layer of alumina is laid upon the substrate at a rate of 100 Å mn^{-1}. Then SiO is evaporated by Joule heating from a furnace onto the alumina film forming a 500 nm thick film composed of SiO, SiO_x, SiO_2 $(1 < x < 2)$.[4]

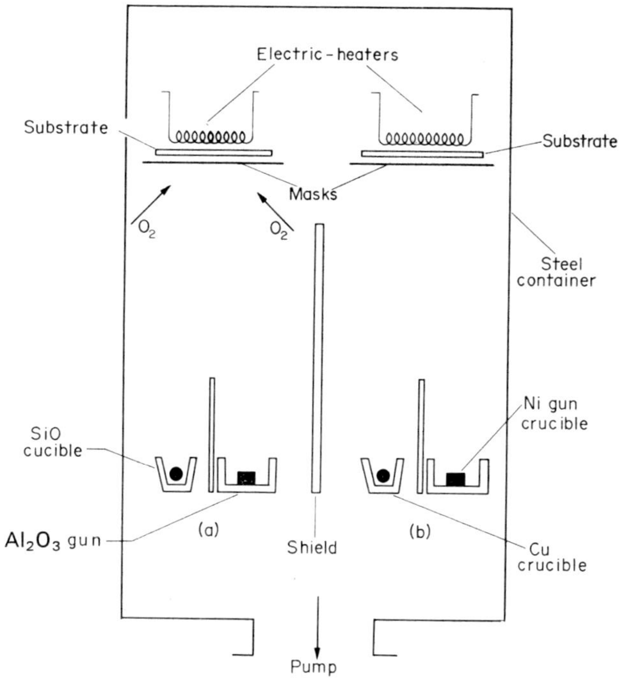

Figure 1. Vacuum apparatus.

An annealing treatment of several hours is done in an oxygen atmosphere of 200 torr, in order to oxidize and stabilize the dielectric film.

3. Preparation of the alloy film[11–13]

Coevaporation. The substrate is then transferred into another part of the jar or in another vacuum apparatus where the pressure is lowered down to 10^{-7} torr (Figure 1b). A mask is being put very close to it, in order to define the geometry of the alloy deposit in the form of a thin rectangular element. The film of alloy (60% at Cu, 40% at Ni) is obtained from two separate sources. The copper is evaporated from a BeO furnace and the nickel is evaporated simultaneously by an electron gun. Two boxes are located close to the substrate, each supporting a vibrating quartz microbalance, controlled in temperature for the purpose of measuring the flux of the evaporating nickel and copper separately, both of them affecting the substrate.[14]

To obtain an alloy-film upon the substrate with the same composition throughout the thickness, we have to maintain the ratio of the flux K as near as possible to K_0 chosen,* while the evaporation is in progress. Since the frequency change of the quartz crystals is proportional to the deposited mass, a frequency to voltage converter gives a signal dV, proportional to dK ($dK = K - K_0$). It acts as an entrance signal of a servo-system, regulating the current through the gun filament which emits bombarding electrons towards the nickel target, in order to adjust the Ni-flux to the Cu-flux and to give the specified composition.

Stabilization treatment. In the vacuum enclosure, the substrate temperature (after a few hours of ageing) is increased steadily up to 200°C over a period of 32 h, is maintained at 300°C for 48 h and is then lowered slowly down to 20°C. The subsequent relative reduction of resistance is about 3%.

Experimental results

1. Insulation of the dielectric. The Figure 2a shows that the insulation-resistance between the gauge and the beam has satisfactorily high values up to 300°C.

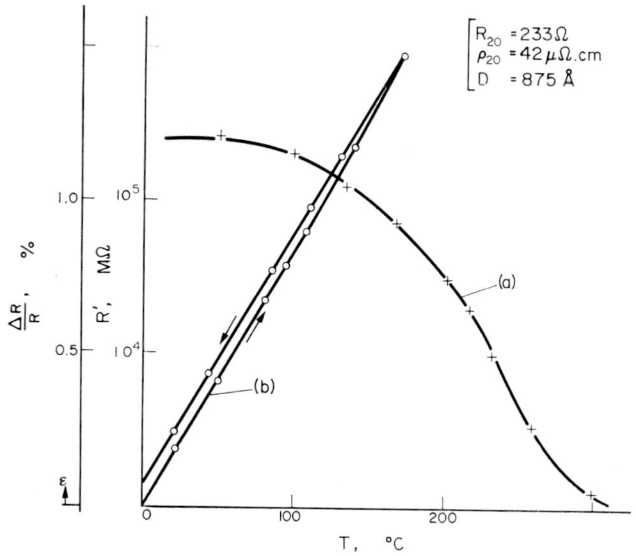

Figure 2. (a) Insulation-resistance of the dielectric: R'. (b) Ohmic behaviour of the gauge: $\Delta R/R$.

2. Ohmic behaviour with temperature of the gauge. A four-terminal method is used for measuring the resistance variation of the alloy-film during the course of a temperature-cycle (20–200°C) the beam being heated in the nitrogen atmosphere of an oven. The curve of the Figure 2b exhibits a thermal zero shift of less than 10^{-5}/°C and a zero balance drift (ε) of less than 10^{-3}, the resistivity being approximately 42 $\mu\Omega$ cm

3. Strain ohmic behaviour of the gauge. The beam (Figure 3) is put in the horizontal position with its part A mounted in a rigid wall.

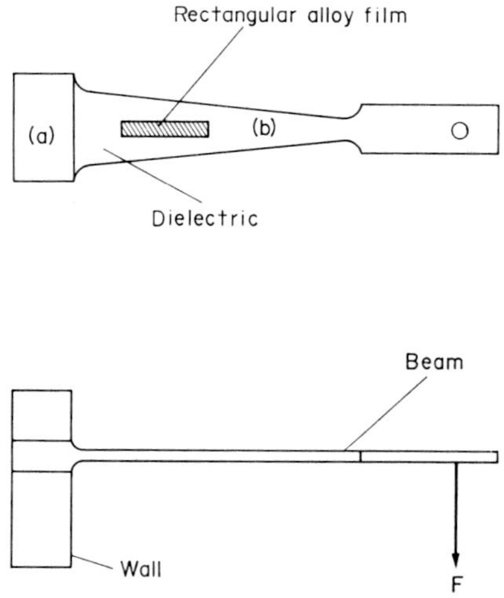

Figure 3. Operational bar.

There is an isoflexion state for the part B (where the full-component is deposited) while a strength F is being exerted vertically, its intensity being constant or varying in a sinusoidal way. It can be observed:

a strong adhesiveness and the keeping of the insulation between the component and the steel up to 4000 micro-deformations (μD) during a static stage.

a convenient holding of the gauge for 10^6 cycles during a sinusoidal stage up to 1500 μD (the value between the lower and upper crests).

a value of about 2.2 for the gauge factor,

a non-linearity and hysteresis lower than 0.5% in the resistance–variation of the alloy-film versus F (Figure 4a),

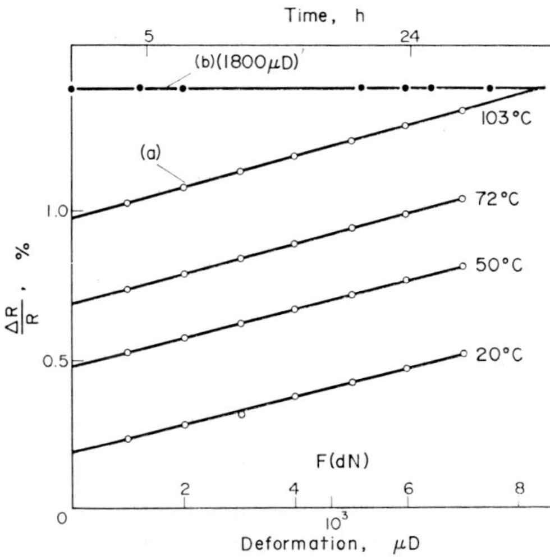

Figure 4. (a) Extensometric behaviour of the gauge. (b) Temporal stability of the gauge under full load at 103°C.

* Where K is the nickel flux and K_0 is the copper flux.

a non-existence of creep at a temperature of 200°C under full load—1800 μD (Figure 4b).

Results

The static stage is obtained by means of a weight hanging on the end of the bar with a rope. The sinusoidal stage is achieved with the aid of a rotating machine which commands a crank-driving rod system acting on a spring fixed on the end of the bar. The linear relationship between the intensity of F and the longitudinal deformation of the part B is well known. The obtained parameters are compared to the characteristics of a bulk strain gauge glued onto the substrate close to the alloy-film.

Conclusion

From these results it appears that the full-component prepared according to the described process exhibits electromechanical properties of insulation, endurance, resistivity, gauge-factor, and stability consistent with the characteristics required for a good quality strain-gauge which is to be deposited on sensing elements of transducers.[11]

References

[1] J O Linde and C H Stade. *Arkiv Fys*, **2**, 1950, 99.
[2] A Colombani, B Laniepce and P Huet, *C R Acad Sci*, **250**, 1960, 3946.
[3] R L Parker and A Krinsky, *J Appl Phys*, **34**, 1963, 2700.
[4] D F A MacLachlan, *Microelectron Reliab*, **3**, 1964, 227.
[5] J H Okimura, *Japan J Appl Phys*, **4**, 1965, 825.
[6] P R Perino, Preprint Inst Soc of Am, 21st on Conf and Ex, New York (1966).
[7] P Malsan and M Granveaud, *Onde Elec*, 1967, 480.
[8] Z H Meiksin, *Thin Solid Films*, **4**, 1970, 241.
[9] B S Verma, G L Malhotra and S K Sharma, *Thin Solid Films*, **6**, 1970, R9.
[10] D W Preston, *Cong Int sur les Couches Minces*, Cannes (1970).
[11] G R Witt, *Thin Solid Films*, **22**, 1974, 133.
[12] J C Lecordier, Thèse de l'Université de Rouen, France (1974).
[13] P H Thureau, A Floch, M F Moley, *Thin Solid Films*, **36**, 1976, 201.
[14] G Richon, J Gouault, J J Herou, *Thin Solid Films*, **36**, 1976, 915.

3.3 Properties of heavily preloaded quartz crystals for thin film thickness monitors

E Söllner, E Benes, A Biedermann and D Hammer, *Institut für Allgemeine Physik, Technische Universität Wien, Karlsplatz 13, A-1040 Wien/Austria*

The validity range of the frequency/mass-relation based on the compound resonator model has been verified with a very sensitive method of mass determination (Atomic Absorption Spectrometry-AAS) up to mass loads $m_f/m_q > 60\%$. The present load range limiting factors are the significant increase of the crystal damping and the occurrence of small frequency jumps in the frequency/mass-curve. The behaviour of the crystal damping has been measured during the deposition process and a strong correlation between the frequency jumps and the damping curve has been discovered. Consequences for the preparation of the crystal electrodes in order to avoid frequency jumps even at high mass loads are introduced.

1. Introduction

The usable mass load range of quartz crystal resonators for thin film thickness monitors depends primarily on a more or less sophisticated measurement-interpretation. The 'frequency measurement technique' of Sauerbrey[1] and Lostis[2] is based on the approximate proportionality between the decrease in resonance frequency of the crystal and the area mass density of the deposited material. This method is well supported by experimental data[3] up to a mass load of $m_f/m_q < 2\%$, where m_f is the area mass density of the deposit and m_q that of the unloaded crystal. The thickness x_f of the deposited film may always easily be calculated from the area mass density

$$m_f = \rho_f \cdot x_f \qquad (1)$$

if the mass density ρ_f of the film material is known.

The so called 'period measurement technique'—introduced by Behrndt[4]—uses the change in period time of the crystal oscillation as measuring parameter. Since the proportionality range between mass load and period time change is much higher than that between mass load and frequency change, the usable load range is extended to $m_f/m_q < 10\%$.

The next approach to a better mathematical model of the quartz crystal microbalance was suggested by Miller and Bolef.[5] They were the first to treat the system quartz plus thin film as an one-dimensional compound acoustic resonator. The obtained relation between the resonance frequency and the mass load was finally transformed[6] to a practicable form, which is applicable up to mass loads of m_f/m_q exceeding 60%.[7,8] This fundamental equation for the long range use of quartz crystal resonators in thin film thickness monitors is here introduced in terms of period time, since this is the measuring parameter of modern digital thickness monitors:

$$\frac{m_f}{m_q} = \frac{\rho_f x_f}{\rho_q x_q} = (1 + \Delta\tau/\tau_q)\frac{1}{\pi z}\arctan\{z\tan[\pi - \pi/(1 + \Delta\tau/\tau_q)]\} \qquad (2)$$

$\Delta\tau$ is defined as difference between the oscillation period time τ of the loaded crystal and τ_q of the unloaded crystal. The factor z is the ratio of the acoustic impedance of the quartz crystal and that of the deposited material:

$$z = \frac{\rho_q v_q}{\rho_f v_f} \qquad (3)$$

v_q and v_f are the transversal acoustic wave velocities in the crystal and the deposit, respectively.

2. Monitoring of large mass loads

Equation 2 indicates that materials with different acoustic impedances will obey different frequency respectively period time versus mass relations. In recent works[6–8] equation 2 was proved using conventional microbalances as reference measurement equipment. We verified it by another, more sensitive method of mass determination using an Atomic Absorption Spectrometer (AAS),[10] which allowed an accuracy of $\pm 0.3\%$ for the m_f value. The great advantage of AAS is the exclusive sensitivity to the interesting element, e.g. to the deposited material Ag, hence there is no registration of contaminations or oxide respectively sulphide formations occurring during the handling of the crystal outside of the vacuum chamber.

The change of the period time $\Delta\tau$ was measured with a digital thickness monitor, which has been developed at our institute.[9] Figure 1 shows the experimental results for silver deposition in comparison with equation 2 and the relations which are based on the frequency and period measurement technique.

The quartz crystals used were plano–convex AT-cuts (14 mm diameter) with a fundamental series resonance frequency of 5 MHz. The exciting Ag-electrodes were approximately 100 nm thick and 14 mm (frontside) respectively 8 mm (backside) in diameter. It should be mentioned here that the use of plane crystals should be discouraged. The plano-convex crystal is superior to the plane one because of its much closer fitting to the theoretical frequency versus thickness curve.[11]

The deposited material was silver, evaporated by boat resistor heating with a deposition rate of about 2. 10^{-3} mg cm^{-2} s^{-1}

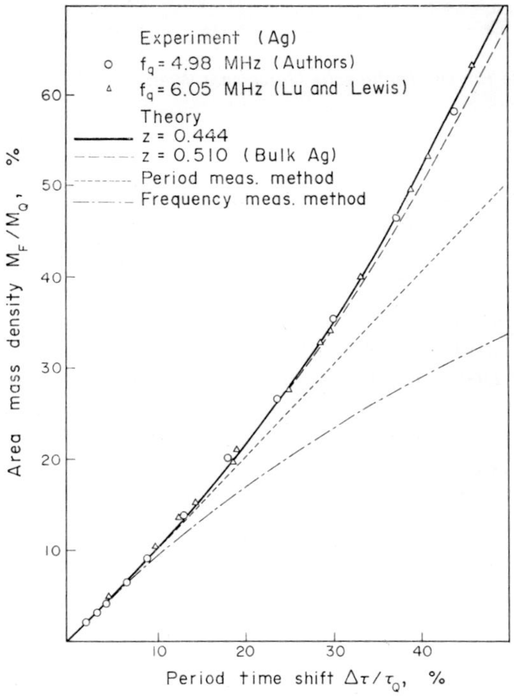

Figure 1. Reduced area mass density m_f/m_q as a function of reduced period time shift $\Delta\tau/\tau_q$ for silver deposition on quartz crystals with different resonance frequencies. Comparison between experimental data and equation 2. The curves representing the 'period measurement method' and the 'frequency measurement method' are also shown for comparison.

in a conventional vacuum system. The value of $z = 0.510$ was calculated from the bulk properties of silver using

$$v = \sqrt{(G/\rho)} \tag{4}$$

where v is the shear wave velocity and G the elastic shear modulus. The 15% deviation of the z-value fitting best the experimental data ($z = 0.444$) from the theoretical value can be understood by the possible differences in the acoustic properties between film and bulk material.[8, 12]

The consideration of the acoustic impedance of the deposit according to equation 2 seems to be rather sophisticated for practical use, but this is not true for state of the art digital thickness monitors, which are able to compute a direct display of the film thickness based on equation 2.[9] Our measurements have shown that such a thickness monitor would allow a thickness determination accuracy of 1% even with heavily preloaded crystals, if the proper thin film ρ_f and z values are known.

3. Limitation in accuracy and deposition range

In earlier works a number of accuracy or deposition range limiting factors like temperature, crystal mounting, electrode configuration, oscillator circuit, etc. has been described, which may be avoided, respectively considered.[11] The present limitations in accuracy or deposition range are treated in the following three sections.

3.1. Crystal damping during deposition—frequency jumps. The damping d of the crystal resonator is defined by the ratio of the dissipated power—which must be supplied by the oscillator—and the reactive power, which is stored within the resonator:

$$d = \frac{1}{Q} = \frac{\text{dissipated power}}{\text{reactive power}} = \frac{I^2 R_s}{I^2 X} = \frac{R_s}{X} \tag{5}$$

Q is the quality factor of the quartz crystal, I the rms-value of the current through the vibrating quartz, R_s the series resonance resistance, and $X = \omega_s L_q$, where ω_s is the series resonance angular frequency and L_q the dynamic inductance in the equivalent electrical circuit of the quartz crystal.

Our measurements of the quality-factors of 5 MHz quartzes showed that they are loosing about 90% of their Q-values by deposition of a mass load of $m_f/m_q = 20\%$. The damping of these loaded quartzes is much higher than expected by the acoustic losses within the evaporated films. This may be due to the high mechanical stresses in these films causing a concave bending of the quartz wafer and the deformation of the resonator produces additional acoustic losses.

To observe the damping curve during the deposition process a simple method has been developed. It is based on the fact that the automatic gain control (AGC) of the sensor oscillator tries to increase the amplification if the crystal damping increases. Hence the amplitude control voltage can be easily recorded during deposition, it may be used to estimate the crystal damping behaviour. Figure 2 shows that the control voltage V_{AGC} decreases approximately in proportion to the increasing series resonance resistance of the quartz crystal.

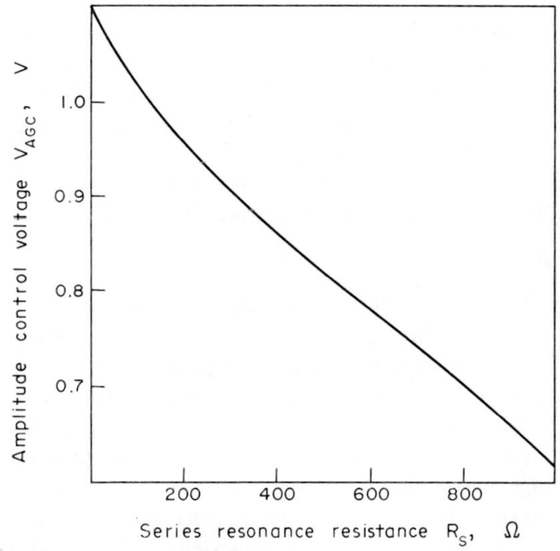

Figure 2. Amplitude control voltage V_{AGC} of the exciting oscillator as a function of the series resonance resistance R_s in the electrical equivalent circuit of the quartz crystal.

On the other hand, the damping is roughly proportional to R_s (equation 5), because the change in the reactance X is very small in comparison with that of R_s, as our measurements of the equivalent electrical circuits of loaded crystals have shown.

We have undertaken several long range deposition runs, while recording the change in crystal frequency as well as the sensor oscillator control voltage. The employed crystals were coated with Ag-electrodes, the deposited material was silver, too. Heavily preloaded crystals showed jumps in the frequency versus mass load curve in spite of continued constant deposition rate. A strong correlation between these frequency jumps and the damping behaviour has been discovered. Figure 3 shows a

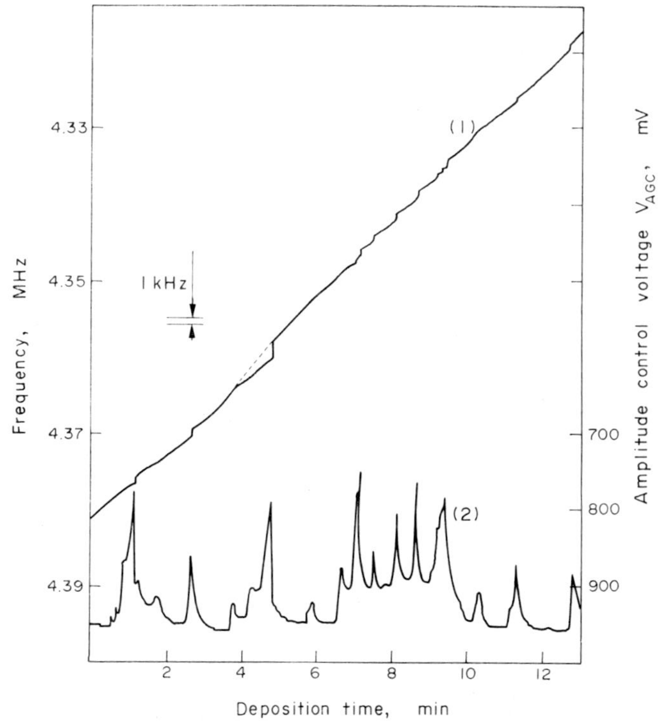

Figure 3. Characteristical section of the frequency (1) and damping behaviour (2) of a heavily Ag preloaded ($m_f/m_q = 12\%$) plane quartz crystal during continued silver deposition. This crystal only has been conventionally cleaned before depositing the exciting Ag-electrodes. The amplitude control voltage V_{AGC} of the oscillator is approximately proportional to the crystal damping (see Figure 2).

typical section of the frequency and damping behaviour obtained with a conventionally prepared plane quartz plate preloaded with $m_f/m_q = 12\%$. Figure 4 shows an example for a plano-convex crystal preloaded with $m_f/m_q = 10\%$. The frequency jumps are in the order of some hundred Hertz to some thousand Hertz and their occurrence may start at mass loads of $m_f/m_q = 8\%$ using plano-convex crystals. With plane crystals the occurrence of frequency jumps starts even much earlier and they are also more numerous.

Every frequency jump is preceded by a significant increase in the damping of the quartz resonator, which is assumed to be due to a local internal stress peak in the deposited film. This stress causes an increase of the resonance frequency,[13, 14]

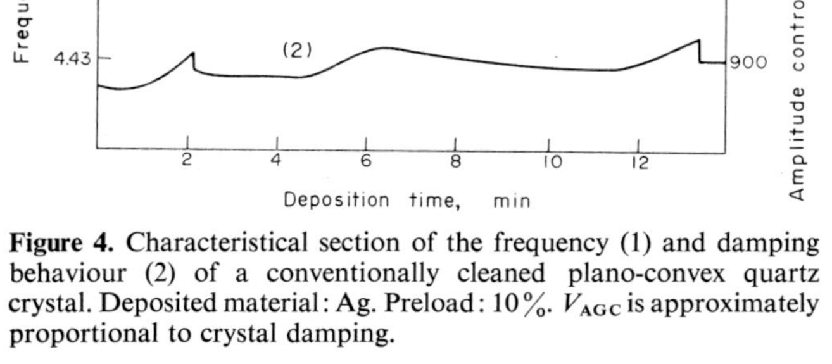

Figure 4. Characteristical section of the frequency (1) and damping behaviour (2) of a conventionally cleaned plano-convex quartz crystal. Deposited material: Ag. Preload: 10%. V_{AGC} is approximately proportional to crystal damping.

which simulates a reduction of the deposition rate (see dotted lines in Figures 3 and 4). At a certain limit the interface electrode–quartz no longer can absorb this stress and a sudden relaxation mechanism occurs due to a microscopic partial removement of the thin film plus electrode from the quartz surface. This results in a frequency and damping jump to a value, which could be expected under regular circumstances at constant deposition conditions.

3.2 Reduction of frequency and damping jumps. If the assumed mechanism is correct, a variation of the sticking coefficient between the electrode and the quartz surface should directly influence the occurrence of frequency jumps. This sticking coefficient chiefly depends on the used electrode material and on the contamination conditions of the quartz surface while depositing the electrodes.

Measurements of the jumping behaviour have been carried out with plane and plano-convex crystals with different surface treatments before the deposition of the electrodes:

A. Cleaning only conventionally (defatting and cleaning with aceton or methanol).

B. Cleaning by glow discharge (defatting, ultrasonic cleaning, cleaning in the vacuum chamber by glow discharge).

C. Predeposition with an adhesion film (after conventional cleaning a Cr–Cr_2O_3-adhesion film has been produced; starting with a pure Cr_2O_3-film, then continuous transition to a pure Cr-film, ending with the Ag-electrode).

D. Special chemical cleaning, this is an eight-step process, which in this work has been developed to improve the jumping behaviour:

1. removing of the Ag-electrodes with HNO_3 (p.a., ca. 50%);
2. washing with distilled water;
3. defatting in boiling methanol (p.a.);
4. washing with deionized H_2O;
5. dipping in warm H_2O_2 p.a. (30%) and addition of 1 part concentrated NH_4OH p.a. to 10 parts H_2O_2 p.a.;
6. heating up to strong foaming;
7. cooling off;
8. washing of the quartz wafer with deionized H_2O.

The characteristic results of these procedures with regard to the jumping behaviour at Ag-deposition for quartzes with Ag-electrodes can be seen from Table 1. Using plane quartz crystals the occurrence of the first frequency and damping jump could actually be shifted by the special chemical cleaning method to twice the frequency shift, but this was still only a mass load of about 3%. Using plano-convex crystals surface treatments like B, C or D improved the jumping behaviour significantly. Especially the predeposition of a Cr–Cr_2O_3-adhesion film or the special chemical cleaning method allowed Ag-deposition up to mass loads of $m_f/m_q < 20\%$ without any frequency jump.

The difference in the jumping behaviour of plane and plano-convex crystals may be due to the different amplitude distributions. The plano-convex crystal exhibits a vanishing amplitude and hence a vanishing sensitivity at the edges of the electrodes.[11] A partial microscopic removement of the electrode edges would not effect the resonance frequency of a plano-convex crystal. It is assumed that microscopic removements of the electrodes

Table 1. Representative results of the jump behaviour of plane (p) and plano-convex (pc) quartzes at silver deposition in dependence on the cleaning methods of the crystal surface before depositing the electrodes.

Quartz	Deliverer	Cleaning method	Electrode material	Start frequency (MHz)	1st f-jump at Δf (kHz)	Magnitude of 1st f-jump (Hz)	1st f-jump $\geq$ 1 kHz at Δf (kHz)
p	Balzers	conventional	Ag	4.97612	74	400	194
p	Balzers	special chemical	Ag	4.99438	145	300	230
pc	Balzers	conventional	Ag	4.97638	393	300	465
pc	Balzers	glow discharge	Ag	5.14704	615	600	1182
pc	Balzers	Cr–Cr_2O_3-adhesion film	Ag	4.94241	872	500	1046
pc	Balzers	special chemical	Ag	4.97284	910	300	1030
pc	Inficon	not specified	Au	6.12105	350	400	855

from the crystal surface are forced primarily at the edge of the electrode, since here is the maximum mechanical load of the crystal–electrode interface.

3.3. Restriction of usable mass range due to film stress. The phenomena mentioned above are only valid for metal films. For dielectric films the intrinsic stresses in the film can reach such high values already at relatively small mass loads that the vibration of the quartz no longer can be maintained by the oscillator due to the rapid increase of crystal damping.

Figure 5 shows the deposition of a MgF_2-film on a conventionally cleaned plano-convex quartz with Ag-electrodes.

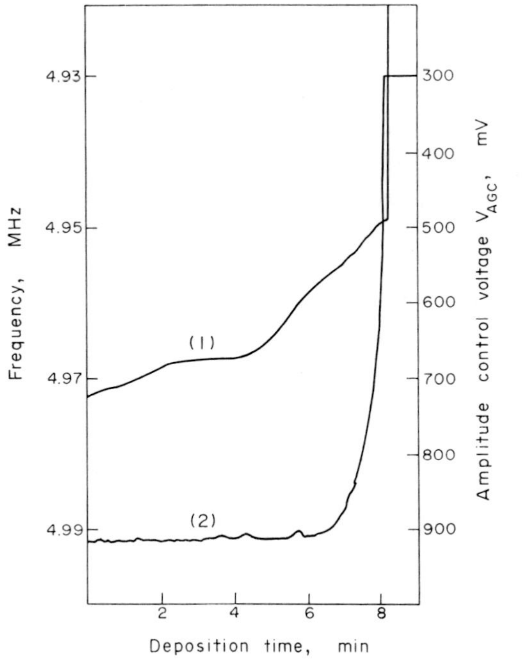

Figure 5. Deposition of a MgF_2-film on a conventionally cleaned plano-convex quartz with Ag electrodes. (1) frequency curve, (2) damping behaviour.

Already at a mass load of $m_f/m_q = 0.4\,\%$ the crystal damping exceeds the oscillator amplitude control range. These high damping values are immediately followed by a complete removement of the electrode plus deposit from the crystal surface.

4. Conclusion

The occurrence of frequency jumps in the frequency/mass-curve of heavily preloaded quartz crystals means a remarkable lack in reliability of this film thickness measurement method, although the probability of an erroneous thickness value caused by a jump event is very small. During deposition, these frequency jumps obviously cannot be noticed on modern digital displays without an analogue recorder. We investigated, therefore, the probable mechanism of the jump events and introduced methods in preparing the crystal surface before depositing the electrodes in order to remove the occurrence of the first jump outside the usable load range.

Only crystals prepared that way allow reliable thickness measurements for metal deposition up to mass loads of $m_f/m_q = 60\,\%$ taken into consideration according to Miller and Bolef. For optical coatings the usable crystal load range is much more restricted on account of the higher mechanical stresses within dielectric layers produced by deposition. In this application field, therefore, the period measurement method which is applicable up to mass loads $m_f/m_q = 10\,\%$ is perfectly sufficient.

Acknowledgements

We would like to thank the initiator of this work, Prof Dr F Viehböck, for all his assistance. We are also indebted to Dr Wernisch of the Institut für Technische Physik at our university for providing the AAS measurements and to Doz Dr Pulker of the Balzers AG for supplying us with various quartz crystals.

This research work was supported by the Austrian 'Fonds zur Förderung der wissenschaftlichen Forschung' under project number F2177 and F2546.

References

[1] G Sauerbrey, *Z Phys*, **155**, 1959, 206.
[2] P Lostis, *Rev Opt*, **38**, 1959, 1.
[3] H K Pulker, *Z Angew Physik*, **20**, 1966, 537.
[4] K H Behrndt, *J Vac Sci Technol*, **8**, 1971, 622.
[5] J G Miller and D I Bolef, *J Appl Phys*, **39**, 1968, 5815.
[6] C S Lu and O Lewis, *J Appl Phys*, **39**, 1972, 4385.
[7] C S Lu, *J Vac Sci Technol*, **12**, 1975, 578.
[8] O Lewis and C S Lu, *Proc 29th Ann Freq Cont Symp*, 1975, 5.
[9] D Hammer, E Benes and H K Pulker, *Thin Solid Films*, **32**, 1976, 47.

[10] H L Kahn, *J Chem Educ*, **43**, 1966, 2.
[11] H K Pulker, E Benes, D Hammer and E Söllner, *Thin Solid Films*, **32**, 1976, 27.
[12] A Wagendristel and E Söllner, *Z Naturforsch*, **30a**, 1975, 44.
[13] E P EerNisse, *Proc 29th Ann Freq Cont Symp*, 1975, 1.
[14] E P EerNisse, *J Appl Phys*, **43**, 1972, 1330.

3.4 Influence of condensation surface temperature on macrostructure in evaporated chromium films

P J Nordlander, K Å B Andersson, H T G Nilsson and S B M Hagström, *Department of Physics and Measurement Technology, Linköping University, S-58183 Linköping, Sweden*

Chromium, with a Vickers diamond hardness of up to 1200 has been vapor deposited on die cast zinc substrates. In each coating cycle up to six specimens with equal initial condition but different thicknesses could be made. A macrostructure of cracks was found in the films. It is very like the cracking network obtained in electrodeposited chromium. The cracks are caused by intrinsic stress in the films. The 'mean distance between cracks' was measured as a function of film thickness, deposition rate and bulk substrate temperature.
The hardness is given as a function of 'mean distance between cracks'. The condensation surface is defined and its temperature dependence on deposition parameters is discussed. Using the concept of condensation surface temperature and the linear expansion coefficient of the film material an atomic model is given to explain the 'mean distance between cracks' dependence on the deposition parameters.

Introduction

Early in the thin film vacuum deposition history it was noticed that the films had a tendency to curl when deliberately peeled off or when film adhesion failed. In the beginning of 1950s Murback and Wilman[1] concluded that the behaviour was caused by tensile stress in the upper parts of the film. This 'stress is explained as arising from the contraction of the upper regions of the growing deposit as they cool after their deposition, which takes place at a temperature generally several hundred degrees higher than the initial substrate temperature'. Later other possible origins for this so-called intrinsic stress were proposed. Halliday, Rymer and Wright[3] suggested surface tension, Story and Hoffman[4] suggested point defects and defect clusters. Finnegan and Hoffman[5] proposed a combined effect of surface tension and growth at the grain boundaries, Klokholm and Berry[6] proposed annealing and shrinkage of disordered material buried under the successively growing film. Sun, Tisone and Cruzan[7] stated that the dominant component of intrinsic stress is a function of the detailed microstructure of the film which in turn is determined by the substrate temperature during deposition.

When doing hard chromium electroplating, a network of cracks is found in the films.[8] The mean distance between these cracks depends on temperature of the electrolyte, current density and film thickness as well as composition of the electrolyte.[9]

In this study chromium was vapor deposited on to die cast zinc. The films show networks of cracks, similar to the ones in electrolytically deposited chromium. These cracks are caused by the intrinsic stress in the film. Crack governing factors are deposition rate, substrate temperature and film thickness.

Experimental equipment and procedure

The geometry of the evaporation apparatus used in this study is shown in Figure 1. The cylindrical chamber, with a diameter of 0.45 m and a height of 0.50 m, is evacuated by an ordinary diffusion pump system with water cooled baffle and liquid nitrogen trap. The deposit is evaporated from a resistance heated zirconium oxide crucible. For chromium a deposition rate of 0.6 μm/min is obtainable at a source–substrate distance of 0.18 m. Between source and substrate three shutters are located. One is used to prevent the substrate from radiation heat as well as deposition while heating up the source to a stable condition. The other two shutters, mounted just below the substrate holder, are used to successively prevent parts of the substrate holder from deposition. This makes it possible to get specimens deposited for different time from the same deposition, with the initial condition equal.

The substrate holder can be heated with a resistant heater, fed from an insulation transformer. The bulk temperature of

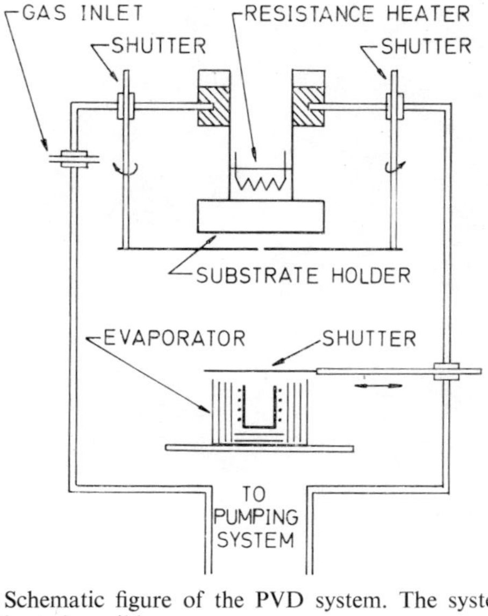

Figure 1. Schematic figure of the PVD system. The system has a resistance heated crucible.

the substrate is measured with a copper–constantan thermocouple. The signal is transmitted via an opto-electric coupling, making it possible to record and regulate the bulk temperature to a predestined value, during all process conditions. A mass-spectrometer, separately pumped by an ion-pump, is connected to the deposition chamber via a needle valve, giving optimum working condition for the mass-spectrometer, independent of the pressure in the evaporation chamber.

As substrates die cast zinc was used (4.03% Al, 2.58% Cu, 0.04% Mg, 0.010% Fe, 0.002% Cd, 0.001% Sn and Zn as the remainder). The specimens had a diameter of 30 mm and a height of 15 mm. After grinding and polishing the specimens were ultrasonically cleaned in trichloroethylene, acetone, and alcohol. In each deposition six specimens were coated. After evacuating to a pressure of 10^{-4} Pa (10^{-6} torr) or lower, the chamber was backfilled with 99.9997% argon to a pressure of 2.6 Pa (20 mtorr). The specimens were sputter cleaned at −2 kV for 20 min. After sputter cleaning the chamber was pumped down to 10^{-3} Pa (10^{-5} torr) or lower before deposition started. The weight gain of the specimens were measured with an accuracy on the order of 10^{-5} g. Thickness and deposition rate were calculated from the measurements. The thicknesses were also controlled with a Talysurf 4. The coatings were examined in optical and scanning electron microscope (SEM) after deposition.

Experimental results

During deposition of chromium on to die cast zinc it was found that the films cracked. The amount of cracks in the films has a strong dependence on deposition rate, substrate temperature and film thickness. Films deposited on to Al, Pb, Cu, and brass were compared with those deposited on Zn and no significant differences were found. Figure 2 shows a cracked chromium surface. The cracks are found to preferentially follow scratches in the substrate if scratches are present.

The cracks are caused by stresses developed in the film. The 'mean distance between cracks' was measured for deposition rates up to 0.65 $\mu m\ min^{-1}$ and three different substrate temperatures, 323 K (50°C), 398 K (125°C) and 473 K (200°C). 'Mean distance between cracks' as a function of deposition rate and substrate temperature are shown in Figure 3. The film thicknesses of all examined specimens are 3–6 μm. For deposition rates less than 0.1 $\mu m\ min^{-1}$ the films are free from cracks.

Figure 2. Chromium film with cracks. When deposited on to die cast zinc, evaporated chromium will crack. The cracks are found to preferentially follow scratches in the substrate.

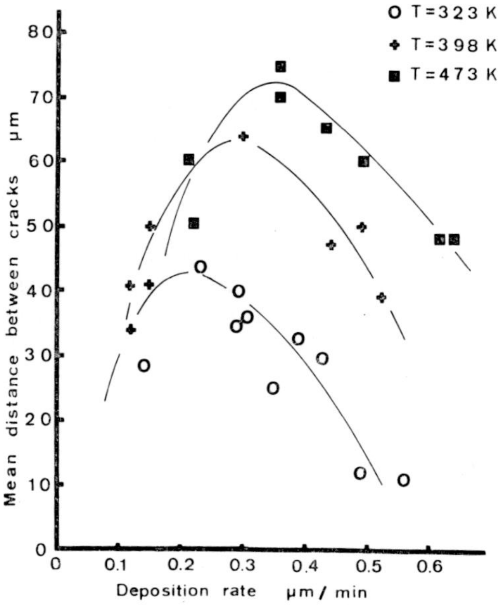

Figure 3. Mean distance between cracks vs deposition rate for different substrate temperatures. The film thickness is 3–6 μm. The cracks are caused by intrinsic stress. For deposition rates lower than 0.1 $\mu m\ min^{-1}$ cracks are not found in the film.

The 'mean distance between cracks' dependence on deposition rate has a maximum when the rate is increased to 0.2 $\mu m\ min^{-1}$, 0.3 $\mu m\ min^{-1}$ and 0.35 $\mu m\ min^{-1}$ respectively.

Figure 4 shows mean distance between cracks as a function of film thickness and substrate temperature. The 'mean distance

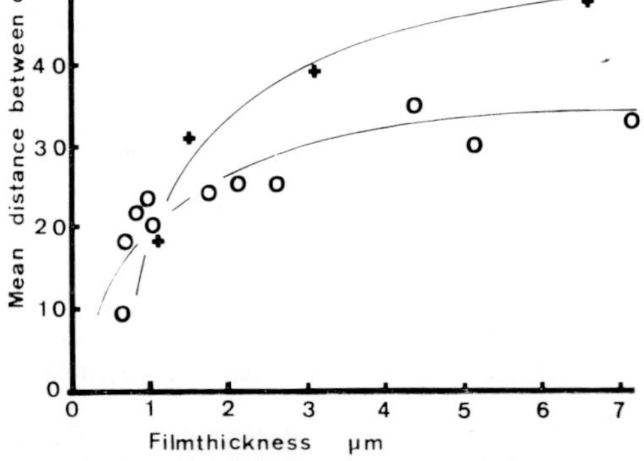

Figure 4. Mean distance between cracks vs film thickness. With bulk substrate temperature (T) 323 K and a deposition rate (v) of 0.3 $\mu m\ min^{-1}$ no cracks were found for a film thickness less than 0.4 μm. For $T = 398$ K and $v = 0.5\ \mu m\ min^{-1}$ this minimum thickness was 0.9 μm.

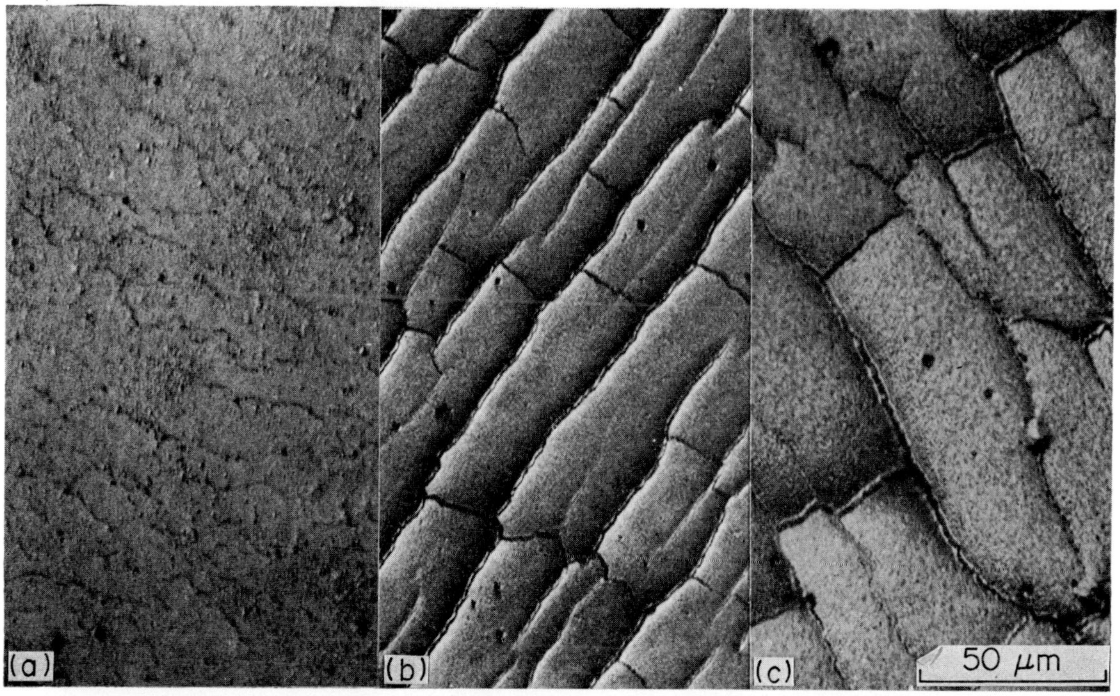

Figure 5. Crack dependence on film thickness illustrating the curves in Figure 4. Substrate temperature 398 K. Deposition rate 0.5 μm min^{-1}.
(a) Thickness (t) 0.6 μm. A large number of cracks can be seen in the film just after cracking has started.
(b) $t = 2.5$ μm.
(c) $t = 4.4$ μm.

between cracks' dependence on film thickness is strongest during the first 2–3 μm after cracking has started. The minimum film thickness, at which the films cracks, rises when the substrate temperature is increased or when the deposition rate is decreased. Just before cracking the surface gets a rough appearance. At 323 K (50°C) and 398 K (125°C) cracking starts at an approximate film thickness of 0.4 μm and 0.9 μm respectively with some changes due to deposition rate. The density of cracks decreases as the film grows thicker. Figure 5 shows photographs of a cracked chromium film at $t = 0.6$ μm, $t = 2.5$ μm, and $t = 4.4$ μm film thickness. The three films are deposited in the same deposition cycle. For thick films the mean distance between cracks approaches a constant value.

The Vickers diamond hardness was measured as a function of deposition rate and substrate temperature (Figure 6). The hardness was measured with a load of 10 g. For loads larger than 10 g it was found that the film collapsed. At deposition rates larger than 0.2–0.35 μm min^{-1} there is a large increase in hardness for increasing deposition rate at all examined substrate temperatures. This shall be compared with Figure 3 where the mean distance between cracks decreases almost linearly above these deposition rates. For deposition rates lower than 0.2–0.35 μm no correlation between increasing hardness and decreasing 'mean distance between cracks' could be found.

SEM observations of the microstructure show that films deposited at a substrate temperature of 473 K (200°C) and 0.2 μm min^{-1} deposition rate have a dense fibric structure. For a substrate temperature of 473 K (200°C) and 0.6 μm min^{-1} deposition rate the microstructure consists of columnar grains having a diameter on the order of 1 μm, Figure 7. Further investigations of the microstructure and its dependence on the deposition rate are in progress.

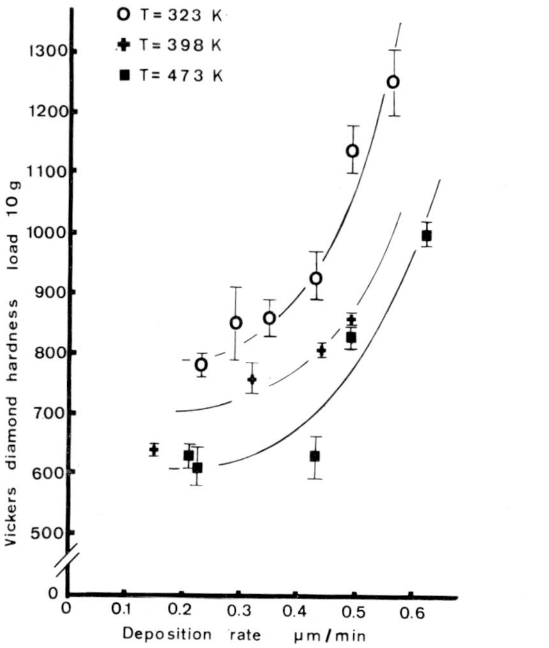

Figure 6. Vickers diamond hardness vs deposition rate for different substrate temperatures. Load 10 g. For deposition rates higher than 0.2–0.35 μm min^{-1} an increase in hardness is found for increasing deposition rate. This corresponds to a decrease in mean distance between cracks.

Discussion

Two observations will be discussed: the appearance of cracks and the crack network dependence of the deposition procedure.

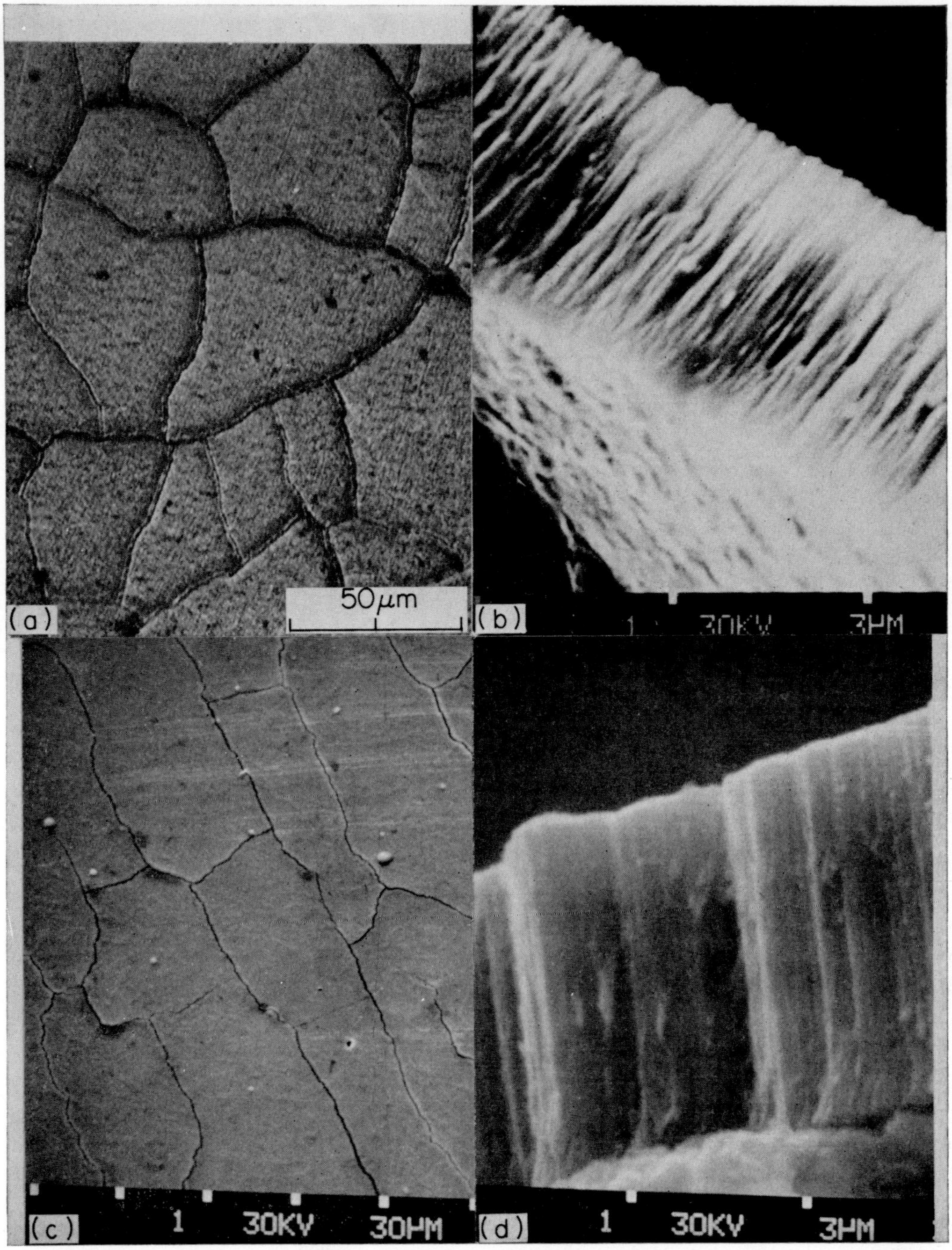

Figure 7. Macro and micro structure in evaporated chromium films.
(a) Optical micrograph showing macrostructures of film coated at $T = 473$ K, $v = 0.2$ μm min^{-1}.
(b) SEM micrograph showing microstructure in film (a). The film has a dense fibric structure.
(c) SEM micrograph showing macrostructure of film coated at $T = 473$ K, $v = 0.6$ μm min^{-1}. The film consists of flakes with some nodules.
(d) SEM micrograph showing microstructure in film (c). The microstructure consists of columnar grains.

Models describing the microstructure dependence on deposition parameters have been given by Movchan and Demchishin[10] and Thornton,[11] and hardness of chromium films have been given by Komiya and Tsuruoka.[12]

Intrinsic stress. The cracks originate from tensile stress in the film. Changing substrate material does not result in any significant altering of the crack behaviour. As zinc has a larger linear expansion coefficient than chromium (28×10^{-6} and 8×10^{-6} respectively) the substrate contracts more than the film when cooled from the same temperature. Cooling from higher temperatures should result in larger compressive stress in the film. The graphs in Figure 3 contradict this. Going from lower to higher substrate temperature during deposition, while keeping deposition rate and film thickness constant, results in less stress in the film. Thus the substrate linear expansion coefficient is a minor component in the crack generation. The networks are thus a consequence of intrinsic stresses in the film and not depending on substrate properties.

The minimum thickness where we found cracks could perhaps

be used to define where thin films should be called thick films, i.e. the limit where the film's own material properties become the major governing factor of the film properties. Using this definition this study is only dealing with thick films.

The origin of cracks. We believe that the first atomic layer condenses at a temperature close to the substrate temperature and the second layer condenses at a slightly higher temperature. This assumption is justified knowing that it takes only a few lattice vibrations until the impinging atoms are in thermal equilibrium with the substrate.[13, 14] Thus the condensation surface temperature gradually increases until it reaches an equilibrium temperature. At this temperature the energy flux from the condensing atoms and the radiating evaporation source, is equal to the energy transport into the substrate and the radiation out from the surface. A temperature gradient is built up in the film from the condensation surface and inwards.

Accurate measurements of the condensation surface temperature have not yet fully been achieved, but indicate a temperature raise on the order of one hundred degrees above the bulk substrate temperature. The tensile stress, due to contraction when cooling one hundred degrees is on the same order as the ultimate tensile stress of chromium. This indicates that this type of force is strong enough to crack the film.

This model can explain the decrease in the quantity 'mean distance between cracks' for high deposition rates, Figure 3. It cannot, however, explain the behaviour to the left of the maxima in Figure 3, where the 'mean distance between cracks' increases with deposition rate. We have to use an atomic model to explain the observed phenomena. When the deposition rate is increased the condensation surface temperature is also increased, as well as the number of impinging atoms. The more atoms migrating on the condensation surface the larger their collision probability. According to nucleation theory a minimum number of atoms have to coincide to obtain a critical nucleus size before the nucleus can grow. We suggest that for low deposition rates this critical nucleus size has not yet been reached. The atoms condense at a very large number of sites with few atoms per site, forming a fibric microstructure, Figure 7b. The crack behaviour in this region is governed by successively burying of disordered material. Going to higher deposition rates the condensation surface temperature is increased, causing enhanced atom migration at the surface and less disorder in the buried material. As the disorder in the buried material decreases the intrinsic stress also decreases and we get less cracks.

Increasing the deposition rate further means a larger number of migrating atoms at the condensation surface. This will give such a high number of atoms that the critical nucleus size for grain growth is reached. The transition from fibric to columnar (Figure 7d) grains appears for deposition rates giving the maxima in Figure 3. The critical nucleus size for columnar grains increases with the condensation surface temperature. This explains the shift of the maxima towards higher deposition rates when depositing on to substrates with higher temperature. This mechanism would also imply that we will find larger columnar grains for higher deposition rates. The crack generation behaviour in this zone of columnar grains, where the atoms are highly ordered, is explained as mentioned earlier by the temperature gradient developed in the film from the condensation surface and inwards. Increasing the deposition rate in this region gives a larger temperature gradient and thus a decrease in 'mean distance between cracks'.

In the zone of columnar structure an increase in intrinsic stress, observed as a decrease in 'mean distance between cracks', corresponds to an increase in Vickers diamond hardness, Figure 6.

Acknowledgement

This is the first paper presented from the surface coating group in Linköping. We wish to express our sincere thanks to all those who have made this group a reality. In particular we acknowledge the support of Mr Hans Ottosson and Mr Börje Ingvarsson and the most valuable contributions to the group of Professor Sven-Erik Karlsson.

References

[1] H P Murback and H Wilman, *Proc Phys Soc Lond*, **B66**, 1953, 905.
[2] D M Evans and H Wilman, *Acta Crystallogr*, **5**, 1952, 731.
[3] J S Halliday, T B Rymer and K H R Wright, *Proc Phys Soc Lond*, **A225**, 1954, 548.
[4] A S Story and R W Hoffman, *Proc R Soc Lond*, **B70**, 1957, 950.
[5] J D Finegan and R W Hoffman, *Trans 8th Nat Vac Symp*, p 947, Pergamon Press, New York (1961).
[6] E Klokholm and B S Berry, *J Electrochem Soc*, **115**, 1968, 823.
[7] R C Sun, T C Tisone and P D Cruzan, *J Appl Phys*, **46**, 1975, 112.
[8] L E Grant and L F Grant, *Trans Electrochem Soc*, **53**, 1928, 509.
[9] H von Wiegand and H-R Kaiser, *Metalloberfläche*, **19**, 1965, 161.
[10] B A Movchan and A V Demchishin, *Fizika Metall*, **28**, 1969, 653. *Phys Met Metallogr (USSR)* **4**, (1969), 83.
[11] J A Thornton, *J Vac Sci Technol*, **12**, 1975, 830.
[12] S Komiya and K Tsuruoka, *J Vac Sci Technol*, **13**, 1976, 521.
[13] J H McFee, PhD Thesis, Carnegie Institute of Technology (1960).
[14] J E Lennard-Jones, *Proc R Soc Lond*, **A163**, 1937, 127.

3.5 Morphologies of rf sputter-deposited solid lubricants

K Å B Andersson, S-E Karlsson and N Ohmae,* *Department of Physics and Measurement Technology, Linköping University, S-581 83 Linköping, Sweden*

In this study molybdenum disulphide films, about 0.5 micrometer thick, were rf sputter deposited on mild steel at ambient temperature and at different substrate bias. Scanning electron microscopy and energy dispersive X-ray analysis were used to investigate and characterize the coatings. Static friction tests were also employed.
The coatings were found to contain various amounts of isolated and clustered growths, surrounded by areas having a very smooth MoS_2 coating. The outgrowths were found to be concentrated around surface defects such as impurities and scratches, but they could also be stimulated by electron bombardment. They showed different structures such as cauliflower-like and worm-like structures, sometimes with a double layer structure. The static friction coefficient was found to be affected by the substrate bias, giving a minimum around −100 V bias.

Introduction

In this study, rf sputter deposition was employed in order to coat lamellar solid lubricant compounds on a steel surface. Solid film lubricants appear to be potentially useful in reducing friction between two surfaces subjected to relative movements, if the film strongly adheres to the substrate. Relatively little has been reported in this field of study, except by Spalvins[1–6] from NASA Lewis Research Center. Using transmission and scanning electron microscopy, Spalvins has shown that rf sputter deposited molybdenum disulphide (MoS_2) films are greatly affected by the substrate temperature[4] and that thick sputtered MoS_2 films exhibit growths with characteristic worm-like and cauliflower-like structures.[5–6] From a tribological point of view, it seems quite natural that MoS_2 films having few irregularities exhibit better friction properties, although the mechanisms of friction and wear are not fully understood yet.

MoS_2 and tungsten disulphide (WS_2), both of which are well-known solid lubricants, were rf sputter deposited at various dc bias voltages. Morphological studies of these solid lubricant films were made using scanning electron microscopy, and the composition of the films were examined by energy dispersive X-ray analysis. In this paper the growth of irregularities and the influence of bias on the properties of rf sputter deposited films are discussed. An attempt to clarify the inter-relationship between micro-topography of the film and static friction property is also given.

Apparatus and experimental procedures

The sputter deposition apparatus used in this experiment is an rf-diode mode as shown in Figure 1. The stainless steel working chamber has a diameter of 0.5 m and a height of 0.3 m. The main components are three 0.15 m diameter disc targets, two titanium shutters and a 2.5 kW rf power supply. The optimum pumping speed of the diffusion pump (Varian Sputtorr 0175) is 350 l. s^{-1} in the pressure range from 1 to 10^{-5} Pa (10^{-2} to 10^{-7} torr) and at high pressure of 6 Pa (5×10^{-2} torr) a high pumping speed of 175 l. s^{-1} is attainable.

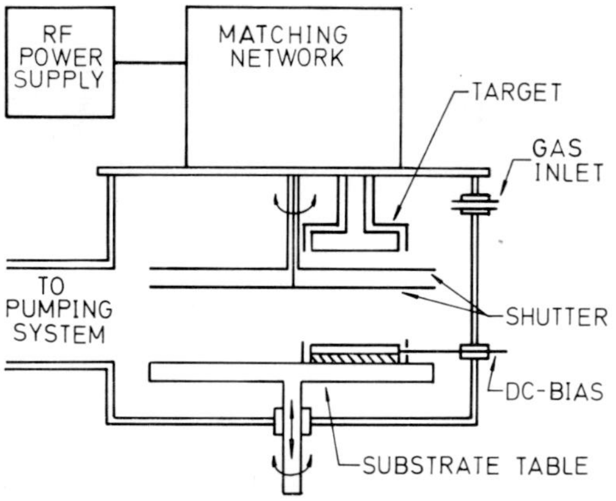

Figure 1. Schematic diagram of radiofrequency diode sputtering apparatus with dc-biased specimen.

Before deposition, the chamber was evacuated to 10^{-4} Pa (10^{-6} torr) or less, and sputter cleaning was done in a backfilled argon atmosphere of 2.6 Pa (2×10^{-2} torr) at −2 kV dc potential for 20 min. The argon gas employed in this study was as pure as 99.9997%. rf sputter deposition was conducted at about 1 Pa (8×10^{-3} torr) argon pressure with 20 kW/m² rf power density and an rf-frequency of 13.56 MHz. The distance between target and substrate was kept constant at 40 mm. A dc negative potential was supplied to a specially designed substrate holder. The final film thickness was about 0.5 μm and the deposition rate about 0.01 μm min^{-1}.

The substrate material employed was a mild steel in the shape of round bars, with a diameter of 10 mm, having a composition of 0.12% C, 0.05% Si, 0.4–0.7% Mn, 0.08% P, 0.06% S and Fe

* On leave of absence from Department of Precision Engineering, Osaka University, Suita, Osaka 565, Japan.

rest. The flat end of the bar was polished by diamond compound down to a surface roughness of 0.1 μm R_{max}. Before deposition, substrates were ultrasonically cleaned with trichlorethylene, acetone and alcohol. Using a tiltable plane, static friction experiments of MoS_2 coated substrates were carried out under controlled atmosphere of 293 K (20°C) temperature and 50% relative humidity. Static friction coefficients were calculated from the angle with respect to horizon when the coated specimen began to move. As counter specimen a freshly abraded high carbon steel was used. The friction experiments as well as tape tests showed that the coatings had strong adherence to the steel substrate.

Results and discussion

Scanning electron microscope observations. The sputtered MoS_2 coatings were examined using a scanning electron microscope, fitted with an energy dispersive X-ray analyser. The MoS_2 coatings were found to contain various amounts of isolated and clustered growths, surrounded by areas having a very smooth MoS_2 coating. The frequency of surface outgrowths had a small but not very pronounced dependence on substrate bias during deposition. Coatings performed at a bias of between −75 and 125 V exhibit a smaller amount of growth than those performed at higher and lower bias. The size of the surface outgrowth varies from less than one micrometer up to several micrometers in diameter, having nodular or cauliflower-like structures, but worm-like structures of tens of micrometers in length were also observed, as shown in Figure 2.

It was also found that the growths often exhibit a double layer structure, as can be seen in Figure 3. In agreement with statements of Spalvins,[6] differences in chemical composition can effect the amount of growths in sputtered coatings, and impurities can act as nucleation sites for preferential growth. X-ray analyses have revealed that at many of the growths a larger amount of impurity elements such as aluminium, silicon and manganese can be found. Examination before deposition of highly polished steel specimens showed that the specimens

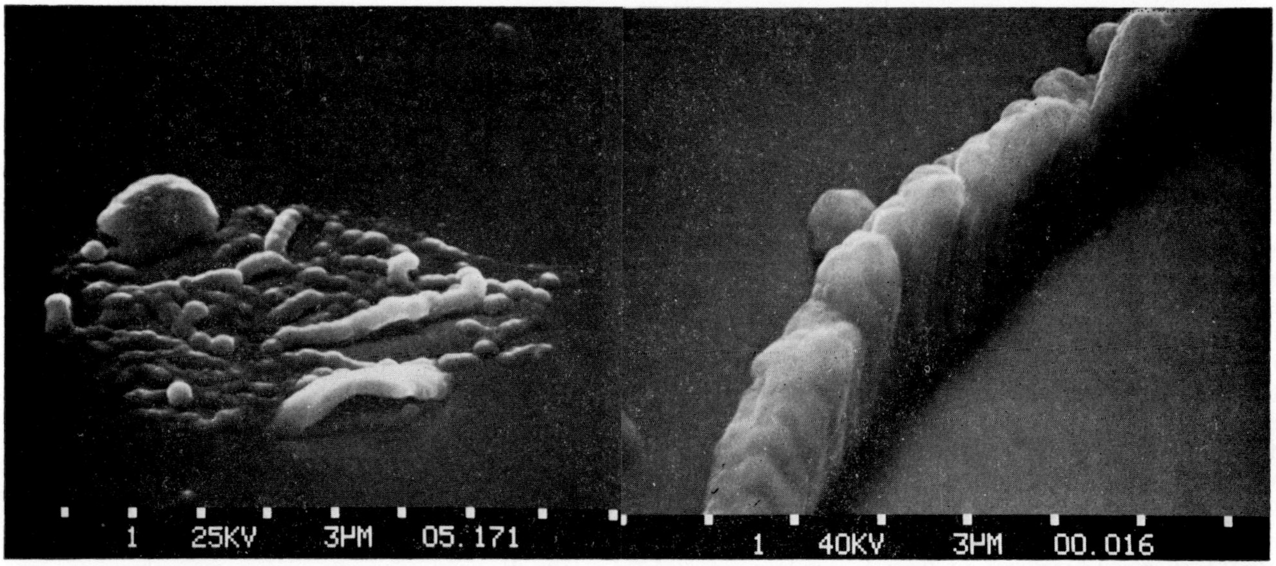

Figure 2. Surface outgrowths with worm-like structures in sputtered MoS_2 film.

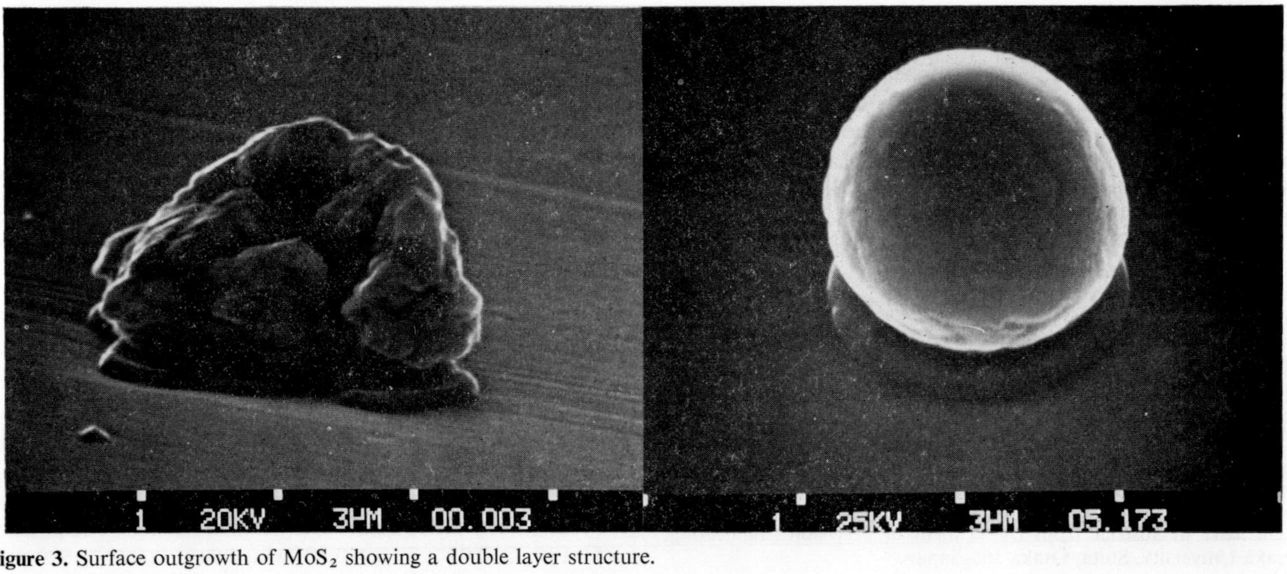

Figure 3. Surface outgrowth of MoS_2 showing a double layer structure.

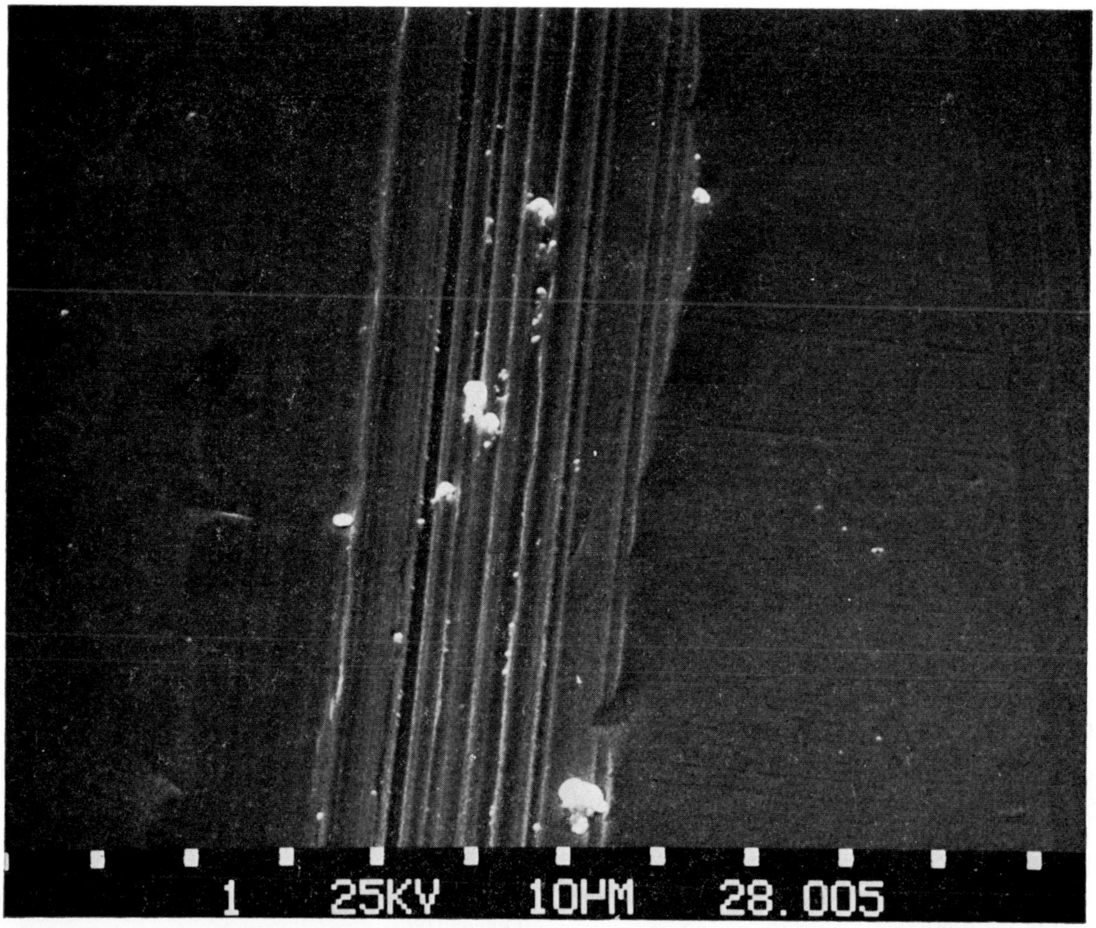

Figure 4. MoS_2 coated steel surface with scratches. A higher amount of surface outgrowth can be seen at the scratches.

contained small particles or inclusions with a high amount of impurity elements such as aluminium, silicon and manganese in addition to sulphur. The amount of particles and inclusions corresponds fairly well to the amount of surface outgrowths observed in the MoS_2 films. Thus it is possible to state that impurities in the specimens can have a pronounced effect in forming surface outgrowths of deposited MoS_2.

In order to study the influence of geometrical surface irregularities on the nucleation and growth of MoS_2 coatings, small scratches and indentations were made on some of the specimens before deposition. The irregularities produce a higher amount of surface outgrowths, as can be seen in Figure 4, but the effect was not very pronounced.

During X-ray point analysis the electron beam usually produced local swelling of the MoS_2 film, thus causing structures very similar to the observed surface outgrowths obtained during deposition. This indicates that electron and probably also ion bombardment during deposition may cause abnormal growth at highly bombarded areas. Local disturbance of the rf field during deposition could be a possible explanation for enhanced electron or ion bombardment, causing surface outgrowths at highly bombarded areas.

Surface topography and tribological property. Figure 5 shows the relationship between static friction coefficient and surface microtopography for MoS_2 coated specimens. Number of peak

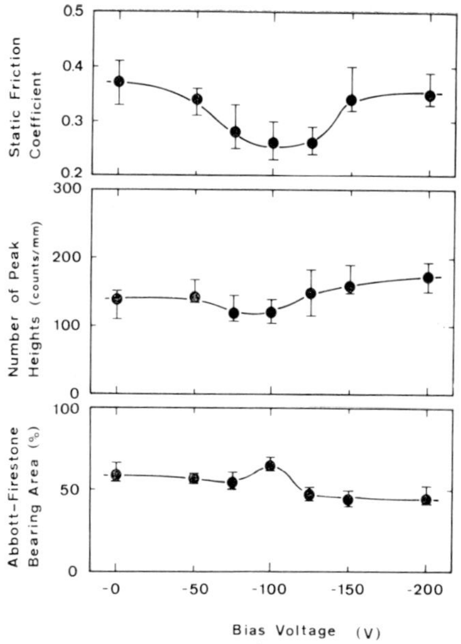

Figure 5. Static friction coefficient, number of peak heights at mean line and Abbott–Firestone bearing area at mean line vs bias potential. The minimum in friction coefficient is strongly connected to minimum in number of peak heights and maximum of Abbott–Firestone bearing area.

heights above mean line and Abbott–Firestone bearing area at mean line[7,8] were measured with a Talysurf 4. We note that the static friction coefficient attains a minimum value for the specimens deposited at -100 V bias. The number of peak heights exhibits the same tendency as the static friction coefficient and the Abbott–Firestone bearing area shows an inverse relation to the number of peak heights. Thus it is possible to say that the static friction property is greatly affected by the surface microtopography. The coated surface becomes smooth at around -100 V bias, and the surface may be characterized by a hillock geometry. In contrast, the coated surface at -200 V bias, for instance, may be represented by rough geometry, because the ratio between bearing area and number of peak heights becomes small, resulting in a mountainous topography with high hills and deep valleys. Therefore, it is reasonable to say that irregularities cause resistance to sliding and produce high friction coefficients. The variation of surface topography with amount of bias, as measured by Talysurf 4, appears to correspond well with the results obtained in the scanning electron microscope studies. It may be concluded that the best tribological properties in MoS_2 are obtained with around -100 V bias potential. The same tendency was found for WS_2 coatings.

Summary of results

Scanning electron microscopy, X-ray analysis and static friction experiments of rf sputtered MoS_2 on steel substrates, deposited at ambient temperature with different substrate bias, revealed the following results:

1. The MoS_2 films were uniformly coated and had strong adherence to the substrate.
2. Impurities were found to act as a starting point for surface outgrowths.
3. Scratches were found to give an enhanced amount of surface outgrowths.
4. Electron bombardment can be responsible for local swelling of MoS_2 coatings.
5. The static friction coefficient showed a minimum for specimens coated at about -100 V bias.

Acknowledgement

We would like to thank Professor S B M Hagström for stimulating discussions, and H O E Ottosson for valuable assistance in the initiation period of the project.

One of us (N O) would like to express his gratitude to Linköping University for the guest research period therein, and his hearty thanks are due to Professor T Tsukizoe of Osaka University for his valuable remarks and encouragement of this study.

References

[1] T Spalvins, *ASLE Trans*, **12**, 1969, 36.
[2] T Spalvins, *ASLE Trans*, **14**, 1971, 267.
[3] T Spalvins, *NASA TN* D-7169, 1973.
[4] T Spalvins, *ASLE Trans*, **17**, 1974, 1.
[5] T Spalvins, *Microstructural Science*, **3**, 1975, 525.
[6] T Spalvins, *NASA TM X*-71742, 1975.
[7] T Tsukizoe and T Hisakado, *Trans ASME, Ser D*, **89**, 1965, 666.
[8] L Miller, *Engineering Dimensional Metrology*, p 30, Edward Arnold, London (1962).

4 Optical properties of thin films

4.1 The monitoring of thin films for optical purposes

H A Macleod, *Department of Physics and Physical Electronics, Newcastle-upon-Tyne Polytechnic, Newcastle-upon-Tyne NE1 8ST, England*

The control of optical thickness is the principal objective of the monitoring of optical thin films. The common methods are first of all optical ones, which have the advantage of directness, and then the measurement of film mass by the quartz crystal microbalance, which has the advantage of simplicity. Accuracy is all important and the desire for greater accuracy has led to many developments in the techniques. A closely related topic is the assessment of allowable tolerances in coating designs.

Nevertheless, in spite of the great advances which have been made, the most characteristic feature of the process remains the difficulty of achieving absolute accuracy and hence high yield in single coating runs, while for a series of identical runs the eventual accuracy and repeatability are amazingly good.

The reasons for this lie in the behaviour of the thin films themselves and especially in their condensation, their structure and their stability after deposition. Accuracy and tolerancing calculations usually assume ideal thin film behaviour and it is only by considering their real behaviour that it is possible to begin to understand the exasperating difficulties which occur when attempting to coat a substrate which is a little different from normal.

Introduction

In a short paper it is impossible adequately to cover the entire field of the monitoring of thin films for optical purposes and so this account deals with just a few aspects of the monitoring of evaporated thin films for optical coatings.

The parameter we are most interested in is optical thickness, and although there are very many ways in which this could be measured, virtually all the common techniques involve the measurement of either optical reflectance or transmittance at one or many wavelengths, or of film mass using the quartz crystal microbalance.[1,2]

Monitoring methods

We can, initially, divide the monitoring methods into two main groups, direct monitoring, where all the layers are monitored directly on the component which is to be produced, and indirect monitoring, where the control is carried out on separate test plates. The original and obvious attraction of direct monitoring is that one is measuring what is actually being deposited while indirect monitoring allows the use of separate test plates for each layer if necessary, avoiding many of the difficulties of direct monitoring. Indirect monitoring with the quartz crystal microbalance is particularly suitable for automatic control.

To produce a given multilayer, how accurately must we control layer thickness? This is a surprisingly difficult question to answer, but with the help of powerful computers much progress has been made recently and we now know that the allowable tolerances in layer thickness depend on the monitoring method. In fact it is much more meaningful to talk in terms of tolerances related to the monitoring signal rather than layer thickness. The most successful technique for studying tolerances has been found to be the simulation by computer of the monitoring process including random errors of prescribed mean and standard deviation in all layers simultaneously. The variation between successive results represents the expected variation in successive production batches.

The earliest versions of this technique[3] simply involved random thickness errors in all the layers, the errors being quite uncorrelated but drawn from the same random population. This reproduces what is found in practice with indirect monitoring when all the layers are monitored on separate test plates. Figure 1 is an example of such a plot. The results for narrow band filters were those most difficult to reconcile with practice. For example a 2.0 nm bandwidth filter centred on 600 nm would require errors in thickness of not greater than 0.3%. The process most often used for these filters consists of turning value monitoring which simply involves monitoring at one single wavelength and terminating the layers at turning values of reflectance or transmittance, corresponding to complete quarter wave optical thicknesses. For individual layers we can immediately demonstrate[4] that the achievable accuracy is at least an order of magnitude poorer than we need. However, it has now been shown[4,5] that in direct turning value monitoring using the first order, that is where the monitoring wavelength coincides with the peak wavelength of the filter, a powerful error compensation process exists which ensures that while errors in thickness even up to 50% in some layers can exist, they are compensated by corresponding errors in the subsequent layers. This question of narrow band filter monitoring was reviewed briefly recently.

Thus an advantage of direct monitoring over indirect

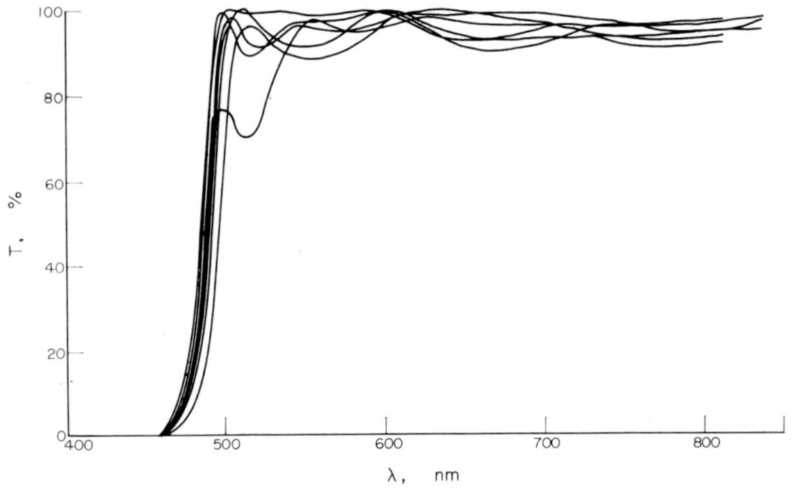

Figure 1. Effect of random thickness errors of zero mean and 5.0% standard deviation on the performance of a 17-layer edge filter of design Glass/H/2$(LH)^7LH$/2/Air with H and L quarter waves of zinc sulphide and cryolite centred on 400 nm.

monitoring is that error compensation processes can exist. This result leads us to rearrange the monitoring classifications by introducing a further classification, that of semi-direct monitoring where although the test plate is separate from the substrates to be coated, all the layers are monitored on the same test plate so that advantage may be taken of any error compensation which may exist.

In indirect monitoring when the index difference between the film and test plate is small, it has been common practice to precoat the test plate in order to increase the magnitude of the monitoring signal. Recently Seeley and colleagues[7] and Ward[8] have described two specific cases of semi-direct monitoring where precoatings make a substantial improvement.

Seeley's primary interest is the production of filters for the far infrared and he monitors all the layers in reflectance on one single germanium test plate at a wavelength which is one quarter of the reference wavelength, i.e. in the fourth order, and terminates the layers at turning values, so that if the coating designs use only eighth wave optical thicknesses or integral multiples, then both the start and finish of every layer will be at exactly the same level of reflectance. A precoating consisting of a single layer anti-reflection coating at the monitoring wavelength usually of zinc sulphide ($n = 2.3$) has been found to give greatly increased monitoring accuracy. Very small consistent overshoots of the turning values improve the accuracy still further.

Ward[8] has described a four layer anti-reflection coating which is shown in Figure 2 and has found that a useful technique for the production of such a coating is semi-direct turning value monitoring using a glass test plate with either a three layer or a five layer precoating of form *HLH* or *HLHLH*, *H* and *L* being quarter waves of zinc sulphide and cryolite ($n = 1.35$) respectively.

The errors in semi-direct monitoring at one single wavelength have recently been analysed theoretically[9] and it has been shown that the reason for the success of these systems is largely the small consistent overshoot which skilled operators always make in determining turning values. The function of the precoating turns out to be not just the sharpening of the turning values by increasing the amplitude of the monitoring signals but also the

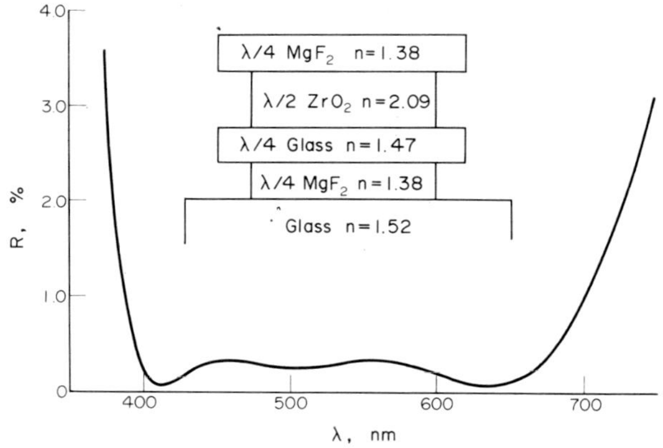

Figure 2. A four layer antireflection coating designed by Ward.[8]

elimination of unfavourable error interactions which in the presence of overshoots become especially important.

It does occur that coatings have to be produced where the layer thicknesses bear no simple relationship to quarter waves at one reference wavelength, and for such designs current commercial practice is to use indirect monitoring because it is much more straightforward than other types of monitoring. Metal layers, too, can cause special problems and most often these are monitored indirectly. Indirect monitoring is reviewed[10] and has not been considered in detail here but it should be noted that it is used with considerable success for the manufacture of a wide range of coatings and especially broad band components. Nevertheless many workers have an almost instinctive feeling that a system of monitoring where all the layers are deposited together on the same test plate, is preferable to one in which separate test plates are used, and still further that a system where the performance of the multilayer is monitored over the entire spectral region of interest during deposition, is better than one where the performance is measured at only one single wavelength. A recently developed system of this type[11] suitable for use with absorbing layers and layers of any thickness is shown in Figure 3. Central to the system is a rapid scanning monochromator which uses a stepping motor to drive the grating. A mini-computer (PDP11) is incorporated, which in the simplest arrangement compensates for the background variation but which can also be used for more complex tasks. There is a number of possible ways in which this system may be used. The transmittance of the coating over the spectral range of interest may be displayed on a cathode ray tube and compared with the performance curve which it should possess at the termination of the layer. After termination the thickness which has actually been deposited is computed and displayed. To eliminate the visual assessment of the performance curves which can be difficult, a merit function can be calculated as the layer is being deposited, using the difference, with suitable weights, between the actual and the desired performance at 10 preselected wavelengths. A scanning monochromator monitoring system has also been developed.[12]

Great progress in the monitoring of optical thin films has undoubtedly been made. Our computer simulations allow us to predict the accuracy which we require and to choose the appropriate monitoring arrangements. Yet, if I may quote from the recent book[13] the second last sentence, which is an admirable summary of what must surely be the experience of anyone who has worked in the optical coating industry, 'Finally we remark that the error scale may shrink to remarkably small values in repeated production of one item (i.e. one system on a given substrate and workholder), whereas even an experienced laboratory may have a rough life if it lends an ear to the diversified wishes of individual customers.' Why is this?

Our computer simulations and error predictions assume ideal thin film behaviour and it is in the real behaviour of the films that the answer is to be found.

Film condensation

The condensation of thin film materials is the first area we should consider. The condensation coefficient, sometimes also referred to as the sticking coefficient, is the fraction of the evaporant incident on the surface which remains and contributes to the growing film. In 1957, Günther[14] published the results of a study he had made of the condensation coefficients of silicon monoxide and of boron trioxide. This is a beautiful paper which seems inexplicably to have gone unnoticed because I have not been able to find any reference to it in the later literature. The substrates used in the investigation were polished iron, but the thicknesses of the films were sufficient to ensure that the initial nucleation on the iron formed a negligibly small part of the process and the condensation could be considered as taking place on the growing film.

The simple model of the condensation of a thin film of one single component involves first of all the adsorption of the arriving atoms or molecules on the surface with adsorption energy U_{ads}. The atoms remain on the surface for a finite time before they either condense with condensation energy U_k and form part of the growing film, or re-evaporate. In the steady state, when the film is beyond the initial nucleation stage, the number of atoms or molecules condensing will be equal to the number being adsorbed. Following Günther we can write for the condensation coefficient.

$$a = \frac{n_+ - n_-}{n_+} = 1 - C\exp\left(-\frac{U_{\text{ads}}}{kT}\right) \qquad (1)$$

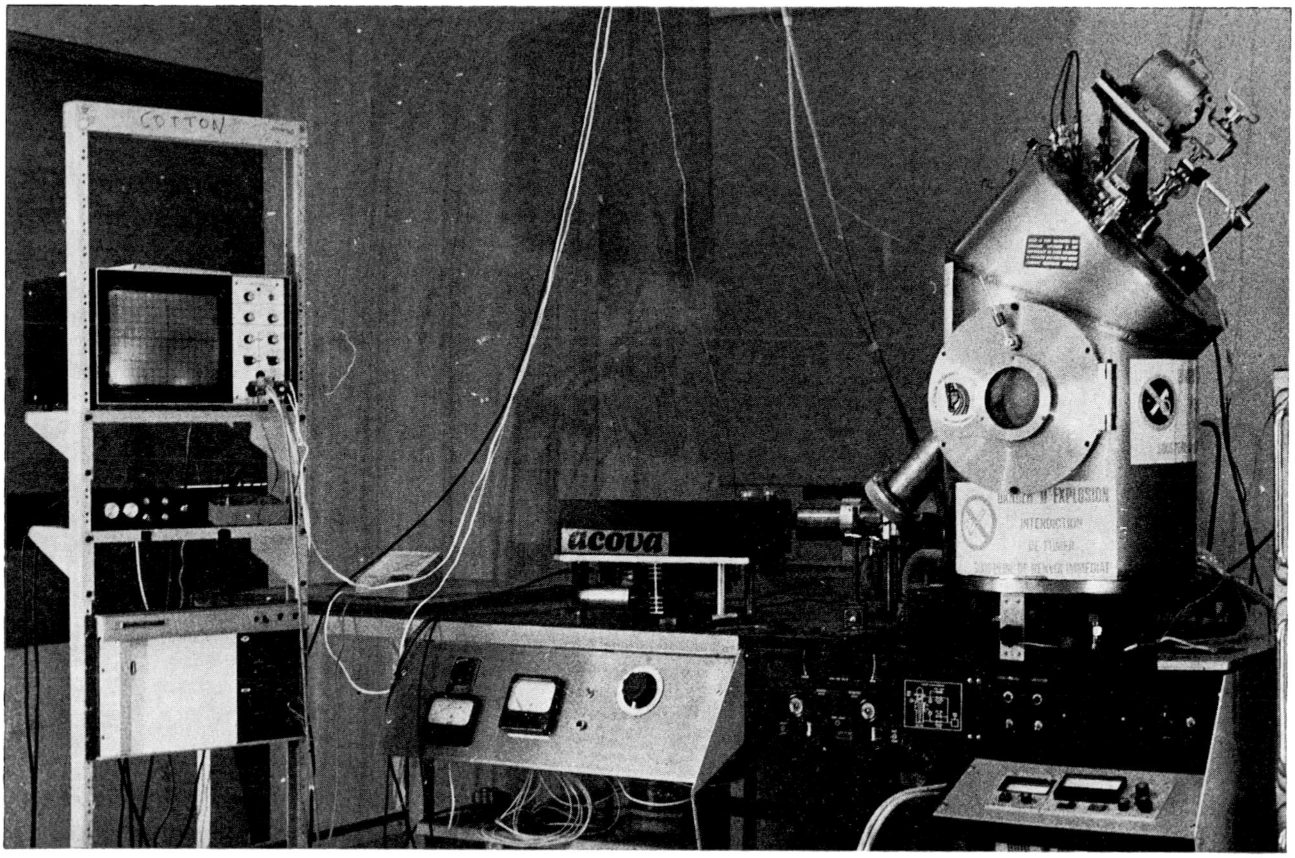

Figure 3. Monitoring system using a rapid scanning monochromator. The chamber is on the right with the light source on top. The rapid scanning monochromator is the black box on the left of the chamber.[11]

where n_+ and n_- are the beam densities of arriving and re-evaporating atoms or molecules respectively, k is Boltzmann's constant, T the substrate temperature (kelvin) and C is a constant. If now we plot log$(1 - a)$ against l/T we should obtain a straight line and be able to measure U_{ads} from the slope. Figure 4 shows Günther's plot of his results for SiO. For temperatures below 350°C the plot is consistent with an adsorption

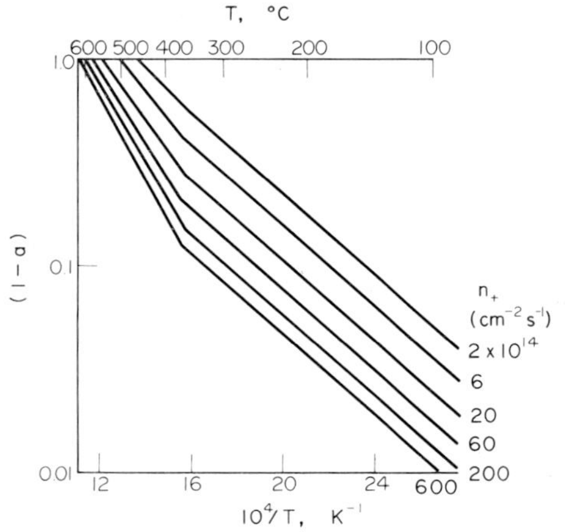

Figure 4. Plot of $(1 - a)$ on a logarithmic scale against l/T for SiO. After Günther.[14]

energy of 4.320 kcal/mol. The constant C is clearly a function of n_+. In the same temperature region the adsorption energy for B_2O_3 was found to be 4.050 kcal/mol.

Ritter and Hoffmann[15] investigated the condensation coefficient of zinc sulphide and magnesium fluoride on glass substrates. Magnesium fluoride appeared to show no variation of condensation coefficient with temperature over the range 40 to 365°C. Zinc sulphide, on the other hand, showed behaviour similar to that of SiO with an adsorption energy of 4.2 kcal/mol, although no variation of C with n_+ over the range 1 to 6 nm s^{-1} was detected. Figure 5 shows the results of

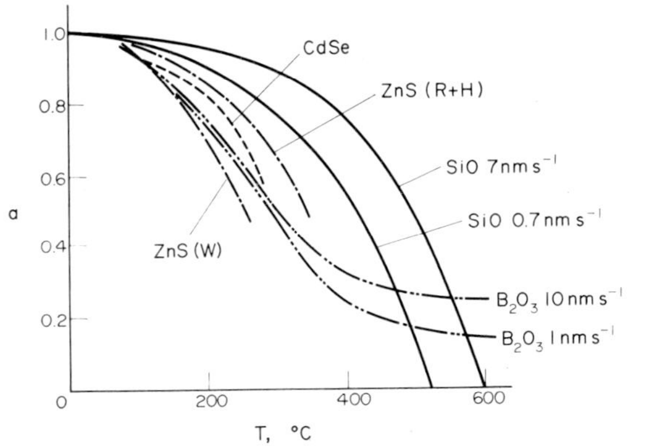

Figure 5. Various results for condensation coefficients showing the variation with temperature. SiO and B_2O_3 results from Günther[14], ZnS $(R + H)$ from Ritter and Hoffmann,[15] and ZnS (W) and CdSe from Warren.[16]

Ritter and Hoffmann plotted together with some results extracted from Günther. Warren[16] obtained similar results for zinc sulphide and cadmium selenide and some of his preliminary results are also shown in Figure 5.

There is a temperature dependence also in the condensation of metal films. Again there is a difference in the behaviour of the metal when it is condensing on an uncoated substrate from when it is condensing on a growing film and many of the published results apply to films less than one monolayer thick showing a rapid reduction in the condensation coefficient with temperature.[17] Recently Dobrev and Boicheva[18] have studied the temperature dependence of the condensation of thicker silver films on to glass. For low rates of evaporation up to 3.7 nm s^{-1} there was no detectable difference in condensation coefficient with temperature but with rates increasing from 15 to 160 nm s^{-1} they found a marked minimum of condensation coefficient between 150 and 200°C and a marked maximum between 400 and 500°C the relative difference being as high as 18%. The reason for this behaviour is not at all clear.

It has long been known that water vapour should be eliminated as far as possible from the coating chamber. Bloor and Olson[19] have shown that small amounts of water vapour can have a considerable effect on the condensation coefficient of metal films. Water vapour of partial pressure 8.1×10^{-7} torr reduced the condensation coefficient of lead on lead from unity to 0.81. With a partial pressure of 6.8×10^{-8} torr the condensation coefficient of lead on SiO was reduced from unity to 0.85. These results have implications for other metal films and also for dielectrics, especially those which decompose on evaporation and recombine on the substrate.

The significance of the condensation coefficient from the point of view of optical monitoring is mainly its variation with temperature. In semi-direct and in indirect monitoring the control is exercised on a test plate separate from the batch and any difference in temperature between them will mean an error in film thickness. We can assess the likely magnitude of such errors by using some of the results of Figure 5. Plotted in Figure 6 is the percentage error in thickness of zinc sulphide and silicon monoxide films for a temperature difference of 1°C between the test plate and batch component. This is clearly one reason for the difficulty of producing accurate narrow band filters using other than direct monitoring.

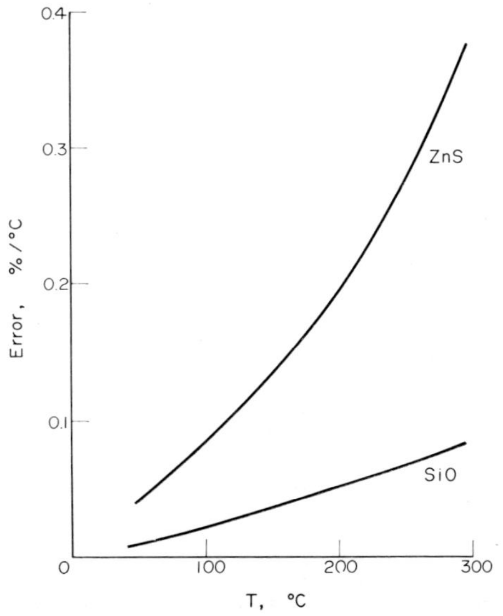

Figure 6. Film thickness errors for a temperature difference of 1°C between test plate and batch component calculated from the SiO (7 nm^{-1}) and ZnS $(R + H)$ curves of Figure 5.

Control of substrate temperature during deposition is not easy, especially with glass substrates which are rotated. Normal commercial practice is to use radiant heaters and simply control the temperature of a probe adjacent to the substrates. The evaporation sources can radiate considerable heat and can cause temperature rises of 10°C or more during the deposition of a single layer.[20] The thicker the substrate the greater is the heat capacity and the smaller the temperature rise due to this effect. Obviously the greatest consistency in results will be obtained when the test plate and batch are identical and mounted identically, easier to achieve with semi-direct or direct than with indirect monitoring. The seriousness of any temperature difference between test glass and batch or from one batch substrate to another will depend on the materials and on the coating specification, but for anti-reflection coatings, taking the SiO results as typical, a temperature difference of 15°C would probably be tolerable while for a dichroic required to have an accuracy of 2.5 nm at 500 nm, say, the difference should be not greater than 5°C. Temperature shifts of this type are even more serious when monitoring is by the quartz crystal microbalance because the crystal is held near room temperature and is often water cooled. This is an area where there is a clear need for measurements of the properties of coating materials.

Film structure

Another important aspect of the behaviour of a thin film is its structure. A considerable amount of work has been carried out in recent years using the transmission electron microscope and it is clear that the vast majority of dielectric thin film materials condense with a columnar structure.[2,22] The micrograph in Figure 7 shows clearly the tops of these columns. The cross section of the columns is roughly circular and so even when they are closely packed there can be spaces left which are unfilled with material and which reduce both the density and refractive index of the film. The term packing density, p, has been introduced as a measure of the amount of space within the film.

$$p = \frac{\text{volume of solid part of film (i.e. the columns)}}{\text{geometrical volume of film (i.e. columns + spaces)}} \quad (2)$$

Figure 7. Transmission electron micrograph of a replica of a ZnS film surface.

The expression usually used to calculate the refractive index of such a film is[23]

$$n_f = pn_0 + (1 - p)n_p \tag{3}$$

where n_f is the resulting film index, n_p is the refractive index of the voids within the film and n_0 is the refractive index of the solid portion of the film. This expression is correct for mixtures of materials but not strictly so for a columnar solid. Unfortunately, exact calculation of refractive index is difficult and so the simple expression is normally used. It introduces only a small error in n_f, probably less than 1% for the normal range of film indices and packing densities, but when the expression is inverted to give a formula for packing density, the error in p can be rather greater.

There are several ways in which this structure is important for monitoring. First of all, packing density varies both with substrate temperature and with film thickness, Figures 8 and 9

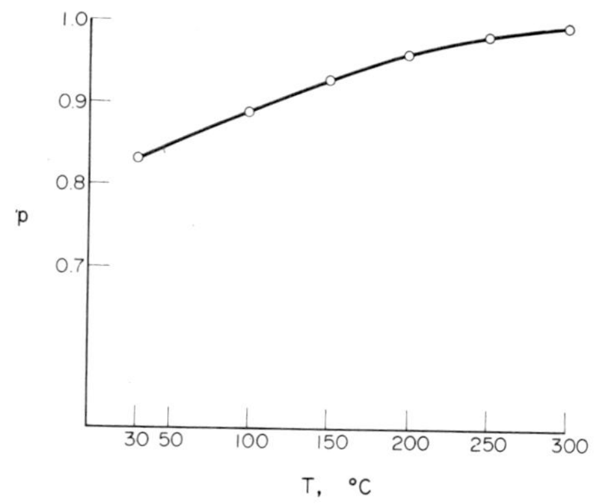

Figure 8. The variation of the packing density of MgF_2 with substrate temperature. After Ogura.[24]

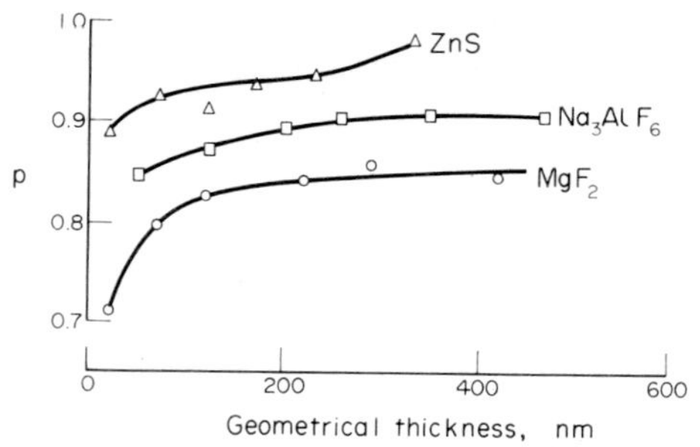

Figure 9. The variation of packing density, p, with film thickness for several materials deposited at room temperature.[24,25]

showing typical behaviour.[24,25] Since the refractive index of the material of the columns remains substantially constant, the refractive index of the films will increase with packing density, and we can use equation 3 to give us some idea of the importance of this effect. If a constant mass of film is deposited then the optical thickness, D_f, can be shown to be given by

$$D_f = \left\{1 + \frac{(1-p)}{p} \cdot \frac{n_p}{n_0}\right\} D_0 \tag{4}$$

where D_0 is the optical thickness which the film would have with a packing density of unity. If we take as typical values 0.85 for p, 1.0 for n_p, and 1.4 for n_0, then the change Δ_p in p required for a 1% change in optical thickness becomes 0.011. If we further assume that the packing density changes from 0.85 to 1.0 over a range of 200°C then a change of 0.011 represents a temperature change of some 15°C. In the case of semi-direct and indirect optical monitoring this represents the tolerable shift in temperature of the batch substrates with respect to the test plate during the course of the run if the thickness error due to this effect is not to exceed 1%, the absolute temperature being less important. For quartz crystal monitoring, where D_0 is what is always measured by the crystal, this represents the stability with which the temperature of the batch substrates must be maintained. This type of error is in the same sense as that due to changes in condensation coefficient. When we recall that packing density also varies with film thickness, we find that the difficulties in achieving absolute accuracy become still less surprising.

Unfortunately, this variation of packing density with temperature is just one of the effects of the columnar structure. Much more serious is the way in which the film is affected by atmospheric moisture. The voids in the film have the form of pores which stretch from the outer film surface through to the inner boundary and when the film is exposed to atmospheric humidity these pores fill with liquid water. This increases n_p from 1.0 to 1.33 which in turn increases the index n_f of the film and the optical thickness. Not all the pores fill with water immediately. Only those up to a certain critical radius, which depends on the relative humidity, fill. The equilibrium vapour pressure above a curved meniscus is lower than that above a plane liquid-surface so that a full pore will remain in equilibrium with vapour pressure below saturated if the meniscus is sufficiently curved. Since the curvature is greater the smaller the pore, pores up to a certain critical radius will fill with water while those above critical will remain empty, although the walls of the empty pores will have the usual monolayer or multilayer covering of adsorbed water which always exists even on a plane surface. This mechanism has been used for the analysis of pore size distributions in porous solids[26] and has been adapted by Ogura[24] to analyse the water adsorption isotherms in a wide range of dielectric thin films. Typical isotherms with their pore size distributions are shown in Figure 10.

The implication of the form of the isotherms is that the equilibrium value of film index will depend on the ambient relative humidity. Figure 11 shows the way in which the index of magnesium fluoride varies, calculated from the adsorption

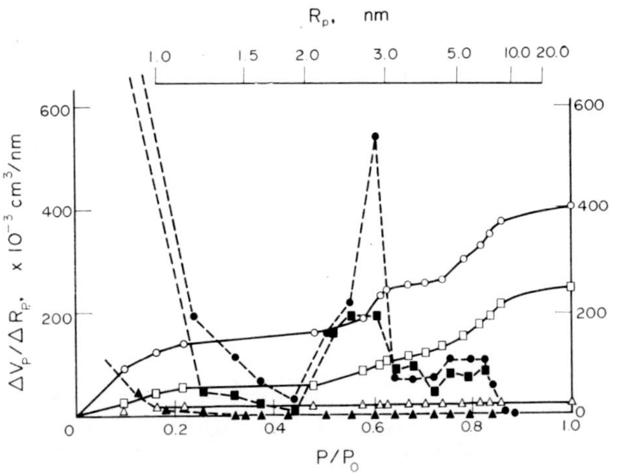

Figure 10. Water vapour adsorption isotherms and corresponding pore size distributions for MgF_2 (top), cryolite and ZnS.[24,25]

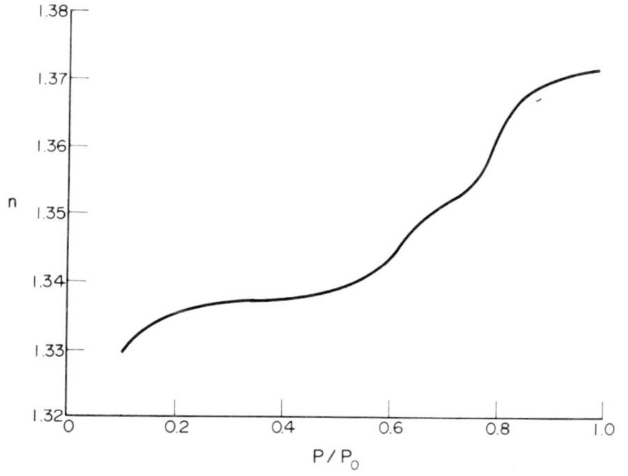

Figure 11. The variation of the refractive index of MgF_2 with relative humidity calculated from the isotherm of Figure 10 assuming $n_0 = 1.38$ and given that the mass of the film before exposure to moisture corresponds to a frequency shift of 6307 Hz.

isotherm in Figure 10 using an analysis based on equation 3. This behaviour means that immediately a coating is exposed to the atmosphere, the refractive indices and, because the geometrical thicknesses remain constant, the optical thicknesses begin to increase. The characteristic of the coating will therefore move away from the monitoring wavelength towards longer wavelengths. As a further complication in multilayers the atmospheric moisture does not penetrate evenly over the entire surface of the coating but through a limited number of pores, spreading out into the layers laterally. The progress of the moisture can be followed by illuminating the coating with collimated monochromatic light either in the neighbourhood of the pass band in a narrow band filter or of the edge in a narrow band filter. The technique is described in reference[27] and has been used in the detailed study of narrow band filters by Richmond.[28] Figure 12 shows moisture penetration patterns in a 27 layer reflector of ZrO_2 and Schott glass no 8329 both deposited at near 300°C. The mottled appearance which coatings sometimes have just after removal from the plant is almost invariably due to this effect. The drift and subsequent stabilization is often referred to as ageing or settling. The rate of moisture penetration varies enormously with film materials, deposition conditions and multilayer design. Zinc sulphide and cryolite multilayers deposited at room temperature may take several weeks to stabilize while ZrO_2 and glass multilayers such as that of Figure 12 take only 1 or 2 h.

For coatings which are to be cemented there is an interest in attempting to prevent this ingress of moisture by exposing the coatings to the atmosphere for as short a time as possible, measured in seconds, before cementing. This is especially true of narrow band filters.[6] Although the cement does not completely prevent the entry of moisture, it slows it down considerably. For coatings, which must be left exposed to the atmosphere, such as anti-reflection coatings, dichroics and laser mirrors, it is best simply to allow them to stabilize and to correct the monitoring wavelength to allow for the drift. Of course the properties will vary slightly with relative humidity. In some cases baking the

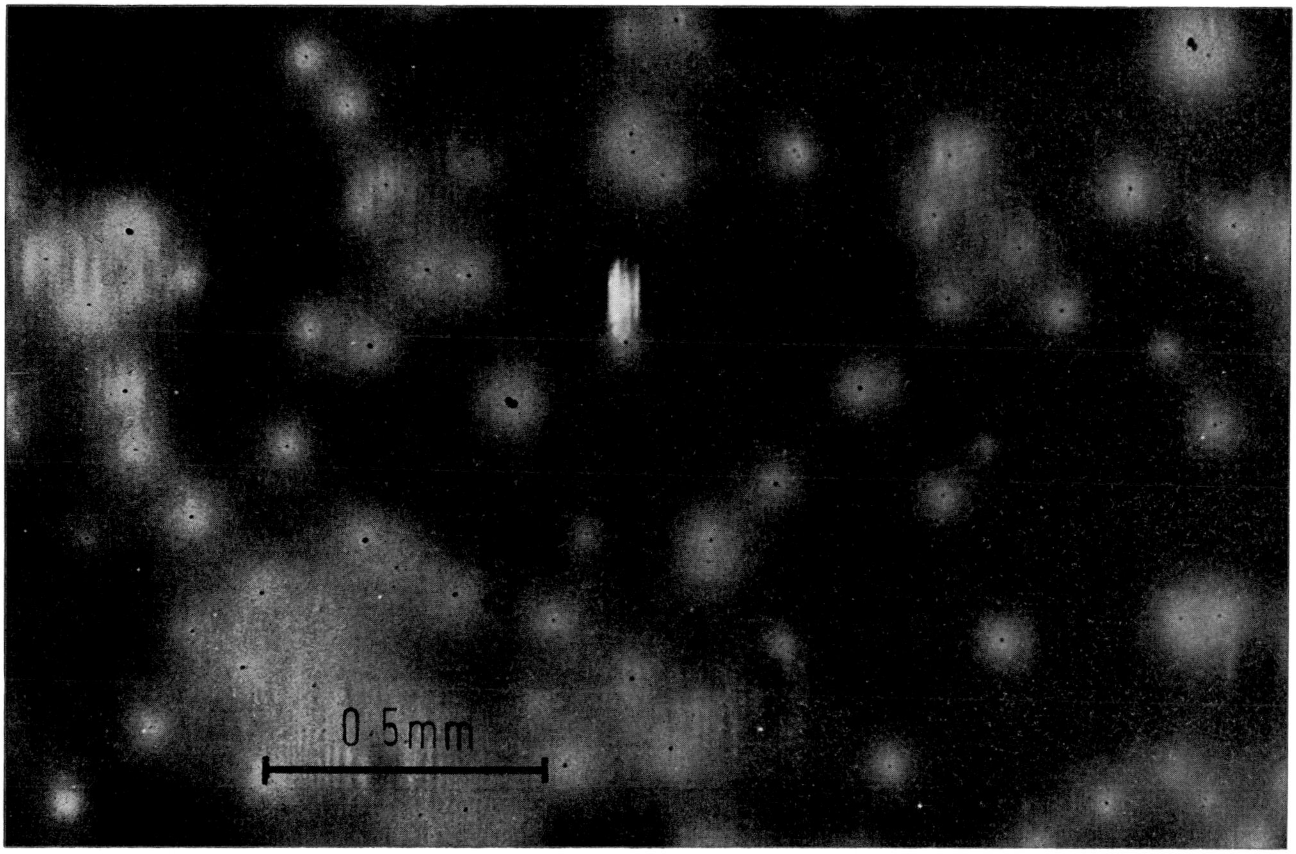

Figure 12. Moisture penetration patterns in a ZrO_2/Schott glass 8329 27-layer reflector deposited at greater than 250°C with $\lambda_0 = 632.8$ nm. The photograph was taken at 572 nm 23 min after deposition.

coating in air may help to inhibit the entry of moisture.[27] Protective coatings formed by the evaporation of extra layers may also help although Richmond in his study of narrow band filters[28] could not find any coatings of this type sufficiently effective to make cementing unnecessary.

This lack of stability due to water adsorption is a major, if not the most important cause of monitoring difficulties. However there are other sources of instability and one interesting effect which has so far not been satisfactorily explained has been found in lead telluride by Evans, Hunneman and Seeley.[29,30] This is a growth in optical thickness which takes place during the 15 min or so immediately after deposition of the layer and while under vacuum. The magnitude of the increase in optical thickness is variable, but of the order of 1 or 2%.

Conclusions

This has been a very abbreviated account of some of the factors which determine the accuracy with which we can control the deposition of optical thin films but nevertheless I hope it has helped to explain why it is that optical monitoring can be so exasperatingly difficult. Of all the parameters which play a part in the process, substrate temperature is clearly the most important and probably also the most difficult to control. A great deal of work remains to be done, much of it involving the painstaking measurement of the properties of thin film materials. And then there is the problem of moisture adsorption which seems common to all thermally evaporated thin films and for which no ideal solution has yet been found.

Acknowledgements

I am indebted to my colleagues at Newcastle upon Tyne Polytechnic and at the University of Aix–Marseille III who have worked with me on optical monitoring and I gratefully acknowledge the support given by the Science Research Council to the work at Newcastle.

References

[1] K H Behrndt, *Physics of Thin Films*, **3**, 1966, 1. (Edited by G Hass and R E Thun), Academic Press, London.

[2] H A Macleod, *Thin-Film Optical Filters*, Adam Hilger, London (1969).

[3] F S Ritchie, PhD Thesis, University of Reading, England (1970) (A brief summary of the work on monitoring errors is given in ref 2).

[4] H A Macleod, *Opt Acta*, **19**, 1972, 1.

[5] P Bousquet, A Fornier, R Kowalczyk, E Pelletier and P Roche, *Thin Solid Films*, **13**, 1972, 285.

[6] H A Macleod, *Thin Solid Films*, **34**, 1976, 335.

[7] J S Seeley, UK Patent Specification 1 425 941 (1972). The technique is also described in refs 29, 30.

[8] J Ward, *Vacuum*, **22**, 1972, 369.

[9] H A Macleod and E Pelletier, *Optica Acta*, **24**, 1977, in press.

[10] A Thelen, Invited paper at International Symposium on Vacuum and Thin Film Technology, Uppsala (1976).

[11] A Fornier, E Pelletier and B Vidal, Laboratoire d'Optique, University of Aix-Marseille III, Marseille, France, private communication.

[12] R Hiraga, N Sugawara, S Ogura and S Amano, *Japan J Appl Phys*, Suppl 2, pt 1, 1974, 689.

[13] Z Knittl, *Optics of Thin Films*, p. 541, John Wiley, New York (1976).

[14] K G Günther, *Z. Phys*, **149**, 1957, 538.

[15] E Ritter and R Hoffmann, *J Vac Sci Technol*, **6**, 1969, 733.

[16] S W Warren, ITT, Harlow, Essex, England, private communication

[17] R F Benck, C Crisco and H L Spies, *J Vac Sci Technol*, **10**, 1973, 562.

[18] D Dobrev and L Boicheva, *Thin Solid Films*, **32**, 1976, 237.

[19] E W Bloore and F A Olson, *J Vac Sci Technol*, **10**, 1973, 144.

[20] D Hacman, *Optik*, S28, Vol (1968/69) 115 (1968).

[21] H K Pulker, *Optik*, Heft 5, 1971, 496.

[22] J M Pearson, *Thin Solid Films*, **6**, 1970, 349.

[23] K Kinosita and M Nishibori, *J Vac Sci Technol*, **6**, 1969.

[24] S Ogura, PhD Thesis, Newcastle upon Tyne Polytechnic, England (1975).

[25] S Ogura and H A Macleod, *Thin Solid Films*, **34**, 1976, 371.

[26] S J Gregg and K S W Sing, *Adsorption, Surface Area and Porosity*, Academic Press, New York (1967).

[27] H A Macleod and D Richmond, *Thin Solid Films*, **34**, 1976, 163.

[28] D Richmond, PhD Thesis, Newcastle upon Tyne Polytechnic, England (1976).

[29] C S Evans, R Hunneman and J S Seeley, *J Phys D: Appl Phys*, **9**, 1976, 321.

[30] C S Evans, R Hunneman and J S Seeley, *Opt Acta*, **23**, 1976, 297.

4.2 Monitoring of optical thin films using a quartz crystal monitor

C J vd Laan and H J Frankena, *Department of Applied Physics, Delft University of Technology, Delft, The Netherlands*

Quartz crystal monitoring in the production of dielectric multilayer stacks is compared to optical monitoring. The stacks are calculated with the aid of an optimizing program. This yields optical thicknesses of the layers differing from multiples of a quarter of the wavelength. Optical monitoring is complicated in this case. Using quartz crystal monitoring, planning and execution of the process is easier and yields stacks with satisfactory reflectance characteristics faster than with optical monitoring.

The reproducibility of the optical thickness of MgF_2 layers is investigated. In general, both methods turn out to be equivalent in performance.

1. Introduction

Optical thin dielectric film stacks, fabricated in the Delft University Optical Group (e.g. broad band mirrors, high- and low-pass filters, anti-reflectance coatings, etc.), are usually optimized with the aid of a specially designed computer program. This optimization is performed by varying the layer thicknesses keeping the refractive indices fixed on experimentally determined values.

This design process yields stacks whose layers, in general, have optical thicknesses differing from exact multiples of a quarter of the wavelength. For monitoring[1] the evaporation process of these stacks an optical method or a quartz crystal monitor can be used. The aim of this paper is to give some practical information on the use of quartz crystal monitoring. After information about the equipment used, we discuss Figure 1 showing the processes leading to a multilayer stack satisfying preset specifications. Finally, we consider the reproducibility of the optical thickness of MgF_2 layers. Optical and quartz crystal monitoring are compared.

2. The equipment used

Two Balzers coating plants are available, viz. a BA 500[2] with the optical monitor GSM 210[3] (measuring the reflectance and transmittance of a layer as a function of the optical thickness) and a BA 510[4] provided with the quartz crystal monitor QSG 201.[5]

The principle of the quartz crystal monitor [6–8] is outlined in Figure 2. An oscillating quartz crystal (frequency about 4.96 MHz) is placed in the bell jar, above the vapour source. The condensing vapour increases its mass leading to a decrease in the oscillating frequency f_c. This frequency f_c is compared with the stable standard frequency f_0, their difference $f_d = |f_c - f_0|$ is supplied to a frequency meter. The standard frequency is variable from 4.8 to 5.0 MHz. The frequency shift Δf in f_d during deposition of the layer is a measure for its thickness t. In first approximation, we may assume a linear relationship between Δf and t.

In the bell jar of the BA 510, also a changer for test substrates is fixed to the domed calotte, in a position with the same distance to the centre of the bell jar and on the same height as the regular substrates. Using a shutter before the test substrates, we can control the exposition during evaporation independently.

The test substrates are used for measurement of the obtained refractive indices after the process following the Abelès method.[9] The changer can contain six test substrates.

Furthermore, we used a Beckman spectrophotometer DB–GT for determination of the reflectance and transmittance of the stack afterwards.

3. The evolution of a thin film stack

In Figure 1 the evolution of a stack is sketched; our discussions will follow this diagram.

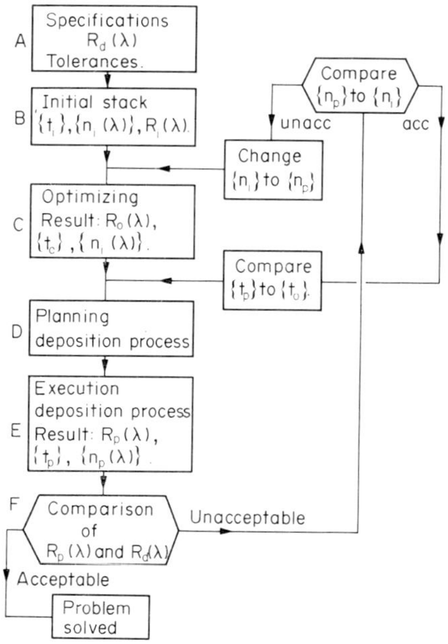

Figure 1. Diagram of the evolution of a multilayer stack.

A. Stating the problem. We shall consider the problems related to producing broadband anti-reflection coatings, low- and high-pass filters, broadband laser mirrors, etc., all composed of dielectric layers and in general for use at visible wavelength. In our optimizing program the desired reflectance $R_d(\lambda)$, where λ is the wavelength, must be specified as an ideal curve (consisting of a set of straight lines in the various wavelength regions that must be specified by the user) plus bands of constant widths around it in which deviations of the ideal curve can be tolerated. The number and positions of the wavelength regions, the place of the straight lines and the width of each of the tolerance regions can be freely chosen. An example, relating to a low-pass filter, is given in Figure 3. The computer input in this example

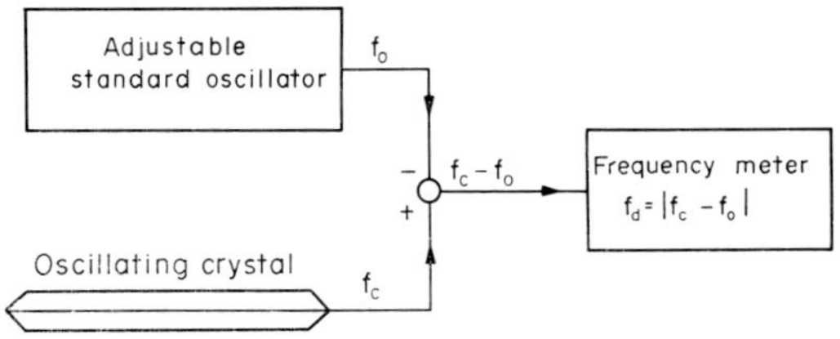

Figure 2. Principle of a quartz crystal thin film monitor.

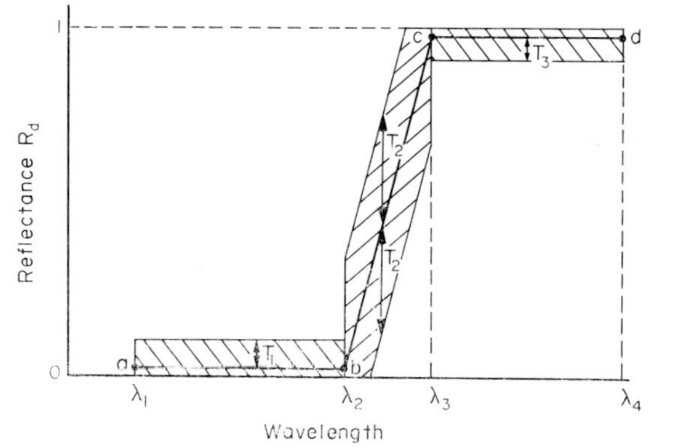

Figure 3. Example of the specification of the reflectance R_d.

will consist of the wavelengths $\lambda_1 \ldots \lambda_4$ (determining the wavelength regions), the reflectances for the ideal curve at these wavelengths (the points $a \ldots d$) and the widths $T_1 \ldots T_3$ (we suppose the tolerance band symmetric around the ideal curve as long as no natural boundaries ($R_d = 0$ or $R_d = 1$) prevent this).

B. The initial stack. Our program optimizes by varying the layer thickness (starting from an initial stack) until the reflectance of the stack is brought completely (if possible) within the given tolerance bands. The used initial stack is often derived from systems described in the literature. The initial stack is supplied to the program by giving the refractive indices $\{n_i(\lambda)\}$ of the layers and their starting thicknesses $\{t_i\}$. Dispersion is taken into account. The refractive indices are obtained from previous experiments by which layers, with optical thicknesses of about $\frac{1}{4}\lambda$, are deposited on test substrates. The conditions (such as: the composition of residual gas, the bell jar pressure, the substrate temperature and the rate of evaporation) under which the layer is deposited, have to be accurately described to make reproduction possible.

C. Optimizing of the stack. The discussion of the optimizing program is beyond the scope of this paper. Important is, that it leads to stacks consisting of layers with optical thicknesses that, in general, differ from multiples of a quarter of a wavelength. This fact influences the choice of the monitoring method. The so-called turning value method[1, 10, 11] which is in use for $\frac{1}{4}\lambda$ stacks, cannot be used in our case.

Another optical method,[1] in which deposition is terminated after (if possible) or before (to avoid) a turning value is reached can be applied. However, as will be discussed later, this method shows certain disadvantages in comparison with quartz crystal monitoring.

Another feature of the program is, that it determines the derivatives of the reflectance R_0 (λ), viz. $\partial R_j/\partial t_k$, where R_j is the reflectance in the jth extremum and t_k is the thickness of the kth layer. They are used for finding the layers, for which errors in thickness have serious consequences for the total reflectance of the stack and also for the calculation of deviations in the thicknesses of a produced stack. In Figure 4 an example is given of the matrix $\partial R_j/\partial t_k$ for an anti-reflection coating consisting of four layers.

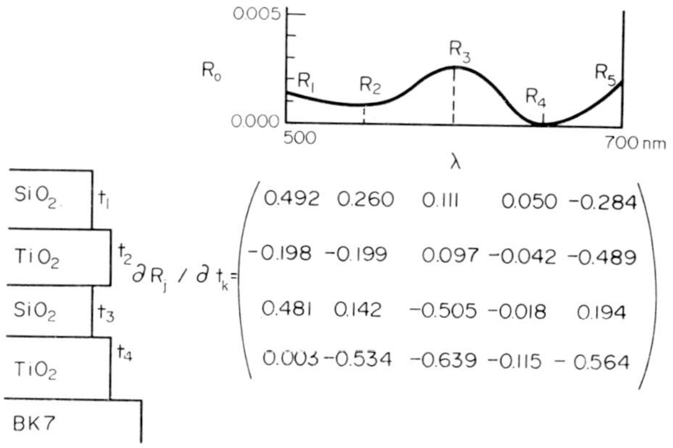

Figure 4. Example of the matrix of derivatives $\partial R_j/\partial t_k$ (dimension μm^{-1}) of an anti-reflectance coating. The refractive indices of TiO_2 and SiO_2 are given in Figure 10. The thicknesses of the layers $t_1 \ldots t_4$ are 94.7, 107.1, 22.7 and 15.2 nm, respectively.

D. Planning the deposition process. The planning for the deposition process of a multilayer stack is, using quartz crystal monitoring, relatively easy. The frequency shift Δf for making a single layer of thickness t of a given material follows from

$$\Delta f = C_f C_p C_m t \tag{1}$$

where

- C_f is the frequency factor, depending only on the oscillating frequency f_c of the crystal,
- C_p is the position factor, depending only on the relative positions of the crystal to vapour source and substrate and
- C_m is the material factor, depending only on the material of the layer.

The factor C_f is introduced since, after use in several measurements, the mass loading of the crystal is increased considerably, and then the assumption of a linear relationship between mass and frequency is no longer valid for the total range in which the frequency is decreased. For a single layer, however, we still can use this relationship provided that C_f is related to the total decrease in frequency. Denoting the total decrease in frequency (starting with a clean crystal) by Δf_t we can use Pulker and Schädler's graphical relation[8] for C_f as is depicted in Figure 5 (where we normalized its value at $\Delta f_t = 0$ to unity). If, starting deposition of a layer, the crystal is already loaded to a frequency shift $\Delta f_t'$, while that layer itself will give an estimated shift Δf, one should determine C_f from Figure 5 for the mean shift $\Delta f_t' + \frac{1}{2}\Delta f$.

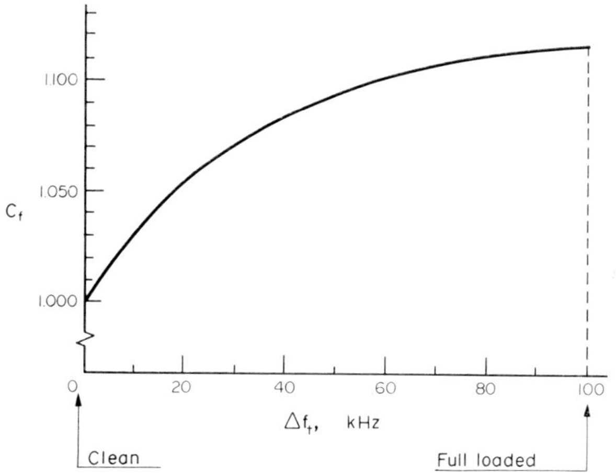

Figure 5. Relation between C_f and the total frequency shift Δf_t of a crystal. We call a crystal 'clean' if, apart from electrodes, it carries no other layers.

The factor C_p takes into account the influence of the difference in position between crystal and rotating substrate and is defined as

$$C_p = \frac{\text{(thickness of the layer on the crystal)}}{\text{(thickness on substrate)}} \tag{2}$$

Assumed is that the sticking properties on crystal and substrate are equal. Then C_p depends only on the relative position of the crystal, substrate, vapour source and the emission characteristic of that source. Many authors[1,12] dealt with this problem. We use the formula[1]

$$C_p = h_c^2[(h^2 + R^2 + \rho^2)^2 - 4R^2\rho^2]^{3/2} / [h^2(h_c^2 + R_c^2)^2(h^2 + R^2 + \rho^2)] \tag{3}$$

for a rotating substrate and a directed surface source. The used symbols are explained in Figure 6.

The factor C_m depends on the material of the layer and the conditions under which deposition takes place. C_m has been determined in the same process whereby the experimentally obtained refractive indices are measured (see paragraph B). This is done as follows:

We deposit a thick layer of several $\frac{1}{4}\lambda$ optical thickness on a substrate (in a known position for determining C_p) which gives a frequency shift Δf of about 5 to 10 kHz. Before doing this the initial total frequency shift of the crystal $\Delta f_t'$ is determined to obtain C_f.

In order to reduce the influence of inhomogenities the layer of a $\frac{1}{4}\lambda$ on a test substrate, necessary for determining the refractive index, is deposited about the middle of the deposition process of the thick layer. The optical thickness of the thick layer follows from the reflectance curve by using the wave-

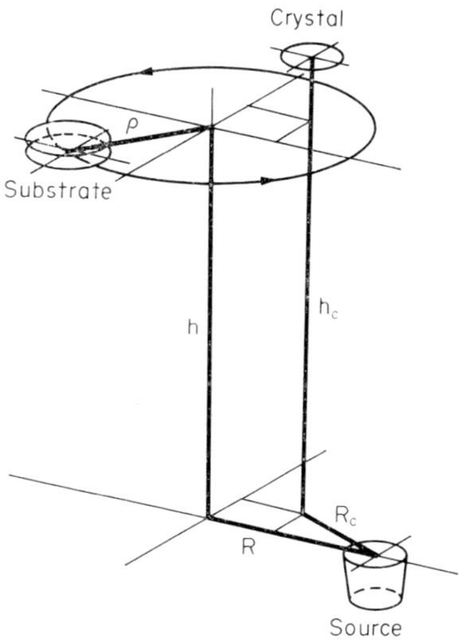

Figure 6. Position of crystal, rotating substrate and directed surface source. For the BA 510 equipment, we have $h_c = 514$ nm, $R_c = 188$ mm, $h = 464$ mm, $R = 165$ mm and $\rho = 91$ mm which gives $C_p = 0.842$.

lengths where $nt = m\lambda/4$ $(m = 1, 2, \ldots)$ and the thickness equals

$$t = m\lambda/4n. \tag{4}$$

The disadvantage of this method is that for obtaining an acceptable accuracy, very thick layers have to be deposited. In order to avoid this, a unit, measuring thicknesses by interferometry, is in preparation which is able to determine thicknesses within an error of about 1 nm. If Δf, t, C_p and C_f are known C_m follows from equation 1.

An example of the process planning of the anti-reflection coating of Figure 4 is given in Table 1.

Table 1. Calculation of the deposition process of a four layer anti-reflection coating using quartz crystal monitoring.

Order of deposition —	Material —	C_m Hz/nm	C_p —	t nm	$\Delta f_t'$ kHz	C_f —	Δf Hz
1	TiO_2	10.69	0.842	15.2	30.4	1.070	146
2	SiO_2	9.73	0.842	22.7	30.5	1.070	149
3	TiO_2	10.69	0.842	107.1	30.7	1.071	1032
4	SiO_2	9.73	0.842	94.7	31.8	1.073	832
1	2	3	4	5	6	7	8

In order to determine $\Delta f_t'$ for the first layer (row 1, column 6), before planning the process, the crystal, on which the process will be carried out, has to be chosen. The factor C_f (column 7) is now found from

$$\Delta f_t = \Delta f_t' + C_m C_p t/2. \tag{5}$$

In the given simple example it appeared to be acceptable to use the same C_f for all layers. Only for a stack of many layers the change in C_f is important.

Comparing optical monitoring with quartz crystal monitoring. In the case of optical monitoring, by which deposition is terminated before or after a turning value is reached, the process has to be planned with the aid of a computer, since it is rather complicated. Then, problems arise concerning the number of layers that must be deposited on a single monitor glass, which interference filters must be used and what refractive indices the materials used will have in vacuum. Much skill is necessary in meeting these problems, since the performance of the resulting stack will strongly depend on the way in which the process is planned. If we compare this with the plan using quartz crystal monitoring, the simplicity of the latter method is obvious; the calculations can be carried out with the aid of a pocket calculator and not much experience is required.

E. Execution of the deposition process. In order to obtain a stack with a reflectance $R_p(\lambda)$ close to the calculated reflectance $R_0(\lambda)$ reproduction of the deposition conditions by which $\{n_i\}$ and the factors C_m are determined is very important. In our case the temperature of the substrate and the rate of evaporation are controlled manually. Our methods yield satisfactory results but, in our opinion, usage of automatic control units is better.

The temperature of the substrate is indicated by a copper–constantan thermocouple which is in continuous contact with the rotating calotte. Heating takes place by radiation from a heat element.

The rate of evaporation is controlled by handling the vapour source such that the (analogue) signal Δf follows a straight line on the recorder. The slope of that line determines the rate of evaporation and must be calculated at beforehand.

In order to lower the reactivity of the residual gas a liquid N_2 cooled-baffle is used. The pressure in the bell jar can be controlled by bleeding in gas (N_2, Ar, O_2 or air) by an automatic gas-inlet system.

We prefer to use the same measuring crystal for all layers of a stack, since then the sticking on crystal and substrate are as near as possible equal. Moreover, then no effects occur from small differences in sensitivity of the crystals. Given a certain Δf, the deposition of a layer is carried out as follows. We preset the adjustable standard oscillator (Figure 2) on a frequency which is $\Delta f + 50$ Hz lower than the crystal frequency. This gives by increasing layer thickness a decreasing frequency difference.

The frequency difference is obtained in digital and in analogue form. The analogue signal is used for controlling the rate of evaporation. The digital readings, before and after deposition of each layer, determine the resulting frequency shift Δf. The readings appear to be easily influenced by mechanical vibrations from the rotations of the calotte. To avoid this, we stop the rotation when the digital meter is read. In Figure 7, the operating order is illustrated by means of the analogue signal for the case that $\Delta f = 800$ Hz.

The handlings to be carried out at the various times $t_0 \ldots t_7$, are listed below:

- t_0 By means of the digital frequency meter the adjustable oscillator is set on a frequency 850 Hz lower than the crystal (monitor range is 1 kHz). The substrate is not rotating.
- t_1 The rotation is started and the vapour source shutter is opened. The rate of evaporation is controlled.

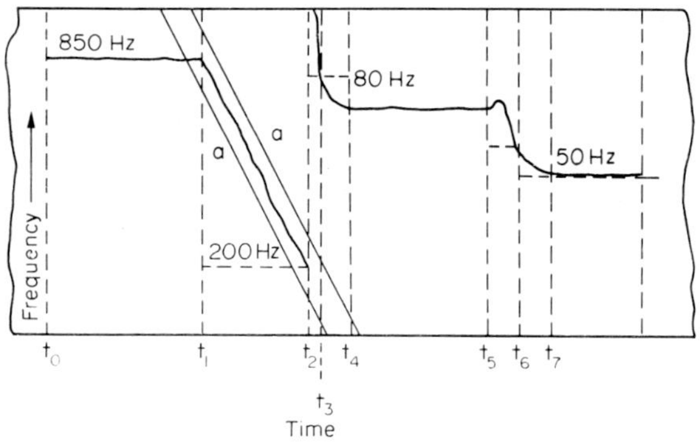

Figure 7. Recorder sheet obtained from deposition of a quartz crystal monitored layer ($\Delta f = 800$ Hz). The slope of the lines *a* is calculated and the lines are drawn on the sheet before deposition of the layer.

- t_2 Reaching about 200 Hz, the monitor is switched to the range of 0.1 KHz. This is done since the terminating point (50 Hz) is better defined in this range.
- t_3 When reaching a frequency of about 80 Hz, the shutter is closed. Due to the slowness of the frequency meter overshoot occur which must be experimentally determined. The rotation is stopped.
- t_4 The digital frequency meter is read.
- t_5 If now the wanted terminating frequency (50 Hz) is not yet reached, the shutter is opened again for a short time. The substrate rotates.
- t_6 Shutter closed and rotation stopped.
- t_7 The digital frequency meter is read and $\Delta f = 800$ Hz is reached.

  In this way, a certain terminating frequency can be obtained within an accuracy of better than 5 Hz.

This procedure is repeated for each layer until the stack is completed. Furthermore, for each stack test substrates are evaporated with layers of all used materials, making it possible to verify the obtained refractive indices.

Comparing optical monitoring with quartz crystal monitoring. In comparison with optical monitoring the execution of the process is somewhat easier, because the terminating points are sharply defined. In the case of optical monitoring the obtained reflectance of the turning value mostly differs from the calculated value and the terminating point is influenced by this difference. Hence, the operator has to determine this point in a very short time making optical monitoring an uneasy job. Another advantage of quartz crystal monitoring is the ease with which the rate of evaporation is controlled. This gives a better reproducibility of the refractive indices.

F. Detection of errors in the stack. When the deposition process is performed the reflectance $R_p(\lambda)$ of the stack is measured and compared to the desired reflectance $R_d(\lambda)$. If the difference does not agree with the tolerances we have to determine the reason before restarting the process. Firstly, the obtained refractive indices are compared to those for which the stack has been optimized. Deviations can be caused by:

Deviations from the assumed deposition conditions. This is verified from recorder sheets and measurements readings during the process.

Deviations in the used evaporation materials (other type or mark).

In the latter case, the new refractive indices are supplied to the program and the cycle restarts (Figure 1). If the refractive indices are correct, the thicknesses of the layers must differ from the calculated values. Supposing that the deviations of the reflectance are known accurately enough, the differing layers can be detected from the matrix $\partial R_j/\partial t_k$. For small deviations, however, the (unavoidable) absorbtions and inhomogenities become important and the previous method is not applicable. In general, for incorrect layers $\partial R_j/\partial t_k$ is relatively high; assuming this, a trial-and-error method can be used. If the source of the failure cannot be found in this way, the previous process is repeated; in many cases this second run leads to a more satisfactory result.

Comparing optical monitoring with quartz crystal monitoring. Not much difference exists in the error detection using optical or quartz crystal monitoring. However, there is a difference in the way in which errors in previous layers influence the errors in later layers. In the case of optical monitoring, normally more than one layer is put on the same monitor glass in order to avoid influences of the difference in sticking properties between the substrate and the monitor glass. Now, an error in the terminating point of a foregoing layer also changes the reflectance at the point where the calculated thickness of the next layer is reached (Figure 8). It is not possible to correct this

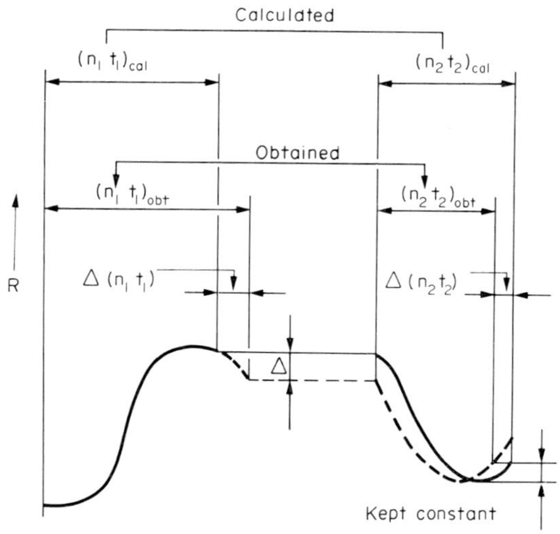

Figure 8. Explanation of the error in optical thickness $\Delta(n_2t_2)$ caused by an error in the optical thickness of a foregoing layer $\Delta(n_1t_1)$. $(n_1t_1)_{cal}$ and $(n_2t_2)_{cal}$ are the calculated optical thicknesses. $(n_1t_1)_{obt}$ and $(n_2t_2)_{obt}$ are the obtained optical thicknesses. Δ is the terminating error in the first layer.

complicated influence during the execution of the process. So, in general, an error in a foregoing layer introduces errors in the following layers which are deposited on the same monitor glass.

In the case of $\frac{1}{4}\lambda$ stacks, using turning value monitoring, Macleod[10, 11] proved that these errors compensate, however, for a certain wavelength only. In our case compensation effects are not observed for the whole wavelength range of interest.

Using quartz crystal monitoring, errors in previous layers have no influence in the later layers and this is possibly one of the reasons for our favourable results with this method.

In Figure 9, an example is given of the measured transmittance of a high-pass filter obtained after the first process.

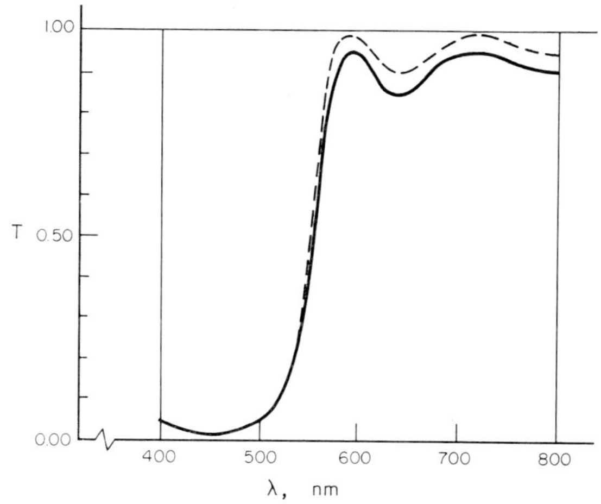

Figure 9. Transmittance curve of a high-pass filter. The dotted line gives the calculated transmittance, the solid line the measured transmittance after the first process.

The materials and the calculated thicknesses are given in Table 2. The refractive indices of the used materials as a function of the wavelength are depicted in Figure 10.

Table 2. Materials and calculated thicknesses of the high-pass filter of Figure 9.

Material	Thickness nm
BK7	—
TiO_2	36.3
SiO_2	70.8
TiO_2	48.3
SiO_2	82.2
TiO_2	46.0
SiO_2	76.3
TiO_2	46.8
SiO_2	83.3
TiO_2	46.8
SiO_2	64.3
TiO_2	36.7
SiO_2	84.1
Air	—

The difference between calculated and measured transmittance is mainly caused by the 4% reflectance of the uncoated reverse of the substrate. Furthermore, the measured curve is shifted over a few nm to longer wavelengths. Nevertheless, the performance of the fabricated stack is already acceptable for most applications. Mostly, also other stacks, like broad-band anti-reflectance coatings and broad-band beam-splitters give satisfactory results after the first process. If we use optical monitoring for the same stacks, in general, more processes are required.

4. The reproducibility of the optical thickness

An argument in use against quartz crystal monitoring is that the reproducibility of the optical thicknesses is worse than with optical monitoring. In order to verify this statement we compared the deviations in 10 layers, monitored by quartz crystal

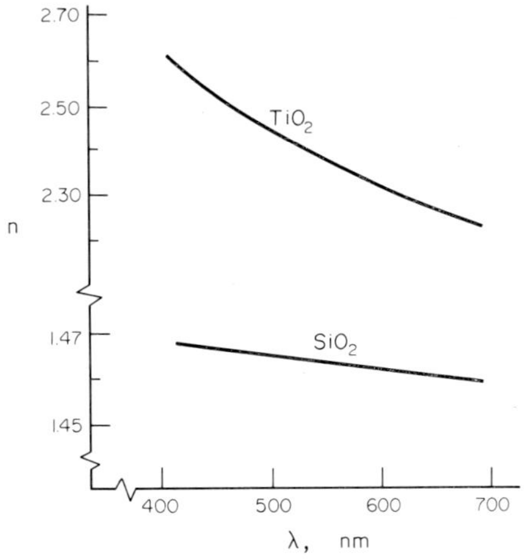

Figure 10. Refractive indices n of TiO_2 and SiO_2 as a function of the wavelength λ.

with 10 comparable optically monitored layers. The used material was MgF_2 and the optical thickness of all layers was about 430 nm. Optical monitoring was performed at such a wavelength λ_m that the evaporation process terminated at $5\lambda_m/8$ (Figure 11). The frequency shift Δf using quartz crystal

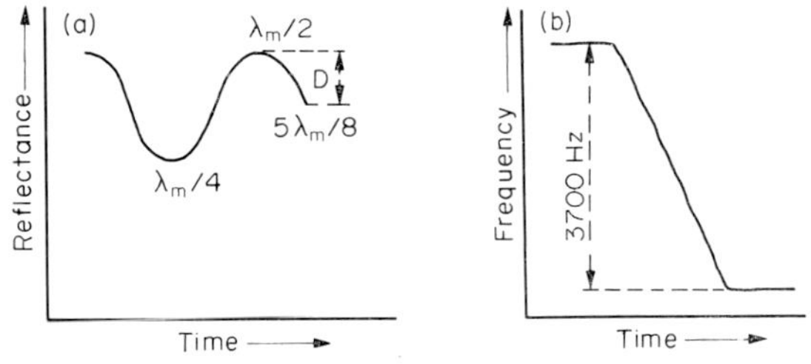

Figure 11. Monitoring signals: (a) optically. In order to correct the influence of the small fluctuations of the $\lambda_m/2$ turning value, the distance D was made the same for all layers. (b) quartz crystal.

monitoring was 3700 Hz. In both cases the deposition conditions were carefully controlled by skilled operators. After the processes the transmittances were measured. From the mutual shift of their curves the mean optical thickness $\langle nt\rangle$ and the standard deviation $\overline{nt}$ defined as

$$\overline{nt} = \{\textstyle\sum_{j=1}^{k}[(nt)_j - \langle nt\rangle]^2/(k-1)\}^{1/2} \qquad (6)$$

was determined. The results are given in Table 3.

Table 3. Standard deviation and relative deviation of the optical thicknesses of MgF_2 layers using optical and quartz crystal monitoring.

Monitoring Method	Standard deviation $\overline{nt}$ nm	Relative deviation $\overline{nt}/\langle nt\rangle$ —
Optically	2.15	0.0050
Quartz crystal	2.32	0.0054

As can be seen, the standard deviation using a quartz crystal monitor is higher in this case. However, in the case of optical monitoring, the terminating point was chosen such that the best accuracy was obtained. Now we, calculate the relative deviation as a function of the optical thickness, assuming that in both methods the absolute error in the terminating point of the monitoring signals will remain constant. The optical monitoring signal is approximated by

$$R = a\cos(\phi) + b, \qquad (7)$$

in which the phase thickness ϕ equals

$$\phi = \pi x = \text{constant} \times nt, \qquad (8)$$

where x denotes the fractional number of $\frac{1}{4}\lambda_m$ positions (Figure 12a). From (7) follows the terminating error ΔR as a function of the error in the phase thickness $\Delta\phi$ as:

$$\Delta R = -a\sin(\phi)(\Delta\phi) - \tfrac{1}{2}a\cos(\phi)(\Delta\phi)^2 + 0((\Delta\phi)^3) \qquad (9)$$

Neglecting third and higher orders of $\Delta\phi$, the relative deviation in the optical thickness equals

$$\overline{nt}/\langle nt\rangle$$
$$= \{[\sin^2(\phi) + |\pi P\cos(\phi)|]^{1/2} - |\sin(\phi)|\}/|\phi\cos(\phi)|, \qquad (10)$$

where

$$P = 2\Delta R/(\pi a). \qquad (11)$$

The absolute values in equation 10 originate from the fact that ΔR must be positive in the minima and negative in the maxima. From the experimentally determined point $x = 5/2$, $\overline{nt}/\langle nt\rangle = 0.0050$ follows $P = 0.025$, after which the curve can be computed to other values of x and nt (see equation 8).

In the case of quartz crystal monitoring the relation

$$\overline{nt}/\langle nt\rangle = 0.0054 \times 430/(nt) \qquad (12)$$

holds (Table 3).

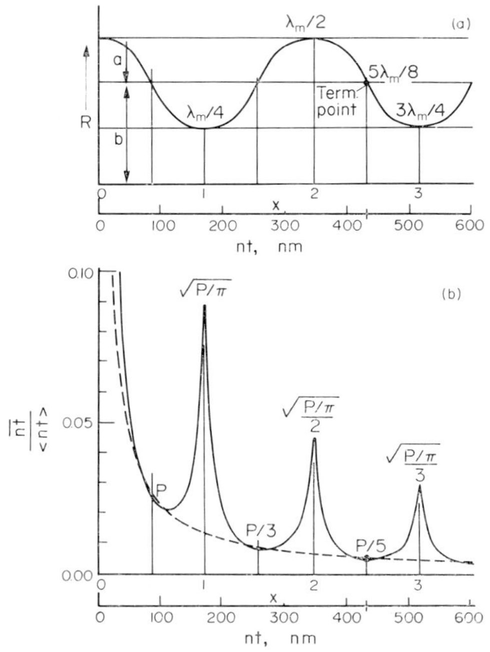

Figure 12. (a) Approximated monitoring signal (equations 7 and 8). (b) Relative deviation $\overline{nt}/\langle nt\rangle$ as a function of the optical thickness nt.
Solid line: using optical monitoring (equation 10).
Dotted line: using quartz crystal monitoring (equation 12).

For both cases, the graphs are given in Figure 12b. From that, we conclude, that in the case of MgF_2 layers, quartz crystal monitoring is in general as good as optical monitoring and for layers with optical thicknesses smaller than $\lambda_m/8$, quartz crystal monitoring is more accurate than optical monitoring. Experiments with materials of higher refractive index (ZnS and TiO_2) will be carried out in the near future.

5. Conclusion

We emphasize that our experiences and results with quartz crystal monitoring are obtained under laboratory conditions. In the case of large-scale production, factors like differences in position, shape and temperature of the substrates can spoil the results. However, this is also the case when optical monitoring instead of quartz crystal monitoring is used.

We remark that in the case of quartz crystal monitoring it is necessary to reproduce the deposition conditions under which the material factors C_m and the refractive indices of the used materials are determined. Most problems with this method arise from insufficient care by the determination, the description and the reproduction of those conditions. Further, we restrict our conclusions to dielectric stacks, used in the visible.

The main advantages of quartz crystal monitoring above optical monitoring are:

The planning of the deposition process is easier.

The execution of the process is easier because the terminating points are better defined. Furthermore, the rate of evaporation is more easily controllable which results in a better reproducibility of the refractive index.

The errors in foregoing layers have no influence on the errors in the subsequent layers. This gives less deviations of the reflectance characteristic from the calculated curve, and consequently, quartz crystal monitoring yields stacks with satisfactory performance faster than optical monitoring.

Concerning the reproducibility of the optical thickness of MgF_2 layers, the performance is comparable to that of optical monitoring. For layers with optical thickness smaller than $\lambda_m/8$, quartz crystal monitoring is more accurate.

Acknowledgements

We are indebted to A Kuntze and A Looyen for execution of the deposition processes and to A Thelen, W Kindli and H A Macleod for stimulating discussions.

References

[1] H A Macleod, *Thin-Film Optical Filters*, Adam Hilger, London (1969).

[2] *Balzers Betriebsanleitung A* 11–01 d. Hochvakuum-Bedampfungsanlage BA 500.

[3] *Balzers Betriebsanleitung A* 11–3995 d. Optische Schichtdicken-Messeinrichtung GSM 210.

[4] *Balzers Betriebsanleitung A* 11–3918 d. Hochvakuum-Bedampfungsanlage BA 510 Automatic.

[5] *Balzers Betriebsanleitung A* 11–4000 d. Schwingquartz-Schichtdickenmessgerät QSG 201.

[6] H K Pulker, *Z Angew Phys*, **20**, 1966, 537.

[7] H K Pulker and H Hilbrand, *Z Angew Phys*, **23**, 1967, 15.

[8] H K Pulker and W Schädler, *Il Nuovo Cimento*, Serie X, **57B**, 1968, 19.

[9] G Hass, R E Thun, *Physics of Thin Films*, **2**, 1964, 210, Academic Press, London.

[10] H A Macleod, *Opt Acta*, **19**, 1972, 1.

[11] H A Macleod, *Opt Acta*, **20**, 1973, 493.

[12] L Holland, *Vacuum Deposition of Thin Films*, Chap 5, Chapman & Hall, London (1956).

4.3 Vacuum evaporated thin films for solar collectors

L E Flordal and R Kivaisi,* *Institute of Optical Research, S-100 44 Stockholm, Sweden*

Selectively absorbing surfaces are important for the efficient use of solar energy. The main characteristics aimed at for this type of coatings are the highest possible absorptivity for solar radiation and the lowest possible emissivity for thermal radiation. (Temperatures up to 100–200°C). There are several ways to create surfaces with these properties. In the present work evaporated semiconductor films with suitable anti-reflection coatings were used on different metals. For the laboratory tests, the metals were vacuum evaporated on glass substrates. The evaporations were performed with an electron beam gun in a diffusion pumped system. The evaporated thicknesses and the rate of evaporation were controlled by a quartz crystal monitoring system. The selective coatings were studied on flat and rough metal surfaces. Sinusoidal, metal covered gratings with different periods were used as the rough absorber surfaces. This simplified the measurements of absorptivity and emissivity and gave a more thorough understanding of the basic thin film phenomena. It is possible to use gratings because the general rough surface can be looked upon as a superposition of different sinusoidal gratings. A comparison is made between the properties of the layers on flat and rough surfaces. The coatings were subjected to high humidity and high temperature tests. They were also cycled between the intended working temperature and ambient temperature to check possible degradation. Results concerning film adherence and stability during these environmental tests are given.

Introduction

The use of solar energy in Sweden might seem very optimistic during cloudy and dark autumn and winter days. However, the total insolation per year on to a horizontal surface is 900 kWh/m^2, which means a mean value over the year of 100 W/m^2. This is more than 35% of the insolation at the most intensively illuminated desert areas of the earth.[1] The insolation on to an oblique area in Sweden is thus during spring and summer quite comparable to that in e.g. Sahara. The radiation we receive is, however, rather different from that which is falling on most of the sunny areas. The main differences are that our insolation is diffuse to a high degree and that it is unevenly distributed over the year. The use of solar energy therefore puts somewhat special demands in our country. Thus, concentrating systems can hardly be used because of the many cloudy days. What we need are absorbers which work well even with diffuse light. The construction should be made in such a way that they work efficiently without tracking mechanisms. Another very important demand for the practical use of solar energy in Sweden is cheap and efficient systems for energy storage, preferably from one season to another. This is not dealt with in the present work. However, the primary element in all systems for direct use of the Sun as a continuous source of energy is an efficient absorber of some type. The present study has the goal to find some possibilities to absorb solar energy thermally in thin film coatings.

Physical background

The spectral distributions of the incident sunlight at the surface of the earth and of the emitted black-body radiation at different temperatures are shown in Figure 1. This shows clearly the separation in wavelength between the two spectra. The boundary wavelength lies roughly at 2.5 μm for temperatures below 500 K. Because of this separation, it is possible to make a surface which absorbs sunlight efficiently, but which does not re-emit the absorbed energy at low temperatures. Such a surface, a selective absorber, should have the highest possible absorptivity for wavelengths below 2.5 μm but a low emissivity for longer wavelengths.[2] A selective absorber placed in sunlight will thus reach a much higher temperature than a common black surface. This is important, since thermal energy is more valuable at high temperatures than at low. A system consisting of a thin semiconductor layer with a simple anti-reflection coating on a metal with high reflectivity in the infrared was found to fulfil the demands.[3] The high metal reflectivity

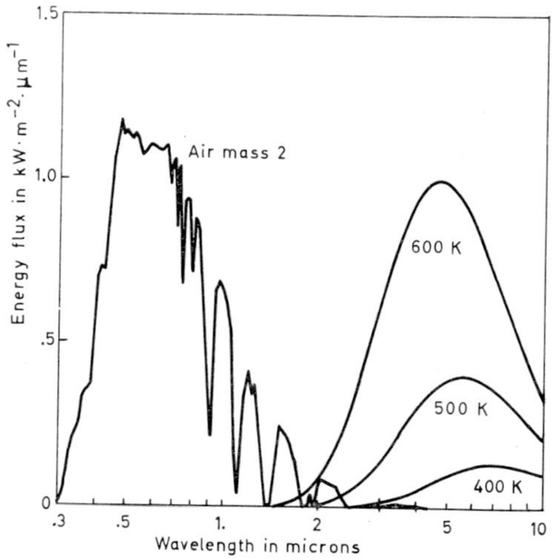

Figure 1. Solar irradiance at the Earth's surface and blackbody radiation at different temperatures.

* Permanent address: University of Dar es Salaam, P.O. Box 35063, Dar es Salaam, Tanzania.

corresponds directly to a low infrared emissivity. Aluminium, nickel, chromium and copper were investigated as base metals. Copper was later abandoned due to oxidation even at moderate temperatures. The metal was covered with a thin semiconductor coating in order to increase the absorptivity of solar radiation. If the semiconductor has a band-gap of about 2.5 μm, this can be done without increasing the thermal emissivity, since the semiconductor is transparent for infrared radiation. Germanium and lead sulphide were chosen as suitable semiconductors. The semiconductor coating absorbs incident solar radiation, but has a rather high reflection loss because of its high refractive index. The absorptivity can be increased by covering the semiconductor with a simple antireflection coating. Silicon monoxide was chosen for this purpose since it is rather resistant and easy to evaporate. This gives the basic design: metal–semiconductor–anti-reflection coating.

Theoretical calculations

Different combinations of metal substrates (aluminium and nickel) and coatings were investigated theoretically. The absorptance versus wavelength at different angles of incidence was calculated for many different layer thicknesses. One example is shown in Figure 2. The solar absorptance was then calculated using the irradiance curve (Figure 1) as a weighting function. The method with selected wavelengths was used for this purpose (20 wavelengths were used). Figure 3 shows the variation of solar absorptance when the semiconductor thickness is varied. It is obvious from these curves that the solar absorptance is not very sensitive to changes in coating thicknesses around the optimum values. This is very important for practical use. Figure 4 shows together with Figure 2 that this type of selective absorber works efficiently in diffuse light. The solar absorptance is almost constant from normal to 60° angle of incidence.

The theoretical calculations also showed that the absorption in the systems is due to two different mechanisms. An example of this is shown in Figure 5 where the absorption is divided between the metal substrate and the coating. In the wavelength range where the semiconductor is non-transparent, it is anti-reflected by the single silicon monoxide layer. Thus, short wavelength radiation is absorbed mainly in the semiconductor.

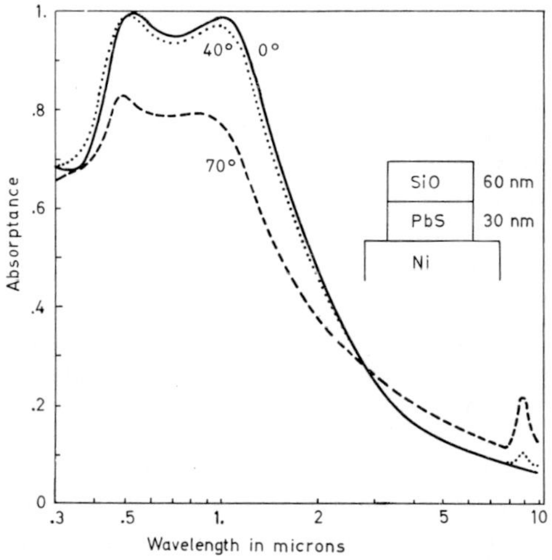

Figure 2. Theoretically calculated absorptance at different angles of incidence for one example of the two layer coatings.

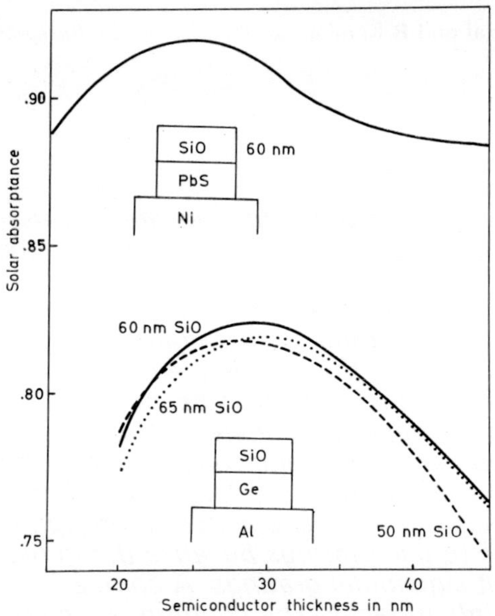

Figure 3. Theoretically calculated solar absorptance versus semiconductor thickness at normal incidence.

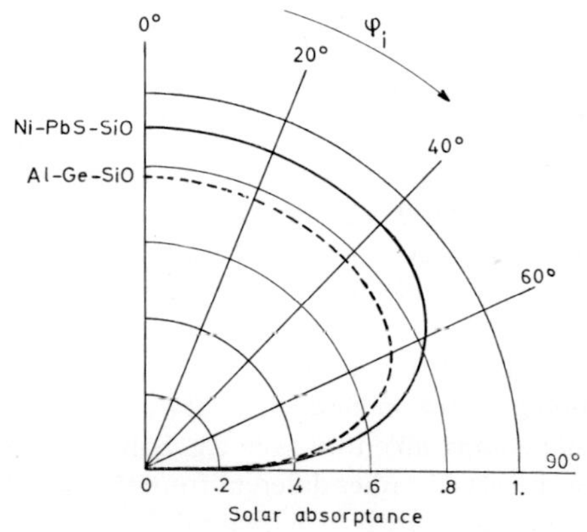

Figure 4. Theoretically calculated solar absorptance versus angle of incidence for two optimized coatings. Semiconductor thickness 30 nm, silicon monoxide thickness 60 nm.

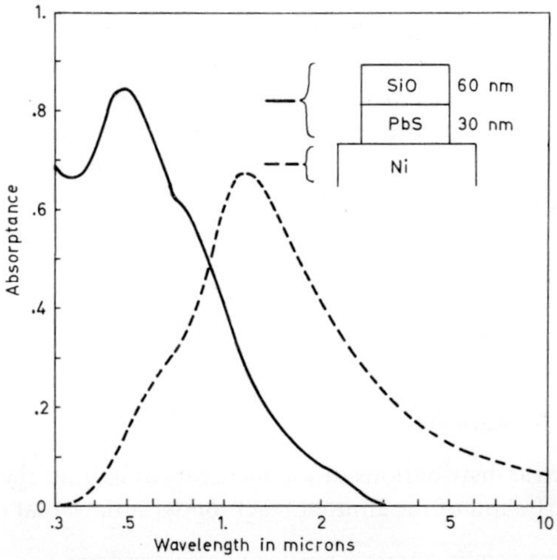

Figure 5. Theoretically calculated absorptance in the coating and the metal substrate at normal incidence. Fifty-six per cent of the incident energy is absorbed in the semiconductor layer, 35% in the metal.

The wavelength of maximum absorptivity can be chosen at will by adjusting the single anti-reflection layer thickness. In the region where the semiconductor is transparent, it gives an anti-reflection coating of the base metal together with the silicon monoxide layer. Here, the radiation is absorbed in the base metal. The wavelength of maximum absorptivity can be chosen to a certain degree, and is given by the compound optical thicknesses of the layers. The optimum anti-reflection layer for all the combinations was 60 nm silicon monoxide, which gives anti-reflection at roughly 550 nm. The optimum thicknesses for the semiconductor were between 20 and 40 nm, depending on base metal and semiconductor. These thicknesses are well suited for vacuum evaporation, which was used to make the selective absorber samples.

Experimental

For measurement purposes, both the two layers and the base metals were produced by vacuum evaporation. The metal was evaporated on a smooth glass to get an even surface. From the optical point of view this is equivalent to a semi-infinite metal substrate, as soon as the metal film is thick enough to be non-transparent for all radiation under consideration. The flat, mirror-like surfaces simplified the subsequent practical measurements. The evaporations were performed in a conventional, diffusion-pumped system with a working pressure of roughly 10^{-5} torr. The substances were evaporated from an electron beam gun with maximum 5 kW power. The evaporation rates differed from 1 to 1.5 nm s^{-1} for germanium and silicon monoxide. Lead sulphide films were obtained at very low rate, about 5 to 6 nm min^{-1}, to get good film properties. A quartz crystal detector, which was earlier calibrated with the actual substances, was used to control rate and thickness. During the evaporations, where four samples were coated in each run, the samples were rotated to get even thicknesses of the coatings. Several series with different thicknesses were produced to be able to study the influence on performance caused by thickness variations.

The reflectivities of the finished samples were measured in spectrophotometers at close to normal incidence between 0.35 and 40 μm wavelength. Since the samples were non-transparent, these measurements gave the absorptivity or emissivity directly as unity minus the reflectivity. The solar absorptivity of each sample was calculated by weighting the measured absorptivity at different wavelengths with incident solar irradiation at two airmasses. The method with 20 selected wavelengths was again used for this purpose. The thermal emissivity was calculated in a similar manner using black-body radiation curves at 400 and 500 K. The measurements were made at room temperature, under the assumption that the optical properties were not changed considerably at elevated temperatures. However, this assumption has to be investigated by measurements. The measurements showed reasonable agreement with the theoretical predictions. One example of this is given in Figure 6. The deviations were probably caused by the uncertainty of the optical constants of the materials used in the calculations. Thus,

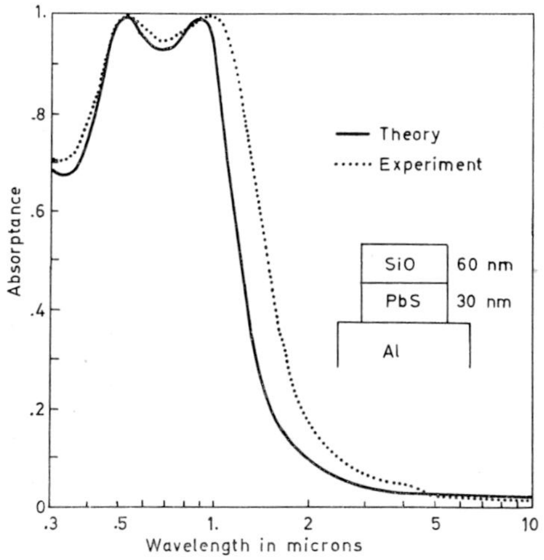

Figure 6. Comparison between calculated and measured absorptance at normal incidence.

Table 1. Solar absorptances α_s and thermal emittances ε of some selective absorbers. Silicon monoxide thickness 60 nm.

Design	Semiconductor thickness nm	α_s theory	α_s exp	ε_{400K}	ε_{500K}	$\alpha/\varepsilon_{400K}$	$\alpha/\varepsilon_{500K}$
Al–Ge–SiO	20	0.77	0.75	0.012	0.014	63	54
	30	0.84	0.79	0.012	0.014	66	56
	40	0.79	0.74				
Al–PbS–SiO	20	0.75	0.77	0.018	0.018	43	43
	30	0.83	0.89	0.018	0.02	52	44
	40	0.84	0.87				
Ni–Ge–SiO	20	0.91	0.88	0.035	0.044	25	20
	30	0.89	0.85	0.04	0.052	21	16
	40	0.85	0.82				
Ni–PbS–SiO	20	0.91	0.92	0.041	0.054	22	17
	30	0.92	0.93	0.043	0.055	22	17
	40	0.89	0.93				
Cr–Ge–SiO	20		0.93	0.11	0.12	8	8
	30		0 90				
	40		0.86				
Cr–PbS–SiO	20		0.93	0.11	0.13	8	7
	30		0.94	0.12	0.14	8	7
	40		0.94				

systems with germanium showed 3 to 4% lower absorptivity in practice, while the designs with lead sulphide gave somewhat higher absorptivities than the calculated. These results are summarized in Table 1. The differences were mainly due to changes in the absorption edges of the semiconductors. The highest measured solar absorptivities were for Al–Ge–SiO 79%, Al–PbS–SiO 89%, Ni–Ge–SiO 88%, Ni–PbS–SiO 93%, Cr–Ge–SiO 93% and Cr–PbS–SiO 94%. The samples with chromium as the base metal showed the highest infrared emissivities 10–15%, while these were only 1–2% with aluminium. Nickel showed intermediate emissivities of 4 to 5% at 400 K. The difference in emissivity means that chromium is suited for the low working temperatures, while aluminium is more efficient for high temperature applications. The ratio between solar absorptivity and emissivity at 400 K was 50–60 for aluminium, about 20 for nickel but only 8 for chromium as the base metal.

Stability test

A simple environmental test of the coatings was also performed. The samples were subjected to high relative humidities and were heated to 250°C in air several times to check possible deterioration. The performances of the absorbers were measured before and after the treatment and were compared. The investigations are summarized in Table 2 and showed that the absorbers with chromium as the metal base were not affected at all. The combination nickel–lead sulphide lost 4% in absorptivity. The loss was of the same magnitude for the combination germanium–aluminium, but as high as 10% for lead sulphide and the same metal. The test was by no means complete, but gives some hints about the stabilities. A far more extensive investigation with homogeneous metal substrates is needed to give the long term stability of the coatings.

Table 2. Stability of selective coatings. Semiconductor thickness 30 nm, silicon monoxide thickness 60 nm.

Design	α_s untreated	After heating in air, 240°C α_s	$\Delta\alpha_s$	After humidity test, 90–95% α_s	$\Delta\alpha_s$
Al–Ge–SiO	0.79	0.76	0.03	0.77	0.02
Al–PbS–SiO	0.79	0.69	0.10	0.70	0.09
Ni–Ge–SiO	0.85	0.85	0.00	0.84	0.01
Ni–PbS–SiO	0.92	0.88	0.04	0.88	0.04
Cr–Ge–SiO	0.90	0.93	−0.03	0.90	0.00
Cr–PbS–SiO	0.94	0.94	0.00	0.94	0.00

Rough surfaces

The two layer coating systems were also studied on rough metal surfaces. This study was simplified by using gratings with different periods as prototypes for rough surfaces. The gratings were made by exposing holographic plates in interfering laser beams. After development, the gelatin on the plates had a sinusoidally varying thickness. With the gelatin side as a substrate metals were evaporated to give rough metal surfaces with known roughness. The two coatings were evaporated on those rough metals and measured almost as the flat surfaces. The measurements showed that these rough selective absorbers had higher solar absorptivities than the flat ones. The thermal emissivity, however, was the same. The increase in absorptivity was a few per cent, depending on the grating period and the coating system. A possible use of rough surfaces could be to simplify the selective absorber to one single coating. Work is going on in this direction.

Discussion

This investigation shows that it is possible to make good selective absorbers by vacuum evaporation of coatings. A practical use of these coatings to capture thermal energy from the Sun requires very large areas. These coatings may be possible to produce at a competitive cost in large-scale vacuum plants in the future.

Acknowledgements

One of us, R Kivaisi, was a participant to the International Seminar in Physics sponsored by SIDA (Swedish International Development Authority) during this work. The work was also supported by NFR (Swedish Natural Science Research Council).

References

[1] H E Landsberg, *Solar Energy*, **5**, 1961, 95.
[2] D M Mattox and R R Sowell, *J Vac Sci Technol*, **11**, 1974, 793.
[3] J Jurisson, R E Peterson and H Y Mar, *J Vac Sci Technol*, **12**, 1975, 1010.

4.4 Co-sputtered optical films

C Misiano and E Simonetti, *Selenia SpA, Via Tiburtina Km. 12,400, 00131 Roma, Italy*

We have investigated the co-sputtering of two dielectric materials with indices of refraction as widely different as possible with the aim of obtaining both homogeneous films with any intermediate index of refraction and inhomogeneous films with predetermined profiles. An rf sputtering module will be described which has been especially designed, with two separate cathodes and two independently tunable rf generators. The substrates are placed on a circular anode rotating underneath the two cathodes. So far mainly CeO_2, TiO_2 and SiO_2 targets have been used. The deposition rate from each cathode and the total film thickness are determined by means of two quartz thickness monitors, sputtering compatible. Values obtained for the refractive index and optical thickness will be reported, as well as repeatability, mechanical and chemical characteristics, reliability and high power optical radiation resistance. Finally, results obtained on optical components of practical interest will be discussed.

Introduction

The present increasing interest in the design of thin film optical components has brought in turn an increasing need for a wider range of optical materials with different refractive indices.

More recently a demand has also developed for inhomogeneous films with pre-determined refractive index profiles.

The best known approaches proposed so far include the following:

(a) Co-evaporative from two or more carefully controlled thermal sources[1,2];

(b) evaporation of a mixture from a single thermal source[3,4];

(c) deposition of an alternating sequence of thin layers of two different materials[5,6];

(d) reactive deposition of SiOx under controlled oxidation conditions.[7]

However, all of these techniques either involve complicated equipment and elaborate processes, or are not easily repeatable or do not allow the processing of many substrates at the same time. Therefore their application is not very widespread outside research laboratories. As an example we show in Figure 1 the set up proposed by Jacobson for the co-evaporation from four thermal sources. It is apparent that an involved automatic system is required for the system to be sufficiently reliable. It was our aim to develop a simple technology which would enable us to deposit repeatably both homogeneous films with pre-determined refractive index and inhomogeneous films with pre-determined refractive index profile. This could be achieved by depositing a mixture of two materials with high and low refractive index respectively. We decided to choose the sputtering technique, by co-sputtering from two separate targets onto substrates placed on a rotating anode underneath. In practice there is no actual co-sputtering since the substrates never 'see' the two sources simultaneously. However, as long as the alternatively deposited layers have a thickness smaller than 1 Å, a perfect mixture of the two materials is assured. Our experiments were carried out in a home made system but co-sputtering systems are now commercially available which are perfectly suitable for this purpose.

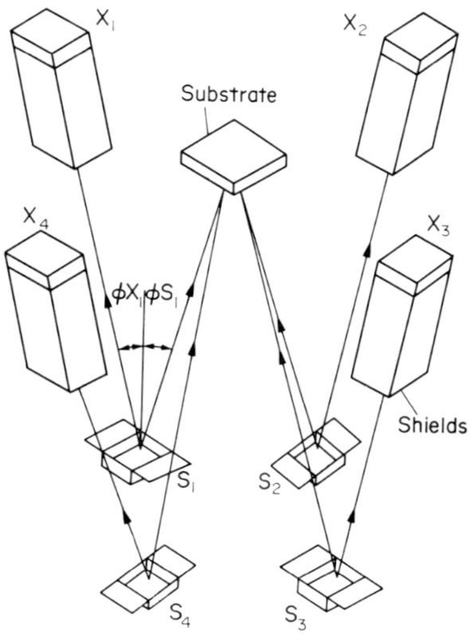

Figure 1. Experimental set up for the co-deposition from four thermal sources (after R Jacobson).

Experimental

A sketch of the sputtering system is shown in Figure 2: two 5 in. diameter sources are powered at the same time by two separate 13.5 MHz, 1.3 kW supplies. The matching networks are also separate. The anode, which is not cooled and has a 40 cm diameter, is rotating with a speed of 20 rpm at a distance of 6 cm from the targets. Two separate rf sputtering compatible quartz measuring heads are used as rate and thickness monitors. They are located one under each of the targets in such a way as not to see the material emitted from the other target.

The pumping system is of the conventional diffusion pump type. The sputtering atmosphere was pure Ar throughout. Most of the experiments were performed by using either the couple CeO_2–SiO_2 or the couple TiO_2–SiO_2 as source materials. The obvious choice of MgF_2 as a low index material was prevented by the excessively high absorption typically obtained when sputtering MgF_2 in pure Ar. Each measuring head was calibrated for all materials of interest. The effective deposition rate on the rotating substrate was determined as a function of

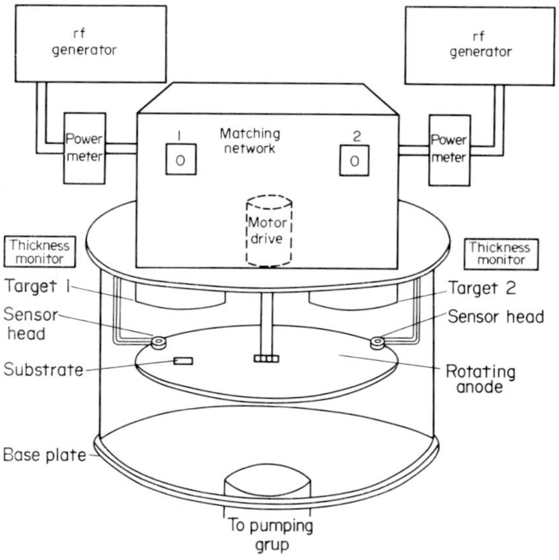

Figure 2. Sketch of the co-sputtering system.

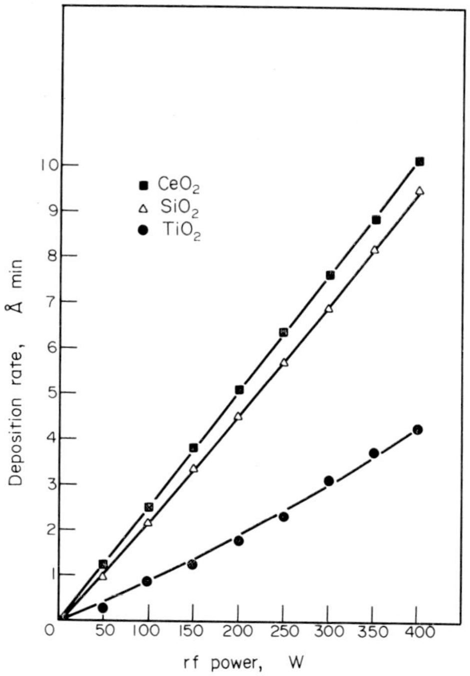

Figure 3. Sputtering rate as a function of the rf power at the cathode for TiO_2, CeO_2 and SiO_2.

the power at the cathode by comparison with spectrophotometric measurements performed on the deposited films.

As an example we show in Figure 3 the calibration curves for TiO_2, CeO_2 and SiO_2.

Results

Homogeneous films. In the case of homogeneous films with predetermined refractive index, our attention was focused on the repeatability attainable under realistic working conditions. We felt this was the real test for the usefulness of the proposed technology. Because of the dispersion of all materials in question (particularly TiO_2 and CeO_2) the refractive index must be determined always at the same wavelength λ, in our case generally 5500 Å. Films with a targeted optical thickness equal to $\lambda/4$ were deposited on Corning microsheets 0211, BK7 and sapphire. Both refractive index and actual optical thickness were determined by measuring the transmittance of the samples before and after deposition as a function of the wavelength.

Shown in Figure 4a and b is the refractive index of the mixtures SiO_2–TiO_2 and SiO_2–CeO_2 as a function of the volume percentage of SiO_2. Marked with the shaded area is the scattering of results obtained for the refractive index. It is apparent from Figure 4 that the scattering of values is higher for mixtures SiO_2–CeO_2 than for mixtures SiO_2–TiO_2. This is likely

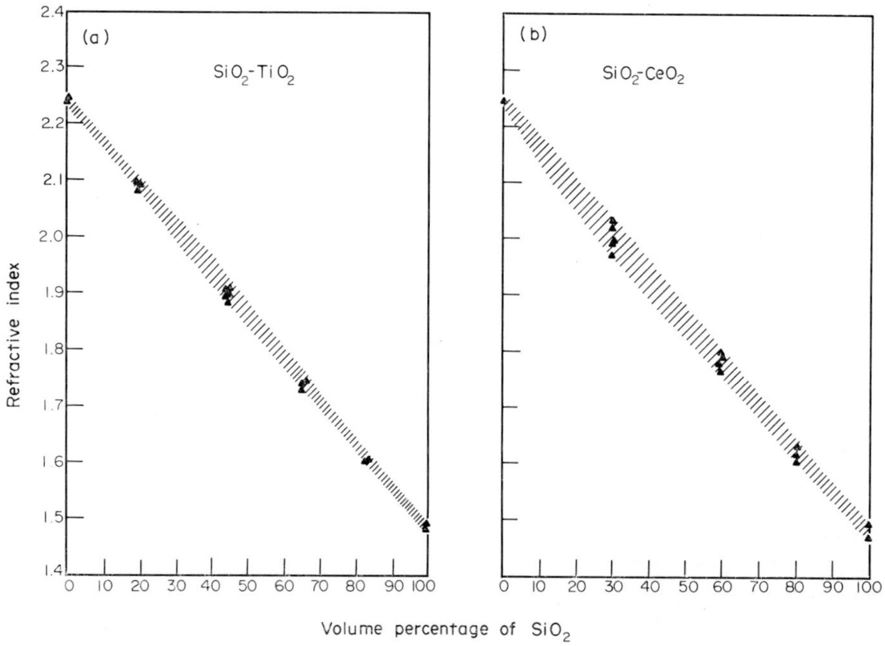

Figure 4. Refractive index of co-sputtered films as a function of the volume percentage of SiO_2: (a) SiO_2–TiO_2; (b) SiO_2–CeO_2. The shaded area points to the range of values obtained.

to be due to the strong dependence of the refractive index of CeO_2 films on the substrate temperature. Although thickness monitors were actually used in this experiment, it was found that the same reproducibility could be obtained for the refractive index by controlling time and rf power at the cathode, provided all other parameters were kept constant. However, in this case, reaching a good vacuum ($<10^{-6}$) before deposition and keeping the discharge atmosphere free from reactive gases is especially essential because even small percentages of such gases (particularly O_2) could affect in an appreciable way the deposition rate of all materials in question. The conditions of our experiments are summarized in Table 1 which give also the maximum dispersion obtained for the refractive index.

Inhomogeneous films. In the case of inhomogeneous films, because of difficulties involved in determining the refractive index profile on the deposited film, we chose to start from the simplest case of a linear variation. Starting from the refractive index of the substrate (1.53) the refractive index of the film was made to change linearly up to a given higher value with a total thickness of the film much higher than the wavelength of interest.

The apparent refractive index of the substrate was thereby modified. We made use of the experience gained with homogeneous films to determine the values to be used for the process parameters. Displayed in Figure 5 is the variation with time of the power delivered to the SiO_2 and CeO_2, cathodes together with a list of the values used for the main parameters. The total physical thickness of the film was ~2 μm. In Figure 6 we have

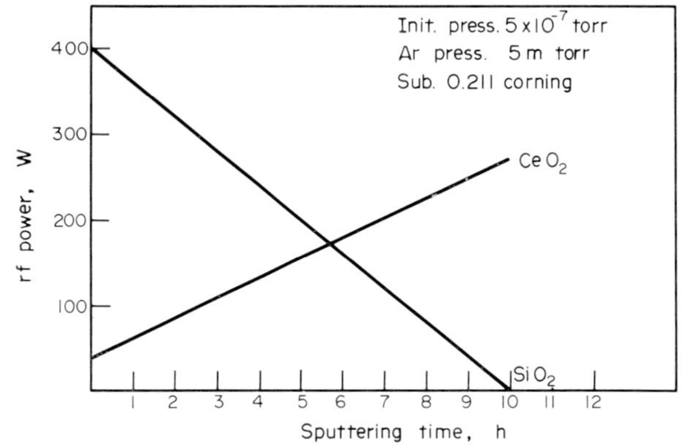

Figure 5. Time dependence of the power delivered to SiO_2 and CeO_2 cathodes during the deposition of an inhomogeneous film.

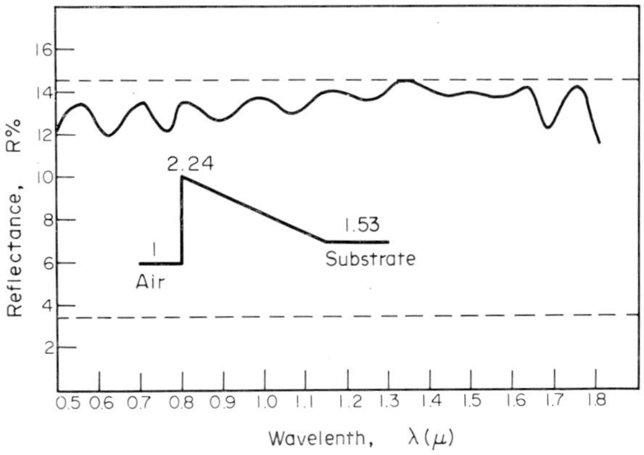

Figure 6. Reflectivity of an inhomogeneous film. The upper dotted line marks the reflectivity of a bulk material with refractive index equal to 2.24. The lower one the reflectivity of the uncoated substrate.

reported the profile targeted for the refractive index, ranging from 1.53 to 2.24. Marked with dotted lines are the reflectivity at normal incidence for a bulk material with refractive index equal to 2.24 and the reflectivity of the untreated substrate. While the apparent refractive index of the treated sample is clearly changing, oscillations due to interface effects are still clearly visible. We feel these could be reduced by modifying the profile as follows: the corners at the two ends of the sloping profile should be smoothened, the refractive index should be kept at its high value for an appreciable thickness (of the order of 1 μm) and the total thickness of the film should be increased.

Film properties. Both homogeneous and inhomogeneous films obtained by co-sputtering either SiO_2–CeO_2 or SiO_2–TiO_2 underwent the following tests: adhesion, abrasion resistance and resistance to humidity cycles. In all cases the results were quite good. The radiation resistance was tested with a multimode 1.06 μm pulsed laser. Films of SiO_2–CeO_2 started to deteriorate at powers above 500 mW/cm^2 while films of SiO_2–TiO_2 could stand about 150 mW/cm^2.

Applications. In order to evaluate the feasibility of practical applications two anti-reflective coatings were considered:

The first one was made of three homogeneous films, one of them co-sputtered with a $\lambda/4 - \lambda/2 - \lambda/4$ structure as shown in Figure 7.

Table 1. Co-sputtered optical films.

Couple	% of SiO_2	SiO_2 power W	H index power W	Total rate Å/min	$\bar{n}$ average	n_{MIN}–n_{MAX}	
	0	0	400	4.1	2.24	2.23–2.24	
	30	85	400	5.8	2.00	1.99–2.00	
SiO_2–TiO_2	50	175	400	8	1.87	1.86–1.88	
	66	360	400	12.5	1.75	1.74–1.76	
	85	400	175	11.2	1.62	1.61–1.62	Ultimate pressure $<10^{-6}$ torr
	100	400	0	9.5	1.49	—	Sputtering atmosphere 6 mtorr
	0	0	260	6.6	2.23	2.21–2.24	of pure Ar
SiO_2–CeO_2	30	130	260	9.4	2.00	1.97–2.02	
	60	400	260	16	1.82	1.79–1.83	
	80	400	85	12	1.65	1.63–1.67	
	100	400	0	9.5	1.49	—	

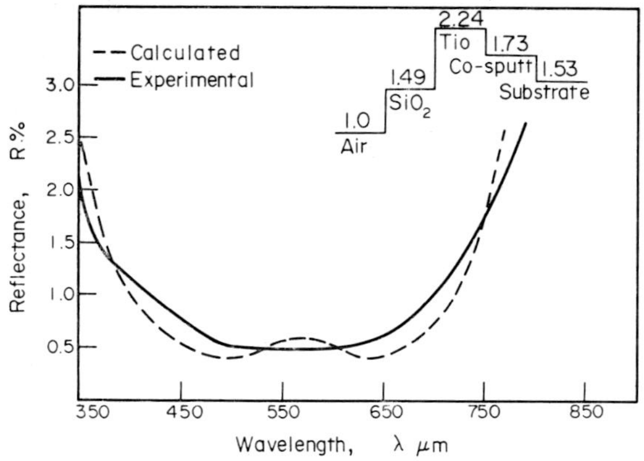

Figure 7. Reflectivity as a function of wavelength of a three-layer anti-reflection coating on glass ($n_s = 1.53$).

The other one was made up of a 2 μm thick inhomogeneous film followed by a homogeneous film with refractive index equal to the square root of the new apparent refractive index of the substrate as shown in Figure 8. As seen from the reflectivity

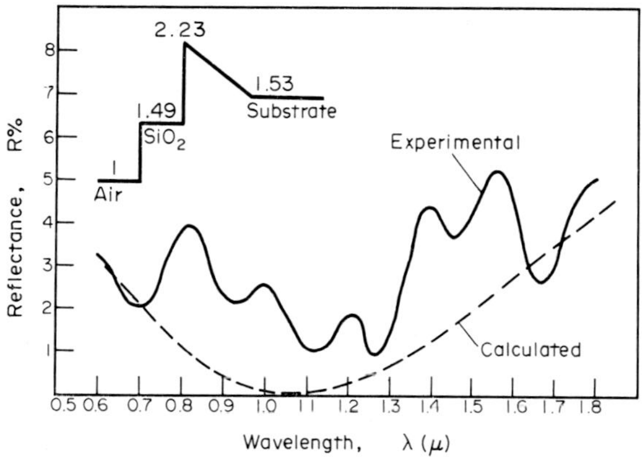

Figure 8. Reflectivity of an inhomogeneous anti-reflection coating on glass ($n_s = 1.53$). The theoretical curve refers to a mono-layer with refractive index equal to 1.49 on a substrate with refractive index 2.23.

curves, also shown in Figures 7 and 8, while the homogeneous film is quite close to the target, so far the oscillations present in the inhomogeneous case very seriously hinder practical applications.

Conclusions

From our experiments so far we can state that homogeneous optical films with index of refraction repeatability of ± 0.01 can be readily obtained by co-sputtering. This is comparable with the repeatability obtainable by depositing single components.

Mechanical characteristics and radiation resistance were excellent. The advantages of the proposed technology with respect to other methods of deposition can be summarized as follows:

(a) Reduced control instrumentation.

(b) Higher reproducibility of the values obtained for the refractive index of the deposited mixture.

(c) A superior performance both concerning hardness and resistance to environment.

The main disadvantage is found to be the long deposition time. Inhomogeneous films of pre-determined refractive index profile can be also easily deposited. However, the practical applicability is still to be tested. Finally, we would like to recall that the range of values obtainable for the refractive index can be further extended in the ir by using Ge or Si as high refractive index material.

Acknowledgements

We are grateful to Dr F Marcelia for the fruitful computer simulations and to Dr M Bujatti for helpful advice.

References

[1] R Jacobson and J O Martensson, *Japan J Appl Phys*, **4**, 1965, 333.
[2] R Jacobsson, *Phys Thin Films*, **8**, 1975, 51.
[3] V N Yadava, S K Sharma and K L Chopra, *Thin Solid Films*, **17**, 1973, 243.
[4] V N Yadava, S K Sharma and K L Chopra, *Thin Solid Films*, **22**, 1974, 57.
[5] K L Chopra, S K Sharma and V N Yadava, *Thin Solid Films*, **20**, 1974, 209.
[6] V N Yadava, S K Sharma and K L Chopra, *Thin Solid Films*, **21**, 1974, 297.
[7] W J Coleman, *Appl Opt*, **13**, 1974, 946.

4.5 Use of lateral waves in detecting surface absorption

O S Heavens, *Physics Department, University of York, Heslington, York, YO1 5DD, England*

For light incident on an interface in the neighbourhood of the critical angle, some energy is coupled into a lateral wave which is propagated along the interface region, close to the surface. Light is coupled back into the medium of incidence at all points and can be detected after an appreciable length of lateral wave path. This provides a sensitive method for detecting weak absorption, e.g. in very thin films or in cases where the absorbing region is confined to the immediate neighbourhood of the surface. The method is illustrated by applying it to the study of small concentrations of F-centres produced on a KCl crystal surface by electron bombardment.

1. Introduction

A few observations have been made of the lateral wave which is propagated along the interface region between two media when radiation is incident in the neighbourhood of the critical angle. As with the case of total reflection, the field of such a wave penetrates into the second medium to a depth of the order of the wavelength. Energy from the lateral wave is coupled back into the medium of incidence at all points along the surface and may easily be detected, in a typical experimental arrangement, at distances of the order of centimeters from the point of entry. Such waves can potentially be extremely useful for probing the surface region of a material and for detecting whether there is significant absorbing region close to the surface. Preliminary results in the use of such waves have been given by Heavens and Sharma[1] who have shown that very weak absorption extending to a depth of only a few tens of nm is readily detectable.

2. Theory

Calculation of the lateral wave flux, even at a plane interface between uniform media, is a somewhat difficult diffraction problem. The incident wavefield is expanded as a set of inhomogeneous plane waves and the field at the receiver involves the Fresnel reflection coefficient $R(\theta)$ for the boundary. The inhomogeneous wave may be specified by allowing the angle of incidence θ to take on complex values and the field at the receiver takes the form:

$$\psi_{\text{ref}} = \frac{i\kappa}{2} \int_{-\pi/2+i\infty}^{\pi/2-i\infty} R(\theta) H_0{}^1(u) \sin\theta \exp\{ik(z+z_0)\cos\theta\} \mathrm{d}\theta \tag{1}$$

where $H_0{}^1(u)$ is the Hankel function of the first kind and $u = \kappa r \sin\theta$.

The necessary contour integration is performed with due note taken of poles or branch cuts. Since the Fresnel coefficient $R(\theta)$ involves the radical $\sqrt{(n^2 - \sin^2\theta)}$ (where $n = n_1/n_2$, the ratio of refractive indices in the first and second media), a two-sheeted Riemann surface is involved. The double surface intersects in a branch cut of locus $n^2 - \sin^2\theta = \text{constant}$ and in certain circumstances the path of integration needs to be taken around the branch cut, resulting in a contribution to the receiv e wavefield from a region very close to the critical angle θ_c This contribution corresponds to the lateral wave. For the geometry shown in Figure 1, the lateral wave flux at P is given by:

$$F_{\text{lat}} = \frac{n^2\lambda^2}{\pi^2 m^2 (n^2-1)^2 r L^3} \tag{2}$$

where $m = 1$ for light polarized with the electric vector parallel to the plane of incidence and $m = n^2$ for the perpendicular polarization.

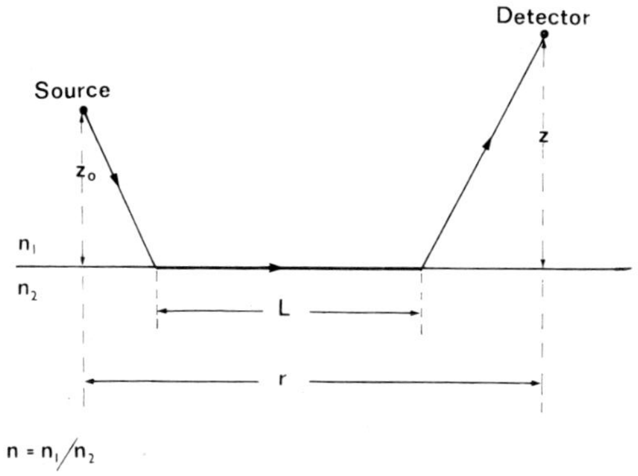

Figure 1. Lateral wave along an interface.

If the lateral wave is generated in a surface carrying an absorbing film, then the above flux will be attenuated by a factor depending on the extinction coefficient k in the film. An exact calculation for the general case cannot as yet be made, but the special case of a film of thickness small compared with the wavelength[3] indicates that the form of the attenuation, for a film of given thickness, is represented by:

$$F_{\text{film}} = \text{const} \times \exp\left(-\frac{4\pi kl}{\lambda}\right) \tag{3}$$

where l is the length of the lateral wave path in the film. Since lateral waves are easily detected for values of l of the order centimetres, then for visible radiation $l/\lambda \sim 10^4$. Thus significant attenuation will occur for values of k as low as 10^{-6} so that measurement of the lateral wave flux offers a powerful method of obtaining information on very weak absorption.

3. Studies of colour centres formation in KCl

In investigating the influence of *F*-centres, formed by electron bombardment on KCl, on the nucleating properties of the KCl crystal surface, the need arose for a method of detecting very small concentrations of such defect sites. The energy range of the bombarding electrons was 1–5 keV, for which the depth to which *F*-centres form lies in the range 50–400 nm. At the lower end of the range, the condition $d \ll \lambda$ applies, so that equation 3 should be appropriate for the determination of the extinction coefficient k. The value of k increases with the duration of electron bombardment and is found[4] to show saturation effects. For a crystal with refractive index n and with a concentration N_0 of colour centres, N_0 is related to k by the expression:

$$N_0 = \left(\frac{1.6 \times 10^{18} \Delta E}{f \lambda_m (2 + n^2)^2}\right) k \qquad (4)$$

where ΔE is the half-width of the absorption band in eV, λ_m is the wavelength at the absorption maximum and f is the oscillator strength for the transition. Inserting values appropriate to the *F*-centre in KCl ($f = 0.81$, $\lambda_m = 556$ nm, $\Delta E = 0.32$ eV, $n = 1.49$) yields

$$N_0 = 6.4 \times 10^{20} k \text{ centres/cm}^3 \qquad (5)$$

The numerical coefficient is that calculated on the assumption of a Lorentzian lineshape. For a Gaussian profile, this factor would be $4.3 \times 10.^{20}$

If absorption in the *F*-centre layer is determined by direct measurement at normal incidence then since the thickness of the absorbing layer is small, the smallest value of k which can readily be measured is of the order 0.005, corresponding to $N_0 \sim 3 \times 10^{18}/\text{cm}^3$. By ellipsometry, values of k down to perhaps 0.002 may be measured under the best conditions. Thus by 'direct' methods, defect densities below $\sim 10^{18}/\text{cm}^3$ cannot be measured.

The method adopted using lateral waves is as follows. Specimens of KCl were polished by hand on a Selvyt cloth stretched over a slightly curved glass plate, initially with 0.3 μm polishing powder in alcohol, but finally in a dry state. The polishing procedure is crucial. Conventional polishing methods leave fine surface scratches which scatter the lateral waves and make quantitative measurement impossible. Electron bombardment is carried out through a mask (Figure 2) which enables lateral wave measurement to be made along a bombarded region (AA) adjacent to a clear region (BB). If it is assumed that scattering due to residual scratches affects both regions equally, then the effect of *F*-centre absorption is readily obtained from the ratio of lateral wave fluxes for paths AA and BB. A path length of 10 mm was adopted; a filtered high-pressure xenon arc was used as light source. The light was coupled into the crystal by an optically contacted high index prism.

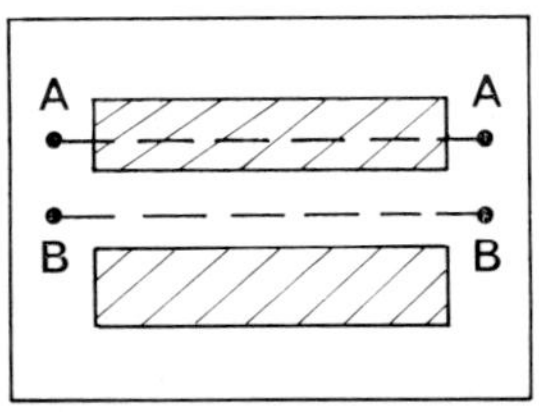

Figure 2. Electron bombardment of the crystal is through a mask to produce the shaded regions shown. Lateral waves are used along paths AA and BB.

Specimens were prepared by bombardment for different times ranging from a few seconds to a few minutes at electron energies of 1 and 2 keV. For the longer periods, the lateral wave absorption was so large that the measurements were adversely influenced by adventitious scattering. In practice, the largest value of k which could be measured was of the order 2×10^{-5}, corresponding to an *F*-centre concentration of $\sim 10^{16}/\text{cm}^3$.

The results obtained are shown in Figure 3. The dependence of the rate of production of *F*-centres on the energy of the incident electrons is as would be expected. The actual electron current at the crystal surface is not known and would be difficult to measure. When the total bombarding current at 2 keV was increased by a factor of 50, absorption by normal transmission was just detected ($k \sim 1.2 \times 10^{-3}$) after bombardment for 4 min. This is reasonably consistent with the result obtained by extrapolating the upper curve of Figure 3 and allowing for the increased current density.

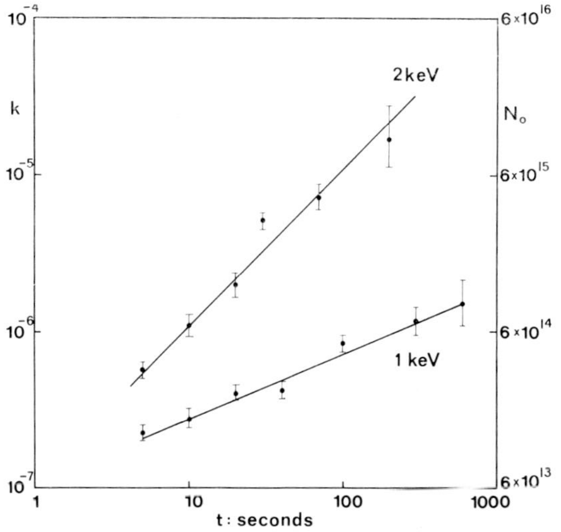

Figure 3. Extinction coefficient, and *F*-centre concentration, in electron-bombarded KCl.

4. Acknowledgement

I am grateful to Dr S K Sharma for evolving the polishing technique crucial for these experiments.

References

1 O S Heavens and S K Sharma, *Opt Comm*, **17**, 1976, 339.
2 L H Brekhovskikh, *Waves in Layered Media*, Academic Press, New York (1960).
3 O S Heavens, *Opt Acta*, **21**, 1974, 1.
4 F Herrman and P Pinard, *J Phys C*, **3**, 1970, 1037.

4.6 Some optical properties of thin Cd_3As_2 films

L Żdanowicz and T Kwiecień, *Institute of Solid State Physics, Polish Academy of Sciences, Zabrze, Poland*

The transmission of vacuum deposited Cd_3As_2 films has been measured in the spectral range 0.6–15 μm. The absorption coefficient α has been calculated from the data for crystalline films obtained at substrate temperature T_s = 440 K and also for amorphous films obtained on substrates at T_s = 300–393 K. The position of the low-energy absorption edge in the spectral range 0.1–0.3 eV is discussed and the value Δ_g^{opt} = 0.06 eV for indirect transion $\Gamma_1 \rightarrow \Gamma_{15}$ is obtained for crystalline films. For amorphous films, a shift of the minimum absorption edge to $\Delta E_g^{opt} \simeq$ 0.4–0.5 eV was observed. For higher energies two direct transitions have been established: the first at E_1 = 0.9 eV in the photon energy range 0.9–1.2 eV and the second at E_2 = 1.2–1.3 eV in the photon energy range 1.3–1.6 eV. The last value is in good agreement with the value calculated by Lin Chung and with reflectivity data; it corresponds to the transition $\Gamma_{15} \rightarrow X_1$ in the hypothetical band model of Lin Chung. Differences in the refractive index values for crystalline and amorphous Cd_3As_2 are also discussed.

Introduction

During the past several years considerable attention has been given to cadmium arsenide—Cd_3As_2 because of its interesting transport properties. This compound is a degenerate n-type semiconductor with extremely high mobility ($\mu m \simeq 2 \times 10^4$ cm^2/V s at 300 K and $n \geq 10^{18}$ cm^{-3}) and very small electron effective mass.[1–4] However, in spite of a continuously increasing number of publications on the subject, there continues to be a lack of agreement on the energy band structure of Cd_3As_2, especially on the nature of the fundamental absorption edge. The major difficulty in a theoretical treatment of this problem originates from the exceptionally complicated crystal structure of this compound-body-centred tetragonal (space group I 4_1 cd) with 32 molecular units per unit cell.[5,6] Recently band structure calculations have been performed by Lin Chung[7] by simplifications of the real crystal structure into a fluorite-like structure.

The energy band structure obtained for this simplified structure is presented partially in Figure 1.

This calculation located the valence-band maxima away from the Γ point, although the minimum of the conduction band was fixed at Γ, and this resulted in a small positive energy gap. Lin Chung's model seems to be reasonable and it provides satisfactory agreement with published experimental data.

Other band models for Cd_3As_2 have also been proposed.[8–10] Numerous detailed investigations of the band structure have been based on reflectivity data in the visible and uv-regions.[11–14] However, only a few papers have been published[3,10,12,15] on studies in the infrared region for energies below 1.2 eV and some disagreement in results has been reported. Accurate experimental investigations of the electrical and especially the optical properties are needed to establish the true energy band structure of this compound and for quantitative comparisons.

In the work described in this paper we have measured the transmission spectra of crystalline and amorphous Cd_3As_2 thin films over the photon energy range 0.05–1.6 eV and have calculated the absorption coefficient in this range. The results are discussed in terms of the Lin Chung hypothetical band structure model and the data are compared to that of other authors.

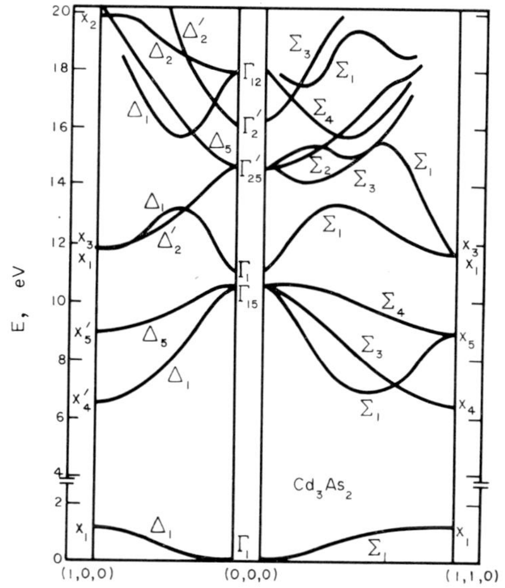

Figure 1. Excerpt of energy band structure of Cd_3As_2 in its hypothetical crystal structure.[7]

Experimental details

Thin Cd_3As_2 films for optical experiments were thermally evaporated in a vacuum of 10^{-6} torr using a method described elsewhere.[15,18,19] The substrates used were optically polished BK-7 glass or air cleaved NaCl, KCl or As_2S_3 plates depending on the spectral range. The substrate temperature was about 445 K for crystalline films and 100 to 400 K for amorphous films. The thicknesses of the films measured by the interference method were in the range 0.3–10 μm. The structure of the films was determined by electron diffraction or by X-ray analysis. Optical transmission measurements were performed using a Beckman Acta M-4 spectrophotometer in the range from 0.6 to 3 μm and Zeiss Spectrometer UR-10 or Zeiss Specord 71-IR with NaCl prism in the infrared region up to 15 μm.

For temperature measurements the samples were mounted on a copper cold finger inside a special Dewar fitted with glass or NaCl windows; a resistance heater mounted inside the cold finger was used to control the temperature.

The absorption coefficient, α, was calculated from the transmission data by using the relations[15]:

$$T_{av} = \frac{A \exp(-\alpha_1 \cdot d_1)}{1 + r_1^2 r_2^2 \exp(-\alpha_1 d_1)},$$

where

$$A = \frac{16 n_1^2 n_2}{(1+n_1)^2 \cdot (n_1+n_2)^2},$$

$$r_1 = \frac{n_1 - 1}{n_1 + 1}, \quad r_2 = \frac{n_1 - n_2}{n_1 + n_2}, \quad \text{for} \quad k_1 \ll n_1,$$

or

$$T = \frac{A \exp(-\alpha_1 d_1)}{1 - 2 r_1 r_2 \exp(-\alpha_1 d_1) + r_1^2 r_2^2 \exp(-2\alpha_1 d_1)}$$

for interference maxima.

The refractive index n_1 of the crystalline Cd_3As_2 films was taken from our earlier publication.[15] The refractive index n_1 for amorphous films was calculated from the transmission interference maxima and the absorption coefficient α was calculated for each a-sample using its own values of n_1.

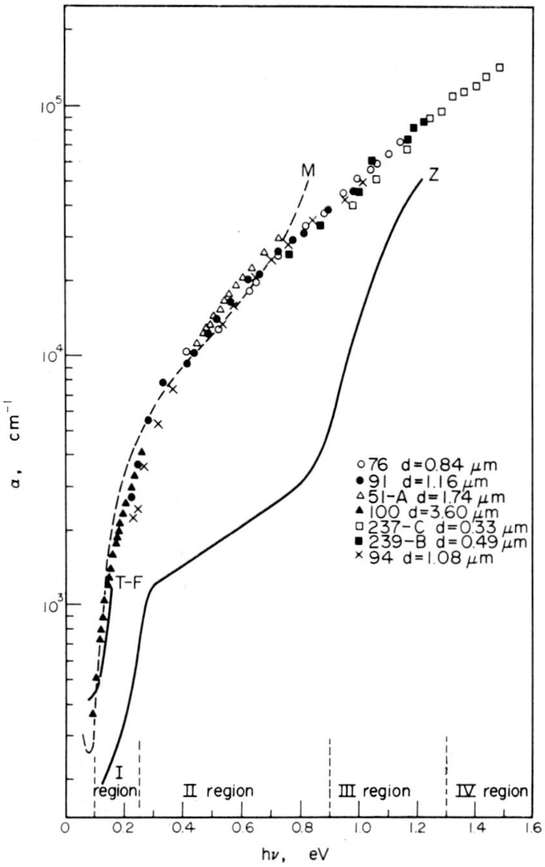

Figure 2. Absorption coefficient for crystalline Cd_3As_2 films. Curve TF—acc. to ref 3, M—acc. to ref 16, Z—acc. to ref 15.

Results and discussion

(a) Crystalline films. The spectral dependence of the absorption coefficient α for a number of crystalline films of different thicknesses in the energy range 0.06–1.6 eV is shown in Figure 2.

For ease of interpretation the data can be divided into four regions. The first region is the photon energy range 0.1 to 0.3 eV, where the absorption coefficient α increases from 2×10^2 to about 3×10^3 cm^{-1}. The second region is the energy range 0.3–0.9 eV, where α changes from about 3×10^3 cm^{-1} to 3×10^4 cm^{-1}. The shape of the α-curve in this region seems to depend on the quantity and kind of contamination in the sample. In the third region (energies 0.9–1.2 eV), the absorption coefficient α reaches values up to 10^5 cm^{-1}. In the fourth region (energies 1.2–1.6 eV) with $\alpha > 10^5$ cm^{-1} it is possible to measure α only for very thin films ($d \sim 0.3$–0.5 μm). In Figure 3, the data of the first region are replotted as $(\alpha h\nu)^{1/2}$ versus energy for the rather thick film no 100 ($d \simeq 3.6$ μm) with good electrical parameters (electron mobility $\mu_H = 4200$ cm^2/V s) and with large-grain microstructure. It is clearly seen, that the experimental points follow the relation $\alpha \sim (h\nu - E_g)^2$ and that absorption in this region indicates the existence of an indirect gap with $E^{opt}_{g\,ind} \simeq 0.06$ eV. However, replotting of the data as $(\alpha h\nu)^2$ versus $h\nu$ also gives a satisfactory fit to the experimental points and yield $E^{opt}_{g\,dir} \simeq 0.14$ eV (Figure 3). This can be explained in terms of the Lin Chung model. The band structure calculations of Lin Chung have shown, that for the simplified model the valence band maxima in Cd_3Qs_2 are located off the Γ point by 12.5% of the distance to the zone boundary in the

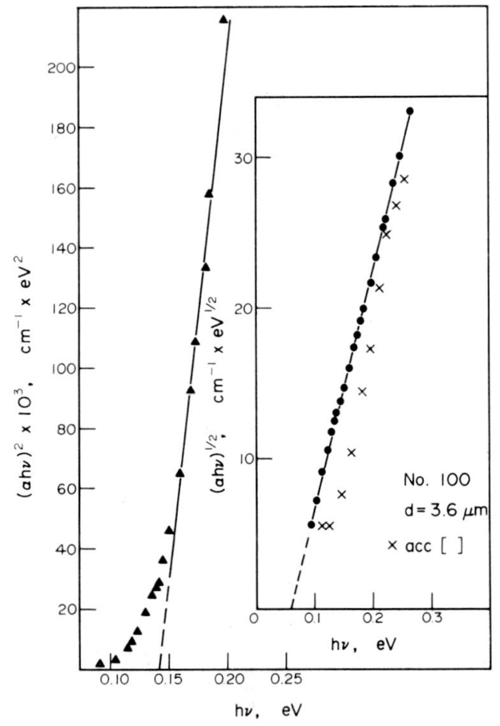

Figure 3. The dependences $(\alpha h\nu)^2$ and $(\alpha h\nu)^{1/2}$ (inset) on photon energy for the first region of crystalline Cd_3As_2 films.

(110) direction, while the bottom of the conduction band is at the Γ point and has an *s*-like representation Γ_1 (Figure 1). It is therefore reasonable to expect that absorption measurements should exhibit the indirect transitions $\Gamma_1 \rightarrow \Gamma_{15}$. This is in agreement with our earlier suggestions[15] and also with the results of Iwami, Yoshida and Kawabe.[16] They obtained the value $E^{opt}_{g\,ind} \simeq 0.11$ eV. However, the direct transition (at or near Γ) which is responsible for the direct optical absorption is located at a slightly higher energy (about 0.05 eV higher[7]).

Evaluation of the exact value of E_g^{opt} in this region is difficult from the optical data only because of the high degeneracy of the material ($n \sim 10^{18}$ cm^{-3}) and a significant Burstein–Moss shift. Small differences in determined values of E_g^{opt} may be explained by different level of contamination for different samples.

Analysis of the absorption data in the third and fourth regions of Figure 2 suggests that direct transitions occur in these regions. The energies associated with these transitions as extrapolated from straight-line plots $(\alpha h\nu)^2$ versus $h\nu$ are $E_1 \simeq 0.9$ eV for the third region and $E_2 = 1.2$–1.3 eV for the fourth region.[1] The value $E_2 = 1.2$–1.3 eV corresponds to the transition $\Gamma_{15} \rightarrow X_1$. Since the *X* and Γ points are equivalent in the hypothetical structure, in the real crystal structure the *X* point becomes a Γ point and this transition is also at the centre of the zone in accordance with the Lin Chung model. Lin Chung has placed this transition at 1.2 eV.[7] This transition corresponds also to the reflectivity peak observed by Bhola[17] at 1.32 eV. The value $E_1 = 0.9$ eV, also observed by Bhola[17] (as 0.91 eV) and in ref 14 (as 0.88 eV), may be attributed to transitions originating from tetragonal splitting of the *p*-like valence band at the Γ point.[21]

The problem of the exact values of the refractive index of Cd_3As_2 and other II–V compounds and solid solutions is also not resolved unambiguously as yet. Thus, our values of n_1 for Cd_3As_2[15] which have been confirmed by Rogers *et al*[4] from interference fringe data in thin-film transmission and reflectivity spectra (Figure 4) are significantly greater than those obtained by Zivitz and Stevenson[6] and also by others[3,9,10,13] from reflectivity measurements on bulk samples (Figure 4). This fact is probably due to unsatisfactory surface conditions resulting from mechanical polishing of bulk samples and to the usually poor quality of bulk samples; these problems do not exist for as-deposited films.

From the interference data we obtain mainly bulk properties of material. On the other hand, the value of *n* derived from thin film measurements contains relatively large errors owing to the poor accuracy of film thickness measurements (of the order of 5 to 8%) and to different techniques of film preparation. For example, Figure 4 shows data calculated for film no 100 of thickness 2.6 μm deposited at slightly different conditions than those in ref 15. We found the unexpecting result that the value n_1 was $n_1 \sim 6$ at $\lambda = 14.5$ μm in this sample (Figure 4).

However, this value seems to be confirmed by the results obtained by Cisowski and Bodnar[20] from a study of transport phenomena in Cd_3As_2 at low temperatures, assuming the strong degeneracy and nonparabolicity of the conduction band.

The computed curves give the best fit to experimental data for the value $\varepsilon_0 = 42$. Subsequent calculations[21] assuming spin–orbit and tetragonal field splitting gave the best fit to experiment for $\varepsilon_0 = 34$.

This last value is in very good agreement with our value $n_1 = 6$. It is also surprising that we really did not observe a

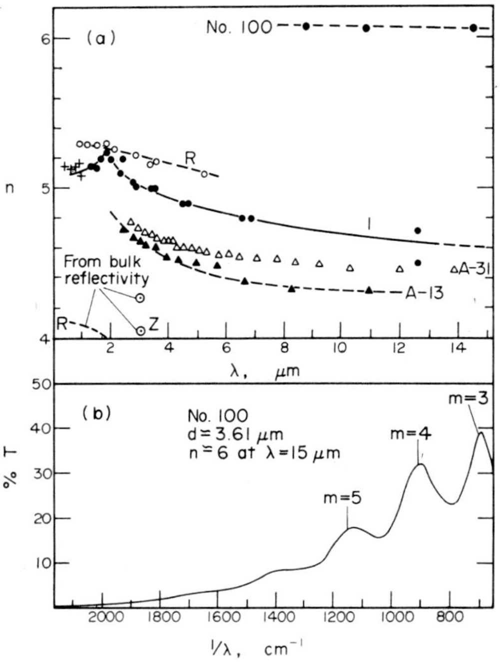

Figure 4. Refractive index for crystalline and amorphous Cd_3As_2 films calculated from the interference data. The continuous curve 1—acc. to ref 15, *R*—acc. to ref 8, Z—acc. to ref 12, A—for amorphous films.

temperature shift of the absorption edges in regions III and IV (Figure 2), this follows from the data presented in Figure 5 (sample no 55-B) for the temperature range 130–423 K. The temperature dependence of the low energy edge in region I was not investigated by us. However, the measurements by Turner *et al*[3] in this region on bulk samples indicate a temperature coefficient $\beta = -2 \times 10^{-4}$ eV/K.

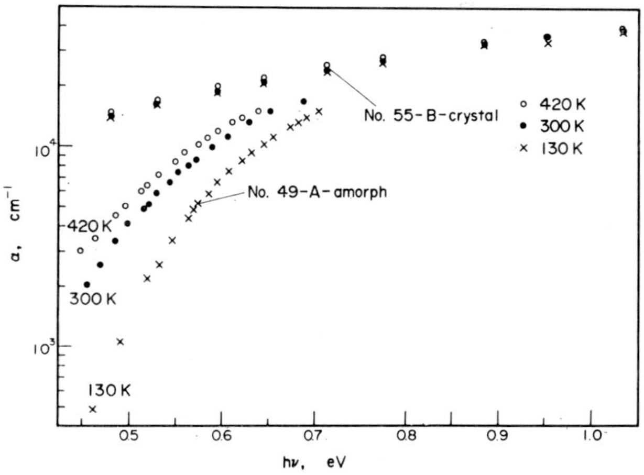

Figure 5. The absorption coefficient α as function of photon energy for several temperatures. 55-B—crystalline film: 49-A—amorphous film.

(b) Amorphous films. The spectral dependence of the absorption coefficient for a number of amorphous films obtained at different temperatures is presented in Figure 5. A comparison of the absorption coefficient of crystalline and amorphous materials reveals that the absorption coefficient is smaller in the amorphous films at all measured photon energies below the photon energy 0.9 eV. Then, the higher transparency of amorphous films in the infrared region should be noted. This is also the case for *a*-films of other materials.[27]

Moreover, the values of the absorption coefficient data for different amorphous films obtained under different conditions on substrates cooled to various temperatures in the wide range from liquid nitrogen[26] up to 120°C lie exactly on the same α-curve, if the values of refractive index n_1 measured for each sample are used in the calculation. This reproducible optical behaviour of these amorphous films confirms our earlier assertion that the amorphous Cd_3As_2 structure is a stable and homogeneous phase and that it transforms irreversibly to the crystalline one at a well defined temperature ($T_t \simeq 130$°C in our vacuum conditions). Significant differences in the values of refractive index n_1 for different films shown in Figure 4 and also in ref 26 can be explained by small deviations from the stoichiometry, as observed for other films with small variations of composition[23] and also by changes in 'space packing' of film. For interpretation of results, we relied on the concept of independently distorted bands of the Cd_3As_2 crystal structure when it becomes amorphous; power-law adsorption edge of the following form was assumed:

$$\alpha \sim [h\nu - E_g^{opt}]^n,$$

where n is believed to have the value 2.0 (see for example ref 24). Replot of the data of Figure 6 in the form $(\alpha h\nu)^{1/2}$ versus photon energy is presented in the inset of Figure 6. A good fit of experimental points is observed and the value $\Delta E_{g\,ind}^{opt} = 0.4$–o.5 eV can be estimated for the absorption edge of amorphous Cd_3As_2.

A considerable shift of the minimum absorption edge to higher energies is observed for amorphous Cd_3As_2 as compared to the crystalline material; this would have been predicted by the 'Verlicek's rule'.[25] From the temperature shift of the absorption edge shown in Figure 5 the temperature coefficient

$$\beta \cong -3.4 \times 10^{-4} \text{ eV/K}$$

was estimated.

A more detailed discussion of the experimental data presented here will be published later. The primary objective of this paper has been to make a comparative study of amorphous and crystalline Cd_3As_2 films.

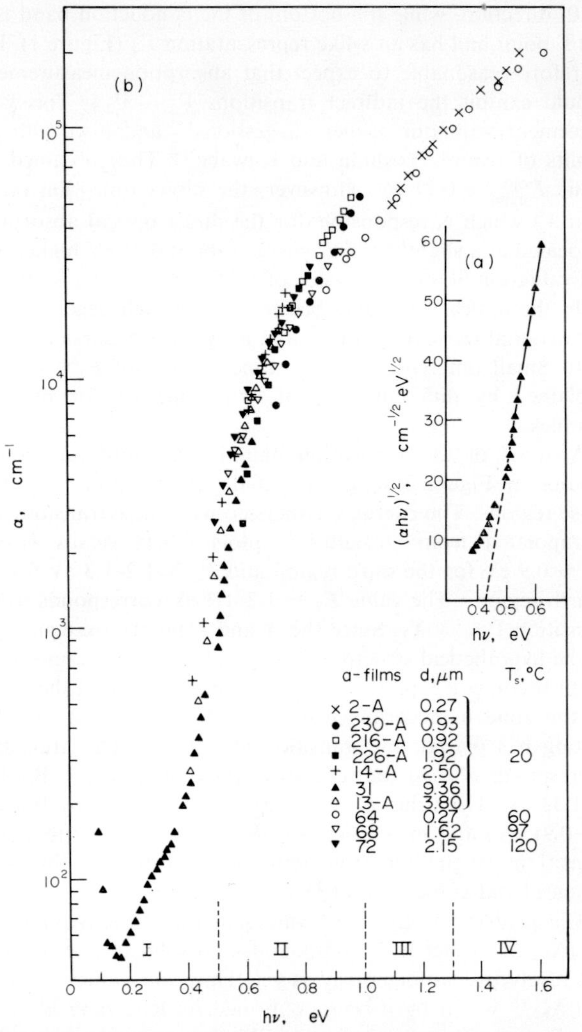

Figure 6. Absorption coefficient for amorphous Cd_3As_2-films. Inset (a) $(\alpha h\nu)^{1/2}$ as function of energy for *a*-films.

Acknowledgements

We are much indebted to Professor W Żdanowicz for continuous help and the valuable discussion of results and to Professor J J Loferski for reading the manuscript. We also wish to thank Mr G Pocztowski, and Mrs M Korona for preparing the films. The work was sponsored by the Institute of Physics, Polish Academy of Sciences.

References

[1] W Żdanowicz and L Żdanowicz, *Annual Review of Materials Science*, **5**, 1975, 301. (Edited by R Huggins).
[2] W Żdanowicz, *Proc IX Int Conf Semicond Phys*, p 1095, Prague (1960).
[3] W J Turner, A S Fischler and W E Reese, *Phys Rev*, **121**, 1961, 759.
[4] F A P Blom and J Schrama, *Phys Lett*, **A30**, 1969, 245.
[5] W Żdanowicz, K Lukaszewicz and W Trzebiatowski, *Bull Acad Pol Sci Ser Sci Chim*, **12**, 1964, 169.
[6] G A Steigmann and J Goodyear, *Acta Crystalogr*, **B24**, 1968, 1062.
[7] P J Lin Chung, *Phys Rev*, **188**, 1969, 1272.
[8] L M Rogers, R M Jenkins and A J Crocker, *J Phys D Appl Phys*, **4**, 1971, 793.
[9] P L Radoff and S G Bishop, *Phys Rev, B*, **5**, 1972, 442.
[10] R J Wagner, E D Palik and E M Swiggard, *J Phys Chem Solid*, **32**, 1971, 471.
[11] V V Sobolev, N N Syrbu and S D Shutov, *Chemical Bonds in Semiconductors and Thermodynamics*, p 165. (Edited by N N Sirota), Consult, Burcan, New York (1968).
[12] M Zivitz and J R Stevenson, *Phys Rev B*, **10**, 1974, 2457.
[13] H W Ellis and J R Stevenson, *J Appl Phys*, **46**, 1975, 3066.
[14] M J Aubin and P J Cloutier, *Can J Phys*, **53**, 1975, 1642.
[15] L Żdanowicz, *Phys State Solid*, **20**, 1967, 473.
[16] M Iwami, M Yoshida and K Kawabe, *Japan J Appl Phys*, **12**, 1973, 1276.
[17] V P Bhola, *Phys State Solid*, (a), **34**, 1976, K179.
[18] L Żdanowicz, *Acta Phys Polon*, **31**, 1967, 1021.
[19] L Żdanowicz and S Mietkowska, *Thin Solid Films*, **29**, 1975, 177.
[20] J Cisowski and J Bodnar, *Phys State Solid* (a), **28**, K49, 1975.
[21] J Bodnar, *Proc Nat Symp Phys II–VI Semicond*, Jaszowiec, April 1977, Poland, to be published.
[22] T S Moss, *Proc Phys Soc B*, **63**, 1950, 174.
[23] R Jacobsson, in, *Physics of Thin Films*, **8**, 1975, 51. (Edited by G Hass, M Francombe and R W Hoffman).
[24] M Zavetova and B Velicky, in, *Optical Properties of Solids, New Developments*, Chap 8. (Edited by B O Seraphin), Elsevier, New York (1976).
[25] V Vorlicek, *Chechosl. J Phys*, **B25**, 1975, 1042.
[26] L Żdanowicz, G Pocztowski, C Weclewicz, M Niedzwiedz, and T Kwiecien, *Thin Solid Films*, **34**, 1976, 00.
[27] S K Bahl and K L Chopra, *J Appl Phys*, **40**, 1969, 4940.